COLLAGEN

Volume II

Biochemistry and Biomechanics

Editor

Marcel E. Nimni, Ph.D.
Professor of Biochemistry, Medicine, and Orthopaedics
University of Southern California School of Medicine
and
Director, Bone and Connective Tissue Biochemistry Laboratory
J. Vernon Luck Research Center
Orthopaedic Hospital of Los Angeles
Los Angeles, California

CRC Press
Taylor & Francis Group
Boca Raton London New York

CRC Press is an imprint of the
Taylor & Francis Group, an **informa** business

First published 1988 by CRC Press
Taylor & Francis Group
6000 Broken Sound Parkway NW, Suite 300
Boca Raton, FL 33487-2742

Reissued 2018 by CRC Press

Library of Congress Cataloging-in-Publication Data

Collagen.

Includes bibliographies and indexes.
Contents: v. 1. Biochemistry -- v. 2. Biochemistry and biomechanics -- v. 3. Biotechnology -- [etc.]
1. Collagen--Collected works. I. Nimni, Marcel E.
[DNLM: 1. Collagen. 2. Connective Tissues. QU 55 C6965]
QP552.C6C66 1988 591.19'245 87-20946
ISBN 0-8493-4600-2 (set)

A Library of Congress record exists under LC control number: 87020946

Publisher's Note
The publisher has gone to great lengths to ensure the quality of this reprint but points out that some imperfections in the original copies may be apparent.

Disclaimer
The publisher has made every effort to trace copyright holders and welcomes correspondence from those they have been unable to contact.

ISBN 13: 978-1-315-89168-2 (hbk)
ISBN 13: 978-1-351-07078-2 (ebk)

Visit the Taylor & Francis Web site at http://www.taylorandfrancis.com and the
CRC Press Web site at http://www.crcpress.com

PREFACE

This series was conceived with the idea of integrating current aspects of ongoing research in the collagen field. It originated with the idea of stimulating cross-fertilization of ideas among scientists who with different goals in mind require a thorough understanding of the physiochemical properties of the fundamental molecules of the connective tissue — in particular, collagen, its biosynthetic pathway, and the characteristic surface properties of the collagen fibril.

We have assembled a spectrum of papers which discuss diverse aspects such as X-ray structure, the thermodynamics and mechanism of fibrillogenesis, and the use of collagen as a biomaterial for the manufacturing of many implantable, sometimes life saving, devices. I believe that a greater understanding of the collagen molecule and its properties will enable the applied scientist to encounter new applications for this unique biological material. On the other hand, the basic scientist as he becomes more aware of the biodynamics of the connective tissues and the industrial applications of collagen, will not only enjoy the satisfaction of seeing his efforts put to practical use, but may be stimulated to interact with the applied scientist and technologist.

Information on the various collagen types, their distribution in the animal kingdom, their fundamental physico-chemical and structural properties, the way the collagen molecules are processed and deposited in tissues, and how they are degraded by specific enzymes provides the bridge between the molecular biologist and biomechanical engineer. As we all tend to become specialists in restricted areas, we retain the need to understand the implications of our observations and a desire to relate to the process of human life. In a field that has expanded so rapidly, and currently involves X-ray cystallographers, physicists, chemists, biochemists, biologists, engineers, molecular biologists, physicians with various trainings, and more recently manufacturers of drugs and medical devices, it is essential that we communicate with each other. It is part of my own fascination with what is going on in this very active field and the frustration of wanting to keep up with the tremendously interesting world of collagen that prompted me to edit this series. Although obviously incomplete, and possibly a wishful attempt, I believe that some of these ideals have been achieved and hope that the readers and contributors will endeavor to communicate with each other.

I must end by thanking the contributors who made this series possible. I hope their effort will prove to be of value to them and that they will derive the same satisfaction that I did from reading such outstanding papers. I also want to thank Dr. Bjorn Olsen for co-editing the final volume on the molecular biology of collagen.

THE EDITOR

Marcel Nimni is professor of Biochemistry, Medicine, and Orthopaedics at the University of Southern California School of Medicine and a director of the Bone and Connective Tissue Biochemistry Laboratory, J. Vernon Luck Research Center, Orthopaedic Hospital of Los Angeles. He obtained his Baccalaureate and doctoral degrees from the University of Buenos Aires, and his masters in Biochemistry from the University of Southern California. His early training in pharmacology and protein metabolism, his work with Ernest Geiger on the mechanism of action of anabolic steroids, and his work with Lucien Bavetta on the biosynthesis of collagen led to his interest in the structure and metabolism of connective tissue macromolecules.

Dr. Nimni is a member of the American Society for Biological Chemists, American Institute of Nutrition, the American Rheumatism Association, the Society for Bone and Mineral Research, the Society for Experimental Biology and Medicine, the Society for Biomaterials, the Sigma Xi Society, and the Orthopaedic Research Society. He is a fellow of the American Association for the advance of Science and a medical staff consultant at the Los Angeles County and Orthopaedic Hospitals in Los Angeles. He is on the editorial board of several journals, a past member of the Pathobiological Chemistry Study section and current member of the Orthopaedics and Musculo-Skeletal Study Section of the NIH, and serves on many advisory committees. His discovery of the inhibitory effects of penicillamine on collagen cross-linking provided significant clues as to the nature of the native cross-links in collagen and led to its clinical use in the treatment of fibrotic disease (scleroderma, cirrhosis, post-surgical adhesions). In 1986 he received the Founders Award from the Society for Biomaterials for his work on cross-linking of collagen, which led to the implantation in humans of tissue-derived bioprostheses (such as the chemically stabilized porcine heart valve) and for his observation and characterization in the early 1960s of toxic plasticizers which leached out of implantable devices (this subsequently led to the development of heavy-metal-free plastics). He has published more than 100 papers and several chapters in books. His current research relates to the cross-linking of collagen, the characterization and identification of collagen types (primarily in bone and cartilage), and the mechanism of normal and pathologic calcification.

VOLUME II CONTRIBUTORS

Robert Alper, Ph.D.
Assistant Director
Connective Tissue Research Institute
University City Science Center
Philadelphia, Pennsylvania

Eric Baer, Ph.D.
Leonard Case Professor
Macromolecular Science Department
Case Western Reserve University
Cleveland, Ohio

Robert J. Boucek, M.D.
Professor
Division of Cardiology
Connective Tissue Laboratories
Loma Linda University School of Medicine
Jerry L. Pettis VA Hospital
Loma Linda, California

H. Bouissou
Professor and Director
Anatomie-Pathologique
C.H.U. Rangueil
Toulouse, France

Neil D. Broom, Ph.D.
Senior Research Fellow
Mechanical Engineering Department
University of Auckland
Auckland, New Zealand

James J. Cassidy, M.S.
Research Fellow
Department of Macromolecular Science
Case Western Reserve University
Cleveland, Ohio

Milos Chvapil, M.D., Ph.D.
Professor and Head
Section of Surgical Biology
Department of Surgery
College of Medicine
University of Arizona
Tucson, Arizona

I. Kelman Cohen, M.D.
Chairman and Professor
Department of Surgery
Medical College of Virginia
Richmond, Virginia

Alan S. Craig, Ph.D.
Research Scientist
Biotechnology Division
DSIR
Palmerston North, New Zealand

Robert F. Diegelmann, Ph.D.
Associate Professor
Department of Surgery
Medical College of Virginia
Richmond, Virginia

Anne Hiltner, Ph.D.
Professor
Department of Macromolecular Science
Case Western Reserve University
Cleveland, Ohio

L. C. U. Junqueira, Ph.D.
Senior Researcher
Ludwig Institute for Cancer Research
Sao Paulo, Brazil

Nicholas A. Kefalides, M.D., Ph.D.
Professor, Department of Medicine, and Director, Connective Tissue Research Institute
University of Pennsylvania
Philadelphia, Pennsylvania

Shigeru Kimura, Ph.D.
Associate Professor
Laboratory of Biochemistry
Tokyo University of Fisheries
Tokyo, Japan

William J. Lindblad, Ph.D.
Assistant Professor
Department of Surgery and Pharmacology
Medical College of Virginia
Richmond, Virginia

Gregorio Santiago Montes, Ph.D.
Professor
Department of Histology and Embryology
University of Sao Paulo
Cidade Universitaria
Sao Paulo, Brazil

Marisabel Mourelle, Ph.D.
Professor
Phamacology and Toxicology Department
CINVESTAV
Mexico City, Mexico

Van C. Mow, Ph.D.
Professor
Department of Orthopaedic Surgery and Mechanical Engineering
Columbia University
New York, New York

Elizabeth R. Myers, Ph.D.
Department of Orthopaedic Surgery
Charles A. Dana Research Institute
Beth Israel Hospital and Harvard Medical School
Boston, Massachusetts

David A. D. Parry
Professor
Department of Physics and Biophysics
Massey University
Palmerston North, New Zealand

M. Th. Pieraggi
Chargé de Recherche CNRS
C.H.U. Rangueil
Toulouse, France

Marcos Rojkind M.D., Ph.D.
Professor
Department of Biochemistry
CINVESTAV, IPN
Mexico City, Mexico

Terry J. Sites
Division of Orthopaedics and Rehabilitation
University of California at San Diego
La Jolla, California

Marvin L. Tanzer, M.D.
Professor and Head
Department of BioStructure and Function
University of Connecticut Health Center
Farmington, Connecticut

Savio L. Y. Woo, Ph.D.
Professor
Division of Orthopaedics and Rehabilitation
University of California at San Diego
La Jolla, California

Wen bo Zhu, M.S.
Department of Orthopaedic Surgery
Columbia University
New York, New York

VOLUME I TABLE OF CONTENTS

Chapter 1
Molecular Structures and Functions of Collagen 1
Marcel E. Nimni and Robert D. Harkness

Chapter 2
Energetics and Thermodynamics of Collagen Self-Assembly 79
George Némethy

Chapter 3
X-Ray Diffraction as a Tool for Studying Collagen Structure 95
Barbara Brodsky, Shizuko Tanaka, and Eric F. Eikenberry

Chapter 4
Collagen Fibrillogenesis 113
Arthur Veis and Kevin Payne

Chapter 5
Collagen Types: Structure, Distribution, and Functions 139
Edward J. Miller

Chapter 6
Cross-Linking of Collagen 157
Mitsuo Yamauchi and Gerald Mechanic

Chapter 7
Interactions between Structural Glycoproteins and Collagen 173
J. Labat-Robert and L. Robert

Chapter 8
Biochemistry and Physiology of Mammalian Collagenases 187
George P. Stricklin and Margaret S. Hibbs

Chapter 9
Molecular Anatomy of the Vertebrate Eye: Distribution of Collagen in
Ocular Tissues 207
Eunduck P. Kay

Chapter 10
Cartilage Collagens 225
Natsuo Yasui and Marcel E. Nimni

Chapter 11
Collagens in the Nervous Tissue 243
H. Mei Liu

Chapter 12
Collagens in Heart Valves 257
Reza I. Bashey and Sergio A. Jimenez

Chapter 13
Collagens of Normal and Diseased Blood Vessel Wall 275
Michael John Barnes

Index 291

VOLUME II TABLE OF CONTENTS

Chapter 1
Collagen Fibrils during Development and Maturation and their Contribution to the Mechanical Attributes of Connective Tissue 1
David A. D. Parry and Alan S. Craig

Chapter 2
Phylogenetic Aspects of Collagen Structure and Function 25
Marvin L. Tanzer and Shigeru Kimura

Chapter 3
Histochemical Localization of Collagen and of Proteoglycans in Tissues 41
G. S. Montes and L. C. U. Junqueira

Chapter 4
Structure and Organization of Macromolecules in Basement Membranes 73
Nicholas A. Kefalides and Robert Alper

Chapter 5
Changes in the Morphology and Chemistry of Connective Tissues during Aging 95
H. Bouissou and M. Th. Pieraggi

Chapter 6
Fibrogenic Processes during Tissue Repair 113
R. F. Diegelmann, W. J. Lindblad, and I. K. Cohen

Chapter 7
The Liver as a Bioecological System: Modifications during Regeneration and Repair 137
Marcos Rojkind and Marisabel Mourelle

Chapter 8
Method of Treatment of Fibrotic Lesions by Topical Administration of Lathyrogenic Drugs 161
Milos Chvapil

Chapter 9
Hierarchical Structure of Collagen and its Relationship to the Physical Properties of Tendon 177
Eric Baer, James J. Cassidy, and Anne Hiltner

Chapter 10
Contributions of Elastin and Collagen Organization to Passive Mechanical Properties of Arterial Tissue 201
Robert J. Boucek

Chapter 11
Current Advances on the Study of the Biomechanical Properties of Tendons and Ligaments 223
Savio L-Y. Woo and Terry J. Sites

Chapter 12
The Collagen Framework of Articular Cartilage: Its Profound Influence on Normal and Abnormal Load-Bearing Function 243
Neil D. Broom

Chapter 13
Viscoelastic Properties of Articular Cartilage and Meniscus 267
Elizabeth R. Myers, Wenbo Zhu, and Van C. Mow

Index 289

VOLUME III TABLE OF CONTENTS

Chapter 1
Bioprosthesis Derived from Cross-linked and Chemically Modified Collagenous Tissues 1
Marcel E. Nimni, David T. Cheung, Basil Strates, M. Kodama, and Khalid Sheikh

Chapter 2
The Chemistry of Tanning 39
E. Heidemann

Chapter 3
Collagen Graft Copolymers and Their Biomedical Applications 63
K. Panduranga Rao and K. Thomas Joseph

Chapter 4
Regeneration of Skin and Nerve by Use of Collagen Templates 87
Ioannis V. Yannas

Chapter 5
Injectable Collagen for Tissue Augmentation 117
Donald G. Wallace, John M. McPherson, Larry Ellingsworth, Linda Cooperman, Ross Armstrong, and Karl A. Piez

Chapter 6
Morphology of Collagen in Bioprosthetic Heart Valves 145
Victor J. Ferrans, Stephen L. Hilbert, Yoshifumi Tomita, Michael Jones, and William C. Roberts

Chapter 7
Cell Growth on Collagen-Hema Hydrogels....191
Carl Franzblau, Barbara Faris, Diane M. Dunn-Lanchantin, Peter J. Mogayzel, Jr., and Paul Toselli

Chapter 8
Use of Collagen Solutions in Surgery and as Adjuvant to Prosthetic Implants....209
Shmuel Shoshan and Avinoam Yaffe

Chapter 9
Biochemistry of Tendon and Ligament....223
David Amiel and J. B. Kleiner

Chapter 10
Collagen and Vascular Prosthesis....253
Shu-Tung Li

Chapter 11
Calification in Cardiovascular Tissues and Bioprostheses....273
Basil Strates, Jane Lian, and Marcel E. Nimni

Index....293

Chapter 1

COLLAGEN FIBRILS DURING DEVELOPMENT AND MATURATION AND THEIR CONTRIBUTION TO THE MECHANICAL ATTRIBUTES OF CONNECTIVE TISSUE

David A. D. Parry and Alan S. Craig

TABLE OF CONTENTS

I. Introduction 2

II. Growth Patterns of Collagen Fibrils in Connective Tissues 2

III. Modes of Growth of Collagen Fibrils 9

IV. Mechanical Role of Connective Tissue 14

V. Conclusions 19

Acknowledgment 20

References 20

I. INTRODUCTION

Connective tissues are complex biological structures assembled from a variety of distinct chemical components. Such tissues contain not only collagen, but glycosaminoglycans (GAG), minerals, elastic fibers, cellular components, and water (Table 1). The relative amounts of each of these vary significantly with age and with the particular connective tissue. Although the noncollagenous components have a significant effect in determining the structure and mechanical role of connective tissue, this review will concern itself primarily with the collagenous component. Reference, however, will be made to the other components of connective tissue where appropriate.

Collagen, a truly ubiquitous fibrous protein, is the major protein constituent of all vertebrate species and has been estimated to account for about 25% of the body protein in mammals.[32] It occurs in a variety of macroscopic forms in tissues as diverse as skin, tendon, blood vessels, cartilage, bone, cornea, vitreous humor, and basement membrane, and is present in the framework and interstices of all other tissues and organs with the exception of blood, lymph, and the keratinous tissues. In most cases the prime role of the collagen is to provide the tissue with its structural integrity; the most singular feature which permits collagen to play this supportive role is the high tensile strength of its fibrils. This property is due to its unique molecular conformation which is bestowed upon it by the regular repeating units in the amino acid sequence, the highly specific alignment and packing of the molecules in the fibrils, and the axial and lateral cohesion afforded by the formation of intermolecular covalent crosslinks. Knowledge of the structure of the collagen molecule is therefore of considerable importance if the mechanisms of fibrillogenesis, the modes of fibril growth, and the physical properties of collagenous tissues are to be understood. Many extensive reviews of collagen structure and function have been published (see for example Traub and Piez,[75] Miller,[51] Fietzek and Kuhn,[19] Ramachandran and Ramakrishnan,[65] Glanville and Kuhn,[28] Bornstein and Traub,[5] Fraser et al.,[24] Bornstein and Sage,[4] Bailey and Etherington,[2] Brodsky and Eikenberry,[8] and Parry and Craig[59].

Not all of the genetically distinct collagen types that have been characterized to date form fibrils. Type IV collagen, for example, forms a lattice of side 800 nm; this is designed to prevent movement of cells and migration of high molecular weight macromolecules. Types I, II, and III collagen molecules, however, do form fibrils and are the most widely studied of the collagens. Such fibrils are easily recognizable in the electron microscope from their periodic banding structure seen in longitudinal section, the so-called D period being 67 nm (Figure 1). This banding, seen in both positively stained sections and negatively stained fibrils has also been observed indirectly in the X-ray diffraction patterns of the native hydrated tissue and has been interpreted in terms of the relative axial staggering of similarly directed collagen molecules in a polar fibril. Considerable efforts have been made to understand this aspect of the fibril structure. In contrast, less effort has been devoted to analyzing the transverse structure of fibrils, especially as a function of age. New data have become available in the past few years, however, and these are now beginning to reveal how fibrils grow and how they provide the connective tissue with an important part of its mechanical attributes. This review will therefore concentrate on transverse aspects of fibril structure. Portions of this work have been taken from the thesis of Craig[13] and have thus been previously published in a modified and much extended form.

II. GROWTH PATTERNS OF COLLAGEN FIBRILS IN CONNECTIVE TISSUES

Before discussing the growth pattern of collagen fibrils it is important to specify the basis on which the electron microscope data have been interpreted.[59] First, since collagen fibrils

Table 1
COMPONENTS OF CONNECTIVE TISSUES IN MATURE ANIMALS

Tissue	Percentage of composition[a] Collagen	Elastin	GAGs[b]	Water
Tendon/ligament	30	1.5	0.03—0.3	65
Skin	30	0.2	0.03—0.35	60—72
Fibrocartilage	20	0.1—0.2	0.6	75
Elastic cartilage	16	5—7	3—4	70
Hyaline cartilage	5—18	<0.1	5—11	75
Bone[c]	5—20	—	0.4	30—50
Cornea	12—15	—	0.2—1.0	80
Aorta	5—15	7—15	0.2—2.5	70—75
Elastic ligament	9	35	—	55
Wharton's jelly	12	—	0.3	88
Vitreous humor	0.25	—	0.02	99

[a] All expressed as percentages wet weight of tissue. Total percentage does not always add up to 100% due to lack of inclusion of cellular components and noncollagenous proteins. No value quoted means that no quantitative data are available.

[b] Glycosaminoglycans.

[c] From the ranges of values cited in the literature it would appear that bone contains ~45% mineral.

Modified and extended from Parry, D. A. D. and Craig, A. S. in *The Ultrastructure of the Connective Tissue Matrix*, Ruggeri, A. and Motta, P. M., Eds., Martinus Nijhoff, The Hague, 1984, 34. With permission.

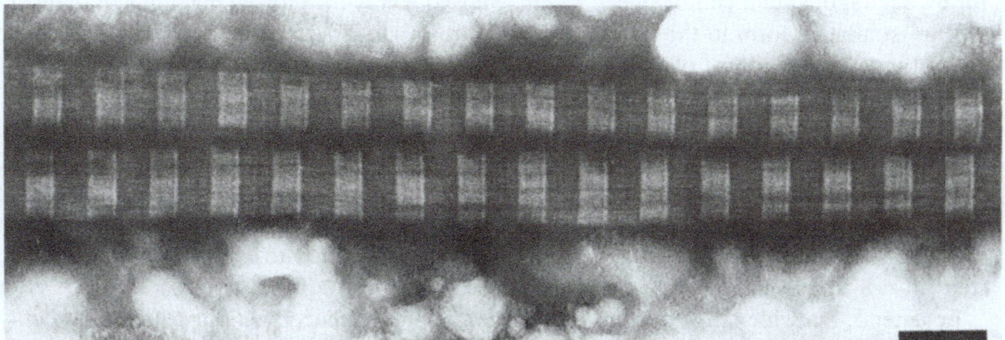

FIGURE 1. Whole mount preparation of collagen fibrils negatively stained to show the D-periodic banding pattern (D ~67 nm). Each D-period is made up of a lighter-staining "overlap" region and a darker-staining "gap" region. Bar = 0.1 μm.

are approximately circular in cross section and show no convincing evidence of tapering, the measured collagen fibril diameter distributions have been considered as arising from a population of infinitely long, nonanastomosing, cylindrical fibrils of constant diameter, rather than from a sampling of cross sections through a homogeneous population of tapering fibrils. Second, although the absolute diameters of the collagen fibrils recorded by electron microscopy are likely to be significantly less than occur in vivo, it is assumed that the relative diameters of the collagen fibrils at different stages of development will be maintained.

Fetal or immature connective tissues, which have high cellular contents, often contain widely spread bundles of small, relatively constant diameter collagen fibrils (Figures 2 and 3). Within such oriented bundles the relative polarities of the polar fibrils vary in a random manner. Subsequent development of the tissue depends markedly on the kind of animal involved.[59,60] In the *altricial* animals (i.e., those animals which have a low body mass at birth compared to that at maturity and which require a considerable period of parental supervision and support after birth relative to their natural life span) the collagen fibril diameter distribution remains sharp throughout fetal development. At later stages of development, however, the distribution broadens in line with the increased tensile requirements of the tissue. In contrast, in the *precocial* animals (i.e., those animals which have a relatively high body mass at birth compared to that at maturity, and which are able to run and fend for themselves soon after birth) the collagen fibril diameter distribution broadens considerably prior to birth. It is not unexpected, of course, that the development of the fibril diameter distribution should parallel the tensile requirements of the tissue.

With further development the mean fibril diameter increases substantially (except in a few highly specialized tissues such as cornea) and either a unimodal or bimodal distribution of fibril diameters occurs at maturity depending on the mechanical properties required (Figure 4). Beyond this stage the mean fibril diameter usually decreases, sometimes rather spectacularly, as in horse tendon[61] (Figure 5). In such senescent tissues fibril morphology can also become markedly altered from normal. Observations made for many tissues from a variety of animals over the period mid-fetal to old age have provided the raw data on which the growth, development, and breakdown of collagen fibrils may be understood in a quantitative manner. Such data are vital if an understanding is to be achieved of the mechanism of fibrillogenesis and of the relationship between fibril diameter and mechanical properties.

X-ray diffraction studies of oriented specimens of rat tail-tendon[52,53] revealed the presence of Bragg reflections in the equatorial and near-equatorial regions that could be accounted for by a tetragonal unit cell of side 7.55 nm. Subsequent work showed that units of diameter 3.8 nm similar in form to the microfibril structure originally proposed by Smith[72] could be packed together in groups of four such that they were related to one another by a 4_3 screw axis. Fraser et al.[24] pointed out that if the structure of crystalline collagen fibrils is indeed based upon this lateral (7.55 nm) repeating unit then the smallest fibrils observed in fetal tissues would be likely to be single groups of four microfibrils. Such a group of microfibrils, if they were to appear circular in cross section, would be expected to have a diameter of about $7.55 \times 2/\sqrt{\pi}$ ($\sim$8.5 nm) and the diameter of all other fibrils would be multiples of this value. In other words collagen fibrils would be expected to "grow" in incremental steps of $\sim$8.5 nm, and their diameters could be said to be "quantized".

In general, the wide range of fibril diameters in mature connective tissues does not readily allow this postulate to be tested. However, in cornea and in fetal connective tissues the collagen fibrils have been shown to have extremely uniform sizes, the mean diameters of which can be measured with some accuracy. Jackson[44] measured collagen fibril diameters in developing chick metatarsal tendons and showed them to have diameters of 8, 12, 17, 25, 31, and 40 nm at 8, 11, 11 to 13, 14, 16, and 20 days fetal respectively, while Greenlee and Ross[30] in their study of the rat flexor digital tendon measured fibril diameters of 16, 20 to 25, and 40 nm in 15 - to 16-day fetal, 18-day fetal, and neonate animals. Further, the collagen fibril diameter measurements made by Parry and Craig (see Parry and Craig[59] for a summary of the data and the original references) and those measurements made for small diameter fibrils from corneal stromata and embryonic tissues, have strengthened the hypothesis formulated that collagen fibrils do indeed increase in diameter by about 8-nm increments (Figure 6 and see also Parry and Craig[57] Parry et al.,[62] Craig and Parry,[14,15] and

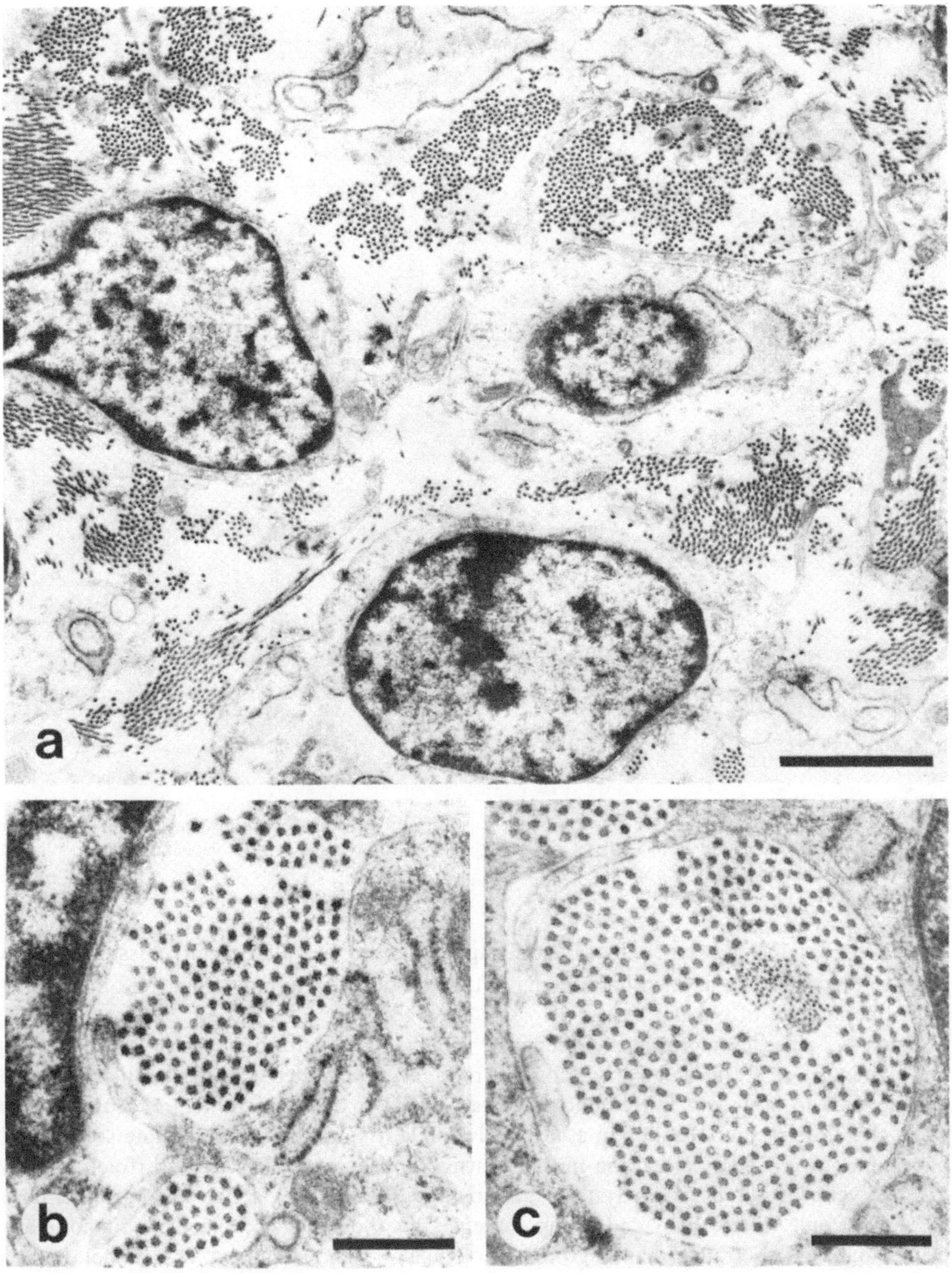

FIGURE 2. Transverse sections of collagen fibrils from 60-day fetal sheep tendons. At this stage of development the collagen fibrils have a distribution of diameters that is both sharp and unimodal. (a) Low power electron micrograph showing loosely organized embryonic fibroblastic tissue and newly synthesized collagen fibrils. Bar = 2 μm. (b and c) Higher power electron micrographs emphasizing the uniformity of fibril diameter. Bars = 0.5 μm. In (b) cytoplasmic processes delineate two discrete collagen fibril bundles (fibers) while in (c) a newly formed bundle is completely enveloped by the cytoplasmic processes of two adjacent tenocytes. A group of elastic microfibrils can be seen within the collagen fibril bundle.

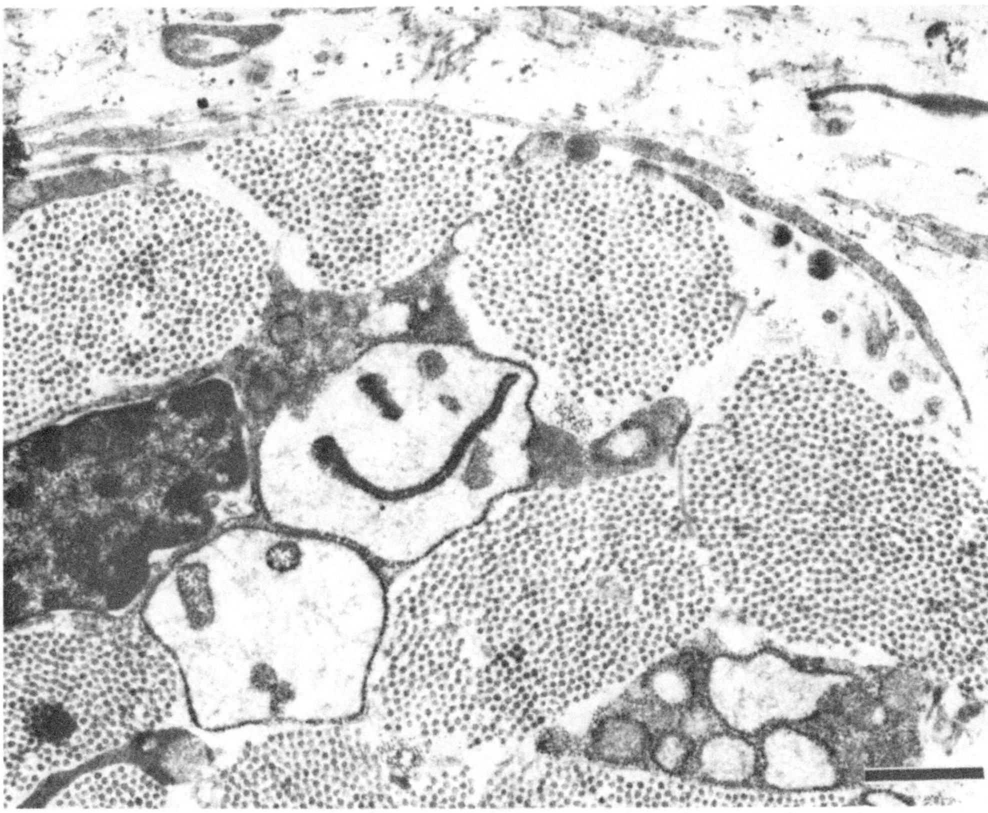

FIGURE 3. Transverse section of a bundle of collagen fibers from the tail-tendon of a neonate rat. Each fiber contains an axially ordered array of fibrils of constant diameter. At this stage of development the tendon still has a high cellular content. Bar = 1 μm.

Parry and Craig[58]). In essence, Parry and Craig[57] proposed that all connective tissue collagen fibrils are constructed in a similar manner with each fibril being built from ~8-nm units or a peripheral layer of ~4-nm units.

Extensive electron microscope evidence in support of a microfibril type of structure (3.8 nm diameter; see Craig and Parry[16] for full references) together with the published X-ray data seemed to provide a fairly conclusive case in favor of such a structure. However, at about this time hydrated tendon in a near native state (though soaked in a dilute solution of phosphotungstic acid) yielded the most extensive and high quality X-ray diffraction data that had yet been obtained. Analysis of these data by Fraser and MacRae[23] and of other data by Hulmes and Miller[39] provided strong evidence that the collagen molecules in tendon are packed individually (rather than in groups) on a quasi-hexagonal lattice. This model did not eliminate the possibility that a microfibrillar structure, delineated by some specific pattern of intermolecular crosslinking, may exist — but it did not require a subfibrillar structural element greater in lateral size than that of a single molecule. The question then arose as to the interpretation of the electron microscope observations.

The analysis of all measurements pertaining to small collagen fibrils clearly shows that the diameters do not have a normal distribution of sizes but indeed are grouped into populations with preferred sizes close to a multiple of 8 nm. The data supporting this statement are shown graphically in Figure 6a which has been constructed by plotting a frequency distribution (histogram) of all of the mean diameters of the sharp unimodal populations of

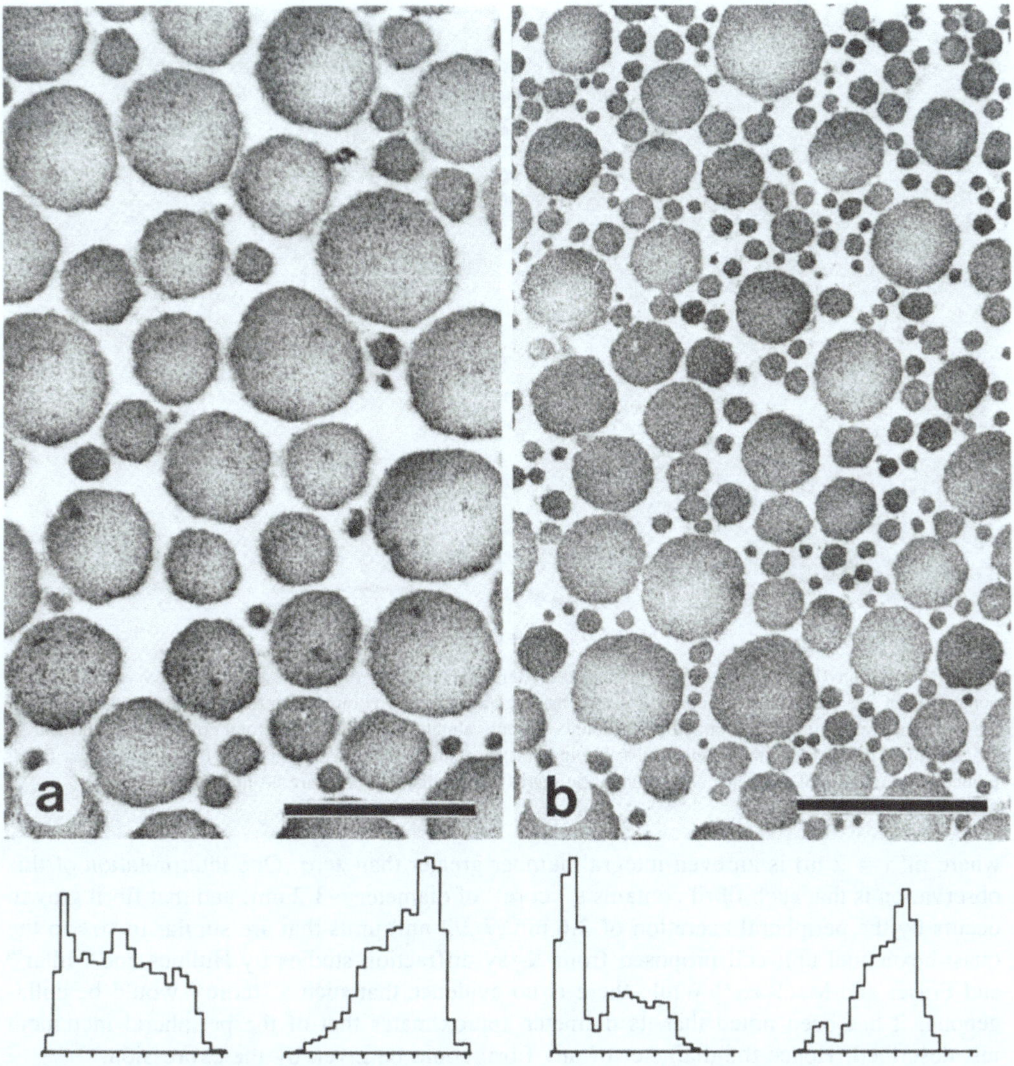

FIGURE 4. Transverse sections of the collagen fibrils from (a) the common digital extensor tendon of a mature horse showing a unimodal distribution of fibril diameters, and (b) the suspensory ligament of a mature horse showing a bimodal distribution of fibril diameters. Bars = 0.5 μm. Each pair of histograms (below) shows the number distribution (left) and the mass distribution (right) of the collagen fibrils plotted as a function of diameter. The graduation marks on the diameter axis are 50 nm apart.

collagen fibrils measured. Further, this distribution has been analyzed using a computer program to resolve multimodal distributions into discrete populations. No constraints were imposed on the mean diameters of the resolved populations or on their standard deviations. To a first approximation the separation of the mean is ~8 nm and thus the mean diameters determined at 17.9, 25.2, 32.0, and 39.8 nm correspond to about 2, 3, 4, and 5 of these ~8-nm increments, respectively. When these data are plotted (Figure 6b) a straight line graph is obtained which intercepts the diameter axis at 3.24 nm and has a slope of 7.28 nm. The equation of the line is thus given by the expression:

$$\begin{aligned} \text{diameter} &= 7.28\ \text{m} + 3.24 \\ &= 3.64\ \text{m}' + 3.24 \end{aligned}$$

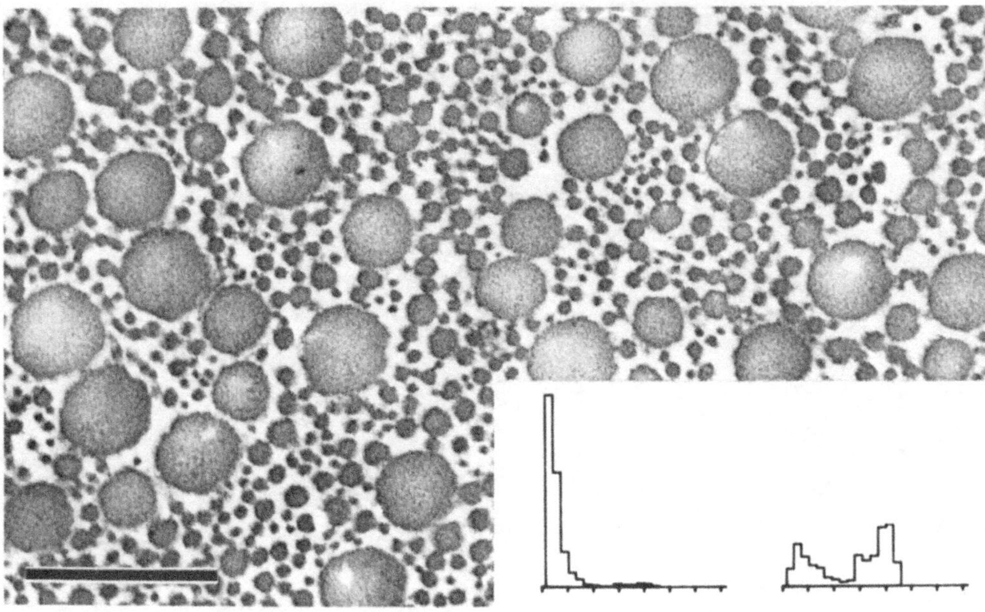

FIGURE 5. Transverse section of the superficial digital flexor tendon from a senescent horse (19 years old). The percentage of the small diameter collagen fibrils has increased greatly compared to that of a mature horse though the collagen present in these small fibrils accounts for only about 30 to 40% of the total collagen present. Bar = 0.5 μm. The inset shows the number distribution (left) and the mass distribution (right) of the collagen fibrils plotted as a function of diameter. The graduation marks on the diameter axis are 50 nm apart.

where m′ (= 2 m) is an even integral number greater than zero. One interpretation of this observation is that each fibril contains a "core" of diameter ~3.2 nm, and that fibril growth occurs by the peripheral accretion of 3.6 nm (7.2/2 nm) units that are similar in size to the quasi-hexagonal unit cell proposed from X-ray diffraction studies by Hulmes and Miller[39] and Fraser and MacRae.[23] While there is no evidence that such a "core" would be collagenous, it has been noted that its diameter approximates that of the peripheral increment just described. Hence the diameter of any fibril could be given by the expression:

$$\text{diameter} = 3.6(m' + 1)$$
$$= 3.6(2m + 1)$$

Prior to these studies it had been suggested that the peripheral accretion of 4-nm units gave rise to fibrils whose diameters were multiples of ~8 nm (the diameters thus being *even multiples* of ~4 nm). However, the interpretation of all the data now available suggests that fibril growth occurs by the peripheral accretion of 3.6-nm units around a core of similar size — the diameter of the fibrils thus being an *odd multiple* of 3.6 nm. In either case the fibrils will grow by increments of about 7 to 8 nm as previously reported. The difference, however, lies in the absolute diameters that the fibrils may attain.

Direct support for the concept of 4-nm peripheral accretion around a collagen fibril comes from optical diffraction studies on the electron micrographs of thin sections of tendon.[41] In that work, evidence was presented that fibrils had tangentially oriented layers of this spacing. Further studies using content-dependent, anisotropic spatial frequency filtering have confirmed these conclusions.[43] The model for fibril growth proposed by Hulmes,[38] in which accretion of staggered concentric layers of nonstaggered collagen molecules is postulated,

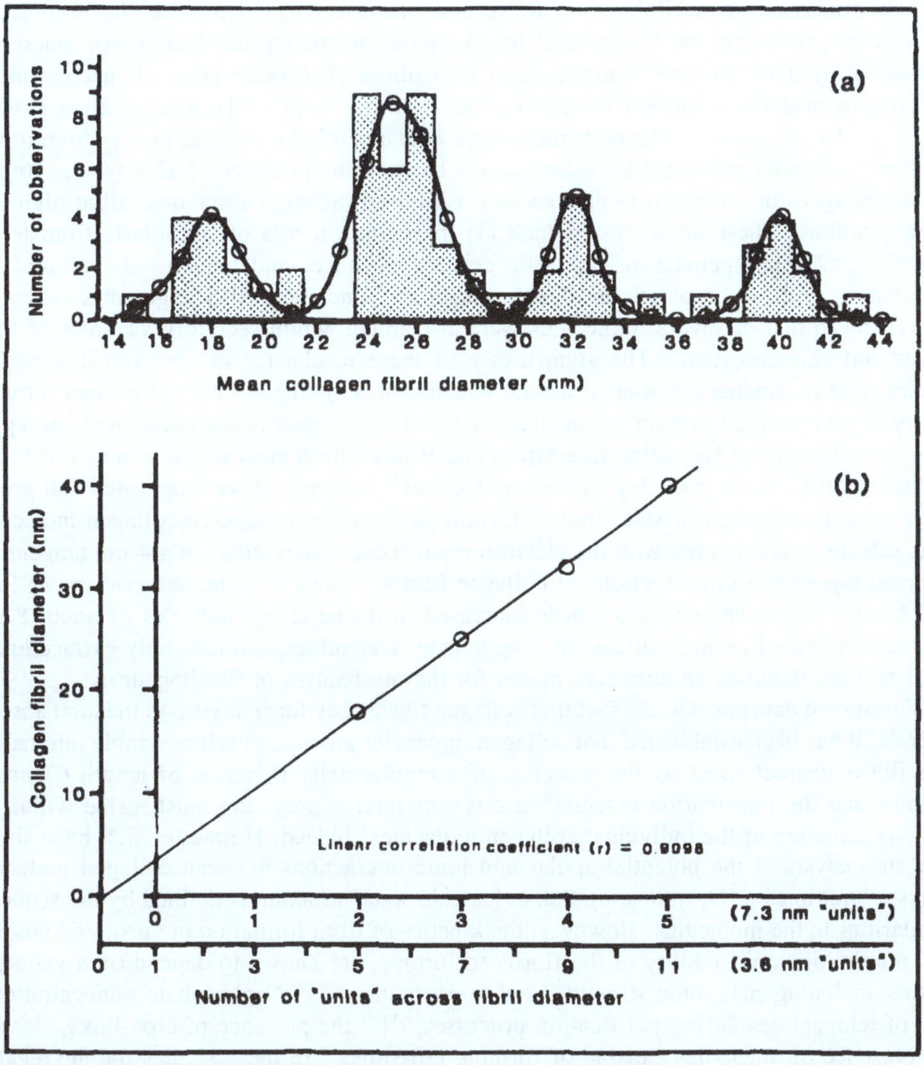

FIGURE 6. (a) Histogram showing the distribution of the mean diameters of collagen fibrils measured from sharp unimodal populations (standard deviations < 3.5 nm). These data represent all those obtained from corneas and from fetal/immature tissues. Using a computer program for the deconvolution of multimodal populations these data were resolved (smooth continuous line) into four populations with means of 17.9, 25.2, 32.0, and 39.8 nm. (b) Mean fibril diameters plotted against the integers representing possible number of "units" traversing a fibril diameter (see text for details).

thus remains a satisfactory explanation of the quantization of fibril diameter. Further, coordinated X-ray and electron microscope studies by Chew and Squire[11] have also confirmed that the data previously cited as indicating a structural unit containing a number of molecules, are interpretable in terms of the quasi-hexagonal packing of individual molecules.

III. MODES OF GROWTH OF COLLAGEN FIBRILS

The sequence of events between the synthesis of procollagen molecules and the appearance of morphologically distinct collagen fibrils is unknown. The mechanisms postulated, however, fall into two main classes. In the first, the procollagen molecules synthesized by the cell are processed extracellularly; in the second, stable aggregates of procollagen molecules formed intracellularly are transported by exocytosis to the extracellular space where they subsequently undergo further aggregation into fibrils. The latter class of mechanism has received considerable support in recent years from the work of Trelstad et al.,[77] Bruns et al.,[10] and Hulmes et al.[42] Electron microscope data and other evidence, mostly from in vitro studies, have been presented by these authors to show the presence of SLS type aggregates of procollagen (or collagen) molecules in a variety of different situations, all at high local concentrations. These situations include (1) the culture media of fibroblasts from tendon and skin, (2) homogenates of lathyritic cornea, cartilage, and tendon, (3) cultured skin fibroblasts, and (4) preparations of procollagen.[10] Bundles of SLS aggregates of lengths equivalent to that of several molecules were also found within secretory vacuoles of fibroblasts and chondrocytes.[10] The significance of these results for in vivo fibrillogenesis is unclear and it remains a matter of debate whether SLS aggregates are (1) present intracellularly in vivo and (2) present extracellularly in a form subject to rearrangement and aggregation to allow fibril formation (see Gross and Bruns[31] for a clear exposition of this topic). In the model for fibril growth proposed by Hulmes[38] however, it was suggested that growth occurs via an accretion of staggered concentric layers of nonstaggered collagen molecules. This scheme is compatible with the electron microscope observations of a 4-nm tangentially oriented layering in cross sections of collagen fibrils[41,43] and with the observations of Parry and Craig[57] which showed that fibrils increased in diameter by multiples of about 8 nm. The scheme based on intracellular SLS aggregates and subsequent assembly extracellularly must be considered as an attractive model for the mechanism of fibrillogenesis.

Widespread data attest to the fact that collagen fibrils may form in vitro in the total absence of cells. It has been established that collagen molecules are able to self-assemble into native-like fibrils characterized by the presence of a regular axial D period of length 67 nm. It follows that the information essential to this structural aggregation must reside within the primary structure of the individual collagen molecules. Indeed, Hulmes et al.[40] have shown that an analysis of the potential apolar and ionic interactions between collagen molecules peaks at multiples of D, indicating that at least the axial structure is defined by the sequence regularities in the molecules. However, the kinetics of fibril formation in vitro, and possibly the morphology and stability of the fibrils so formed, are known to depend on a variety of factors including pH, ionic strength, and temperature,[27,67,81,84] phosphate concentration,[81] loss of telopeptides during purification processes,[7,33,81] the presence of crosslinks, [12,27] and the presence of aldehydes capable of forming crosslinks.[6] In these studies the aggregation of molecules into intact fibrils appears to be a multistep process involving at least an initiation step, a linear growth phase in which filaments are formed, and a lateral growth phase in which the filaments associate.[27] After the initiation step it is believed that individual molecules assemble into very long thin filaments and that in the final phase these fine filaments aggregate to form normal size fibrils.[27,77,81]

Collagen molecules from dense connective tissues are produced by the cleavage of amino- and carboxy-terminal peptides from the procollagen molecules synthesized by the cells. Each of these peptides is liberated by the action of a specific enzyme known as the procollagen amino-terminal proteinase and the procollagen carboxy-terminal proteinase respectively. Studies of the effects of these enzymes on pN-collagen (collagen with the amino propeptide attached) and pC-collagen (collagen with the carboxy-terminal propeptide attached) have shown that although their rates of action are similar the order in which the terminal propeptides are cleaved in vitro does have a significant effect on the diameter of the collagen fibrils formed.[55] It follows that similar biochemical processing in vivo could provide some degree of control on the diameters and lengths of the collagen fibrils. It was pointed out by these authors that the concentration of the procollagen amino-terminal proteinase bore no simple relationship to the ultimate fibril diameter though it did affect the rate of enzymatic cleavage of pN-collagen and hence of fibril assembly. Other studies by Miyahara et al.[54] have shown that it is most probably the cleavage of the carboxy-propeptide that is the crucial step in initiating fibril formation in the in vitro conditions employed in their experiments.

The possibility that fibril growth could be modified by the presence of uncleaved amino propeptides on collagen molecules was suggested from the results of work on the disease known as dermatosparaxis. In this condition pN-collagen accumulates in skin (or other tissues) leading to poorly organized fibrils of smaller diameters than those found in normal tissue. From these considerations it is possible that it is the subsequent processing of the amino propeptides that regulates fibril growth as regards both rate of fibril formation and ultimate size. Fleischmajer et al.[20,21] have shown that small diameter collagen fibrils (~40 nm diameter) contain type I, D. periodically disposed, pN-collagen molecules, whereas fibrils of larger diameter do not. They also showed that the carboxy-propeptides, although in near proximity to the small diameter collagen fibrils, were not bound periodically. It is therefore possible that they are cleaved from procollagen prior to pN-collagen molecules being incorporated in the fibrils. It remains a possibility that these small fibrils are aberrant forms and are not precursors of the large diameter fibrils found at later stages of development. However, labeling with antibodies against type III collagen amino propeptides shows that this propeptide is present in small collagen fibrils with diameters up to 60 nm, from the embryonic state right through to maturity.

Fibrils formed by the action of procollagen carboxy-terminal proteinase on pC-collagen have indicated the presence of a 4-D period. This observation is consistent with the ideas of Silver et al.[70,71] who postulated that 4-D staggered linear dimers and trimers represent an early step in the collagen assembly mechanism. Subsequent lateral aggregation of about five trimers followed by linear and lateral aggregation of these assemblies has also been proposed.[70]

Trelstad and Hayashi[76] have shown that fibrillogenesis in chick tendon appears to occur at the surface of the fibroblast. Goldberg[29] has also shown that type I collagen binds to the membrane of the fibroblast and that secretion of procollagen, as well as the initiation and control of fibrillogenesis, may occur at that site. Evidence was also presented by Trelstad and Hayashi[76] that subassemblies of collagen molecules occur intracellularly, that these are excreted by exocytosis, and that they are built onto the ends of existing fibrils that are in close association with the cell surface, thus allowing the cell to regulate, in part at least, the growth and morphology of the collagen fibrils. An elegant study on cornea by Birk and Trelstad[3] has revealed that fibrillogenesis occurred within small surface recesses on the corneal fibroblast. These recesses each contained about ten mature collagen fibrils of diameter ~25 nm, and as corneal stromal development proceeded these small recesses fused into larger ones resulting in well-defined cell-surface pockets containing up to 100 fibrils. In turn, these pockets fused, the cell surface retracted, and large compartments were formed in which bundles of collagen fibrils combined to form a corneal stromal lamella (Figure 7). The cellular processes were shown to have near orthogonal axes approximately parallel to the large surface compartments. Thus, even the nature of the lamellar structure of corneal stroma appears to be specified at the cellular level. Cornea, of course, is unique in that all of the fibrils in the mature tissue have the same (small) diameter. Nonetheless, the observations of Birk and Trelstad[3] are clearly of major importance in understanding the mechanism of fibrillogenesis in an in vivo situation.

It has also long been established that proteoglycans and GAGs bind to collagen in vitro and that they have an effect on the ultimate sizes of the collagen fibrils so formed (see for example Wood,[83] Keech,[46] Toole and Lowther,[74] Oegema et al.[56]). Those GAGs which delay fibrillogenesis and which give an extended nucleation phase tend to allow collagen fibrils to grow to only relatively small diameters, whereas those GAGs which increase the rate of fibrillogenesis tend to allow collagen fibrils of larger diameter to form.[46,83] Scott et al.[68,69] have suggested that proteoglycans, and a chondroitin-sulfate-rich proteoglycan in particular,

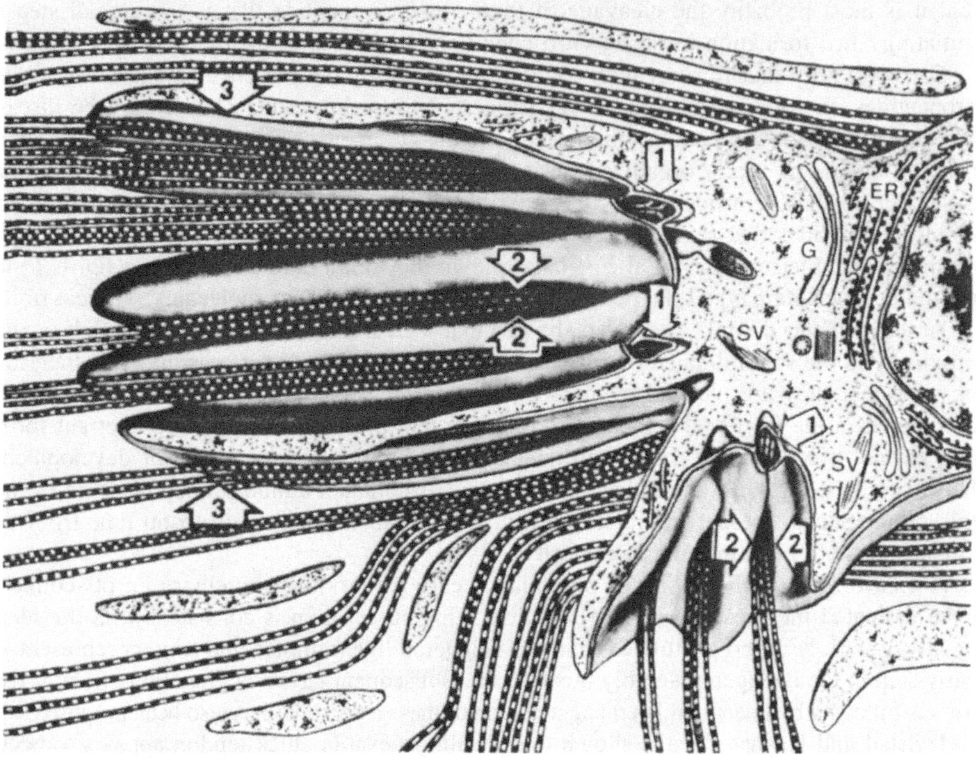

FIGURE 7. Schematic representation of the model proposed by Birk and Trelstad[1] to describe the compartmentalization of the extracellular space by the corneal fibroblast. Procollagen molecules after post-translational modification undergo preliminary aggregation. These steps involve the endoplasmic reticulum (ER), the Golgi apparatus (G), and secretory vacuoles (SV). The latter fuse with the cell surface and give rise to small surface recesses (1) which typically contain about 10 collagen fibrils of mature diameter. Lateral fusion of these recesses produces larger recesses and surface folds (2) in which 50 to 100 collagen fibrils form a fiber bundle. Further fusion of these surface foldings and retraction of the cell surface leads to large surface-associated compartments (3) in which the fibril bundles coalesce to form the even larger bundles which constitute the stromal lamellae. The orthogonal arrangement of the lamellae is specified by the orientation of the surface-associated compartments. (Modified from Birk, D. E. and Trelstad, R. L., *J. Cell Biol.*, 99, 2024, 1984. With permission.)

may have an inhibitory effect on the lateral growth of collagen fibrils in developing rat tail tendon, possibly through interference with the crosslinking between collagen molecules. A more detailed scheme in which specific GAGs are associated with a specific range of fibril diameters has been proposed by Parry et al.[63] From an extensive analysis of fibril diameter and GAG composition they were able to suggest (1) that lateral growth of fibrils beyond a diameter of about 60 nm would be inhibited by an excess of hyaluronic acid, (2) that this inhibition could be removed by increasing concentrations of chondroitin sulfate and/or dermatan sulfate, (3) that high concentrations of chondroitin sulfate would inhibit growth beyond diameters of about 150 nm, and (4) that this inhibition could be removed by increasing concentrations of dermatan sulfate such that it became the dominant GAG in the tissue. It was thus postulated that collagen fibril stability is closely related to the composition of the GAGs in the matrix (Figure 8).

Fibronectin, a high molecular weight glycoprotein that occurs in the extracellular matrix and on the surface of cells binds to denatured collagen and, to a lesser extent, native collagen.[47] It has been reported that fibronectin inhibits the rate at which collagen fibrils

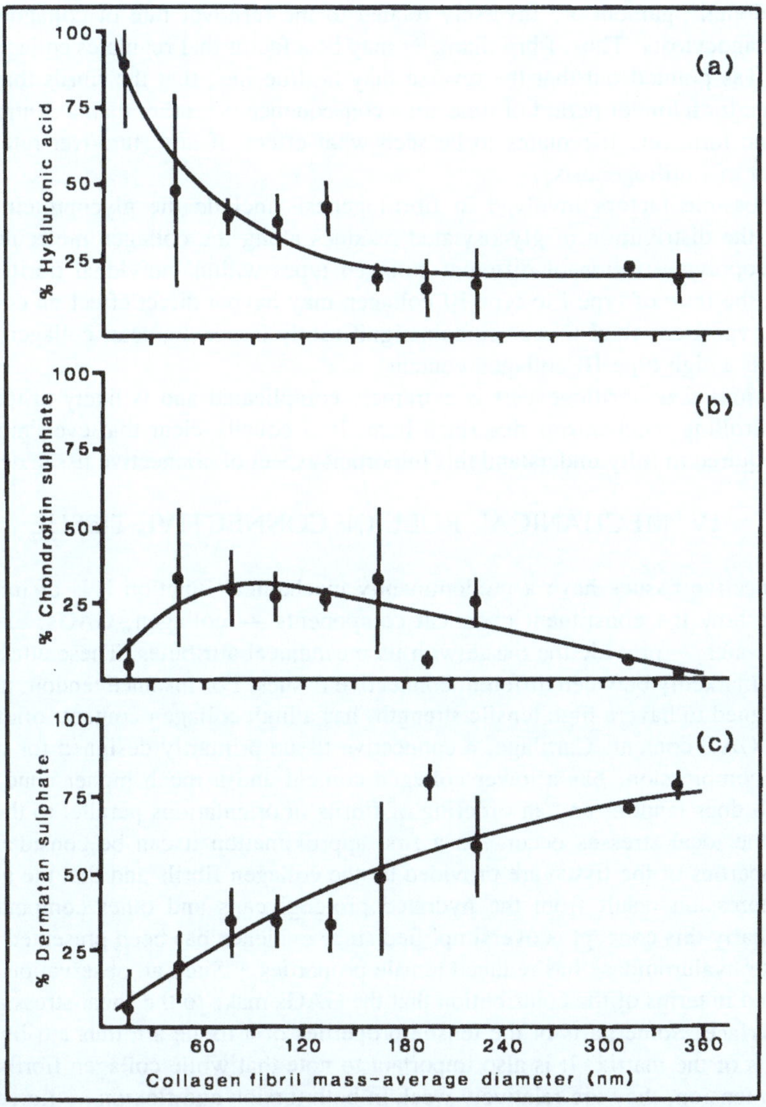

FIGURE 8. The mean percentage contents of (a) hyaluronic acid, (b) chondroitin sulfate, and (c) dermatan sulfate have been measured from a variety of connective tissues at different ages and have been plotted here as a function of the mass-average diameters of the collagen fibrils present in those tissues. The GAG contents are shown with limits of ± 1 SD and the collagen fibril diameters have been grouped into 30-nm intervals (see Parry et al.[63] for full data). In (a) it can be seen that the hyaluronic acid level drops significantly with increasing fibril diameter, whereas in (c) the dermatan sulfate level increases significantly with increasing diameter. The chondroitin sulfate levels in (b) peak for tissues having collagen fibrils of mass-average diameter 60 to 150 nm.

are formed though it apparently does not change the extent of fibril formation. Thus, since fibronectin delays fibrillogenesis it is conceivable that it could regulate the size distribution of collagen fibrils in vivo, possibly in conjunction with the proteoglycans. Recent work by Brokaw et al.[9] suggests that fibronectin has an effect on the kinetics of collagen self-assembly and that this may result from its ability to prevent molecular collisions.

Svoboda et al.[73] have noted that the mean sizes of the collagen fibrils in rat skin, gingiva, and periodontal ligament are inversely related to the turnover rate of collagen and to the level of phagocytosis. Thus, fibril diameter may be a factor that regulates collagen turnover, though it was pointed out that the reverse may be true i.e., that the fibrils that have been in existence for a longer period of time are a consequence of, rather than a controlling factor in, collagen turnover. It remains to be seen what effect, if any, turnover rate has on the mechanism of fibrillogenesis.

Other possible factors involved in fibrillogenesis include the glycoprotein content of collagen,[1] the distribution of glycosylated residues along the collagen molecule,[50] and the effect of copolymerization of different collagen types within individual fibrils.[34,49] In the latter case the ratio of type I to type III collagen may have a direct effect on collagen fibril size since type I enriched tissues contain significantly larger diameter collagen fibrils than tissues with a high type III collagen content.

It is obvious that fibrillogenesis is extremely complicated and is likely to include many of the controlling mechanisms described here. It is equally clear that even more research effort is required to fully understand this important aspect of connective tissue development.

IV. MECHANICAL ROLE OF CONNECTIVE TISSUE

As connective tissues have a predominantly mechanical function it is of importance to understand how the constituent chemical components — collagen, GAGs, elastic fibers, minerals, water — provide the tissue with its mechanical attributes. These attributes, however, vary markedly between different connective tissues. For instance tendon, a connective tissue designed to have a high tensile strength, has a high collagen content, oriented fibrils, and a low GAG content. Cartilage, a connective tissue primarily designed for its ability to withstand compression, has a lower collagen content and a much higher concentration of GAGs than does tendon, and an ordering of fibrils in orientations parallel to the directions in which the local stresses occur. To a first approximation it can be considered that the tensile properties of the tissue are provided by the collagen fibrils and that the properties to resist compression result from the hydrated proteoglycans and other components of the matrix. Clearly this concept is oversimplified since evidence has been presented that tendon treated with hyaluronidase has reduced tensile properties.[64] Such an observation can readily be explained in terms of the contribution that the GAGs make to the shear stress at the fibril-matrix interface. Some parts of the tensile properties of a tissue are thus attributable to the components of the matrix. It is also important to note that while collagen fibrils have great strength in tension, they are relatively weak in both torsion and flexion — i.e., they can be easily twisted or bent.

Other specialized tissues, such as the *ligamentum nuchae* (prominent in grazing animals), are rich in elastic fibers and their mechanical properties are greatly influenced by this component. Its dominance in the *ligamentum nuchae* and its high levels in arterial walls imply that its importance lies with its ability to give a tissue both resilience and extensibility. These factors should not be confused with elasticity in the physical sense. The moduli of elasticity of elastin and collagen have been recorded as $\sim 6 \times 10^5$ and $\sim 8 \times 10^7$ N/m^{-2} respectively, the latter being two orders of magnitude greater. Collagen is therefore more elastic than elastin, but less extensible, breaking at a strain of about 8 to 10%.[66] Elastin shows long-range reversible extensibility and does not creep when loaded for extended periods of time; purified samples of *ligamentum nuchae* break at about twice their resting length.[80] Elastic ("pliant") composites may be considered as a three-phase system, with the collagen fibrils and the proteoglycan components arranged in parallel with the elastic fibers. The strain rate of this three-phase system, however, will arise mainly from the interactions of the collagen fibrils and the proteoglycan components. If the viscous interactions of the latter components with the material as a whole are not too large then the system will have resilience properties similar to those of a pure "protein rubber" and the composite will be able to function as an energy-storing system for relatively high speed cyclic processes as is the case in the vertebrate arterial wall.[80] Most of the connective tissues with high tensile strengths (e.g., skin and tendon) have high collagen and low elastic fiber contents.

It is of interest in this context to note the experimental values of the resilience of collagen fibrils and elastic fibers. Resilience, defined as the percentage of the energy recovered from an elastic recoil, has a value of about 75% for tendon[18,80] and something in excess of 90% for elastin.[80] Since the strain energy of tendon is about 12 MJ m^{-3} (i.e., 8 to 9 J g^{-1} of tissue) it follows that about 2 J g^{-1} are dissipated in each stretching cycle. In comparison the energy dissipated in each cycle of elastic fiber deformation is negligibly small.

In terms of their mechanical role, connective tissues must have some appropriate combination of two basic attributes — an ability to withstand high tensile or compressional stresses and an ability to recover shape and form when stresses are removed. While every collagen fibril in a tissue will contribute something to each of these properties it is not unlikely that their relative contributions could differ with diameter. This idea has been supported by the observations that tissues which sustain long-term high-stress levels have broad bimodal distributions of collagen fibril diameter at maturity, whereas tissues which suffer short-term or intermittent low-stress levels have sharper, unimodal distributions of fibril diameter. As a result of many such observations it has been suggested that the larger diameter fibrils in the bimodal distributions are responsible for maintaining the tensile properties of the tissue, whereas the smaller diameter fibrils play an important part in determining the creep-inhibition properties.[59,60] These conclusions were primarily based on two considerations. First, the percentage of covalent crosslinks between collagen molecules is predicted to increase rapidly with fibril diameter (up to $\sim$100 nm) and hence lead to fibrils of greater tensile strength.[62] Molecules which lie within the peripheral layer of a fibril are unlikely to be able to form the full complement of covalent crosslinks, in contrast to those molecules lying within the body of the fibril. Large diameter fibrils which have a smaller peripheral layer per unit mass of collagen will thus have higher crosslink densities than small fibrils where the peripheral layer forms an appreciable portion of the mass. Second, since the surface area of fibrils per unit mass of collagen increases as the fibril diameters decrease it follows that the possible number of interactions between the matrix and the fibrils can be maximized by decreasing the diameter (and hence increasing the number) of the fibrils. It is considered that the maximization of fibril-matrix interactions would help to inhibit permanent creep and thus allow a tissue to regain its shape after removal of stress.

As a small fraction of large diameter fibrils can account for a large part of the mass of collagen in a tissue having a bimodal distribution of collagen fibril diameters, it is often useful to consider the mass-average fibril diameter (rather than the mean diameter) as an index of collagen fibril size.[22] On the basis of the hypothesis previously discussed this mass-average diameter should be related to the ultimate tensile strength of a tissue and to the change in strength of a tissue measured as a function of age. Indeed good qualitative correlations can be recognized by combining the data presented by Vogel[78,79] and Parry and Craig[59] (Figure 9). Thus a long-term high-stress tissue has a bimodal distribution of collagen fibril diameters, with the largest fibrils present providing the high tensile strength and the large percentage of smaller diameter fibrils providing the creep-inhibition properties. A narrow distribution of fibril diameters would be unable to provide both the strength and creep-inhibition attributes simultaneously. Further, a polydisperse distribution of fibril sizes, as commonly occurs in tendon, allows better packing of fibrils and gives rise to a greater density of collagen per unit mass of tissue. In contrast, a short-term low-stress tissue has a narrow distribution of fibril diameters sufficient to give its tensile and creep-resistance requirements.

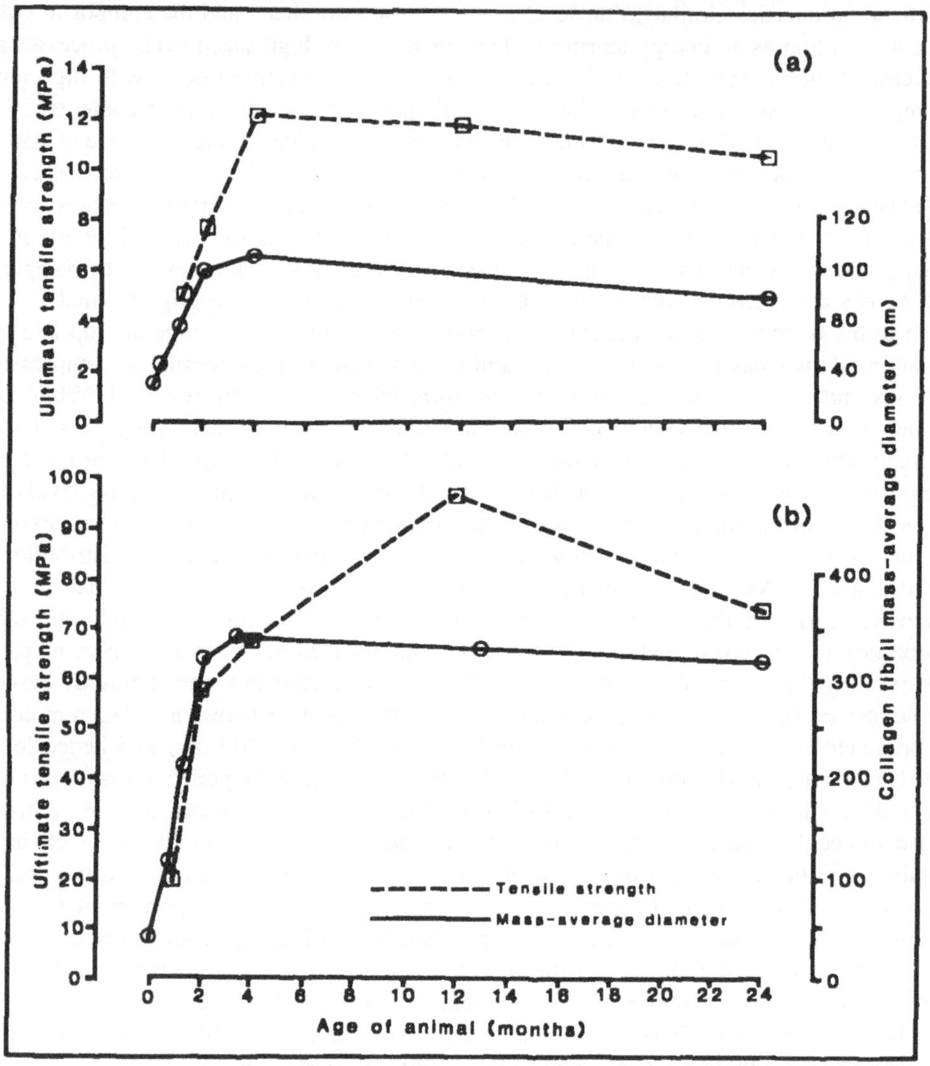

FIGURE 9. Graphs of tensile strength and mass-average collagen fibril diameter vs. age for (a) rat skin and (b) rat-tail tendon. Tensile strength measurements have been taken from Vogel[78,79] and collagen fibril mass-average diameters from Parry and Craig.[59] The tensile strength and mass-average diameter graphs are of similar form.

Thus far the lengths of the collagen fibrils have not been considered, since it has been a tacit assumption that these are significantly greater than the so-called critical length (l_c), where l_c is given by the expression:[36,45,48]

$$l_c = \frac{\text{radius} \times \text{breaking stress of a collagen fibril}}{\text{shear stress exerted on that fibril by the matrix}}$$

When a connective tissue is stretched the fibril-matrix interface is sheared causing the collagen fibrils to become strained, thus supporting the applied load. The shear stress rises from zero at the ends of a fibril to a maximum value at distances of $l_c/2$ from either end. Providing that the length of a collagen fibril is appreciably greater than l_c it follows that there is no absolute mechanical requirement that collagen fibrils should extend from one end of a tissue to the other. Experimentally there is very little data on the length of collagen fibrils in any connective tissue. Parry and Craig[59] noted that they were able to follow the course of 1000 fibrils in mature tendon over a distance of 1 μm through the use of serial sections. Since no fibril end was observed it was concluded that the average length of the fibrils must be close to 1 mm, and hence that the fibrils have an axial ratio of 10^4 (or even greater). Since the critical length in tendon is believed to be about 100 μm,[37] it is clear that the lengths of fibrils in tendon are as much as an order of magnitude long enough for them to be able to act as high tensile elements. The interactions between the GAGs in the matrix and the collagen molecules in the fibrils affect the value of the shear stress and this, together with the collagen fibril diameter distribution, explains the different elastic and viscous mechanical properties exhibited by connective tissues.

There are many factors other than the collagen fibril length and diameter distribution which contribute to the mechanical attributes of a tissue. For example, the fibril-matrix structure of connective tissue provides it with a mechanism to resist crack propagation. Such design is paralleled in other biological tissues (e.g., wool fibers) and is mimicked by man in his construction of reinforced concrete and carbon fiber composites. The orientation of the collagen fibrils within a specific tissue is also of importance.[35] Examples of extreme cases are viscoelastic fluids such as vitreous humor, in which the collagen fibrils are randomly oriented, and tendon, a tissue subjected to unidirectional loading, where the collagen fibrils are all strictly aligned with their long axes parallel to the direction of the applied load. In the case of a fibril-matrix composite, the Young modulus (E_c) is related to the moduli of the matrix (E_m) and fibril (E_f) by the expression

$$E_c = E_f/V_f + E_m(1 - V_f)$$

where V_f is the volume fraction of the fibrils.

There exist many architectures intermediate between these extremes. The most commonly occurring of these is the layered arrangement of fibrils seen in cornea, tapetum, skin, and some bones and cartilages. The orientation of collagen fibrils in a connective tissue is clearly related to the direction(s) in which the tissue is normally stressed. Fibrils lying in these directions are necessarily the ones which will primarily support the load applied. Fibrils whose long axes do not lie close to the direction of applied stress will contribute little mechanically, though by adding weight to the tissue will reduce the strength to weight ratio.[80] Since collagen fibrils in connective tissues have such a diversity of orientations it is necessary to assess their effectiveness as reinforcing elements in the tissue. Krenchel[48] has defined a term known as the efficiency of reinforcement (η) as

$$\eta = \Sigma a_n \cos^4\phi$$

where a_n = fraction of a group of fibrils lying in a given plane and ϕ = angle between the various fibril groups and the loading axis. The expression for the Young modulus of the composite then becomes

$$E_c = \eta E_f V_f + E_m(1 - V_f)$$

For a parallel fibril assembly (stress direction parallel to the fibril axis) $\eta = 1.0$; for fibrils uniformly distributed in a plane (stress direction anywhere within the plane) $\eta = 0.375$; for fibrils randomly distributed in three dimensions (stress in any direction) $\eta = 0.2$.

Many connective tissues normally subject to rapid application of high stress levels show a macroscopic crimp structure at the fibril level (crimp length and angle typically ~1 to

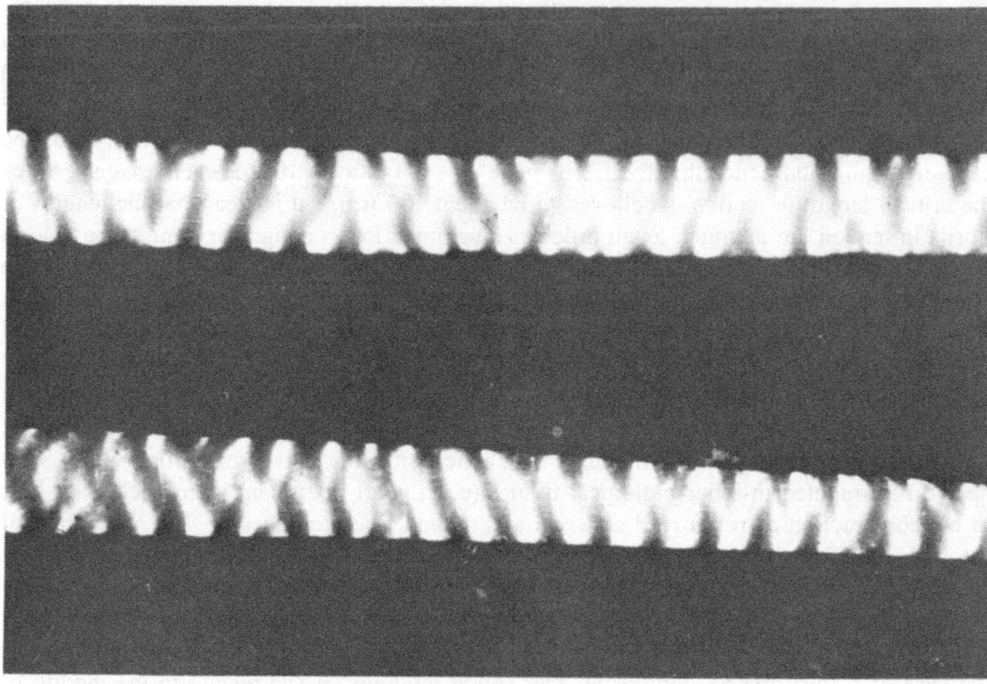

FIGURE 10. Light micrograph of two collagen fibers from mature rat-tail tendon. The crimp period (~100 μm) can readily be seen. It is believed that the planar crimp provides a compliance mechanism for those connective tissues subjected to application of rapid stress. (Magnification ~60 X.)

100 μm and ~5 to 25° respectively, Figure 10). This, together with local reorientation of fibrils under stress, provides a compliance mechanism enabling the tissue to undergo a 3 to 4% length change prior to the individual molecules being stretched, thus helping to prevent the tissue from sustaining permanent damage.[17,26] It should be noted, however, that both the crimp and the orientation of the collagen fibrils gradually change with increasing age and that these changes have, in the senescent years, an adverse effect on the compliance mechanism. Further, structural studies on the conformation of the collagen molecule[25] have shown that it consists of axial regions of alternating "order" (overlap region) and "disorder" (gap region); it follows that the application of stress in a fibril may result in a preferential unraveling of the regions of disorder. Thus a second level compliance mechanism may be present in connective tissue.

In general, the mechanical behavior of connective tissues can be explained in terms of the relationship between tensile force and extension. A stress-strain curve for most tissues may be divided into three distinct regions, each of which may be attributed in part to the different levels of structural hierarchy present in the tissue: (1) a low stress region where nonlinear elastic extension is affected by the angle of crimp — the toe region extension of a fibre will increase with an increase in crimp angle; (2) a linear elastic region in which the slope of the curve is a measure of the Young modulus of the tissue; (3) a high stress "yield and rupture" region in which the yield behavior and rupture strength are determined by progressive breaking of interfibrillar crosslinks at high stress levels, and intrafibrillar cross-links at still higher stress levels.

In the linear elastic region of the force-extension curve the stress which a tissue may sustain increases with an increase in the ratio of collagen to matrix. This ratio is directly dependent on the density per unit volume of collagen in the tissue. A greater density can be achieved in tissues with a bimodal rather than a unimodal distribution of collagen fibril diameters because smaller diameter fibrils can be fitted into the voids created by the packing of the larger ones. This would suggest that the linear elastic modulus of tendons with a unimodal distribution of collagen fibril diameters may be smaller than those which have bimodal distributions of fibril diameters.

The yield behavior of the tissue is determined by the density of interfibrillar crosslinks in the tissue. This is governed, to a large extent, by the form of the distribution of collagen fibril diameters — i.e., by the surface area of collagen fibrils presented to the matrix per unit volume of tissue. The rupture behavior, however, is dependent on the absolute content of collagen per unit volume of tissue. It is therefore probable that the stress which would cause rupture in a large diameter fibril would be greater than for a smaller one.[62] By a similar argument a tissue having a unimodal distribution of fibril diameters would be expected to have a lower rupture strength than one with a bimodal distribution.

V. CONCLUSIONS

Considerable success has been achieved in recent years in furthering our understanding of the mechanism of collagen fibril growth and maturation and also the relationship between collagen fibril size distribution and mechanical attributes of connective tissues. It is therefore appropriate to reflect upon aspects of those areas of research that must be studied in yet greater detail if a better understanding is to be achieved.

Although electron microscopy as a biophysical technique has proved a very powerful method of investigating aggregates of collagen molecules, both in transverse and longitudinal section, several important problems have not yet been addressed by this method; such problems include the quantitative determination of the lengths of the collagen fibrils and the number density of collagen fibrils in connective tissue as a function of age. It can also be argued that greater use should be made of electron microscopy during the earliest stages of fibrillogenesis, where quantitative data are often lacking. New preparative techniques in electron microscopy such as rapid freezing, low-temperature embedding, and ultracryotomy may provide a view of collagen fibril structure that is closer to that occurring in vivo. It is envisaged that techniques such as these will play a significant and increasingly important part in connective tissue research during the coming years.

The difficulties of recording in vivo mechanical moduli from in vitro studies of connective tissues have been discussed many times in the past. Evidence suggests that the experimental values may be as much as an order of magnitude lower than the in vivo values. Since it is clearly important that such moduli be measured as accurately as possible, the use of non-destructive techniques, such as the Brillouin scattering method, may provide a partial answer. It also remains a matter of concern that age-related studies of mechanical properties of connective tissues have rarely been undertaken. Such studies are important if correlations are to be confirmed between the chemically diverse components of the tissue and its mechanical attributes. Indeed, it is worth emphasizing this latter point since it is not uncommon that individual research workers tend to investigate only particular components of connective tissues. Such tissues, however, are complex multicomposite biological structures and it is important to understand the interactions of one component with another. It is the result of these synergistic interactions that endow connective tissues with those special properties which enable them to function in the diversity of ways encountered in vivo.

Connective tissue remodeling, although not considered specifically in this review, has provided important clues in the mechanism of collagen fibril development. It has been established, for instance, that significant changes occur in the relative composition and state of aggregation of the various chemical components in connective tissue when the tensile and compressional stresses to which the tissue is normally subjected are altered. Further coordinated biochemical and structural studies on connective tissue remodeling are needed if the feedback mechanisms existing between the cellular and the fibril-matrix components of the tissue are to be established.

ACKNOWLEDGMENT

The authors would like to acknowledge the photographic assistance rendered by Mr. R. J. Bennett.

REFERENCES

1. **Anderson, J. C., Labedz, R. I., and Kewley, M. A.,** The effect of bovine tendon glycoproteins on the formation of fibrils from collagen solutions, *Biochem. J.*, 168, 345, 1977.
2. **Bailey, A. J. and Etherington, D. J.,** Metabolism of collagen and elastin, in *Comprehensive Biochemistry*, Vol. 19b, Florkin, M. and Stotz, E. H., Eds., Elsevier/North-Holland, Amsterdam, 1980, 299.
3. **Birk, D. E. and Trelstad, R. L.,** Extracellular compartments in matrix morphogenesis: collagen fibril, bundle, and lamellar formation by corneal fibroblasts, *J. Cell Biol.*, 99, 2024, 1984.
4. **Bornstein, P. and Sage, H.,** Structurally distinct collagen types, *Ann. Rev. Biochem.*, 49, 957, 1980.
5. **Bornstein, P. and Traub, W.,** The chemistry and biology of collagen, in *The Proteins*, Vol. 4, Neurath, H. and Hill, R. L., Eds., Academic Press, New York, 1979, 411.
6. **Brennan, M. and Davison, P. F.,** Role of aldehydes in collagen fibrillogenesis *in vitro*, *Biopolymers*, 19, 1861, 1980.
7. **Brennan, M. and Davison, P. F.,** Influence of the telopeptides on Type I collagen fibrillogenesis, *Biopolymers*, 20, 2195, 1981.
8. **Brodsky, B. and Eikenberry, E. F.,** Characterization of fibrous forms of collagen, *Meth. Enzymol.*, 82, 127, 1982.
9. **Brokaw, J. L., Doillon, C. J., Hahn, R. A., Birk, D. E., Berg, R. A., and Silver, F. H.,** Turbidimetric and morphological studies of Type I collagen self-assembly *in vitro* and the influence of fibronectin, *Int. J. Biol. Macromol.*, 7, 135, 1985.
10. **Bruns, R. R., Hulmes, D. J. S., Therrien, S. F., and Gross, J.,** Procollagen segment-long-spacing crystallites: their role in collagen fibrillogenesis, *Proc. Natl. Acad. Sci. U.S.A.*, 76, 313, 1979.
11. **Chew, M. W. K. and Squire, J. M.,** Cryosections of X-ray monitored collagen fibrils provide support for quasi-hexagonal molecular packing, *Int. J. Biol. Macromol.*, 8, 27, 1986.
12. **Comper, W. D. and Veis, A.,** The mechanism of nucleation for *in vitro* collagen formation, *Biopolymers*, 16, 2113, 1977.
13. **Craig, A. S.,** The Fibrillar Organization of Collagen in Connective Tissue, Ph.D. thesis, Massey University, Palmerston North, New Zealand, 1984.
14. **Craig, A. S. and Parry, D. A. D.,** Collagen fibrils of the vertebrate corneal stroma, *J. Ultrastruct. Res.*, 74, 232, 1981.
15. **Craig, A. S. and Parry, D. A. D.,** Growth and development of collagen fibrils in immature tissues from rat and sheep, *Proc. R. Soc. London Ser. B.*, 212, 85, 1981.
16. **Craig, A. S. and Parry, D. A. D.,** The sub-structure of collagen fibrils, in *The Chemistry and Biology of Mineralized Connective Tissues*, Veis, A., Ed., Elsevier/North-Holland, Amsterdam, 1981, 107.
17. **Diamant, J., Keller, A., Baer, E., Litt, M., and Arridge, R. G. C.,** Collagen; ultrastructure and its relation to mechanical properties as a function of ageing, *Proc. R. Soc. London Ser. B.*, 180, 293, 1972.
18. **Dunn, M. G. and Silver, F. H.,** Viscoelastic behaviour of human connective tissues: relative contribution of viscous and elastic properties, *Connect. Tissue Res.*, 12, 59, 1983.
19. **Fietzek, P. P. and Kuhn, K.,** The primary structure of collagen, *Int. Rev. Connect. Tissue Res.*, 7, 1, 1976.
20. **Fleischmajer, R., Timpl, R., Tuderman, L., Raisher, L., Weistner, M., Perlish, J. S., and Graves, P. N.,** Ultrastructural identification of extension aminopropeptides of Types I and III collagens in human skin, *Proc. Natl. Acad. Sci. U.S.A.*, 78, 7360, 1981.
21. **Fleischmajer, R., Olsen, B. R., Timpl, R., Perlish, J. S. and Lovelace, O.,** Collagen fibril formation during embryogenesis, *Proc. Natl. Acad. Sci. U.S.A.*, 80, 3354, 1983.

22. **Flint, M. H., Craig, A. S., Reilly, H. C., Gillard, G. C., and Parry, D. A. D.,** Collagen fibril diameters and glycosaminoglycan content of skins — indices of tissue maturity and function, *Connect. Tissue Res.*, 13, 69, 1984.
23. **Fraser, R. D. B. and MacRae, T. P.,** The crystalline structure of collagen fibrils in tendon, *J. Mol. Biol.*, 127, 129, 1979.
24. **Fraser, R. D. B., MacRae, T. P., and Suzuki, E.,** The molecular and fibrillar structure of collagen, in *Fibrous Proteins: Scientific, Industrial and Medical Aspects*, Vol. 2, Parry, D. A. D. and Creamer, L. K., Eds., Academic Press, New York, 1979, 179.
25. **Fraser, R. D. B., MacRae, T. P., Miller, A., and Suzuki, E.,** Molecular conformation and packing in collagen fibrils, *J. Mol. Biol.*, 167, 497, 1983.
26. **Gathercole, L. J. and Keller, A.,** Light microscopic waveforms in collagenous tissues and their structural implications, in *Structure of Fibrous Biopolymers. Colston Papers No. 26*, Atkins, E. D. T. and Keller, A., Eds., Butterworths, London, 1975, 153.
27. **Gelman, R. A., Williams, B. R., and Piez, K. A.,** Collagen fibril formation, *J. Biol. Chem.* 254, 180, 1979.
28. **Glanville, R. W. and Kuhn, K.,** The primary structure of collagen, in *Fibrous Proteins: Scientific, Industrial and Medical Aspects*, Parry, D. A. D. and Creamer, L. K., Eds., Academic Press, New York, 1979, 133.
29. **Goldberg, B.,** Binding of soluble Type I collagen molecules to the fibroblast plasma membrane, *Cell*, 16, 265, 1979.
30. **Greenlee, T. K. and Ross, R.,** The development of the rat flexor digital tendon, a fine structure study, *J. Ultrastruct. Res.*, 18, 354, 1967.
31. **Gross, J. and Bruns, R. R.,** Another look at fibrillogenesis, in *Role of Extracellular Matrix in Development*, Trelstad, R. L., Ed., Alan R. Liss, New York, 1984, 479.
32. **Harkness, R. D.,** Biological function of collagen, *Biol. Rev.*, 36, 399, 1961.
33. **Helseth, D. L. and Veis, A.,** Collagen self-assembly *in vitro*, *J. Biol. Chem.*, 256, 7118, 1981.
34. **Henkel, W. and Glanville, R. W.,** Covalent crosslinking between molecules of Type I and Type III collagens. The involvement of N-terminal, non-helical regions of the α1(I) and α1(III) chains in the formation of intermolecular crosslinks, *Eur. J. Biochem.*, 122, 205, 1982.
35. **Hukins, D. W. L.,** Collagen orientation, in *Connective Tissue Matrix, Topics in Molecular and Structural Biology*, Vol. 5, Hukins, D. W. L., Ed., MacMillan, New York, 1984, 211.
36. **Hukins, D. W. L. and Aspden, R. M.,** Composition and properties of connective tissues, *Trends Biochem. Sci.*, 10, 260, 1985.
37. **Hukins, D. W. L., Aspden, R. M., and Yarker, Y. E.,** Fibre reinforcement and mechanical stability in articular cartilage, *Eng. Med.*, 13, 153, 1984.
38. **Hulmes, D. J. S.,** A possible mechanism for the regulation of collagen fibril diameter *in vitro*, *Coll. Rel. Res.*, 3, 317, 1983.
39. **Hulmes, D. J. S. and Miller, A.,** Quasi-hexagonal molecular packing in collagen fibrils, *Nature (London)*, 282, 878, 1979.
40. **Hulmes, D. J. S., Miller, A., Parry, D.A.D., Piez, K. A., and Woodhead-Galloway, J.,** Analysis of the primary structure of collagen for the origins of molecular packing, *J. Mol. Biol.*, 79, 137, 1973.
41. **Hulmes, D. J. S., Jesoir, J. C., Miller, A., Berthet-Colominas, C., and Wolff, C.,** Electron microscopy shows periodic structure in collagen fibril cross-sections, *Proc. Natl. Acad. Sci. U.S.A.*, 78, 3567, 1981.
42. **Hulmes, D. J. S., Bruns, R. R., and Gross, J.,** On the state of aggregation of newly secreted procollagen, *Proc. Natl. Acad. Sci. U.S.A.*, 80, 388, 1983.
43. **Hulmes, D. J. S., Holmes, D. F., and Cummings, C.,** Crystalline regions in collagen fibrils, *J. Mol. Biol.*, 184, 473, 1985.
44. **Jackson, S. F.,** The morphogenesis of avian tendon, *Proc. R. Soc. London Ser. B.*, 144, 556, 1956.
45. **Jeronomidis, G. and Vincent, J. F. V.,** Composite materials, in *Connective Tissue Matrix, Topics in Molecular and Structural Biology*, Vol. 5, Hukins, D. W. L., Ed., MacMillan, New York, 1984, 187.
46. **Keech, M. K.,** The formation of fibrils from collagen solutions. IV. Effect of mucopolysaccharides and nucleic acids: an electron microscope study, *J. Biophys. Biochem. Cytol.*, 9, 193, 1961.
47. **Kleinman, H. K., Wilkes, C. M., and Martin, G. R.,** Interaction of fibronectin with collagen fibrils, *Biochemistry*, 20, 2325, 1981.
48. **Krenchel, H.,** *Fibre Reinforcement*, Akademisk Forlag, Copenhagen, 1964.
49. **Lapiere, C. M., Nusgens, B., and Pierard, G. E.,** Interaction between collagen Type I and Type III in conditioning bundles organization, *Connect. Tissue Res.*, 5, 21, 1977.
50. **Mathews, M. B.,** *Connective Tissue Macromolecular Structure and Evolution*, Springer-Verlag, Berlin, 1975.
51. **Miller, A.,** Molecular packing in collagen fibrils, in *Biochemistry of Collagen*, Ramachandran, G. N. and Reddi, A. H., Eds., Plenum Press, New York, 1976, 85.

52. **Miller, A. and Parry, D. A. D.,** Structure and packing of microfibrils in collagen, *J. Mol. Biol.*, 75, 441, 1973.
53. **Miller, A. and Wray, J. S.,** Molecular packing in collagen, *Nature (London)*, 230, 437, 1971.
54. **Miyahara, M., Njieha, F. K., and Prockop, D. J.,** Formation of collagen fibrils *in vitro* by cleavage of procollagen with procollagen proteinases, *J. Biol. Chem.*, 257, 8442, 1982.
55. **Miyahara, M., Hayashi, K., Berger, J., Tanzawa, K., Njieha, F. K., Trelstad, R. L., and Prockop, D. J.,** Formation of collagen fibrils by enzymatic cleavage of precursors of Type I collagen *in vitro*, *J. Biol. Chem.*, 259, 9891, 1984.
56. **Oegema, T. R., Laidlaw, J., Hascall, V. C., and Dziewiatkowski, D. D.,** The effect of proteoglycans on the formation of fibrils from collagen solutions, *Arch. Biochem. Biophys.*, 170, 698, 1975.
57. **Parry, D. A. D. and Craig, A. S.,** Electron microscope evidence for an 80 Å unit in collagen fibrils, *Nature (London)*, 282, 213, 1979.
58. **Parry, D. A. D. and Craig, A. S.,** An 80 Å unit in collagen fibrils, in *Structural Aspects of Recognition and Assembly in Biological Macromolecules*, Balaban, M., Sussman, J. L., Traub, W., and Yonath, A., Eds., Balaban ISS, Rehovot, 1981, 377.
59. **Parry, D. A. D. and Craig, A. S.,** Growth and development of collagen fibrils in connective tissue, in *The Ultrastructure of the Connective Tissue Matrix*, Ruggeri, A. and Motta, P. M., Eds., Martinus Nijhoff, The Hague, 1984, 34.
60. **Parry, D. A. D., Barnes, G. R. G., and Craig, A. S.,** A comparison of the size distribution of collagen fibrils in connective tissues as a function of age and a possible relation between fibril size distribution and mechanical properties, *Proc. R. Soc. London Ser. B.*, 203, 305, 1978.
61. **Parry, D. A. D., Craig, A. S., and Barnes, G. R. G.,** Tendon and ligament from the horse: an ultrastructural study of collagen fibrils and elastic fibres as a function of age, *Proc. R. Soc. London Ser. B.*, 203, 293, 1978.
62. **Parry, D. A. D., Craig, A. S., and Barnes, G. R. G.,** Fibrillar collagen in connective tissue, in *Fibrous Proteins: Scientific, Industrial and Medical Aspects*, Vol. 1, Parry, D. A. D. and Creamer, L. K., Eds., Academic Press, New York, 1980, 77.
63. **Parry, D. A. D., Flint, M. H., Gillard, G. C., and Craig, A. S.,** A role for the glycosaminoglycans in the development of collagen fibrils, *FEBS Lett.*, 149, 1, 1982.
64. **Partington, F. R. and Wood, G. C.,** The role of non-collagen components in the mechanical behaviour of tendon fibres, *Biochem. Biophys. Acta*, 69, 485, 1963.
65. **Ramachandran, G. N. and Ramakrishnan, C.,** Molecular structure, in *Biochemistry of Collagen*, Vol. 1, Ramachandran, G. N. and Reddi, A. H., Eds., Plenum Press, New York, 1976, 103.
66. **Rigby, B. S., Hirai, N., Spikes, J. D., and Eyring, H.,** The mechanical properties of rat tail-tendon, *J. Gen. Physiol.*, 43, 265, 1959.
67. **Ripamonti, A., Roveri, N., Braga, D., Hulmes, D. J. S., Miller, A., and Timmins, P. A.,** Effects of pH and ionic strength on the structure of collagen fibrils, *Biopolymers*, 19, 965, 1980.
68. **Scott, J. E. and Orford, C. R.,** Dermatan sulphate-rich proteoglycan associates with rat tail-tendon collagen at the d band in the gap region, *Biochem. J.*, 197, 213, 1981.
69. **Scott, J. E., Orford, C. R., and Hughes, E. W.,** Proteoglycan-collagen arrangements in developing rat tail-tendon, *Biochem. J.*, 195, 573, 1981.
70. **Silver, F. H.,** Type I collagen fibrillogenesis *in vitro*, *J. Biol. Chem.*, 256, 4973, 1981.
71. **Silver, F. H., Langley, K. H., and Trelstad, R. L.,** Type I collagen fibrillogenesis: initiation via reversible linear and lateral growth steps, *Biopolymers*, 18, 2523, 1979.
72. **Smith, J. W.,** Molecular pattern in native collagen, *Nature (London)*, 219, 157, 1968.
73. **Svoboda, E. L. A., Howley, T. P., and Deporter, D. A.,** Collagen fibril diameter and its relation to collagen turnover in three soft connective tissues in the rat, *Connect. Tissue Res.*, 12, 43, 1983.
74. **Toole, B. P. and Lowther, D. A.,** Dermatan sulphate-protein: isolation from and interaction with collagen, *Arch. Biochem. Biophys.*, 128, 567, 1968.
75. **Traub, W. and Piez, K. A.,** The chemistry and structure of collagen, *Adv. Prot. Chem.*, 25, 243, 1971.
76. **Trelstad, R. L. and Hayashi, K.,** Tendon collagen fibrillogenesis: intracellular subassemblies and cell surface changes associated with fibril growth, *Dev. Biol.*, 71, 228, 1979.
77. **Trelstad, R. L., Hayashi, K., and Gross, J.,** Collagen fibrillogenesis: intermediate aggregates and suprafibrillar order, *Proc. Natl. Acad. Sci. U.S.A.*, 73, 4027, 1976.
78. **Vogel, H. G.,** Correlation between tensile strength and collagen content in rat skin. Effect of age and cortisol treatment, *Connect. Tissue Res.*, 2, 177, 1974.
79. **Vogel, H. G.,** Influence of maturation and age on mechanical and biochemical parameters of connective tissue of various organs in the rat, *Connect. Tissue Res.*, 6, 161, 1978.
80. **Wainwright, S. A., Biggs, W. D., Currey, J. D., and Gosline, J. M.,** *Mechanical Design in Organisms*, Edward Arnold, London, 1976.
81. **Williams, B. R., Gelman, R. A., Poppke, D. C., and Piez, K. A.,** Collagen fibril formation, *J. Biol. Chem.*, 253, 6578, 1978.

82. **Williams, I. F., Craig, A. S., Parry, D. A. D., Goodship, A. E., Shah, J., and Silver, I. A.,** The development of collagen fibril organization and collagen crimp patterns during tendon healing, *Int. J. Biol. Macromol.*, 7, 275, 1985.
83. **Wood, G. C.,** The formation of fibrils from collagen solutions. 2. A mechanism of collagen fibril formation, *Biochem. J.*, 75, 598, 1960.
84. **Wood, G. C. and Keech, M. K.,** The formation of fibrils from collagen solutions. 1. The effect of experimental conditions: kinetic and electron-microscope studies, *Biochem. J.*, 75, 588, 1960.

Chapter 2

PHYLOGENETIC ASPECTS OF COLLAGEN STRUCTURE AND FUNCTION

Marvin L. Tanzer and Shigeru Kimura

TABLE OF CONTENTS

I. Introduction 26
A. Invertebrate Collagen Structure 26

II. Coelenterata (Cnidaria) 27

III. Aschelminthes 29

IV. Annelida 32

V. Mollusca 33

VI. Arthropoda 34

VII. Echinodermata 35

VIII. Vertebrata (Fish) 36

IX. Discussion 37

References 37

I. INTRODUCTION

Collagen has the distinction of being the commonest protein in the animal world and it provides an extracellular framework for all multicellular animals. This framework assumes a variety of architectural forms and indeed, the protein molecule itself shows remarkable variation in size, shape, and chemical nature albeit within the bounds of a characteristic wide angle X-ray diffraction pattern and somewhat restricted amino acid composition (1/3 glycine). Only in the last decade has the diversity of collagen and its fine-tuning at the molecular level begun to be appreciated. Indeed, the hypothesis might now be made that not only is collagen a family of structural proteins[1,2] but that the collagen conformation is utilized by nature in other molecules as a common theme, analogous to the alpha helix and beta-pleated sheet. Although it may be premature to generalize in this way, three distinct examples are pertinent: (1) the circulating complement protein C1q, which has a lengthy collagenous domain;[3] (2) the acetylcholinesterase of electric eel membrane, which also has substantial collagenous mass;[4] and (3) the glycoprotein from the pulmonary alveoli of normal individuals and those patients with alveolar proteinosis.[5] In these three examples, it should be cautioned that although primary and secondary structural data support the presence of lengthy collagenous regions, the rigorous criterion of conformational identity by X-ray diffraction has not been met. However, at this juncture in time it seems plausible to speculate that stretches of collagen conformation may turn out to be a more common component in "noncollagenous" proteins than previously suspected.

A. Invertebrate Collagen Structure

It has been noted by Adams that while invertebrates are estimated as 95% of all animal species most collagen studies have centered on vertebrates.[6] He further points out that comparative studies involving nonvertebrate collagens can expand our view of collagen and can reveal new functions over a great range of organismic needs. We subscribe to this study of invertebrate collagens and provide examples of the diversity of information that has ensued from detailed examination of many such collagens. As will be seen, some surprises arose from the data, probably reflecting the phylogenetic specialization of two major animal lines, the Deuterostomia and the Protostomia. In this regard, it should be noted that the vertebrates comprise only one phylum of the Deuterostomia and therefore are hardly representative of the entire animal kingdom.

In seeking to determine whether type I collagen, the most abundant form in vertebrates, is the major collagen throughout the animal kingdom several points may be raised. Type I collagen is abundant in all animals examined and appears to be present in both large, banded fibers and in non- or poorly striated structures which can form the bulk of some tissues. The difference in the ability of type I collagen to produce different sorts of polymerized architectures may be related to the degree of post-translational modifications of the collagen subunits. This concept is related to the fact that the type I collagen of the Protostomia and of the coelenterates have high levels of hydroxylysine and its glycosides; in these animals the collagen is found primarily in poorly striated forms. Thus, if one accepts that the Deuterostomia are a branch of the main Protozoa-Porifera-Coelenterate-Protostomal line then the argument might be made that the type I collagen in that branch is a specialized form of a more ancestral collagen in the main line. The ability of vertebrate type I collagen to form large, banded fibers may be related to some special, as yet undetermined requirement in the development of large terrestrial animals. Another point which emerges from the comparative studies is that type V collagen is quite similar to type I collagen in amino acid composition and that the post-translational modification patterns of hydroxylysine and its glycosides indicate that type V may retain more of the features of a primordial collagen wherever it occurs in the animal kingdom.

In comparing the type I collagens from various animals it becomes apparent that some variation in properties occurs. For example, the precipitation by NaCl at acidic pH encompasses a wide range, with molluscan and annelid collagens precipitating at 0.4 to 0.5 *M* NaCl while coelenteran, arthropodal, and vertebrate collagens precipitate at 0.7 to 0.9 *M* NaCl. In making this comparison, it is assumed that type I collagen is defined as either $(\alpha 1)_3$, $(\alpha 1)_2\alpha 2$, or $\alpha 1\alpha 2\alpha 3$. In addition, these chains are each about 100 kdaltons and have a composition which clearly places them in the type I category, as defined by R values.[7]

II. COELENTERATA (Cnidaria)

Eumetazoa of the animal kingdom may be subdivided into three groups, Coelenterata, Protostomia, and Deuterostomia. Coelenterata belong to one of the most primitive phyla which is characterized by the absence of mesoderm when compared to the other two groups. The structural feature of coelenterate organization is the presence of a mesenchymatous layer, the mesogloea, which is delimited on both sides by epithelia. From the viewpoint of phylogeny, a collagenous protein appears in the mesogloea as one of the earliest interstitial collagens; clues to the phylogenetic development of collagen structure may exist in this phylum. Thus, most of the studies on coelenterate collagens have focused on the well-developed mesogloea as a main connective tissue. There are three classes: Hydrozoa, Anthozoa, and Scyphozoa. In Hydrozoa, the mesogloea of hydra *Pelmatodydra pseudoligactis* has been shown to contain a collagenous protein, as suggested by a significant amount of hydroxyproline and a high level of glucosylgalactosylhydroxylysine.[8]

Anthozoan mesogloeas were studied extensively as early as 1969 to 1973; the mesogloeal collagen of sea anemones was the first of the most primitive interstitial collagens from invertebrates to be fully characterized with respect to its subunit structure (see Table 1). The pepsin-solubilized mesogloeal collagens of *Actinia equina* and *Metridium dianthus* reveal the existence of only a single type of α chain with a mol wt of ~100,000, indicating a structure of $(\alpha 1)_3$.[9—11] This finding is also seen with acid-soluble collagen from *Actinia*. The compositional features of mesogloeal collagens are the considerable amounts of hydroxylysine (24 to 28 residues per 1000) and its glycoside as well as 3-hydroxyproline (6 to 20 residues per 1000). In addition to the glucosylgalactosyl units present in all collagens, except for cuticle collagens, a peculiar carbohydrate unit consisting of a heteropolysaccharide containing fucose, mannose, and glucosamine is found to covalently link to a *Metridium* α chain through an asparagine residue.[11] Electron microscopic observations show that the general pattern, as well as the length (280 to 300 nm) of the segment long spacing (SLS) crystallites from mesogloeal collagens are indistinguishable from that for SLS from vertebrate type I collagens. This result strongly suggests a close resemblance between both invertebrate and vertebrate collagens not only in molecular dimensions but also in primary structure. In the case of *Actinia* collagen, the reconstituted fibrils exhibit an axial periodicity of about 65 nm, characteristic of vertebrate type I collagen, while such periodicity is rarely observed in the tissue form of mesogloeal fibrils. Recently the mesogloea and axial skeleton of another anthozoan, the sea pen *Veretillum cynomorium*, has been shown by electron microscopy to demonstrate structural variability of collagen fibrils.[12]

More recently, the mesogloeal collagen of a scyphozoan, jellyfish, has been characterized with respect to its subunit structure. A large edible jellyfish, *Stomolophus nomurai*, was chosen as a representative species; more than 60% of its mesogloeal tissue components are rich in hydroxylysine and its glycoside.[13] An electron microscopic study indicates the presence of intact mesogloeal fibers consisting of very thin fibrils, 10 to 30 nm in diameter, with no apparent cross striations.[14] The pepsin-solubilized mesogloeal collagen has been isolated by selective salt precipitation using 0.9 *M* NaCl at acidic pH; the existence of molecules comprising three distinct α chains (α1, α2, and α3) is revealed by biochemical

Table 1
DISTRIBUTION OF COLLAGEN IN INVERTEBRATE PHYLA WHOSE SUBUNIT CHAINS HAVE BEEN BIOCHEMICALLY STUDIED

	Subunit	Molecular form	Class (genus)	Ref.[a]
Interstitial collagen				
Coelenterata	$\alpha 1$	$(\alpha 1)_3$[b]	Anthozoa (sea anemone)	1, 2
	$\alpha 1$, $\alpha 2$, $\alpha 3$	$\alpha 1 \alpha 2 \alpha 3$	Scyphozoa (jellyfish)	3
Annelida	$\alpha 1$, $\alpha 2$	$(\alpha 1)_2 \alpha 3$	Echiuroidea (*Urechis*)	4
Mollusca	$\alpha 1$	$(\alpha 1)_3$	Gastropoda (top shell, abalone)	5, 6
	$\alpha 1$, $\alpha 2$	$(\alpha 1)_2 \alpha 2$	Cephalopoda (octopus, squid)	7—9
Arthropoda	$\alpha 1$	$(\alpha 1)_3$	Insecta (locust, cockroach)	10, 11
	$\alpha 1$, $\alpha 2$	$(\alpha 1)_2 2$	Crustacea (crab, spiny lobster)	6
Basement membrane collagen				
Aschelminthes	145K	$(145K)_6$[c]	Nematoda (*Ascaris*)	12
Cuticle collagen				
Aschelminthes	A, B, C (52K)	Not established	Nematoda (*Ascaris*)	13
	75K, 65K (main chains)	Not established	Nematoda (*Caenorhabditis*)	14, 15
Annelida	A, B (47OK)	$A(B)_2$	Polychaeta (*Nereis*)	16

[a] Reference list for Tables 1 and 2 follows text reference list.
[b] A composition of three identical α chains, $(\alpha)_3$, is designated as $(\alpha 1)_3$; α chains: ~100 kilodaltons.
[c] The main collagenous domain is an 850 kilodalton component and consists of six 145-kilodalton chains of very similar size linked together by disulfide bonds.

analyses.[15] The chemical composition is characterized by high levels of hydroxylysine (32 to 38 residues per 1000) and its glycosylation products (about 70%). The same type of collagen is also found in jellyfish skin which covers the mesogloea.[14] These findings appear to be the first example of such an α1α2α3 heterotrimer collagen in invertebrates.

There has been some discussion concerning the close similarity in chemical composition between coelenterate mesogloea and vertebrate basement membranes.[8,11,16] Such discussion emphasizes the relatively high levels of 3-hydroxyproline and glucosylgalactosylhydroxylysine as well as the epithelial origin of both types of extracellular matrices. Judging from the R values calculated from the amino acid data (Table 2), however, the mesogloeal collagens apparently differ from basement membrane (type IV) collagens, but are similar to type V collagen, and both fall into the category of interstitial collagens. When compared with type IV collagen, coelenterate mesogloeal collagens are unique in containing a high level of polar amino acids. From these composite results, the mesogloeal collagen of coelenterates is classified as type I.

It is clear that the coelenterate mesogloeal collagen has two forms of molecules, $(\alpha 1)_3$ for sea anemone and α1α2α3 for jellyfish, which contrast with $(\alpha 1)_2\alpha 2$ for vertebrate type I collagen. In general, the interstitial collagens of primitive animals are considered to consist of three identical α chains;[17] the occurrence in sea anemone of $(\alpha 1)_3$ homotrimer collagen was supposed to be a typical example for this hypothesis. Thus, the recent finding of an α1α2α3 heterotrimer collagen in jellyfish is quite interesting from the viewpoint of collagen phylogeny. It is reasonable to speculate that the α1α2α3 heterotrimer collagen arose with divergence of Anthozoa and Scyphozoa, as a result of duplication and subsequent mutation of a single gene coding for the collagen α chains of an ancestral coelenterate. In addition to the mesogloea, collagenous proteins have also been detected in organs such as the nematocysts of the hydranth *Hydra littoralis* and the sea anemone *Apitasia pallida*, the float of *Physalia*, the spicules of coral *Briareum asbestinum*, and the axial skeleton of coral *Muricea muricata;* these observations were mostly based on amino acid analyses of the tissue components. The *Apitasia* nematocyst collagen is assumed to be stabilized by disulfide bonds and the coral skeleton collagen by tannage or sclerotization in a manner analogous to insect cuticles. By electron microscopy, the coral skeleton collagen displays a repeat periodicity of 33 to 35 nm which is half the width of that described as typical.[18]

III. ASCHELMINTHES

Although there are five classes within this phylum, only the collagen of the Nematoda has been characterized to any significant degree, with several representative animals having been examined. These include the large parasitic worms, *Ascaris suum* and *Ascaris lumbricoides* and the small worms, *Caenorhabditis elegans* and *Panagrellus silusiae*. Since *C. elegans* has been used extensively as a model system of development, much interest has been paid to its cuticle collagen as a marker for developmental programming.

In contrast to the cuticle collagen of *A. lumbricoides*, which is composed of three separate and genetically distinct subunits of 52 kdaltons,[19] the cuticle collagen of *C. elegans* is much more complex. At least 7 major and several minor components, ranging from 50 to 200 kdaltons, have been found in that cuticle.[20—22] This contrast in composition between the two cuticles was unexpected since their ultrastructural appearances were quite similar. Further insight into the collagen complexity was obtained by genomic information which suggests that more than 50 collagen genes are found in *C. elegans* DNA.[23] Two of the genes were structurally characterized and seem to have one or two introns but not in the collagen repeat coding domain. There are, however, coding domains which do not code for the collagen sequence interrupting those which do code for such sequences. Recent data suggest that not only does the ultrastructure of the cuticle change with stages of development, but that different collagen subunits are present at various stages during the life cycle of *C. elegans*.[24]

Table 2
R VALUES OF INVERTEBRATE COLLAGENS

Tissue source and collagen subunit			R_{Φ}	R_{OH}	R_{P}	Ref.[a]
Interstitial collagen[b]			138—224	284—395	431—565	
Porifera	Sponge	Spongin A	177	364	459	17
		Spongin B	154	325	521	17
Coelenterata	Sea anemone	Mesogloea α1	165	339	496	1
	Jellyfish	Mesogloea α1	167	314	519	3
		Mesogloea α2	170	339	491	3
		Mesogloea α3	146	305	549	3
Platyhelminthes	*Fasciola*	Whole body	219	329	452	18
	Cysticercus[c]	Whole body	257	182	561[c]	19
Aschelminthes	*Macracanthorhynchus*	Subsurface layer	224	284	492	20
	Ascaris	Muscle	212	357	431	21
Annelida	*Urechis*	Skin α1	140	304	556	4
		Skin α2	138	297	565	4
	Neanthes	Intestine	222	330	448	27
Mollusca	Top shell	Muscle α1	162	357	481	5
	Octopus	Muscle α1	168	371	461	22
		Muscle α2	206	343	451	22
Arthropoda	Locust	Ejaculatory duct α1	166	375	459	10
	Crab	Subcuticular membrane α1	202	358	440	6
		Subcuticular membrane α2	221	320	459	6
Echinodermata	Sea cucumber	Body wall	169	346	485	23
	Starfish	Body wall	168	362	470	23
Protochordata	Ascidian	Muscle	174	395	431	24
Basement membrane collagen						
Aschelminthes	*Ascaris*	Intestine	253	358	389	12
Cuticle Collagen						
Aschelminthes	*Ascaris*	Cuticle	177—209	156—181	636—642	25
		Cuticle A chain	224	263	513	13
		Cuticle B chain	173	192	635	13
		Cuticle C chain	152	152	696	13
	Caenorhabditis[d]	Cuticle	134	360	506	14, 15

Annelida	*Nereis*	Cuticle	59—120	440—450	440—450	26
		Cuticle A chain	102	439	459	16
		Cuticle B chain	44	540	416	16
	Lumbricus	Cuticle	114—157	507—574	309—341	26

Note: R values are mostly calculated by Matsumura's method[25] using amino acid data which have been collected since Matsumura's paper (1972).

[a] Reference list for tables follows text reference list.
[b] The R values of most collagens fit the "S range" for interstitial collagens compiled by Matsumura.
[c] *Cysticercus* collagen is listed for comparison because the absence of hydroxyproline is quite unusual in an interstitial collagen.
[d] *Caenorhabditis* collagen is extremely rich in hydroxyproline, 103 residues/1000, when compared with 10 to 20 residues/1000 for *Ascaris* collagen and thus it is rather similar in its R values to interstitial collagen.

The collagens found elsewhere in the nematodes, aside from the cuticle, have also been investigated and most information is available on the intestinal basement membrane of *A. suum*. This membrane, similar to the vertebrate basement membranes, does not have any striated collagen fibrils. Three separate collagens, of 2200, 850, and 420 kdaltons, have been characterized.[25] Each contains three distinct subunits, each with an average size of 150 kdaltons, connected by disulfide bonds. The largest collagen appears to consist of five interlinked triple helical molecules, the middle sized one of two interlinked collagens and the smallest one comprises a single triple helix. The dimeric molecule appears to predominate in pepsin extracts of the tissue and accounts for about 90% of the collagenous constituents of such extracts.

When the classifications of these various collagens obtained from nematodes is analyzed by the method of Matsumura,[7] substantial differences are seen among them,[24] including differences between cuticle collagens themselves and between cuticle collagens and basement membrane collagens. Thus, within one class of a single invertebrate phylum a high degree of collagenous protein diversity occurs.

IV. ANNELIDA

Three major classes comprise this phylum, but only the cuticle collagen from the Polychaetes has been extensively characterized. Periodically banded collagen fibers, similar in ultrastructural appearance to vertebrate types I, II, and III collagens, are seen in a number of locations in the annelid body. In the case of the gut wall of the polychaete, *Nereis japonica*, a collagen was extracted by pepsin which has a molecular structure similar to type I collagen.[26] Recent studies of body wall collagen from *Lumbricus* have shown that a material which primarily precipitates at 3.5 *M* NaCl can be obtained, yielding upon amino acid composition a collagen which has a high content of hydroxylysine and its glycosides.[27] Gel electrophoresis showed primarily one or two chains migrating between $\alpha 1$ and $\alpha 2$ vertebrate reference standards. Comparison of segment long spacing crystallites with type I crystallites of calf skin showed both similarities and differences in distribution and intensity of the stained patterns.

Annelid cuticle collagens are among the longest extended molecules known. Segment long spacing crystallites of *Nereis virens* cuticle collagen measure up to 2400 nm in length[28] and rotary shadowed preparations of these molecules have confirmed this length and shown that the molecules are uniformly flexible with a persistence length of about 80 nm.[29] There do not seem to be any preferentially flexible regions. The close agreement between the two different methods of electron microscope imaging lends credibility to the concept that the cuticle collagen is about six to eight times larger than most vertebrate and invertebrate collagens. This point is also supported by the earlier measurements of size of the cuticle collagen and its subunits;[30] such measurements showed a molecular size of about 1400 kdaltons with subunits of about 470 to 513 kdaltons.

Characteristic features which set the annelid cuticle collagens apart from other classes of collagen are a very high ratio of hydroxyproline to proline, an absence of aldehyde-derived crosslinks and of cysteine and hydroxylysine, a low content of hydrophobic amino acids, and extensive glycosylation by unique *o*-glycosidically linked substituents attached to serine and threonine. The common presence of hydroxyproline in position 2 of the gly-x-y triplet may account for the anomalously low denaturation temperatures when these collagens are compared to other collagenous proteins.[31] Recent data show that peptides containing hydroxyproline in position 2 occur in *Nereis* cuticle collagen as in Oligochaete collagen.[32,33]

One of the peptides from the A subunit of *Nereis* cuticle collagen served as a prototype for constructing a synthetic oligonucleotide.[34] A genomic library was initially screened with a *C. elegans* cuticle collagen probe, col-1, and isolated genomic clones then re-examined with the synthetic oligonucleotide. One major restriction enzyme fragment, at 23 kb, hybridized to the probe under low and high stringency conditions. Two major genomic clones have been further characterized by comparison of restriction maps using various enzymes and the composite results show that the two contain different segments of the same collagen gene or they encode for dissimilar collagen genes. This approach, using recombinant DNA methods, should allow for characterization of the cuticle collagen genes, their relationship to other collagen genes in the annelids, and their relationship to the well-characterized genes of vertebrate collagens.

V. MOLLUSCA

Mollusca, as well as Annelida and Arthropoda, are members of the highly advanced phyla of Protostomia. Recent research on molluscan collagens has been centered on Gastropoda and Cephalopoda, because both classes include a number of species, important as food resources, which have well-developed muscle and/or mantle tissues rich in collagen. The body of gastropods is protected by a relatively large shell, an epithelial calcareous derivative of mantle, and a major matrix component of the shell which is a noncollagenous protein, conchioline. However, it is evident that collagen plays an important role as the main connective tissue component of muscle, mantle, and internal organs. The foot muscle of gastropods, the abalone *Haliotis discus* and the top (or turban) shell *Turbo cornutus*, contain abundant collagen; it amounts to as much as 30 to 40% of the total tissue protein. The pepsin-solubilized muscle collagen of both gastropods is composed of only one type of α chain, indicating the occurrence of $(\alpha\ 1)_3$ homotrimer molecules homologous to vertebrate type I trimer.[35,36] These α chains are relatively rich in hydroxylysine (14 residues per 1000) and 50 to 60% of the hydroxylysines are linked to glucosylgalactose. Such a homotrimer collagen has been found in other tissues, such as the mantle of the top shell and the mantle skin of the snail *Helixpomatia.*[14,37] Morphologically, cross-striation patterns of the fibrils, 60 nm for top shell muscle and 53 to 57 nm for snail mantle skin, display a close relationship to those of vertebrate type I collagen.

Cephalopod collagens so far examined are different from gastropod collagens in having two dissimilar α chains. The body of cephalopods is mainly supported by the collagenous tissues of mantle skin and cartilage, while there are two kinds of shell, cuttlebone in Sepioidea and gladius in Teuthoidea, as a chitinous exoskeleton. Biochemical characterization of mantle skin collagen from the octopus *Octopus vulgaris* demonstrates that its major molecular species has a structure of $(\alpha 1)_2\alpha 2$ homologous to vertebrate type I collagen.[38] The amino acid composition, comparing the α1 and α2 chains of octopus collagen, follows a typical pattern similar to that observed for type I collagen; the α2 is characterized by increased contents of hydroxylysine, histidine, valine, isoleucine and a decreased content of alanine as compared with the α1. This type of collagen is widely distributed in cephalopod tissues such as the arm muscle and sucker of octopus, the cranial cartilage of squid, the neural sheath of cuttlefish, and the skin, tunic, and mantle muscle of squid and cuttlefish.[39-43] The cephalopod collagens contain 10 to 24 residues per 1000 of hydroxylsine, including 3 to 8 residues per 1000 of glucosylgalactosylhydroxylysine, comparable to the gastropod collagens. By electron microscopy, the collagen fibrils of octopus mantle skin as well as octopus aorta, squid mantle skin, and cuttlefish neural sheath are shown to exhibit an axial periodicity of 60 to 67 nm. These combined results provide evidence for characterization of these cephalopod collagens as type I. Direct comparison of octopus type I collagen α chains indicates no intraspecies differences in amino acid and carbohydrate composition or in CNBr-peptide maps between mantle skin and arm muscle tissues.[39] In squid type I collagen, however, tissue-specific differences in proline hydroxylation and thermal stability are observed; the cranial cartilage collagen exhibits more resistance to heat denaturation and has greater hydroxylation than the mantle skin collagen.[41] The difference of 0.8°C between the denaturation temperatures of both tissue collagens is probably a consequence of increased (5 to 6 residues per 1000) hydroxyproline. Moreover, there is an apparent interspecies difference between octopus and squid in CNBr-peptide maps of type I collagen α chains, although the chemical compositions resemble each other except for the high degree of hydroxylysine glycosylation in squid α2 chain when compared to octopus α2.[41] Concerning the glycosylation of hydroxylysine, exactly the same sequences of (Gly-X-Hyl-Gly-Y-Arg) as found for vertebrate collagens are shown by sequence analyses of glycopeptides from the mantle skin collagen of cuttlefish *Sepia subaculeata*.[44] This finding implies that the specific enzymes catalyzing carbohydrate unit formation during collagen biosynthesis are very similar in invertebrates and vertebrates. As for crosslinks, a major reducible crosslink in the skin and tunic collagens of squid and cuttlefish is derived from dihydroxylysinonorleucine obtained after reduction with borohydride.[43]

Invertebrate cartilage is distributed in some phyla such as Mollusca, Annelida, and Arthropoda, but little is known about the molecular properties of its collagenous component. On the other hand, the hyaline cartilage of vertebrates is well established, comprising several types of collagen, with type II predominating; it has a subunit structure of $[\alpha 1(II)]_3$. From a phylogenetic point of view, the cranial cartilage of squid is quite interesting, because it resembles vertebrate hyaline cartilage in its microscopic appearance. However, a protein homologous to type II collagen appears to be a minor component, if any, in the squid cartilage.[41] In addition to type I collagen, some minor molecular species of collagen are present in the cranial cartilage of squid as well as in the mantle skin and arm muscle of octopus. Only the minor collagen fraction of octopus arm muscle was isolated and partially characterized; it contains a unique γ chain-sized component crosslinked by disulfide bonds and is similar to basement membrane collagen (type IV) in its extremely high content of glucosylgalactosylhydroxylysine (49 residues per 1000) and a low content of alanine (28 residues per 1000).[39]

From these composite results, it is tempting to speculate that the most abundant and ubiquitous collagen (type I) of molluscs has two molecular forms, $(\alpha 1)_3$ for gastropods and $(\alpha 1)_2\alpha 2$ for cephalopods. One of the important characteristics of molluscan collagens is their precipitation by salt at acidic pH; type I collagens from molluscs as well as an annelid *Urechis* are readily precipitated at 0.4 to 0.5 *M* NaCl which is about one half the NaCl concentration (0.7 to 0.9 *M*) required for precipitating type I collagens from coelenterates, arthropods, and vertebrates. In addition, the minor collagen fraction of molluscs containing noninterstitial collagens is precipitated at 0.9 *M* NaCl, while that of vertebrates containing types IV and V collagens at 1.2*M* NaCl. On the other hand, little information is available concerning the chemical properties of collagens from Bivalvia except for their amino acid composition from the byssus thread of sea mussel, *Mytilus edulis*, and the mantle of pearl oyster, *Pinctada morterisii*.[45,46] The possible presence of an $(\alpha 1)_2\alpha 2$ heterotrimer collagen has recently been reported in *Mytilus* byssus threads; the pepsin-solubilized collagen, however, is heavily contaminated with a crosslinking or tannage protein thought to be involved in the formation of byssus threads; its α1 and α2 chains remain to be characterized.[47]

VI. ARTHROPODA

The most advanced phylum in the protostome line is assumed to be Arthropoda and is comparable to Vertebrata in the deuterostome line. Biochemical research on arthropod collagens is concerned with two classes, Crustacea and Insecta. The body of arthropods is mainly supported and protected by cuticles as a chitinous exoskeleton, but collagen provides supporting tissues around organs and under the cuticles. Among arthropod collagens, the subcuticular membrane collagens of crustaceans, the spiny lobster *Panulirus japonicus* and the swimming crab *Portunus tribuberculatus*, have first been isolated by limited pepsin digestion. Both collagens are selectively precipitated using 0.7 to 0.9 *M* NaCl at acidic pH and contain two distinct α chains, α1 and α2, in a molar ratio of about 1:2, suggesting the possible presence of an $(\alpha1)_2\alpha2$ heterotrimer molecule.[35] These α chains, particularly α2, are unique in containing a high level of hydroxylysine (36 to 43 residues per 1000) and a low level of alanine (38 to 39 residues per 1000) when compared to other interstitial α chains of invertebrates. It is noteworthy that the majority of hydroxylysine in these α2 chains is linked to glucosylgalactose and that the degree of hydroxylation reaches 84 to86%. Although this chemical feature is reminiscent of vertebrate noninterstitial collagens, the R values of these crustacean collagens apparently fit in with the category of interstitial collagens (Table 2). Further, it is evident that the comparison of amino acid composition between the α1 and α2 chains follows the general pattern for various type I collagens. However, the possibility that the α1 and α2 chains are derived from two different molecular species of collagen cannot be excluded, because there is no definitive evidence for the presence of $(\alpha1)_2\alpha2$ heterotrimer molecules. In this connection, the pepsin-solubilized muscle collagen of some crustaceans has recently been shown to possess a chemical composition very similar to their subcuticular membrane collagen,[48] and it seems to be a mixture of at least two molecular species.

In Insecta, some morphological studies reveal the occurrence in several organs of collagens which are indistinctly or distinctly banded. Two insect collagens from the ejaculatory duct of locust, *Locusta migratoria*, and the mesenteric sheath of cockroach, *Periplaneta americana*, have been extensively characterized with respect to their morphological and biochemical properties.[49,50] The pepsin-solubilized collagens of both insect tissues are precipitated using 4.0 *M* NaCl at neutral pH and share a subunit structure of $(\alpha1)_3$; the locust collagen also precipitates using 0.9 *M* NaCl at acidic pH. These insect collagens resemble other invertebrate collagens in the considerable amounts of hydroxylysine (16 to 24 residues per 1000) and the presence of dihydroxylysinonorleucine as a major crosslink after reduction with borohydride. The carbohydrate moiety is undefined, but the cockroach collagen contains hexoses accounting for as much as 8.6% of the total collagen by weight. Morphologically, the native and reconstituted collagens of locust and the reconstituted collagen of cockroach display the axial periodicity of 62 to 63 nm typical of vertebrate interstitial collagens. Also, the SLS crystallites exhibit the same banding patterns as vertebrate type I collagen.

These interstitial collagens of arthropods are tentatively defined as type I, while another type of collagen has been isolated and partially characterized. The membrane-associated procollagen of cultured cells of an insect, the fruitfly *Drosophila*, comprises three identical α chains linked together by disulfide bonds and resembles (electron microscopically) vertebrate basement membrane (type IV) procollagen in length and appearance.[51] This *Drosophila* collagen gene has a structure very different from vertebrate interstitial collagen genes; it probably encodes a basement membrane like collagen.[52]

VII. ECHINODERMATA

These animals are commonly found in the sea and have been extensively studied with regard to their body structures and as models for fertilization processes and embryonic development. Although collagenous structures have been initially identified by X-ray diffraction, more recent studies have relied primarily upon electron microscopy and amino acid composition.[53] The main features of the amino acid analyses are low hydroxyproline, low lysine, and high levels of acidic amino acids. Analysis by Matsumura et al.[54] has shown these collagens to resemble vertebrate type I collagens. Suggestive evidence for two different types of collagen occurring in cuvierian tubules of the sea-cucumber has been provided and more recent data, using CNBr maps, indicate that the α chains are unique and not identical to vertebrate type I chains.[53]

A unique feature of the echinoderms is that specific collagenous tissues can undergo a rapid change in their mechanical properties, alternating from an extensible state to one that markedly resists extension. While sparse muscle cells have been identified in the tissues, the molecular basis for the change in mechanical properties of the collagenous domains is not known. Several theories have been advanced to account for the reversible alterations but little evidence substantiates any of these at present.

In summary, the collagen of echinoderms may be fundamentally similar to that of other animals, in the sense of containing subunits the size of conventional α chains. These subunits appear to be genetically distinct but have been incompletely characterized. Since collagens are prominent tissue components of this phylum and some of the tissues undergo remarkable, rapid alterations in their mechanical properties, more detailed studies of the collagens would help understand their role as well as provide further information concerning the phylogenetic diversity of the collagenous family of proteins.

VII. VERTEBRATA (Fish)

Vertebrata is situated at the top of evolution in Deuterostomia and the term fish is generally used for lower vertebrates including cyclostomes, chondrichthyes, and osteichthyes. The collagens of fish are widely distributed in skin, cartilage, bone, and many other organs; two types of interstitial collagen, types I and II, are clearly identified in fish.[55—61] Vertebrate type I collagen is characterized by low levels of hydroxylysine and its glycoside when compared with type II collagen and invertebrate type I collagen.[62] From a phylogenetic point of view, type II collagen as the major structural protein of cartilage and notochord is very interesting; this collagen conserves its primary structure during evolution much more precisely than type I collagen, since α1(II) chain peptide maps from lamprey, shark, sturgeon, and human bear a close resemblance. On the other hand, peptide mapping exhibits the interspecies difference between the homologous α1(I) or α2(I) chains from some fish such as lamprey, shark, mackerel, and carp.[63] This situation is not true for higher vertebrate type I collagens and this is why CNBr-peptide maps are considered to provide one of the essential criteria for the classification of higher vertebrate α chains. Among fish, shark is quite unique in possessing a marked difference in amino acid composition between α1(I) and α2(I) chains; the difference is assumed to be larger than that of human type I collagen.[63] Probably, shark collagen is specialized further, corresponding to divergence of cartilage fish from the vertebrate line.

Higher vertebrate type I collagen is composed of $[\alpha1(I)]_2\alpha2(I)$ heterotrimer molecules. As early as 1965, however, codfish skin collagen was reported to contain three distinct α chains in comparable amounts.[56] This interesting information seems to have been rather neglected. In this connection, the existence of a third α chain, α3(I), has recently been shown in mackerel skin type I collagen.[63] Further, α3(I) chains have been detected in more than ten species of bonyfishes and found by carboxymethyl-cellulose chromatography in most of the skin type I collagens.[64] α3(I) Chains are presumably derived from α1(I) chains, but the exact relationship and relative amounts of α3(I) to α1(I) and α2(I) are not yet clear. The molecular forms of fish type I collagen remain to be resolved; most probably, α3(I) chains are present as α1(I)α2(I)α3(I) heterotrimers. At present, there are few data on other types (III-X) of fish collagen, while the possible existence of specialized forms of collagen has been suggested in scale, tooth enamel, egg case, and elastoidin. Quite recently, a new type of collagen has been isolated from the elastoidin of the great blue shark; the collagen molecule consists of three identical α chains and the elastoidin α chain, designated α1(E), is very similar in electrophoretic and chromatographic behavior and amino acid composition to the shark skin α1(I) chain, but the two chains are genetically distinct on the basis of CNBr-peptide maps.[65]

IX. DISCUSSION

Much of our interest in this article has focused upon collagen at the biochemical level, in the context of its identification and characterization for comparison across phylogenetic lines. In viewing the historical context of this approach, it is apparent that as new, increasingly more powerful methods for probing Nature have become available, investigators correspondingly expand our knowledge. As noted by Mathews,[66] "there are four ways to studystructures and forms in the biological world: (1) comparative morphology, (2) comparative embryology, (3) comparative protein structure, (4) comparative DNA structure." With respect to collagen, we are still in the process of level 3 and have only recently entered level 4. Levels 1 and 2 probably have yielded sufficient information to provide a firm base for now exploring these more molecular aspects of structure, expression, and function. Moreover, the molecular studies are often quantifiable and allow for ready comparisons across broad phylogenetic domains. The limiting factor is that we study contemporary animals and must necessarily extrapolate our findings concerning primordial events and origins. However, it is agreed that collagen is of considerable evolutionary antiquity and this is reinforced by analyses of fossilized bones which still contain sufficient remnants of their organic, collagenous matrix to provide definite identification of the protein.

A common theme in contemporary biochemistry is that molecular structure underlies function. To some extent this dictum has prevailed when the collagenous proteins are examined. However, another level of complexity enters because of the propensity for collagens to form polymers, which is their primary functional form in the extracellular matrix. While collagen self-assembly into polymers is intrinsically related to the molecular structure of the protein, the presence of other components of the matrix may well modulate polymer organization and, consequently, function. In addition, the ability of cell surfaces to recognize collagen and other matrix components which bind collagen provides for a variety of modulating influences of cellular origin. Thus, given a primordial origin and an ancestral protein, one can appreciate that during evolution all manner of variations in the expression of collagenous proteins has been obtained. Such rich diversity is reflected by the accumulated information at the morphological and biochemical levels and is emerging at the genomic level. We can look forward to evolving insights into the function and behavior of this most important structural protein, especially relying upon tissue comparisons and a broad phylogenetic approach.

REFERENCE LIST FOR TEXT

1. **Bornstein, P. and Traub, W.,** The chemistry and biology of collagen, in *The Proteins*, Vol. 4, 3rd ed., Neurath, H. and Hill, R. L., Eds., Academic Press, New York, 1979, 411.
2. **Eyre, D. R.,** Collagen: molecular diversity in the body's protein, *Science*, 207, 1315, 1980.
3. **Reid, K. B. M. and Porter, R. R.,** Subunit composition and structure of subcomponent C1q of the first component of human complement, *Biochem. J.*, 155, 19, 1976.
4. **Anglister, L. and Silman, I.,** Molecular structure of elongated forms of electric eel acetylcholinesterase, *J. Mol. Biol.*, 125, 293, 1978.
5. **Bhattacharyya, S. N. and Lynn, W. S.,** Characterization of the collagen- and non-collagen-like regions present in a glycoprotein isolated from alveoli of patients with alveolar proteinosis, *Biochim. Biophys. Acta*, 625, 343, 1980.
6. **Adams, E.,** Invertebrate collagens, *Science*, 202, 591, 1978.
7. **Matsumura, T.,** Relationship between amino-acid composition and differentiation of collagen, *Int. J. Biochem.*, 3, 265, 1972.
8. **Barzansky, B. L., Lenhoff, H. M., and Bode, H.,** Hydra mesogloea: similarity of its amino acid and neutral sugar composition to that of vertebrate basal lamina, *Comp. Biochem. Physiol.*, 50B, 419, 1975.

9. **Nordwig, A. and Hayduk, U.,** Invertebrate collagens: isolation, characterization and phylogenetic aspects, *J. Mol. Biol.*, 44, 161, 1969.
10. **Nordwig, A., Nowack, H., and Hieber-Rogall, E.,** Sea anemone collagen: further evidence for the existence of only one α-chain type, *J. Mol. Evol.*, 2, 175, 1973.
11. **Katzman, R. L., and Kang, A. H.,** The presence of fucose, mannose, and glucosamine-containing heteropolysaccharide in collagen from the sea anemone *Metridium dianthus*, *J. Biol. Chem.*, 247, 5486, 1972.
12. **Franc, S., Ledger, P. W., and Garrone R.,** Structural variability of collagen fibers in the calcareous axial rod of a sea pen, *J. Morphol.*, 184, 75, 1985.
13. **Kimura, S., Miura, S., and Park, Y.-H.,** Collagen as the major edible component of jellyfish *(Stomolophus nomurai)*, *J. Food Sci.*, 48, 1758, 1983.
14. **Kimura, S.,** unpublished data.
15. **Miura, S., and Kimura, S.,** Jellyfish mesogloea collagen: characterization of molecules as α1α2α3 heterotrimers, *J. Biol. Chem.*, 260, 15352, 1985.
16. **Nowack, H. and Nordwig, A.,** Sea-anemone collagen: isolation and characterization of the cyanogen-bromide peptides, *Eur. J. Biochem.*, 45, 333, 1974.
17. **Pikkarainen, J. and Kulonen, E.,** On the evolution and development of collagen, *Scand. J. Clin. Lab. Invest.*, Suppl. 95, 40, 1967.
18. **Goldberg, W. M.,** Evidence of a sclerotized collagen from the skeleton of a gorgonian coral, *Comp. Biochem. Physiol.*, 49B, 525, 1974.
19. **Evans, H. J., Sullivan, C. E., and Piez, K. A.,** The resolution of *Ascaris* cuticle collagen into three chain types, *Biochemistry*, 15, 1435, 1976.
20. **Ouazana, R. and Herbage, D.,** Biochemical characterization of the cuticle collagen of the nematode *Caenorhabditis elegans*, *Biochim. Biophys. Acta*, 669, 236, 1981.
21. **Ouazana, R., Herbage, D., and Godet, J.,** Some biochemical aspects of the cuticle collagen of the nematode *Caenorhabditis elegans*, *Comp. Biochem. Physiol.*, 77B, 51, 1984.
22. **Cox, G. N., Strapans, S., and Edgar, R. S.,** Cuticle of *Caenorhabditis elegans:* its isolation and partial characterization, *J. Cell Biol.*, 90, 7, 1981.
23. **Kramer, J. M., Cox, G. N., and Hirsh, D.,** Comparisons of the complete sequences of two collagen genes from *Caenorhabditis elegans*, *Cell*, 30, 599, 1982.
24. **Ouazana, R.,** The collagen of aschelminthes, in *Biology of Invertebrate and Lower Vertebrate Collagens*, Bairati, A. and Garrone, R., Eds., Plenum Press, New York, 1985, 217.
25. **Hung, C. H., Noelken, M. E., and Hudson, B. G.,** Intestinal basement membrane of *Ascaris suum*. Physical properties of the collagenous domain, *J. Biol. Chem.*, 256, 3822, 1981.
26. **Kimura, S.,** Studies on marine invertebrate collagens. IV. Structural difference in collagens from cuticle and intestines of *Neanthes diversicolor*, *Bull. Jap. Soc. Sci. Fish.*, 37, 432, 1971.
27. **Vitellaro-Zuccarello, L., Cheli, F., and Cetta, G.,** The interstitial collagen of *Lumbricus* sp. (Annelida), in *Biology of Invertebrate and Lower Vertebrate Collagens*, Bairati, A. and Garrone, R., Eds., Plenum Press, New York, 1985, 259.
28. **Murray, L. W. and Tanzer, M. L.,** Characterization of a large fragment from annelid cuticle collagen and its relationship to the intact molecule, *Coll. Rel. Res.*, 3, 445, 1983.
29. **Murray, L. W., Engel, J., and Tanzer, M. L.,** Annelid cuticle collagen: a very long protein, *Ann. N.Y. Acad. Sci.*, 460, 478, 1985.
30. **Kimura, S. and Tanzer, M. L.,** *Nereis* cuticle collagen: isolation and characterization of two distinct subunits, *Biochemistry*, 16, 2554, 1977.
31. **Rao, N. V. and Harrington, W. F.,** Relationship between the positioning of pyrrolidine residues and the stability of collagen, *J. Mol. Biol.*, 21, 577, 1966.
32. **Sharma, Y. D. and Tanzer, M. L.,** High-performance liquid chromatographic separation of glycopeptides from *Nereis* cuticle collagen, *Anal. Biochem.*, 141, 205, 1984.
33. **Goldstein, A. and Adams, E.,** Glycylhydroxylprolyl sequences in earthworm collagen: glycylhydroxylprolylserine, *J. Biol. Chem.*, 245, 5478, 1970.
34. **Har-El, R., Sharma, Y. D., and Tanzer, M. L.,** unpublished data, 1986.
35. **Kimura, S. and Matsuura, F.,** The chain composition of several invertebrate collagens, *J. Biochem.*, 75, 1231, 1974.
36. **Kimura, S. and Tanaka, H.,** Characterization of top shell muscle collagen comprising three identical α1 chains, *Bull. Jap. Soc. Sci. Fish.*, 49, 229, 1983.
37. **Schmut, O., Roll, P., Reich, M. E., and Palm, W.,** Biochemical and electronmicroscopic investigations on *Helix pomatia* collagen, *Z. Naturforsch.*, 35c, 376, 1980.
38. **Kimura, S., Takema, Y., and Kubota, M.,** Octopus skin collagen: isolation and characterization of collagen comprising three distinct α chains, *J. Biol. Chem.*, 256, 13230, 1981.
39. **Takema, Y. and Kimura, S.,** Two genetically distinct molecular species of octopus muscle collagen, *Biochim. Biophys. Acta*, 706, 123, 1983.

40. **Babu, M. and Chandrakasan, G.,** Characterization of collagen from the suckers of octopus, *Leather Sci.*, 31, 109, 1984.
41. **Kimura, S. and Karasawa, K.,** Squid cartilage collagen: isolation of Type I collagen rich in carbohydrate, *Comp. Biochem. Physiol.*, 81B, 361, 1985.
42. **De Biasi, S., Cheli, F., and Vitellaro-Zuccarello, L.,** Ultrastructural and biochemical study on collagen from the neural sheath of *Sepia Officinalis*, in *Biology of Invertebrate and Lower Vertebrate Collagens*, Bairati, A. and Garrone, R., Eds., Plenum Press, New York, 1985, 309.
43. **Shadwick, R. E.,** Crosslinking and chemical characterization of cephalpod collagens, in *Biology of Invertebrate and Lower Vertebrate Collagens*, Bairati, A. and Garrone, R., Eds., Plenum Press, New York, 1985, 337.
44. **Isemura, M., Ikenaka, T., and Matsushima, Y.,** Comparative study of carbohydrate-protein complexes. I. The structures of glycopeptides derived from cuttlefish skin collagen, *J. Biochem.*, 74, 11, 1973.
45. **Pikkarainen, J., Rantanen, J., Vastamaki, M., Lampiaho, K., Kali, A., and Kulonen, E.,** On collagens of invertebrates with special reference to *Mytilus edulis*, *Eur. J. Biochem.*, 4, 555, 1968.
46. **Kimura, S., Nagaoka, Y., and Kubota, M.,** Marine invertebrate collagens. I. Collagens from crustaceans and molluscs, *Bull. Jap. Soc. Sci. Fish.*, 35, 743, 1969.
47. **DeVore, D. P., Engebreston, G. H., Schachtele, C. F., and Sauk, J. J.,** Identification of collagen from byssus threads produced by the sea mussel, *Mytilus edulis*, *Comp. Biochem. Physiol.*, 77B, 529, 1984.
48. **Kimura, S. and Tanaka, H.,** Partial characterization of muscle collagens from prawns and lobster, *J. Food Sci.*, 51, 330, 1986.
49. **Ashhurst, D. E. and Bailey, A.,** Locust collagen: morphological and biochemical characterization, *Eur. J. Biochem.*, 103, 75, 1980.
50. **Francois, J., Herbage, D., and Junqua, S.,** Cockroach collagen: isolation, biochemical and biophysical characterization, *Eur. J. Biochem.*, 112, 389, 1980.
51. **Bachinger, H. P., Doege, K. J., Petschek, J. P., Fessler, L. I., and Fessler, J. H.,** Structural implications from an electron-microscopic comparison of procollagen V with procollagen I, pC-collagen I, procollagen IV and a *Drosophila* procollagen, *J. Biol. Chem.*, 257, 14590, 1982.
52. **Monson, J. M., Natzle, J., Friedman, J., and McCarthy, B. J.,** Expression and novel structure of a collagen gene in *Drosophila*, *Proc. Natl. Acad. Sci. U.S.A.*, 79, 1761, 1981.
53. **Bailey, A. J.,** The collagen of the Echinodermata, in *Biology of Invertebrate and Lower Vertebrate Collagens*, Bairati, A. and Garrone, R., Eds., Plenum Press, New York, 1985, 369.
54. **Matsumura, T., Hasegawa, M., and Shigei, M.,** Collagen biochemistry and phylogeny of echinoderms, *Comp. Biochem. Physiol.*, 62B, 101, 1979.
55. **Piez, K. A., Eigner, E. A., and Lewis, M. S.,** The chromatographic separation and amino acid composition of the subunits of several collagens, *Biochemistry*, 2, 58, 1963.
56. **Piez, K. A.,** Characterization of a collagen from codfish skin containing three chromatographically different α chains, *Biochemistry*, 4, 2590, 1965.
57. **Miller, E. J. and Mathews, M. B.,** Characterization of notochord collagen as a cartilage-type collagen, *Biochem. Biophys. Res. Commun.*, 60, 424, 1974.
58. **Kimura, S., Kamimura, T., Takema, Y., and Kubota, M.,** Lower vertebrate collagen: evidence for the Type I-like collagen in the skin of lamprey and shark, *Biochim. Biophys. Acta*, 669, 251, 1981.
59. **Kimura, S. and Kamimura, T.,** The characterization of lamprey notochord collagen with special reference to its skin collagen, *Comp. Biochem. Physiol.*, 73B, 335, 1982.
60. **Kimura, S.** Vertebrate skin Type I collagen: comparison of bonyfishes with lamprey and shark, *Comp. Biochem. Physiol.*, 74B, 325, 1983.
61. **Rama, S. and Chandrakasan, G.,** Distribution of different molecular species of collagen in the vertebral cartilage of shark *(Carcharius acutus)*, *Connect. Tissue Res.*, 12, 111, 1984.
62. **Kimura S.,** Determination of the glycosylated hydroxylysines in several invertebrate collagens, *J. Biochem.*, 71, 367, 1972.
63. **Kimura, S.,** The interstitial collagens of fish, in *Biology of Invertebrate and Lower Vertebrate Collagens*, Bairati, A. and Garrone, R., Eds., Plenum Press, New York, 1985, 397.
64. **Kimura, S., Ohno, Y., and Uchida, N.,** unpublished data.
65. **Kimura, S., Uematsu, Y., and Miyauchi, Y.,** Shark (*Prionace glauca*) elastoidin: characterization of its collagen as $[\alpha 1(E)]_3$ homotrimers, *Comp. Biochem. Physiol.*, in press, 1987.
66. **Mathews, M. B.,** Evolution and collagen, in *Biology of Invertebrate and Lower Vertebrate Collagens*, Bairati, A. and Garrone, R., Eds., Plenum Press, New York, 1985, 545.

REFERENCE LIST FOR TABLES

1. **Katzman, R. L, and Kang, A. H.,** *J. Biol. Chem.*, 247, 5486, 1972.
2. **Nordwig, A., Nowack, H., and Hieber-Rogall, E.,** *J. Mol. Evol.*, 2, 175, 1973.
3. **Miura, S. and Kimura, S.,** *J. Biol. Chem.*, 260, 15352, 1985.
4. **Kimura, S., Tanaka, H., and Park, Y.-H.,** *Comp. Biochem. Physiol.*, 75B, 681, 1983.
5. **Kimura, S. and Tanaka, H.,** *Bull. Jap. Soc. Sci. Fish.*, 49, 229, 1983.
6. **Kimura, S. and Matsuura, F.,** *J. Biochem.*, 75, 1231, 1974.
7. **Kimura, S., Takema, Y., and Kubota, M.,** *J. Biol. Chem.*, 256, 13230, 1981.
8. **Kimura, S., and Karasawa, K.,** *Comp. Biochem. Physiol.*, 81B, 361, 1985.
9. **Babu, M. and Chandrakasan, G.,** *Leather Sci.*, 31, 109, 1984.
10. **Ashhurst, D. E. and Bailey, A. J.,** *Eur. J. Biochem.*, 103, 75, 1980.
11. **Francois, J., Herbage, D., and Junqua, S.,** *Eur. J. Biochem.*, 112, 389, 1980.
12. **Hung, C.-H., Neolken, M. E., and Hudson, B. G.,** *J. Biol. Chem.*, 256, 3822, 1981.
13. **Evans, H. J., Sullivan, C. E., and Piez, K.A.,** *Biochemistry*, 15, 1435, 1976.
14. **Ouazana, R. and Herbage, D.,** *Biochim. Biophys. Acta*, 669, 236, 1981.
15. **Ouazana, R., Herbage, D., and Godet, J.,** *Comp. Biochem. Physiol.*, 77B, 51, 1984.
16. **Kimura, S. and Tanzer, M. L.,** *Biochemistry*, 16, 236, 1977.
17. **Piez, K. A. and Gross, J.,** *Biochim. Biophys. Acta*, 34, 24, 1959.
18. **Nordwig, A. and Hayduk, U.,** *J. Mol. Biol.*, 44, 161, 1969.
19. **Torre-Blanco, A. and Toledo, I.,** *J. Biol. Chem.*, 256, 5926, 1981.
20. **Cain, G. D.,** *Arch. Biochem. Biophys.*, 141, 264, 1970.
21. **Fujimoto, D.,** *Biochim. Biophys. Acta*, 168, 537, 1968.
22. **Takema, Y. and Kimura, S.,** *Biochim. Biophys. Acta*, 706, 123, 1982.
23. **Matsumura, T.,** *Comp. Biochem. Physiol.*, 44B, 1197, 1973.
24. **Kimura, S., Kobayashi, N., and Kubota, M.,** *Bull. Jap. Soc. Sci. Fish.*, 38, 1163, 1972.
25. **Matsumura, T.,** *Int. J. Biochem.*, 3, 265, 1972.
26. **Murray, L. W., Waite, J. H., Tanzer, M. L., and Hauschka, P. V.,** *Meth. Enzymol.* 82, 65, 1982.
27. **Kimura, S.,** *Bull. Jap. Soc. Sci. Fish.*, 37, 432, 1971.

Chapter 3

HISTOCHEMICAL LOCALIZATION OF COLLAGEN AND OF PROTEOGLYCANS IN TISSUES

G. S. Montes and L. C. U. Junqueira

TABLE OF CONTENTS

I. Introduction 42

II. Collagen 42
- A. Historical 42
  - 1. Trichrome Stains 42
- B. The Picrosirius-Polarization Method 43
  - 1. Specificity 43
  - 2. Sensitivity, Stoichiometry, and Collagen Quantification 43
  - 3. Staining Procedure 44
  - 4. Collagenase Digestion 44
  - 5. Characterization of the Different Collagen Types 45
    - a. Fibrillogenesis 45
    - b. Theoretical Basis for the Characterization of the Different Collagen Types with the Aid of the Picrosirius-Polarization Method 45
- C. Ultrastructural Correlations 51
- D. Revisiting the Classical Histologic Structures in the Light of the Current Histochemical Concepts 53
- E. Immunohistology 54

III. Proteoglycans 54
- A. General Characteristics of Proteoglycans and GAGs 55
  - 1. Functions 55
  - 2. Distinction between Proteoglycans and Glycoproteins 55
  - 3. Macromolecular Structure 55
- B. Hyaluronic Acid and Sulfated GAGs 57
- C. Histochemical Methods 59
  - 1. Electron Microscopy 59
  - 2. Light Microscopy 60
    - a. Alcian Blue 61
    - b. Periodic Acid-Schiff (PAS) Technique 63

IV. Collagen-Proteoglycan Interaction 63
- A. General Considerations 63
  - 1. Papain Digestion of Tissue Sections 64
- B. Specific Association of the Genetically Distinct Collagen Types with the Different Proteoglycans 65
  - 1. Quantitative Aspects 65
  - 2. Qualitative Aspects 65
  - 3. Differential Distribution of the Specific Collagen-Proteoglycan Interactions 65

C. Histochemical Evidence for the Structural Role of the Collagen-Proteoglycan Interaction in Tissues 65

V. Summary 68

Acknowledgments 70

References 70

I. INTRODUCTION

Our greater understanding of collagen and of proteoglycan biology has promoted considerable change in the histochemistry of these macromolecules. Also, much more is now known about the different mechanisms by which dye molecules attach to these tissue components and the chemical bases for much of this are currently widely accepted. In addition, the combination of the selective staining techniques with the appropriate enzymatic digestion or chemical manipulation have rendered the histochemical reactions more specific. All of the above-mentioned reasons have led us to believe that the need has arisen for producing a text which covers, where applicable, the theoretical basis of the most useful histochemical methods employed in the study of collagen and proteoglycans. This chapter has not been designed as a comprehensive reference work, but rather as an outline of the most simple and accurate methods that have proved useful for the histochemical study of collagen and proteoglycans. Thus, only a few selected procedures have been included here and only the procedural details of the methods which are routinely used in our Laboratory have been recorded in this chapter. Details for other methods discussed in this section are found in standard textbooks of histochemistry.[1,2] The reader is encouraged to consult them for additional information.

II. COLLAGEN

A. Historical

This term relates to a family of proteins which are contained in different histological entities such as *collagen fibers, reticulin fibers,* basement membranes, etc. A frequent fault among histologists and pathologists is to speak of collagen when in reality they mean one of the above-mentioned tissue components itself.

1. Trichrome Stains

Several staining techniques have been devised to differentiate between collagen and muscle fibers, and some are satisfactory for this application. The majority fall into the category of *trichrome stains.*

Still in constant use are Mallory, Masson, and van Gieson methods.[2,3] Although collagen fibers are usually stained intensely by these methods, other collagen-containing structures such as reticulin fibers and basement membranes are not selectively stained by the trichrome methods, the density of color produced being insufficient to resolve these fine structures.

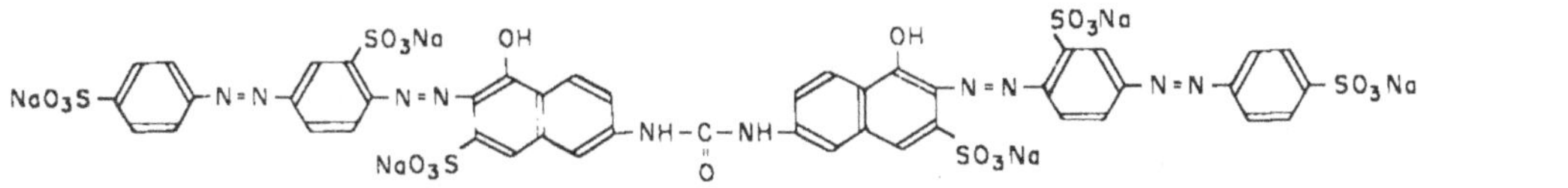

FIGURE 1. Diagram depicting the molecular formula of the dye Sirius Red which has been extensively used in the commercial dyeing of cellulosic fibers (hydrogen atoms on aromatic rings have been omitted). This is an elongated sulfonated azo dye molecule which contains six auxochromic (sulfonic) radicals, by which the dye molecule interacts with the basic groups of collagen and four azo ($-N=N-$) chromophores.

Thus, the various genetically distinct collagen types cannot be distinguished in tissue sections by collagen stains such as van Gieson.[4] Though it was designed for demonstrating collagen, Mallory's method is not especially useful for this purpose either[2] (Plate 1).* In addition, as it has been reported previously,[5,6] and is common knowledge, the trichrome methods do not show consistent results even with collagen fibers themselves which may take on different colors in a same tissue section as shown in Plate 2.

The above-mentioned staining techniques therefore do not readily differentiate between collagen fibers and reticulin fibers. Metal impregnation techniques, while capricious, do provide contrast although they do not distinguish collagen fibers from other tissue components.[7]

Thus, the only specific histochemical procedure so far developed for collagen detection in tissue sections is the Picrosirius-polarization method.

B. The Picrosirius-Polarization Method

1. Specificity

As collagen molecules are orderly disposed in a parallel orientation, a normal birefringence is one of the classic characteristics of collagenous structures (Plate 3A).

Collagen molecules being rich in basic amino acids, strongly react with acidic dyes. Sirius Red, an elongated strongly acidic dye, reacts with collagen and promotes an enhancement of its normal birefringence (Plate 3B) due to the fact that many dye molecules are attached to each collagen molecule in such a way that their long axes are parallel.[8] The enhancement of birefringence promoted by the Picrosirius-polarization method is specific for collagen (Plate 4).

2. Sensitivity, Stoichiometry, and Collagen Quantification

Sirius Red is a strongly acidic azo dye which contains six sulfonic groups (Figure 1) with a mol wt of 1372 that has been much used to stain collagen in tissue sections.[9] It has four azo chromophoric groups and has consequently a high extinction coefficient that permits the detection of small amounts of collagen in tissue sections and in acrylamide gel eletrophoresis to the extent that it can detect 5.76 μg of collagen per milligram of protein.[10]

Based on the selective binding of Sirius Red to collagen and of Fast Green to noncollagenous components, López de León and Rojkind devised a simple and sensitive micromethod for the quantitative estimation of collagen and of total protein in tissue sections.[10] The reliability of the method was checked by comparing the collagen values estimated by the dye-binding technique with the chemical method for collagen estimation (hydroxyproline content) and the results obtained showed a striking correlation between both procedures (Figure 2).

It has been shown that this dye binds to collagen through a strong interaction of its acid sulfonic groups with the basic groups of this protein. Soluble purified collagen types I, II, and III bind, on average, 126 dye molecules per collagen molecule.[8]

Evidence obtained comparing differences in color equivalence found with native or denatured collagen with native fibrous collagen indicate that the degree of polymerization and

* Color plates follow page 54.

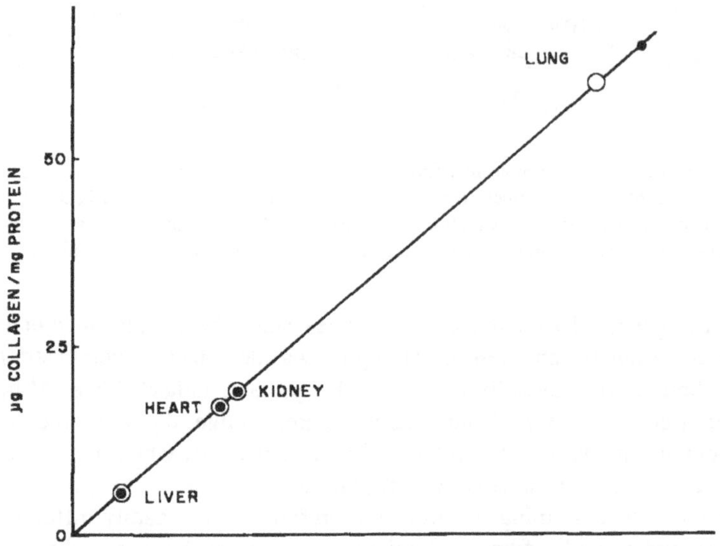

FIGURE 2. Collagen concentration in four rat organs as determined by the chemical and the dye-binding methods.[10] Tissue sections were stained by Sirius Red (which selectively reacts with collagen) and Fast Green (which binds to noncollagenous components). The color equivalence of each dye was determined after destaining the sections and measuring the collagen content by hydroxyproline analysis and the amount of protein by the micro-Kjeldahl procedure. Observe that the dye binding method (○) and the chemical procedure (●) produced comparable results.

three-dimensional organization of the collagen fibers can also be of importance regarding the amount of dye uptake in the collagenous components of tissue sections.[10]

3. Staining Procedure

Due to its high molecular weight, Sirius Red does not stain sections covered with celoidin nor plastic-embedded tissues.

Five-μm sections of formalin or Bouin fixed material are deparaffinized, hydrated to water, and stained for 1 hr in a 0.1% solution of Sirius Red (Sirius Red F 3 B 200, Mobay Chemical Co., Union, N.J.) dissolved in aqueous saturated picric acid. The sections are then rapidly washed in running tap water and counterstained with Harris hematoxylin. A conventional optical microscope with a strong light source (preferably with a halogen lamp) and two polaroid filters should be used to study the birefringence of the stained collagen.

Sirius Red solutions are very stable and can be used for several months. Samples of the stain, however, have deteriorated in our laboratory when kept for more than 4 years on the shelf. Under these conditions the solution loses its specificity and, besides staining collagen, it also stains muscle and epithelia.

4. Collagenase Digestion

Collagen is extremely stable and resists to the action of nonspecific proteases. Collagen turnover, which is necessary for tissue growth and remodeling, requires specific enzymes to initiate collagen degradation. These enzymes, known as animal collagenases, nick the alpha chain at a very specific and limited site promoting denaturation of collagen. Once denatured, nonspecific proteolytic enzymes take over and readily digest the collagen. Collagenases of bacterial origin cut collagen alpha chains at several sites, forming small segments and can, thus, hydrolyze collagen easily. Collagen degradation has been comprehensively reviewed by Pérez-Tamayo.[11]

Combined with the Picrosirius-polarization method, collagenase digestion has been applied to further confirm the specificity of the method. In order to submit histologic material to collagenase digestion, tissue sections are deparaffinized, hydrated to water, and incubated for 24 to 48 hrs at 37°C in a 0.1% solution of collagenase (Collagenase Type V, Sigma Chemical Co., St. Louis, Mo.,), dissolved in a 0.05 *M* Tris buffer solution containing 0.001 *M* $CaCl_2$ and 0.001*M* sodium azide. Control sections should be incubated under the same conditions without the enzyme. Under these conditions collagenase degrades collagen specifically.

5. Characterization of the Different Collagen Types

Interstitial collagens display different interference colors and intensities of birefringence in tissue sections studied by the Picrosirius-polarization method. The comparative study of vertebrate organs by this method disclosed a striking correlation between the localization of the different colors and intensities of birefringence, and the thus far described biochemical distribution of collagen types I, II, and III, which led Junqueira et al.[12] to postulate that these different collagen types could be distinguished in tissue sections by the Picrosirius-polarization method. Their results demonstrated that collagen type I (which corresponds to what has been classically called collagen fibers by histologists) shows up as thick, strongly birefringent, yellow or red fibers; while collagen type III appears under the form of thin, weakly birefringent, greenish fibers that could be identified as reticulin fibers[7] (Plates 6 and 7). Collagen type II, which is present in hyaline and elastic cartilages, does not form fibers and displays a weak birefringence of a variable color[13] (Plate 3).

Although it is generally accepted that basement membrane collagens do not form visible fibrils or fibers, very thin microfibrils can be seen under the electron microscope (Figures 3, 4, and 5). Several authors were of the opinion that collagen molecules were arranged randomly in basement membranes forming a feltwork-like structure.[14] However, a clear birefringence has been detected in all basement membranes studied so far by means of the Picrosirius-polarization method, showing that collagen molecules are orderly disposed in basal membranes[15] (Plate 8).

a. Fibrillogenesis

It is very well known that, during fibrillogenesis, collagen molecules aggregate forming a collagen fibril which can be resolved by the electron microscope. Bunches of these fibrils associate to form the fibers which are visible by optical microscopy. Fibers clumped together in a parallel array result in bundles of fibers.[16] This complete sequence of hierarchical levels of supramolecular organization is only attained in the case of collagen type I (Table 1).

b. Theoretical Basis for the Characterization of the Different Collagen Types with the Aid of the Picrosirius-Polarization Method

The observation that collagen types I, II, and III show different interference colors and intensities of birefringence in a same tissue section,[12] can be explained by the widely known fact that these different interstitial collagens display distinct patterns of physical aggregation.[17-21]

Thus, collagen type I forms thick fibers (collagen fibers), composed of closely packed thick fibrils[7,9,22-24] (Figure 6) and consequently presents an intense birefringence with yellow to red color. Collagen type III forms thin fibers (reticulin fibers) composed of loosely packed thin fibrils[7,9,22-24] (Figure 4) and thus displays a weak birefringence of a greenish color. Collagen type II (which is present in hyaline and elastic cartilages) does not form fibers, and its very thin fibrils are disposed as a loose mesh that strongly interacts with the ground

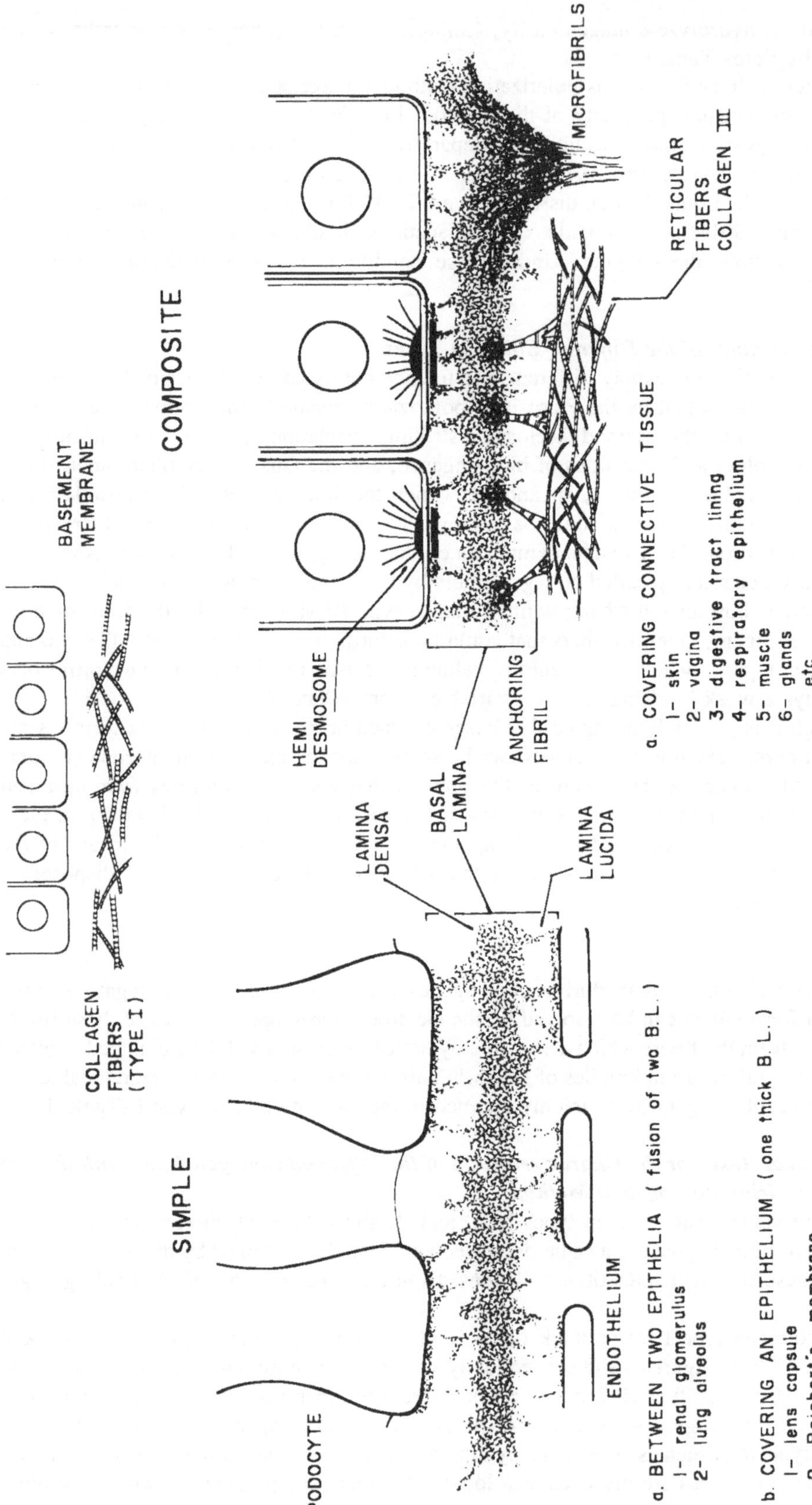
BASEMENT MEMBRANE
SIMPLE
COMPOSITE
COLLAGEN FIBERS (TYPE I)
PODOCYTE
ENDOTHELIUM
LAMINA DENSA
BASAL LAMINA
LAMINA LUCIDA
HEMI DESMOSOME
ANCHORING FIBRIL
MICROFIBRILS
RETICULAR FIBERS COLLAGEN III
a. BETWEEN TWO EPITHELIA (fusion of two B.L.)
1- renal glomerulus
2- lung alveolus
b. COVERING AN EPITHELIUM (one thick B.L.)
1- lens capsule
2- Reichert's membrane
a. COVERING CONNECTIVE TISSUE
1- skin
2- vagina
3- digestive tract lining
4- respiratory epithelium
5- muscle
6- glands
etc.

FIGURE 3. Structure and constitution of basement membranes. The drawing in the upper region represents the light microscopical appearance in which epithelial cells are separated from the subjacent connective tissue by a basement membrane. Electron microscopy of this structure showed that we can distinguish two general types of basement membranes. In the simpler type (left) the basement membrane is composed of a thickened basal lamina (B.L.) formed of central lamina densa and two peripheral laminae lucidae. This type of basement membrane contains more than 20 different proteins. Its best known components are collagen type IV, laminin, nidogen, entactin, amyloid P, and heparan sulfate.[66] In the more complex type (right) the basal lamina is usually thin and is associated with a layer of reticulin fibers (collagen type III) on its connective tissue surface. The so-called anchoring fibrils (probably formed of collagen type VII) form bridges between the reticulin fibers and the basal lamina. Microfibrils, probably of amyloid P, connect the system of elastic fibers to the basal lamina.

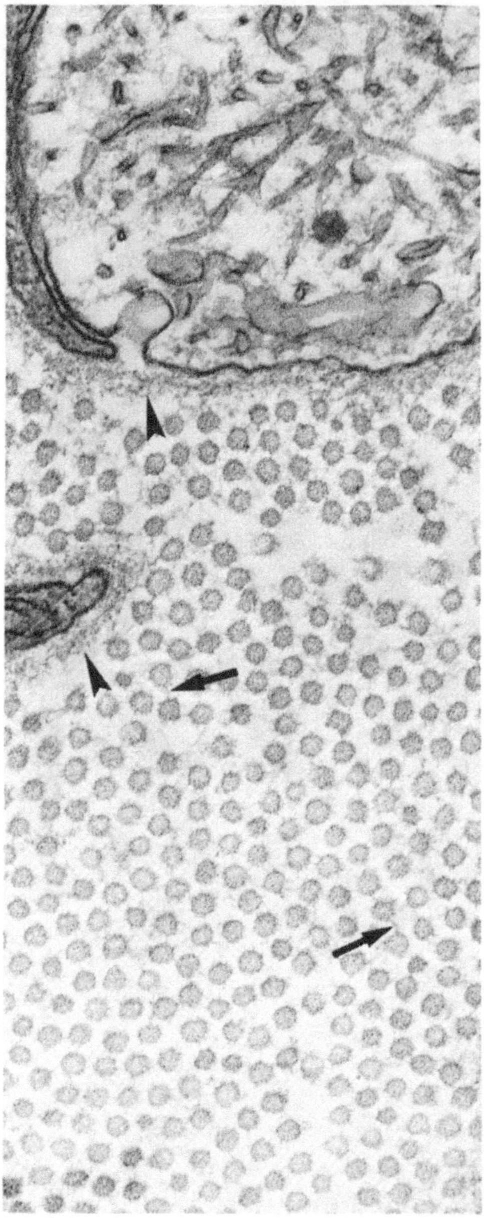

FIGURE 4. Electron micrograph of a reticulin fiber in the endoneurium of a human sural nerve. Compare the loose disposition and uniform diameters of the thin collagen type III fibrils with those of the thicker epineurial fibrils shown at the same magnification in Figure 8. Note the abundant interfibrillary material forming bridges (arrows) that connect the collagen fibrils. Surrounding the cellular portions present in this picture, typical basement membranes can be seen (arrowheads). Magnification × 60,000. (From Montes, G. S. and Junqueira, L. C. U., *Rev. Can. Biol. Exp.*, 41, 143, 1982. With permission.)

substance (Figure 7). This type of physical aggregation results in a weak birefringence, and a variable color, both of which depend on the interaction of collagen molecules with the ground substance.[13]

It is thus evident that, in the 5- to 7-μm sections routinely used in histology, the different colors and intensities of birefringence displayed by collagen types I, II, or III are due to

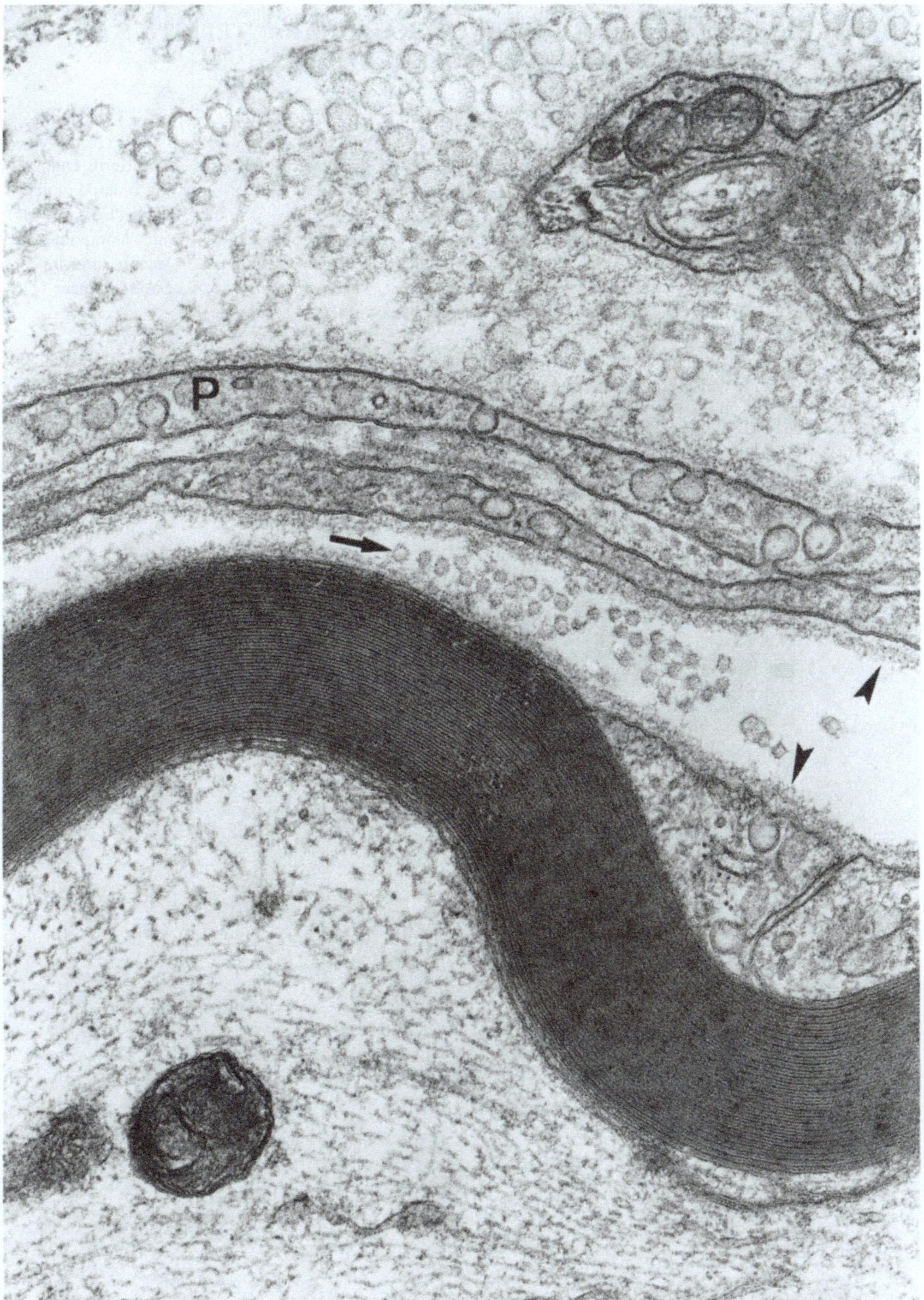

FIGURE 5. Transverse section across a nerve branch supplying the ear pavilion of a mouse. The double-layered perineurium (P) separates a myelinated axon (which is surrounded by thin endoneurial collagen fibrils; arrow) from the epineurium (upper region). Note that a very thin nerve branch composed of unmyelinated axons ensheathed by a Schwann cell, is situated among the thick epineurial collagen fibrils. Surrounding perineurial and Schwann cells are typical basal laminae (arrowheads). Observe that each compartment of the nerve is the site of a different collagen fibril population; fibrils in the epineurium are thicker and show a marked variation in diameter when compared to the thin fibrils of more uniform diameters present in the endoneurium. Magnification × 60,000.

Table 1
HISTOCHEMICAL AND ULTRASTRUCTURAL CHARACTERISTICS OF THE DIFFERENT COLLAGEN TYPES

	Type I	Type III	Type II	Type IV
Maximum level of physical organization attained	Bundles of fibers (2—10 μm thick)	Individual fibers (0.5—1.5 μm thick)	Fibrils (20 to 30 nm thick)	Microfibrils containing ordered molecules (1.5 nm wide)
Histologic features when studied by Picrosirius-polarization method	Closely packed, thick, nonargyrophilic, strongly birefringent, yellow or red fibers	Loose network of thin, argyrophilic, weakly birefringent, greenish fibers	Loose collagenous network displaying a weak birefringence of a variable color	Thin, amorphous weakly birefringent basal laminae
Ultrastructure	Collagen fibers: densely packed, thick (75 nm) fibrils with marked variation in diameter	Reticulin fibers: loosely packed, thin (45 nm) fibrils with more uniform diameters	No fibers, very thin fibrils embedded in abundant amorphous ground substance	Neither fibers nor fibrils; only microfibrils are detected

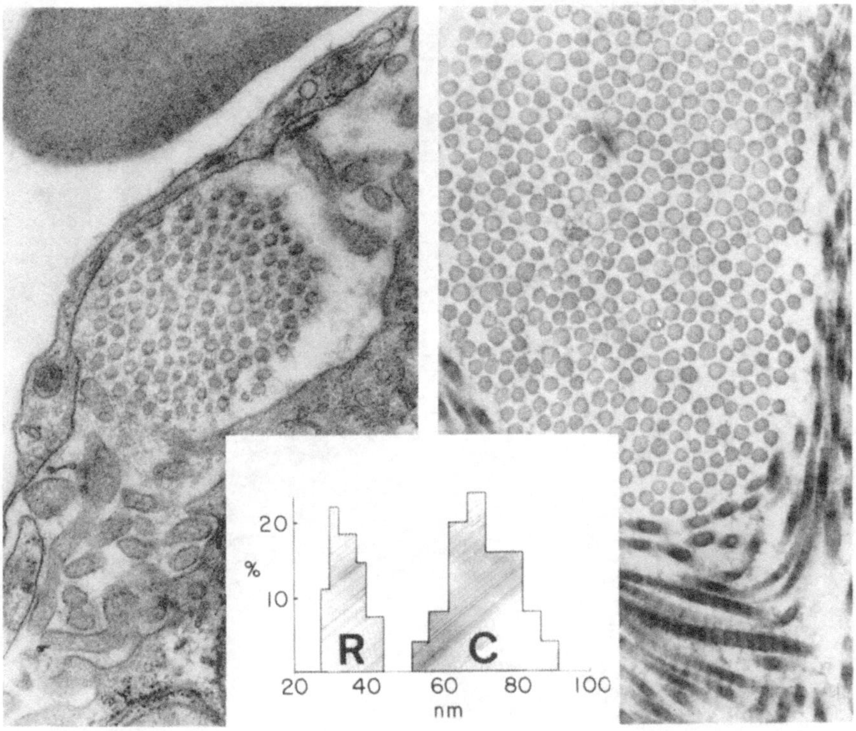

FIGURE 6. Electron micrographs of cross-sectional appearance of reticulin (left) and collagen (right) fibers. Note that each fiber type is composed of numerous smaller collagen fibrils. Each reticulin fiber (R) is composed of a few fibrils of significantly narrower diameter (see histogram inset), and, in addition, the constituent fibrils reveal an abundant surface-associated granularity not present on the many fibrils that form a collagen fiber (C). Magnification × 31,000. (From Junqueira, L. C. U. and Montes, G. S., *Arch. Histol. Jpn.*, 46, 589, 1983. With permission.)

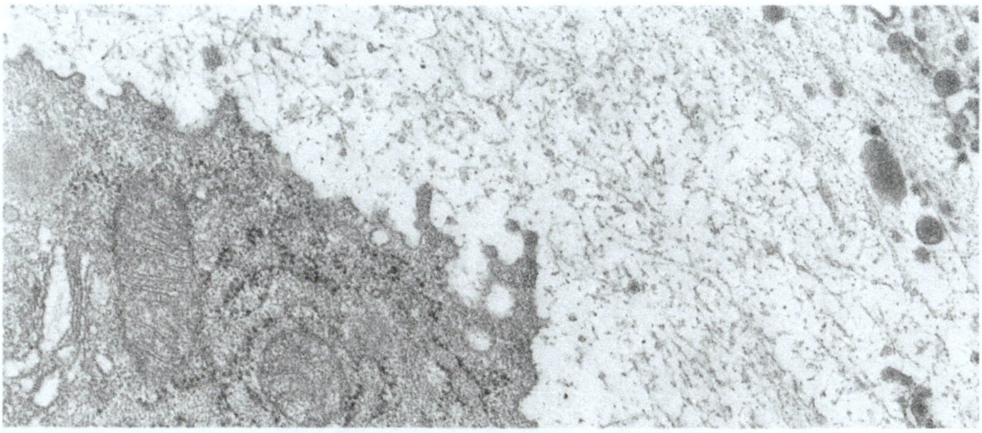

FIGURE 7. Electron micrograph of a mouse hyaline cartilage. Observe that the very thin collagen type II fibrils (in which the characteristic banding pattern is not clearly evident) are oriented in several directions forming a three-dimensional network. Part of a chondrocyte is seen in the lower left corner. Magnification × 30,000.

differences in their patterns of physical aggregation. This is the reason why it has never been claimed that a specific chemical interaction occurs between Picrosirius and the different collagen types.[8,9,12,16,24,25]

This method has proved to be useful for the study of collagen distribution in tissue sections obtained from normal specimens (for review, see Montes et al.)[9] and in various pathological models.[6,26-28] However, in the early stages of the development of this method, it became evident that tissue section thickness is of importance on the color and intensity of birefringence displayed by collagen since folds in tissue sections are more strongly birefringent and present a different color when compared to the surrounding tissue. The influence of tissue section thickness on the color and intensity of birefringence displayed by collagen when observed by the Picrosirius-polarization method, has recently been studied.[25] The results reported clearly show that color and intensity of birefringence are a function of collagen thickness.

The fact that color and intensity of birefringence depend on collagen fiber thickness has also been reported by Pérez-Tamayo and Montfort,[21] and later on confirmed by Vidal et al.[29] and Szendroi et al.[30] when using the Picrosirius-polarization method on different experimental models.

This discussion shows the usefulness of this method for studying the distribution of the different interstitial collagen types in tissue sections from normal adult organs. It also explains why, as has been stated previously,[21] care should be taken when interpreting pictures obtained studying embryonic or pathologic material, where processes involving intense collagen production or degradation are frequent.

However, this method can also be of value for the study of collagen degradation[21,31-32] provided that, under these conditions, one should be cautious regarding the characterization of collagen types.

C. Ultrastructural Correlations

Electron microscopic studies have shown that collagen type I forms coarse fibers (which correspond to the structures that have been classically called collagen fibers by histologists) composed of closely packed thick fibrils, whereas collagen type III forms thin fibers (that have been characterized as reticulin fibers) composed of loosely packed thin fibrils[7] (Figures 4 and 8).

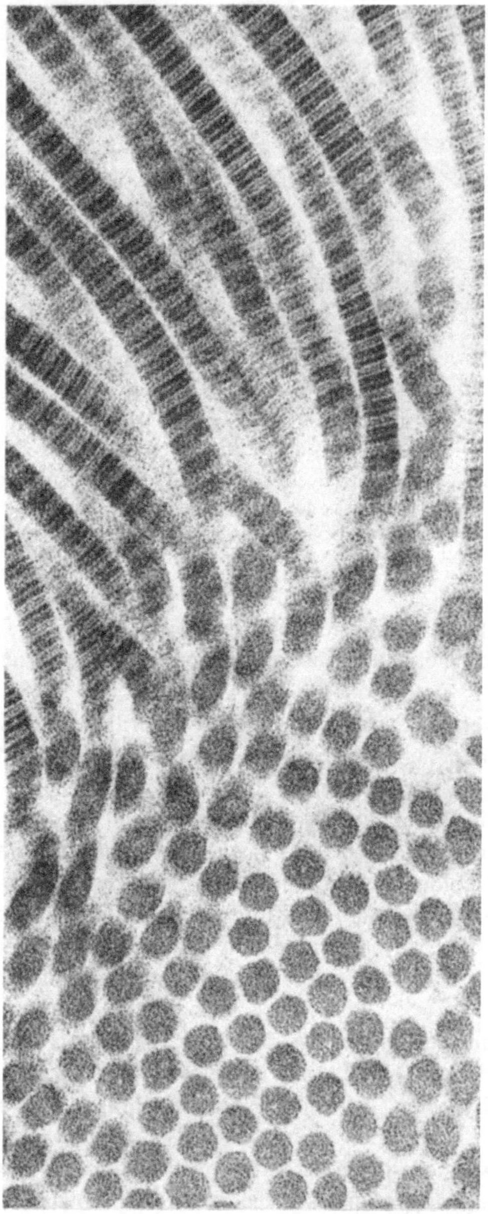

FIGURE 8. This electron micrograph includes part of a collagen fiber in the epineurium of a human sural nerve. Observe the close packing and large diameter of collagen type I fibrils. In the upper region, longitudinal sections of these fibrils show the characteristic banded staining pattern which is particularly evident in this type of collagen. Magnification × 60,000. (From Montes, G. S. and Junqueira, L. C. U., *Rev. Can. Biol. Exp.*, 41, 143, 1982. With permission.)

In certain anatomical locations (for example in nerves,[22,24,33] arteries,[23] smooth muscle,[7] and skin[34]) the above-mentioned distinct fibril populations were found to be segregated into different compartments (Figures 5 and 9). Thus, collagen type I was localized to the epineurium, the adventitial layer of blood vessels, and the deeper layers of the dermis; whereas collagen type III was found in the endonerium, in the tunica media of arteries, in the inner muscular layer of the small intestine, and in the adventitial dermis.

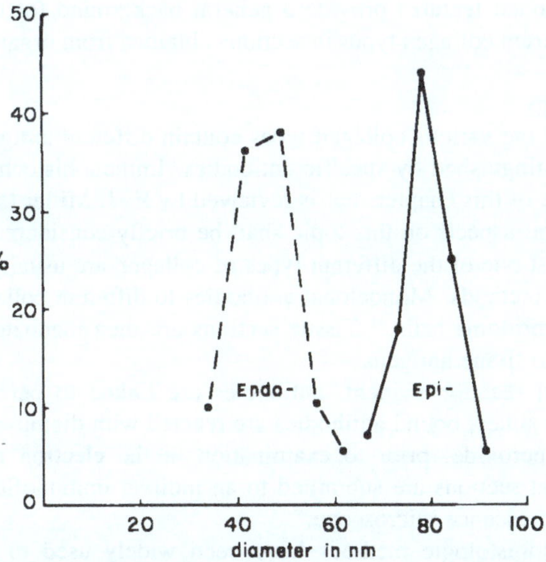

FIGURE 9. Graph illustrating the distribution of collagen fibril diameters in the endo- and epineurium of a human sural nerve showing that each compartment is the site of a different population. Electron micrographs obtained from these same two distinct populations are presented in Figures 4 and 8. (From Junqueira, L. C. U. and Montes, G. S., *Arch. Histol. Jpn.*, 46, 589, 1983. With permission.)

The histochemical results obtained coincide with the electron microscopic observations in showing the presence of two distinct collagen populations segregated into different compartments in nerves,[22,24,33] in arteries,[23] and in skin[34] and these findings are in agreement with immunohistologic[35-37] and biochemical[38-40] findings on the distribution of collagen types I and III in the same organs.

D. Revisiting the Classical Histologic Structures in the Light of the Current Histochemical Concepts

The results discussed so far strongly suggest that collagen type I forms the collagen fibers, and collagen type III the reticulin fibers, of the light microscopists, while collagen type II is identified with hyaline and elastic cartilages and type IV is localized to basement membranes.

Therefore, in all sites where collagen fibers have been described (such as tendons, fasciae, organ capsules, adventitial layer of blood vessels, dermis, and submucous layer of the digestive tube), closely packed, thick collagen fibrils of variable diameters with a clearly evident banded staining pattern, have been observed under the electron microscope (Figure 8). On the other hand, in all structures to which reticulin fibers have been localized (such as endoneurium, arterial tunica media, mucosal lamina propria, smooth muscle layers of the gastrointestinal tract, and the reticular network of parenchymatous organs such as the liver, kidney, and lung), loosely disposed, thin collagen fibrils of more uniform diameters have been detected by means of electron microscopy (Figure 4).

Collagen fibers show up in the form of thick, brilliant (strongly birefringent), yellow or red fibers against a dark background when studied by the Picrosirius-polarization method; whereas reticulin fibers display a weak birefringence and are characterized by their thinness and greenish color (Plate 6).

The above-mentioned features provide a general background for understanding the distribution of the different collagen types in sections obtained from organs of adult vertebrates.

E. Immunohistology

The molecules of the various collagen types contain different antigenic determinants and can therefore be distinguished by specific antibodies. Immunohistochemistry of collagen is not within the scope of this chapter, but is reviewed by E. J. Miller (see Volume I, Chapter 5). However, general aspects on this topic shall be briefly considered below.

Antibodies against one of the different types of collagen are usually raised in rabbits and purified by affinity methods. Monoclonal antibodies to different collagen types can also be produced, using hybridoma cells.[41] Tissue sections are then incubated with the antibodies that can thus bind to tissue antigens.

To make sites of reaction evident, antibodies are linked to peroxidase and, after the incubation of tissue slices, bound antibodies are reacted with diaminobenzidine in the presence of hydrogen peroxide, prior to examination in the electron microscope. For light microscopy, cryostat sections are submitted to an indirect immunofluorescent staining and studied under a fluorescence microscope.

Although immunohistologic methods have been widely used to localize the different collagen types in tissue sections,[41,42] an unambiguous identification of each collagen type has been very difficult to achieve. The limitations of this method include insufficient purification of the antigens (it is difficult to remove all the contaminating noncollagenous proteins) and of the antibodies (all unspecific antibodies must be removed). In addition, localization of a given collagenous antigen requires more than just specific antibodies: fixation conditions suitable for preservation of antigenicity must be tested. Equally important, but often difficult to control or evaluate, is the fact that collagen molecules may be partially masked or may not react completely with the antibody reagents, since it has been shown that each collagen type interacts differently (considering both the quantitative and qualitative aspects) with the ground substance.[43,44] It is interesting that, despite the very well known species specificity of immunological reactions of collagen,[45] it is frequently observed in the reports of the literature that, for the study of human tissue sections, antibodies are obtained from rabbits that are immunized with bovine collagen. Among other artifacts, nonspecific binding of, or failure to wash out unbound, antibodies should also be considered. Besides, immunohistologic methods are not readily adaptable to quantitation.

This technique has been particularly useful to assess collagen type distribution in tissues displaying developmental or disease-associated changes, but its application is still largely empirical and subject to many artifacts; thus, nonimmunological techniques must be used to corroborate the immunohistologic findings.

The above comments explain the discrepancies in results reported by different authors. For example, an immunohistologic study of the collagens of human liver states that reticulin fibers within the liver lobule contain collagen type III and that they are not stained by collagen type I antibodies;[45] these results are not consistent with a more recent report claiming that the same fibers contain both type I and type III collagens in human and bovine livers.[46]

III. PROTEOGLYCANS

The amorphous ground substance that fills the space between cells and fibers of the connective tissue is basically composed of macromolecules and bound water (water of solvation). The macromolecules of the ground substance can be grouped into two main classes: glycosaminoglycans (GAGs) (formerly known as acid mucopolysaccharides) and structural glycoproteins.

PLATE 1A

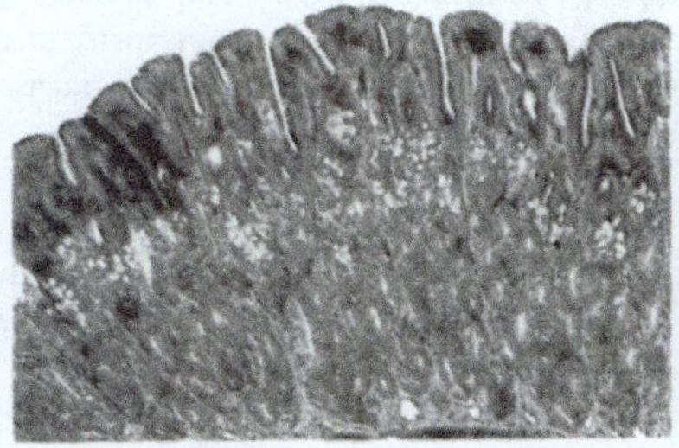

PLATE 1B

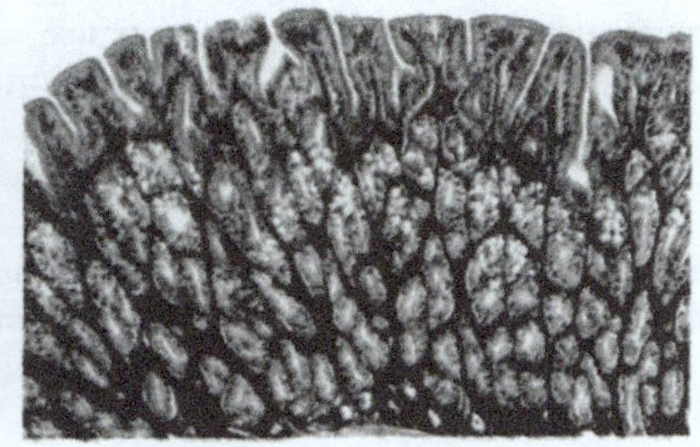

PLATE 1C

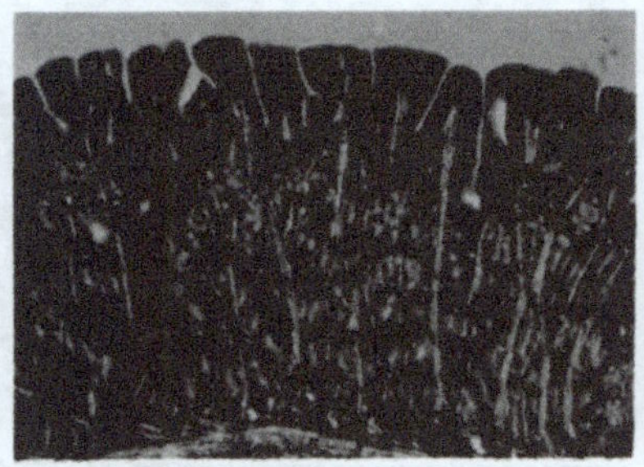

PLATE 2

PLATE 1. Sections of the gastric mucosa of a frog. (A) This histological preparation was stained by Mallory's trichrome technique; observe that this method is not especially useful for collagen which assumes varying shades of grayish blue. (B) An adjacent section stained in the Picrosirius solution (and counterstained with hematoxylin) and viewed by unpolarized light shows that collagen takes on a deeply red color and shows up sharply in the entire lamina propria all the way up to the surface epithelium; compare the crisp localization of collagen by this stain with the picture obtained when using the trichrome technique shown in (A). When the same field of the Picrosirius-stained section is observed under the polarizing microscope (C) the normal birefringence of collagen is enhanced and collagenous structures show up as brightly birefringent structures against a dark background. The enhancement of birefringence in Picrosirius-stained sections studied by polarization microscopy has been shown to be specific for collagen. Magnification X 60.

PLATE 2. Photomicrograph of a human osteosarcoma stained by Masson's trichrome method. Observe that the dense collagen fibers show blue regions interspersed with patches of red color. Magnification X 250.

PLATE 3A

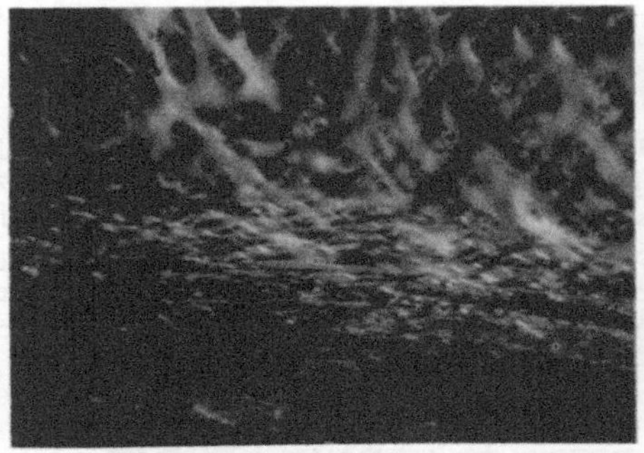

PLATE 3B

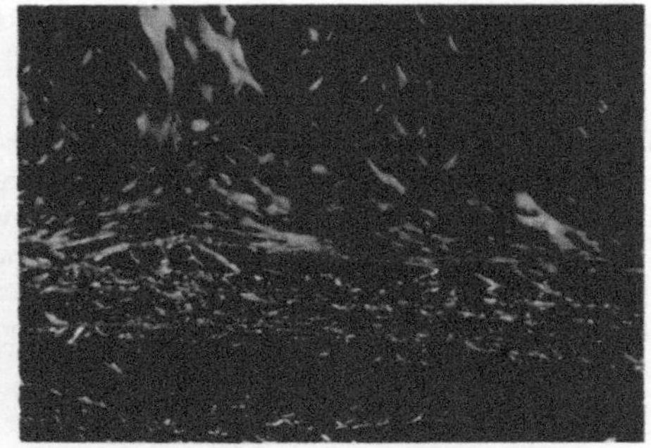

PLATE 4A

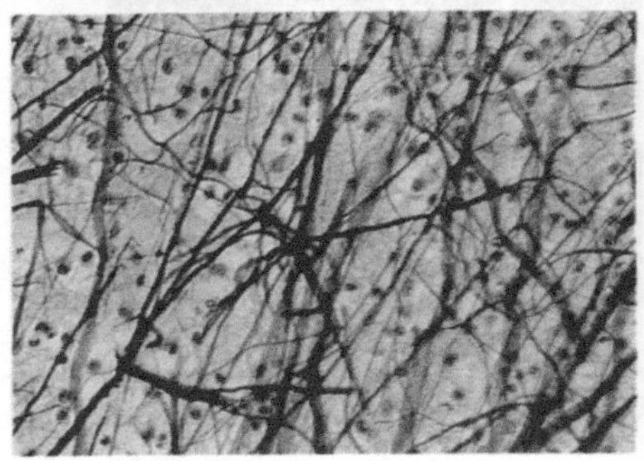

PLATE 4B

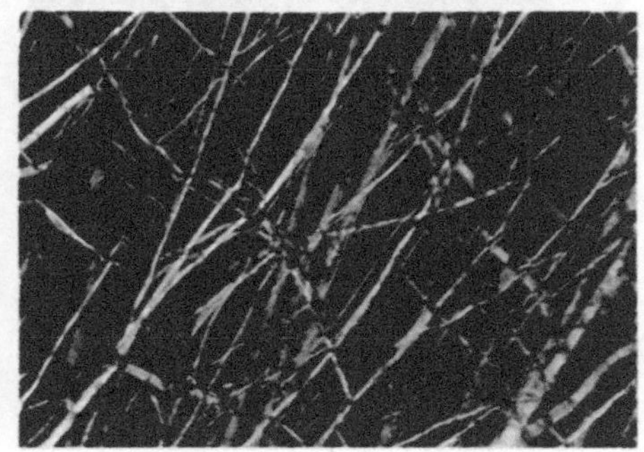

PLATE 3. Adjacent serial sections of hyaline cartilage stained with hematoxylin and eosin (A) and Picrosirius (B) and viewed by polarized light. All birefringent structures which shine against the dark background are of collagenous nature. The perichondrium is in the lower region. Observe that the normal birefringence of collagen (A) is greatly enhanced by Picrosirius staining (B). Magnification × 80.

PLATE 4. Rat dermis stained with Picrosirius and viewed by conventional illumination (A) and by polarized light (B). Picrosirius staining in association with polarization microscopy is specific for collagen detection in tissue sections: note that the structures containing collagen are specifically demonstrated, with the aid of this method, in the form of brightly birefringent structures which shine against a dark background (B). Magnification × 160.

PLATE 5

PLATE 6

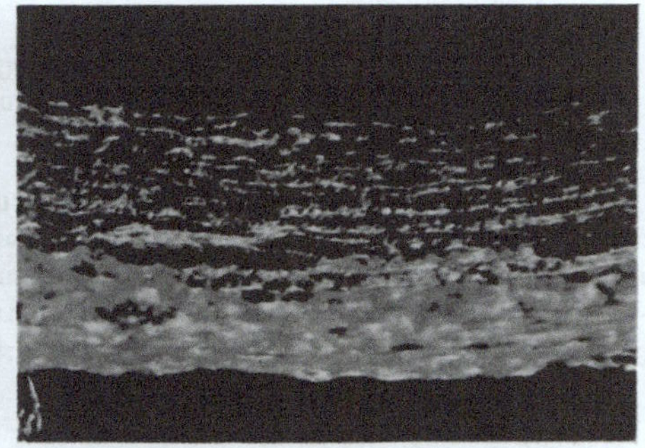

PLATE 7A

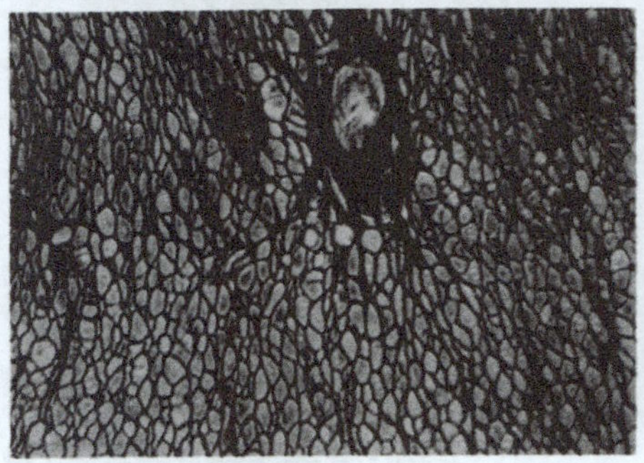

PLATE 7B

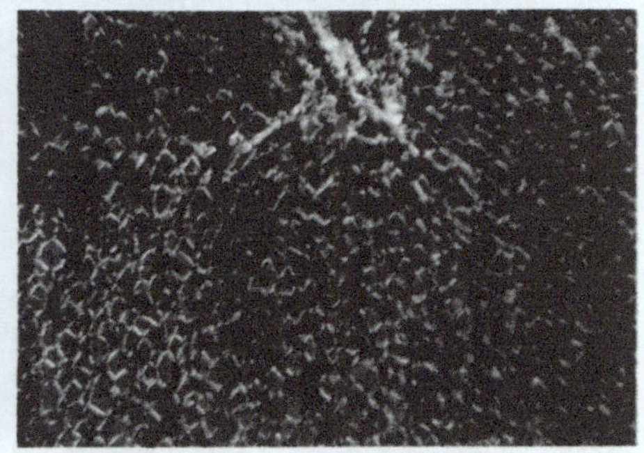

PLATE 5. A spread-out whole mount of guinea pig mesentery observed with the aid of the Picrosirius-polarization method. Note that birefringence brightness and interference colors are directly proportional to the thickness of the collagenous structure, since the strongest birefringence and the red color correspond to the regions where two thick collagen fibers overlap. Single thick collagen fibers show an orange color, while thinner collagen fibers are yellow. The thinnest fibers display a weak birefringence of a greenish color. Magnification × 320.

PLATE 6. Rat aortic wall. Thick, bright red, collagen type I fibers can be seen in the adventitial layer (lower region) whereas the tunica media (upper region) consists of thin, pale greenish, collagen type III-containing reticulin fibers. Picrosirius-polarization method. Magnification × 100.

PLATE 7. Adjacent serial sections of the inner muscular layer of dog small intestine. (A) This histological preparation was obtained by using the double staining method (silver impregnation technique followed by staining in the Picrosirius solution).[7] (B) This section is observed by the Picrosirius-polarization method. Note the striking correspondence in the distribution of the reticulin fibers that surround the smooth muscle cells (A), with the localization of the thin, weakly birefringent, collagen type-III containing fibers disclosed by the Picrosirius-polarization method (B). The adventitia of blood vessels is composed of nonargyrophilic, thick, strongly birefringent, collagen type I fibers (asterisk). Magnification ×300. (From Junqueira, L. C. U. and Montes, G. S., *Arch. Histol. Jpn.*, 46, 589, 1983. With permission.)

PLATE 8A

PLATE 8B

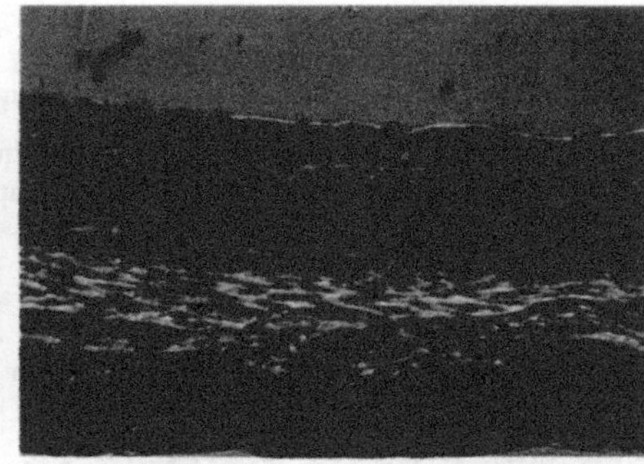

PLATE 9A

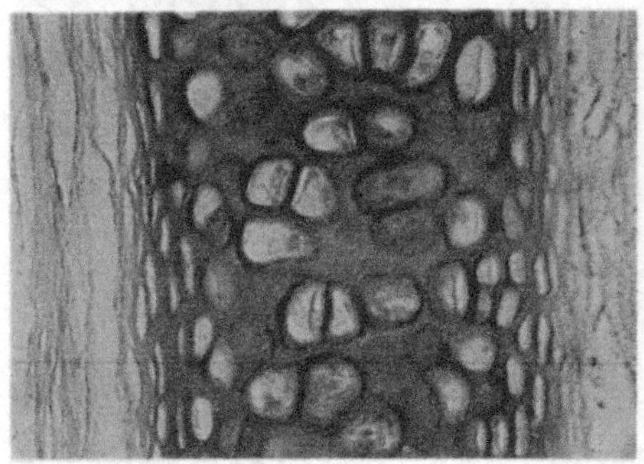

PLATE 9B

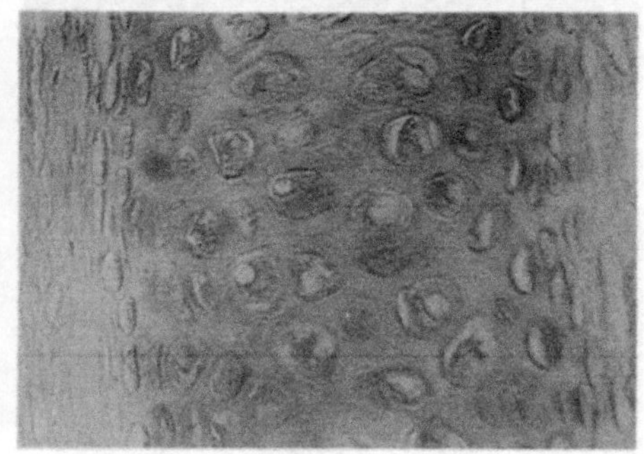

PLATE 8. Mouse pregnant uterus stained with Picrosirius and viewed by conventional illumination (A) and by polarized light (B). Two parallel, horizontally disposed basal laminae can be seen. The upper one is overlying the visceral yolk sac endoderm, while the one below – Reichert's membrane – underlies, and is intimately associated with, the parietal yolk sac endoderm. Observe that while basal laminae (being a continuous sheet each) show up as uninterrupted birefringent lines thoughout the whole length of the photomicrograph, the birefringence displayed by the interstitial collagens in the uterine wall (at the bottom of the picture) denotes the interwoven disposition of fibers that have been cut at the upper and lower surfaces of the tissue section. Magnification X 160.

PLATE 9. Rabbit ear cartilage before (A), and 6 hr after (B), intravenous papain injection. Both sections were stained with Alcian Blue; observe the remarkable loss of alcianophilia of the cartilage matrix as a result of papain injection. Magnification X 70.

A. General Characteristics of Proteoglycans and GAGs

GAGs are a family of heterogeneous polysaccharides, generally composed of repeating disaccharide units consisting of a hexosamine (usually *N*-acetyl-glucosamine or *N*-acetyl-galactosamine) glycosidically linked to either hexuronic (glucuronic or iduronic) acid or galactose. With the exception of hyaluronic acid, all GAGs characteristically possess sulfated esters.

1. Functions

GAG function is primarily related to its polyanionic structure, a consequence of the large number of carboxyl and sulfate groups on the carbohydrate moieties. In solutions containing GAGs, the polysaccharide chains electrostatically repel one another, and as a result they uniformly disperse and "fill" any space available. This behavior directly accounts for one of the principal functions of GAGs, which is to serve as a space filler by physically surrounding and encapsulating connective tissue cells and extracellular fibers.

The high density of anionic charges also influences the functional behavior of the ground substances in other ways. In fact, the electrostatic interaction among the entangled GAGs and their ability to bind to protein leads to the formation of an interlocked supramolecular matrix. In addition, GAGs play an important role in regulating the amount of water in connective tissue. Each molecule of GAG is extremely hydrophilic and combines with a great number of molecules of water. Almost all water present in the amorphous intercellular substance of connective tissue is in the form of solvation water of GAG molecules.

2. Distinction Between Proteoglycans and Glycoproteins

It is well known that sulfated GAGs do not normally occur as free polymers in tissues, but are covalently linked to a protein core forming complexes known as *proteoglycans*.[47]

Proteoglycans differ structurally from glycoproteins; in the latter the carbohydrate exists as heteropolysaccharides rather than homogeneous disaccharide repeating units. In addition, structural glycoproteins are basically all protein, possessing only a minor carbohydrate component (5 to 10% by weight). Fibronectin is the best understood and characterized of the ground substance glycoproteins; laminin and chondronectin are being actively studied at present.[48]

3. Macromolecular Structure

With the exception of hyaluronic acid, most GAGs in tissue are linked covalently to protein, forming complexes known as proteoglycans. In these protein-polysaccharide complexes, the protein moieties always represent a smaller proportion of the molecule than the numerous GAG side chains. Of the proteoglycans, the best characterized in regard to structure and function are those present in cartilage. These proteoglycans possess a central or core protein each, from which extend several GAG chains bound to serine residues of the core protein. The large number of anionic charges on the GAGs cause the polysaccharide chains to become stiff and extend outward from the protein core to occupy a large domain. Proteoglycan molecules, therefore, assume the aspect of a bottle-brush in which the core protein is located centrally and the attached GAGs stick out resembling the bristles.

The GAG chains of the proteoglycan electrostatically bind to extracellular proteins (e.g., collagen). One end of the core protein is relatively devoid of glycosaminoglycan chains and possesses a hyaluronic acid binding site. Through noncovalent interactions and the aid of an accessory "link" protein, approximately 100 proteoglycan complexes bind to each linearly extended hyaluronic acid molecule, forming proteoglycan aggregates[49] (Figures 10 and 11). Chondroitin 4/6-sulfate chains are the predominant GAGs of the cartilage proteoglycans and they specifically bind to collagen type II.[44] As a consequence of electrostatic repulsion, the polysaccharide chains repel one another and ultimately fill all of the space around the collagen

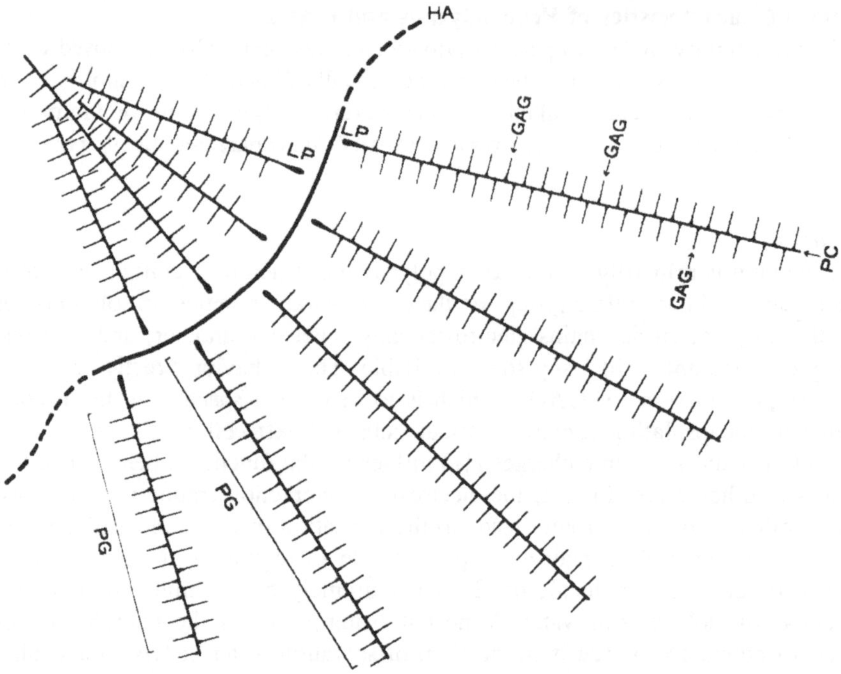

FIGURE 10. Structural and functional characteristics of proteoglycans. This schematic model depicts part of a cartilage proteoglycan aggregate. Observe that several proteoglycans (PG) are associated with a central strand of hyaluronic acid (HA). The chemical structure of a typical cartilage PG is shown schematically: each PG resembles a bottle brush consisting of a central filament (the backbone proteinic core, PC) with numerous projections along its length which correspond to the bristles (these polysaccharide chains which extend out from the core portein are the GAGs). Chondroitin 4/6-sulfate chains are the predominant component of the cartilage proteoglycans which also contain small amounts of keratan sulfate. Linking proteins (LP) covalently bind the PC of PGs to the linear HA molecules. Because of their high carboxylate and sulfate content, GAGs are polyanions possessing arrays of negatively charged groups. Adjacent GAG chains on the PG, therefore, strongly repel each other. As a result of this charge repulsion the PC and GAG chains tend to be fully extended in the cartilage so that each PG, in attempting to occupy the largest possible molecular domain, encompasses a very large volume of solvation water of GAG molecules (stippled area). When cartilage is subjected to a compressive load, solvent is displaced from the molecular domain, however the high negative charge density which is characteristic of the intramolecular interactions between the relatively stiff and polyanionic GAGs is increased (upper left corner). Because of the intermolecular competition between PG molecules for available solvent, when the load is removed, the molecular domains can once again expand. (From Junqueira, L. C. U. and Montes, G. S., *Arch. Histol. Jpn.*, 46, 589, 1983. With permission.)

fibrils. In this interaction, the collagen fibril arrays define the shape and provide the tensile strength of the cartilage matrix, and the hydrated proteoglycans are responsible for the gelation of the matrix, which maintains shape and provides for the compressibility of cartilage.

It is thus evident that the physical organization of extracellular macromolecules is critical: collagen molecules become assembled into fibrils, fibrils aggregate to form fibers, fibers associate into bundles of fibers, and proteoglycans in the ground substance play a major role in the ordering process;[5,50-51] on the other hand GAGs are composed of repeating monomers, GAGs linked to a same protein core form a proteoglycan which, in turn, may bind to a hyaluronic acid molecule to form a proteoglycan aggregate together with other proteoglycans. Further growth of these complex macromolecules at higher hierarchical levels occurs by interaction of collagen with proteoglycans.

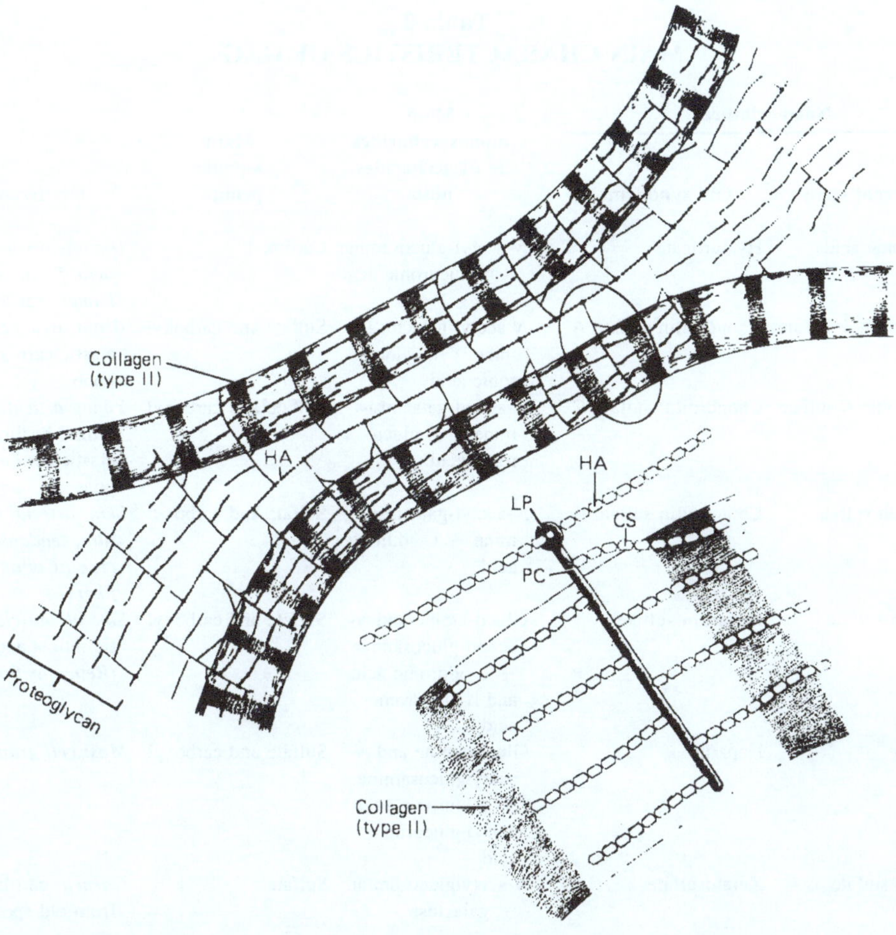

FIGURE 11. Schematic representation of the macromolecular organization of cartilage extracellular matrix. As a result of the polyanionicity and the localization of sulfate and carboxyl groups of proteoglycans, the negative charge concentration within the domain of each molecule is very high. Proteoglycans thus are able to participate extensively in intermolecular electrostatic interactions with adjacent molecules, such as collagen. Thus, cartilage extracellular matrix is a composite structure of collagen type II fibrils with intertwining proteoglycan aggregates. Link proteins (LP) covalently bind the proteinic core (PC) of proteoglycans to the linear hyaluronic acid (HA) molecules. The chondroitin 4/6-sulfate (CS) side chains of the proteoglycan electrostatically bind to the collagen fibrils, forming a crosslinked rigid matrix. (From Junqueira, L. C. and Carneiro, J., *Basic Histology*, 4th ed., Lange Medical, Los Altos, 1983, 129. With permission.)

The principal GAG components of the amorphous ground substance are hyaluronic acid and a closely related family of sulfated polysaccharides which include chondroitin sulfate, dermatan sulfate, and heparan sulfate, among others (Table 2).

B. Hyaluronic Acid and Sulfated GAGs

Hyaluronic acid, the most abundant and ubiquitous of the GAGs, is composed of about 2500 repetitive disaccharide units. Thus, this unbranched linear molecule is large enough to be visualized in the electron microscope. In solution, the overlapping and entanglement of the heavily hydrated hyaluronic acid molecules are responsible, at least in part, for the viscous nature of the amorphous ground substance. Relatively rich solutions of hyaluronic acid serve as lubricants (synovial fluid), space fillers (vitreous humor of the eye), and shock absorbers (nucleus pulposus of the intervertebral disk).

Table 2
MAIN CHARACTERISTICS OF GAGs

Nomenclature: Current term	Nomenclature: Old synonyms	Main monosaccharides in disaccharides units	Main anionic groups	Occurrence[a]
Hyaluronic acid	Hyaluronate	*N*-acetyl-glucosamine + D-glucuronic acid	Carboxyl	*Umbilical cord, synovial fluid, vitreous humor*, cartilage
Chondroitin 4-sulfate	Chondroitin sulfate A	*N*-acetyl-galactosamine + D- glucuronic acid	Sulfate and carboxyl	*Brain*, hyaline and elastic cartilages, bone
Chondroitin 6-sulfate	Chondroitin sulfate C	*N*-acetyl-galactosamine + D-glucuronic acid	Sulfate and carboxyl	*Young articular cartilage*, hyaline and elastic cartilages, bone
Dermatan sulfate	Chondroitin sulfate B	*N*-acetyl-galactosamine + L- iduronic acid	Sulfate and carboxyl	*Skin, arterial adventitia, tendons, fasciae, (Collagen fibers)*
Heparan sulfate	Heparitin sulfate	Glucosamine and *N*-acetyl-glucosamine + L- iduronic acid and D-glucuronic acid	Sulfate and carboxyl	*Smooth muscle*, aortic intima and media *(Reticular fibers)*
Heparin	Heparin	Glucosamine and *N*-acetylglucosamine + L-iduronic acid and D-glucuronic acid	Sulfate and carboxyl	*Mast cell granules*
Keratan sulfate	Keratosulfate	*N*-acetylglucosamine + galactose	Sulfate	*Cornea*, cartilages from old specimens

[a] Italics indicate that the corresponding GAG constitutes the major polysaccharide present in that tissue.

From Junqueira, L. C. U. and Montes, G. S., *Arch. Histol. Jpn.*, 46, 589, 1983. With permission.

In addition to hyaluronic acid, biochemical assays revealed the existence of at least five other closely related GAGs that are present in the amorphous ground substance.[47] These other GAGs are similar to hyaluronic acid in that they consist of repetitive disaccharide units. Unlike hyaluronic acid, the polysaccharide chains are relatively short and generally do not contain more than 40 disaccharide units. Most importantly, in these GAGs 50% or more of the saccharide moieties are sulfated.

The large number of anionic carboxyl and sulfate groups in the sulfated GAGs leads to extensive electrostatic binding to extracellular proteins. Unlike the viscous solutions produced by hyaluronic acid, crosslinked sulfated GAGs in solution form gel-like structures that tend to resist deformation and absorb shock by their ability to withstand compression.

Sulfated GAGs can be distinguished by their sugar composition, by the position and configuration of their glycosidic linkages, and by the location and number of their sulfate groups.

The most abundant of the sulfated GAGs is chondroitin sulfate, which exists in two isomeric forms: chondroitin 4-sulfate and chondroitin 6-sulfate. These forms differ in the position that the sulfate ester occupies on the galactosamine. The sulfated product usually contains about one sulfate group per dissacharide residue although considerable deviations from this ratio have been reported. Individual chondroitin sulfate chains usually contain sulfate groups in either 4- or 6-positions but may also be sulfated in both positions. Partially and totally desulfated chondroitin sulfates have also been detected in normal and pathologic tissues. Chondroitin sulfates constitute the major GAGs present in cartilage and bone; they also occur in soft connective tissues.

Dermatan sulfate differs from chondroitin sulfate in that it contains predominantly L-iduronic acid rather than D-glucuronic acid. Since the purified glycosaminoglycan chain invariably contains some D-glucuronic acid residues which are in disaccharide units identical to those of chondroitin sulfate, dermatan sulfate may be considered a hybrid molecule or a co-polymer of chondroitin sulfate.[47]

Heparan sulfate contains glucosamine as amino sugar and, in addition to the conventional ester sulfate, some of the amino sugars contain an N-sulfate group, which is unique to heparan sulfate and heparin. *Heparin* bears close structural similarity to heparan sulfate, and thus these two GAGs can be regarded as a family of related molecules.[47] The L-iduronic acid: D-glucuronic acid ratio generally increases with increasing sulfate content; low-sulfated, D-glucuronic acid-rich polysaccharides are usually designated heparan sulfate while high-sulfated, L-iduronic acid-rich species are classified as heparin.[52]

Heparin is synthesized and stored in tissue mast cells and, thus, it is not a component of the connective tissue matrix. Occasionally, mast cell granules are secreted into the connective tissue but no structural role of heparin has been reported so far.

All GAGs, except hyaluronic acid (which lacks sulfate groups) and keratan sulfate (which lacks carboxyl groups) bind to collagen by electrostatic interaction at physiological pH and ionic strength.[52]

C. Histochemical Methods

Collagen is embedded in a colorless, transparent, and homogeneous amorphous ground substance with which it interacts. Because of the high polysaccharide and water content of the amorphous ground substance, standard fixation procedures have failed to preserve the integrity and supramolecular organization of its components. During the last decade, modern methods have provided significant insight into the molecular organization and physiologic role of proteoglycans.

1. Electron Microscopy

When adequately fixed, the macromolecular components of the amorphous ground substance aggregate and can be disclosed as coarse granules, or multiform branching structures (frequently bound to collagen fibrils), under the electron microscope.

A series of methods was developed to permit the observation of proteoglycans by aid of electron microscopy. Not only did these methods permit the observation of isolated proteoglycan molecules,[53] but it was also possible to localize structures that could be identified as proteoglycans in connective tissues.

These methods are based on the "stainability" of proteoglycans by Ruthenium Red, Alcian Blue, and Safranin O,[54] for electron microscopy, associated with a previous treatment of the material studied with enzymes that would hydrolyze proteoglycans.

These studies showed that the best results were obtained with Safranin 0 (probably due to its low molecular weight and cationic activity permitting a better penetrability and also the precipitation of proteoglycans in the tissues during fixation)[55] and with papain that promotes the hydrolysis of 90% of the proteoglycans.[56]

The results obtained with the above-mentioned methods showed that, in cartilage, proteoglycans are observed under the electron microscope as coarse granules or multiform branching structures frequently attached to collagen fibrils[55] (Figure 12). This attachment occurs in several, but not in all, regions of the cartilage in a periodic fashion suggesting the

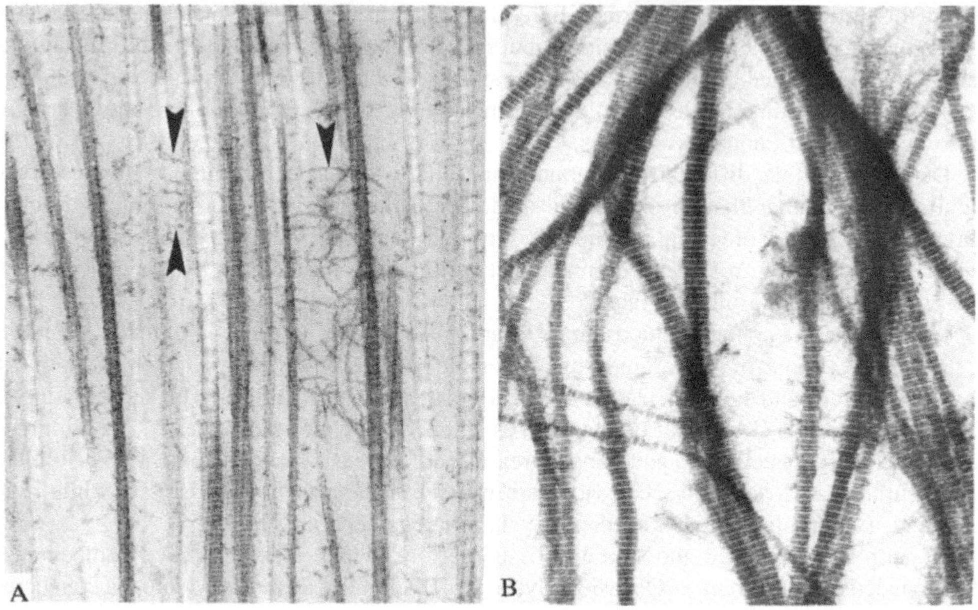

FIGURE 12. Electron micrographs of the body wall of the sea cucumber *Stichopus badionotus:* control (A) and papain-digested (B) specimens. (A) Collagen fibrils are irregularly impregnated by uranyl and lead. The ground substance possesses microfibrils which bind the collagen fibrils to one another (arrowheads). (B) Observe the absence of microfibrillar bridges of proteoglycans: note that removal of proteoglycans by enzymatic activity permits a better visualization of the cross striation of collagen fibrils when compared with the electron microscopic picture of the control tissue in (A). A: Magnification × 31,800; B: Magnification × 60,000. (From Junqueira, L. C. U. et al., *Rev. Can. Biol. Exp.*, 39, 157, 1980. With permission.)

presence of periodicity in the binding sites of collagen to proteoglycans. When tissues were previously incubated with papain, hyaluronidase or chondroitinases (enzymes which are known to degrade proteoglycans, as will be seen below), the absence of stained material was observed in the amorphous component of the matrix, leaving only the naked collagen fibrils[55] (Figure 12).

2. *Light Microscopy*

The presence of some macromolecular components of the amorphous ground substance can be demonstrated in the light microscope by aid of histochemical reactions. By histochemical methods it is possible to distinguish acid from neutral polysaccharides. It is also possible to differentiate acid structures containing sulfated radicals (for instance, heparin or chondroitin sulfates) from others which are known as carboxylated compounds (namely, hyaluronic acid).[57] The use of digestion of tissue sections by specific mucopolysaccharidases associated with histochemical staining methods, provides a good means of localizing the different GAGs in organs under the light microscope.[28,59-60]

Provided that the chromophore in the dye solution used obeys the Lambert-Beer law (which is not always the case and, thus, should be checked by measuring a series of concentrations) these histochemical techniques are readily adaptable to quantitation. By associating either Safranin 0 or PAS staining with microspectrophotometry, Kiviranta and co-workers successfully measured proteoglycans in tissue sections and showed that chondroitin sulfate concentration increases along with depth of articular cartilage.[58,61] Thus, the method provides an accurate technique for quantitation and localization of GAGs in tissue sections. When associated with specific enzyme digestion, this method enables detection of separate GAGs.[58]

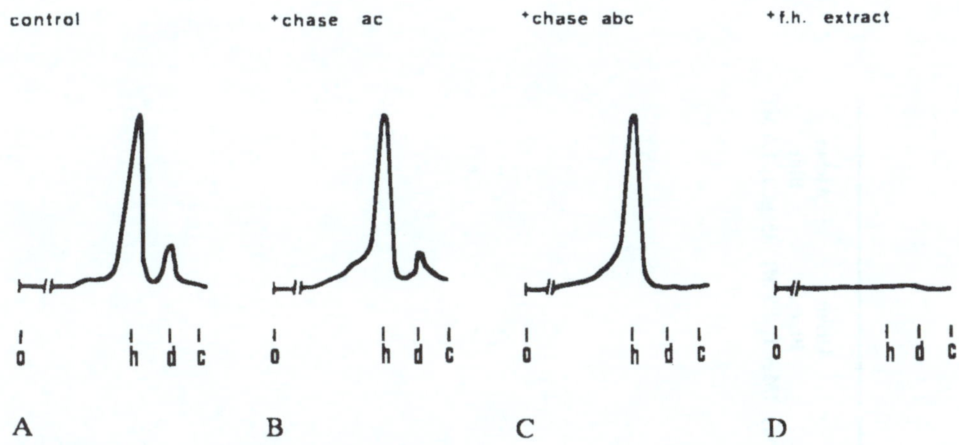

FIGURE 13. Densitometry of the electrophoretic slides from control and enzyme-digested GAGs from the inner muscular layer of dog small intestine. (A) The control curve presents a high peak (87%) with the electrophoretic mobility of standard heparan sulfate (h), and a small peak (13%) corresponding to dermatan sulfate (d). (B) These GAGs are resistant to chondroitinase AC (chase ac) degradation, which shows that they are not affected by the enzyme which degrades chondroitin sulfates AC (c) specifically. (C) The digestion with chondroitinase ABC (chase abc) promotes the disappearance of the small peak, showing that it corresponds to chondroitin sulfate B (dermatan sulfate). (D) All the glycosaminoglycans are susceptible to enzymatic degradation by a crude extract from induced *Flavobacterium heparinum* (f.h.) containing, in addition to chondroitinases, enzymes which degrade heparan sulfate and heparin. (From Montes, G. S. et al., *Histochemistry*, 65, 131, 1980. With permission.)

Proteoglycans and glycoproteins have been demonstrated in cartilage using immunohistologic reactions.[62] Combined with enzymatic treatment, immunohistology has revealed the in situ structure of proteoglycan aggregates.[62]

In addition, specific bacterial mucopolysaccharidases have also been used to identify GAGs by comparing the electrophoretic migration of the GAGs contained in control and digested homogenates prepared for biochemical studies.[44] Among the specific mucopolysaccharidases the most well known are chondroitinase AC (E.C. 4.2.2.5) and chondroitinase ABC (E.C. 4.2.2.4). Chondroitinase AC degrades chondroitin 4-sulfate (A) and chondroitin 6-sulfate (C), and is devoid of proteolytic activity. Chondroitinase ABC degrades dermatan sulfate (B) in addition to chondroitin sulfates A and C. Adapted strains of *Flavobacterium heparinum* produce chondroitinases, besides other enzymes which degrade heparan sulfate and heparin (Figure 13).

a. Alcian Blue

This is a positively charged (cationic) dye which forms electrostatic bonds with certain tissue polyanions containing either carboxyl or sulfate groups.

Alcian Blue at varying pH of solution — A given tissue component is more intensely stained when the dye is used at the pH at which the reacting groups are fully ionized. This phenomenon may be used to employ Alcian Blue solution of varying pH to detect sulfate esters (which react at a lower pH than the carboxylated esters). Thus, hyaluronic acid (which stains well at pH levels ranging from 3.2 down to 1.7) can be separated from sulfated GAGs that react most consistently below pH 1.0 (Table 3). However, the definitive demonstration of the presence of hyaluronic acid is by means of specific bacterial hyaluronidase digestion followed by appropriate staining.[57]

Critical electrolyte concentration — The Alcian Blue technique involving critical electrolyte concentration is a useful separation procedure since the various acid GAGs have

Table 3
STAINING REACTIONS OF POLYSACCHARIDES

	Technique							
	PAS (standard)	PAS (following borohydride)	Alcian Blue (pH 2.5)	Alcian Blue (pH 0.5)	Alcian Blue ($MgCl_2$, 0.3 *M*)	Alcian Blue ($MgCl_2$, 0.65 *M*)	Alcian Blue ($MgCl_2$ 0.9 *M*)	Alcian Blue ($MgCl_2$ 1.0 *M*)
Homoglycans (Glycogen)	+	−						
Neutral glycoproteins	+	−						
Glycosaminoglycans (GAGs)								
Nonsulfated (hyaluronic acid)	−	+	+	−	+	−	−	−
Sulfated[a]								
Chondroitin 4- and 6-sulfates	−	+	+	+	+	+	−	−
Dermatan sulfate	−	+	+	+	+	+	−	−
Heparan sulfate	−	+	+	+	+	+	+	−
Keratan sulfate	−	+	+	+	+	+	+	+

[a] Listed according to increasing degree of sulfation.

Scoring: + denotes positive reaction, − denotes negative reaction.

different critical electrolyte concentration points, and this serves as a potentially useful means of identifying them in tissue sections.[57]

The critical electrolyte concentration is the point at which the amount of an electrolyte, such as magnesium chloride, in Alcian Blue solution is sufficient to prevent staining (due to successful competition of electrolyte cations with the dye cations for binding sites on tissue polyanions). When sections are stained to equilibrium in ascending molarities of magnesium chloride in buffered solution (0.05% Alcian Blue at pH 5.8) there is a diminution or cut off in Alcian Blue staining, depending on the substance present (Table 3).

b. Periodic Acid-Schiff (PAS) Technique

The PAS reaction is a widely used technique in carbohydrate histochemistry. The standard PAS method demonstrates carbohydrate substances which contain glycol groups and usually have neutral electrostatic charge.

The PAS reaction is positive with molecules containing adjacent carbon atoms with hydroxyl groups. Periodic acid breaks the carbon-to-carbon bonds of glycols to form dialdehydes. The Schiff's reagent reacts with the resulting aldehydes to form a magenta-colored compound.[61] Since the GAGs remain unreactive (because the negative electrostatic field of GAGs reduces considerably the rate of periodate oxidation) the PAS method can be used to differentiate between neutral structural glycoproteins and acid GAGs.

Modified PAS method — Since biochemical experiments have shown that uronic acids react slowly with periodic acid when very long oxidation times are used, Scott and Dorling[63] presented a modified PAS reaction to identify uronic acid-containing carbohydrates.

GAGs have an intense electrostatic field which repulses the negatively charged periodate ion and retards the oxidation reaction as compared to the uncharged glycoconjugates (thus the latter molecules, being quickly oxidizable, yield reactive aldehydes long before GAGs do). The early aldehyde groups are then blocked (rendered Schiff negative) with aqueous sodium borohydride. The further periodic acid oxidation makes uronic acid of GAGs reactive with the Schiff solution. This modified PAS method has been successfully used to study the differential distribution of chondroitin sulfate and neutral glycoproteins in articular cartilage.[61]

IV. COLLAGEN-PROTEOGLYCAN INTERACTION

It is well known that collagen interacts with proteoglycans and it has been stated that a "better understanding of these interactions is probably of great biological significance".[64] However, it was not until recently that major advances in our knowledge of the biology of both collagen and proteoglycans considerably facilitated an understanding of their interactions.

Since histochemical and electron microscopic observations in our laboratory have shown that collagen-proteoglycan interaction displays a morphological equivalent and can be quantitated in tissue sections,[5] it appeared expedient to review briefly the biology of collagen-proteoglycan interaction in the light of the current knowledge on the subject.

A. General Considerations

Collagen-proteoglycan interaction occurs mainly due to a electrostatic binding between the acid groups of the GAGs and the basic (amino) groups present in the side chains of collagen lysyl and arginyl residues. The fact that polar amino acid residues are accumulated in clusters along the collagen molecule probably explains why proteoglycans attach to collagen with a periodical pattern.

Although much has been published regarding in vitro studies on collagen-proteoglycan interaction, the papers on proteoglycan morphology, and on their interaction with collagen in tissues, are few and almost limited to cartilage.[49] Recently, however, a method has been

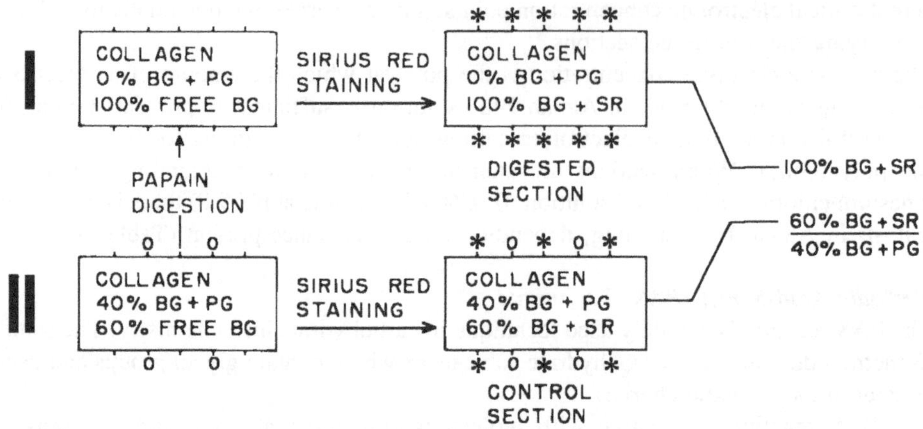

FIGURE 14. Diagrammatic representation of the different steps in the process of quantifying collagen-proteoglycan interaction in tissue sections. Two adjacent serial sections are used: one of the sections (I) is submitted to papain digestion which degrades proteoglycans (PG) and, thus, unmasks the basic groups (BG) of collagen to which the acid radicals of proteoglycans were bound; the other section (II) acts as control: four of the basic groups of collagen are bound to acid radicals of the proteoglycans (0) and the other six basic groups of collagen are free. Both sections are then stained in the Picrosirius solution; Sirius Red (*) being a strongly acidic dye, binds to the available basic groups of collagen. In a next step, the Sirius Red bound to both control and papain-digested sections is eluted in dilute sodium hydroxide and measured in a spectrophotometer: while 100% of the basic groups of collagen were bound to Sirius Red (SR) in the digested section, only 60% of the basic groups of collagen were available for binding to Sirius Red in the control section, thus indicating that 40% of the basic groups of collagen are blocked due to collagen-proteoglycan interaction. (The fact that the unmasking of the basic groups of collagen promoted by papain digestion results in an increased stainability by Sirius Red which, in turn, is responsible for an enhancement in birefringence, is illustrated in Figure 15.) (From Junqueira, L. C. U. and Montes, G. S., *Arch. Histol. Jpn.*, 46, 589, 1983. With permission.)

developed that permits the quantitative estimation of this interaction in tissue sections.[43] This method is based on the quantitation of the collagen basic groups that are blocked by the interaction with the proteoglycan acid radicals. These blocked groups cannot, therefore, bind to the anionic dye Sirius Red, thus reducing the stainability of tissue sections. When adjacent sections are treated with papain, the proteoglycans are degraded, and the previously blocked collagen basic groups are liberated, thus increasing the stainability of this tissue section. In the next step, the Sirius Red bound to both control and papain-treated sections is eluted in dilute sodium hydroxide and measured in a spectrophotometer. Thus, the difference in the amount of dye present in control and papain-digested sections measures the collagen basic (amino) groups originally blocked due to collagen-proteoglycan interaction (Figure 14).

1. Papain Digestion of Tissue Sections

Tissue sections are deparaffinized, hydrated to water, and incubated for 2 hr at 37°C in a 0.5% solution of papain (Papain IV F III, Difco Laboratories, Detroit, Mich.) dissolved in 0.02 *M* phosphate buffer containing 0.005 *M* sodium bisulfite and 0.0005 *M* EDTA. Control slides are incubated under the same conditions without papain.

The unmasking of the basic groups of collagen that is promoted by papain digestion results in an increase in collagen stainability by Sirius Red which, in turn, is responsible for the enhancement in birefringence that can be detected in papain-digested tissue sections when compared with controls (Figure 15).

B. Specific Association of the Genetically Distinct Collagen Types with the Different Proteoglycans

1. *Quantitative Aspects*

When this method was applied to different tissues containing different collagen types from sections of organs obtained from representative species of vertebrates, it showed in all cases studied a close relation between the type of collagen and the intensity of its interaction with proteoglycans. Thus, collagen type II shows high levels of interaction with proteoglycans, whereas collagen type I displays very low levels of interaction with proteoglycans. Collagen type III shows intermediate levels of interaction (Table 4).

2. *Qualitative Aspects*

In another report it was shown that the type and amount of proteoglycans present in tissues also varies according to the type of collagen the sample contains.[44] Thus, tissues composed of collagen type I possess small amounts of proteoglycans which contain almost exclusively dermatan sulfate; while tissues containing only collagen type II have high amounts of 4/6-chondroitin sulfates. Collagen type III is the major fibrillary constituent of tissues that possess intermediate levels of proteoglycans, which contain great amounts of heparan sulfate (Table 5, Figure 16).

3. *Differential Distribution of the Specific Collagen-Proteoglycan Interactions*

These studies show that not only does the amount of proteoglycans vary according to the type of collagen present in the tissues, but also that the type of the constituent GAG is specifically related to the type of interstitial collagen that the tissue studied contains.

These conclusions are important to interpret the morphologic aspects observed in the electron microscopy of the different types of collagen, aspects that can contribute to distinguish them when studying collagen type distribution with the aid of the electron microscope.

Thus, one can easily understand why collagen fibers (composed of collagen type I), that contain a small amount of proteoglycans and interact weakly with these compounds, present their fibrils compactly disposed and show very few branched structures connecting these fibrils (Figure 8). It also explains why hyaline and elastic cartilages that contain collagen type II and are rich in proteoglycans (which strongly interact with the collagen present) show their fibrils widely separated from one another and why the space between them is filled with coarse granules and multiform branching structures that have been identified as proteoglycans.[55]

Collagen type III (present in the reticulin fibers), that shows intermediate levels of interaction with proteoglycans, presents a less compact disposition of its fibrils when compared to that of collagen fibers[7] and shows a net of multiform branching structures binding its fibrils (Figure 4).

The above-mentioned aspects strongly suggest that collagen-proteoglycan interaction has an evident structural role in tissues, lending to them a characteristic elasticity and imparting resistance to pressure that acts like a biological spring.

C. Histochemical Evidence for the Structural Role of the Collagen-Proteoglycan Interaction in Tissues

The importance of collagen-proteoglycan interaction has been observed in several biological examples and experimental models that have been thoroughly reviewed elsewhere.[5] The first evidence for the structural role of this interaction was reported 3 decades ago by Thomas[65] when describing the effect of endovenous papain injection on the ear cartilage in the rabbit. In this experimental model the injection of papain promotes the degradation of the ear cartilage proteoglycans, resulting in a loss of turgidity of the ear which collapses. The ear cartilage, which is strongly alcianophilic before papain injection, contrasts with the

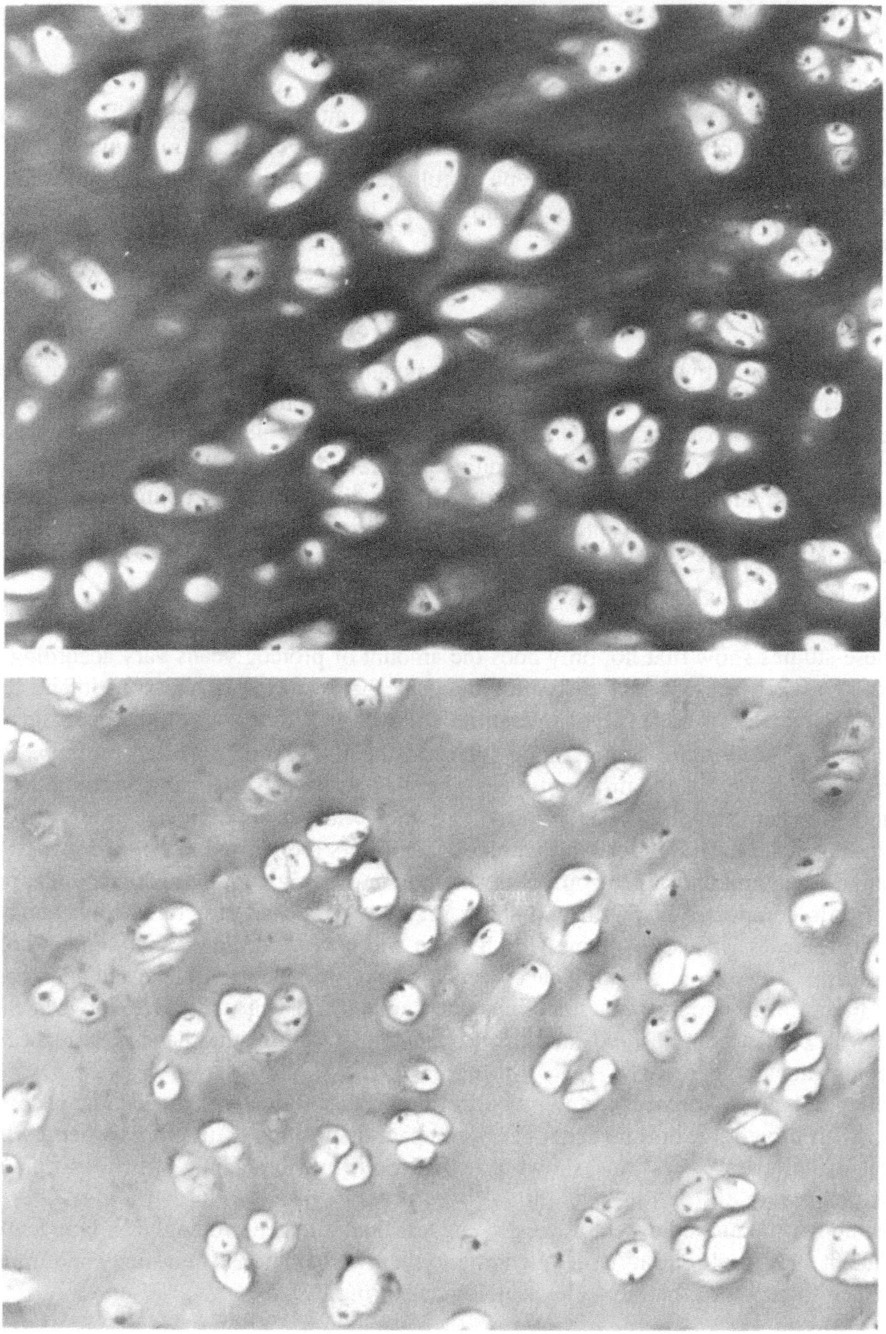

FIGURE 15. Papain-digested (top) and control (lower) sections of hyaline cartilage stained in the Picrosirius solution and observed with conventional (left), and polarizing (right), microscopy. Papain digestion hydrolyzes proteoglycans and, thus, unmasks the basic groups of collagen that were bound to the acid groups of glycosaminoglycans. The unmasking of the basic groups of collagen results in an increase in collagen stainability by Sirius Red which, in turn, is responsible for the enhancement in birefringence observed in the papain-digested section when compared with the control. Magnification × 200.

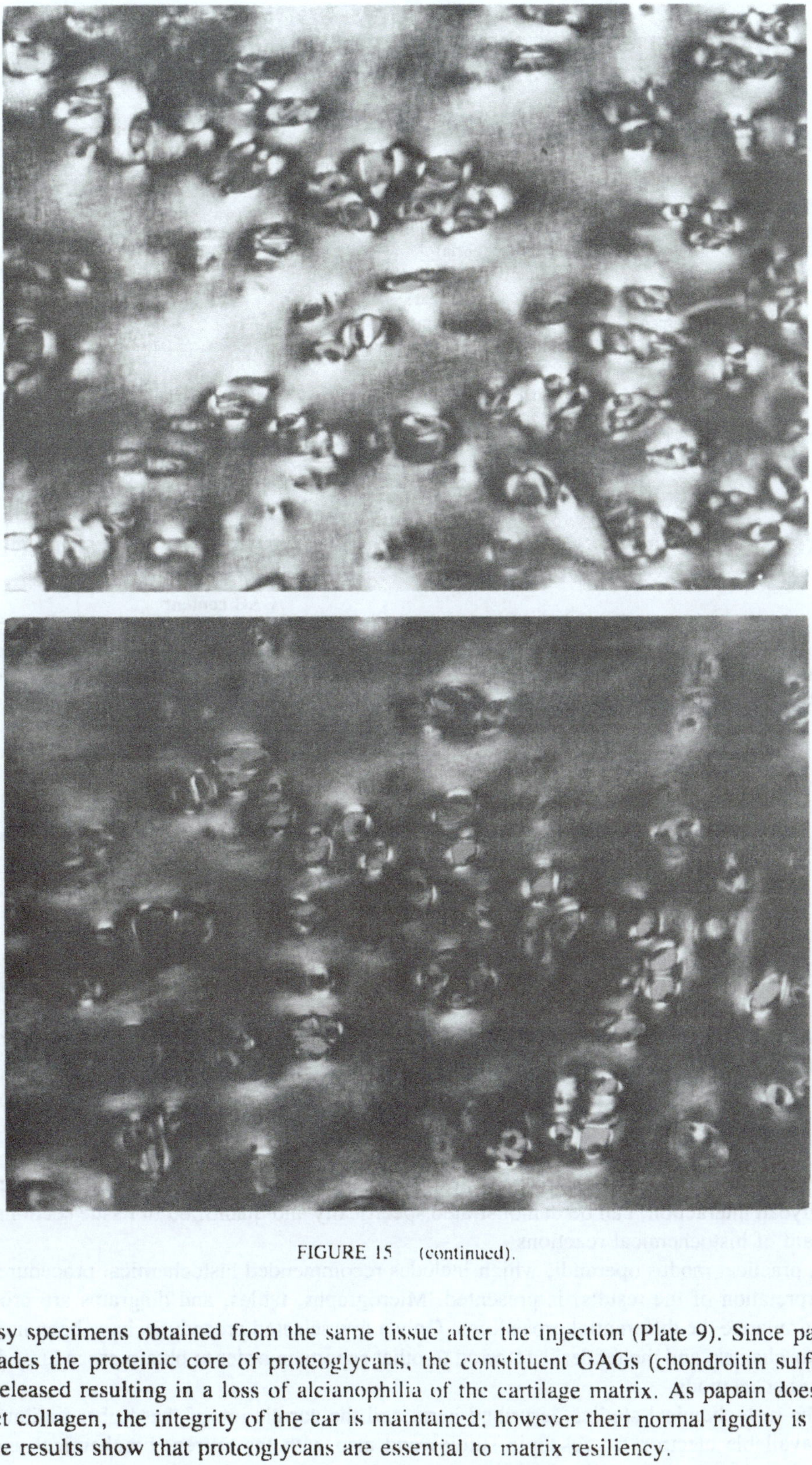

FIGURE 15 (continued).

biopsy specimens obtained from the same tissue after the injection (Plate 9). Since papain degrades the proteinic core of proteoglycans, the constituent GAGs (chondroitin sulfates) are released resulting in a loss of alcianophilia of the cartilage matrix. As papain does not affect collagen, the integrity of the car is maintained; however their normal rigidity is lost. These results show that proteoglycans are essential to matrix resiliency.

Table 4
QUANTITATION OF COLLAGEN-PROTEOGLYCAN INTERACTION IN TISSUES CONTAINING DIFFERENT TYPES OF COLLAGEN

Type of collagen	Species	Tissues and organs	Number of cases	Collagen-proteoglycan interaction[a]	Ref.
I	Human	Dermis (skin)	1	18.00	43
	Human	Adventitia (aorta)	3	17.97	23
II	Pig	Hyaline cartilage (trachea)	4	66.00	43
	Pig	Elastic cartilage (ear)	4	62.00	43
III	Human	Tunica media (aorta)	3	37.26	23
	Dog	Smooth muscle (small intestine)	3	40.00	7

[a] Average; expressed as percentage of collagen basic groups exposed by papain hydrolysis.

From Junqueira, L. C. U. and Montes, G. S., *Arch. Histol. Jpn.*, 46, 589, 1983. With permission.

Table 5
GAGs IN DOG TISSUES CONTAINING DIFFERENT TYPES OF COLLAGEN

		GAG content			
Type of collagen	Tissues and organs	Total mg/g (dry weight)	Chondroitin 4/6-sulfates (%)	Dermatan sulfate (%)	Heparan sulfate (%)
I	Fascia (antebrachial)	1.00	—	100	—
	Central tendon (diaphragm)	2.50	—	100	—
II	Elastic cartilage (ear)	26.00	100	—	—
	Elastic cartilage (epiglottis)	10.52	100	—	—
	Hyaline cartilage (thyroid)	13.20	100	—	—
	Hyaline cartilage (trachea)	10.31	100	—	—
III	Smooth muscle (small intestine)	5.11	—	13	87

From Junqueira, L. C. U. and Montes, G. S., *Arch. Histol. Jpn.*, 46, 589, 1983. With permission.

After papain injection, the resynthesis of the proteoglycans occurs in 72 hr with concomitant recuperation of the ear cartilage turgidity.

V. SUMMARY

The purpose of this chapter is to show that collagen, proteoglycans, and collagen-proteoglycan interaction, can be demonstrated specifically and quantified in tissue sections with the aid of histochemical reactions.

A practical modus operandi, which includes recommended histochemical procedures and interpretation of the results, is presented. Micrographs, tables, and diagrams are provided to summarize the differential procedures. Only a few selected procedures have been included in this chapter, and the reader is referred to other reports in order to obtain procedural details for other methods.

The histochemical studies described here, and the correlation of these observations with the available electron microscopic studies and also with the current biochemical and immunohistologic data on the subject, enable the following conclusions to be drawn.

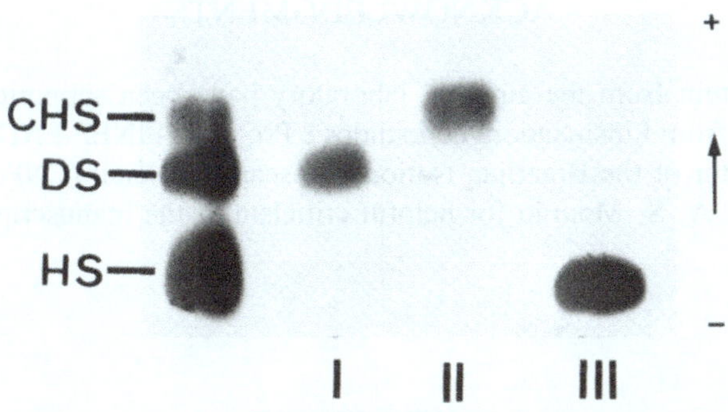

FIGURE 16. Agarose gel electrophoresis of GAGs isolated from tissues containing different collagen types. (I) Dog antebrachial fascia (which contains collagen type I); (II) dog thyroid cartilage (which contains collagen type II); and (III) inner muscular layer of dog small intestine (which contains collagen type III). At left, standard chondroitin 4- and 6-sulfates (CHS), dermatan sulfate (DS), and heparan sulfate (HS). (From Junqueira, L. C. U. et al., *Cell Tissue Res.*, 217, 171, 1981. With permission.)

The Picrosirius-polarization method is a specific histochemical procedure for collagen detection in tissue sections and is also useful for studying the differential distribution of the structurally distinct collagen types in routinely fixed, paraffin-embedded organs.

The results obtained when using the above-mentioned method showed that collagen type I has been localized to all sites where collagen fibers have been described (e.g., tendons, fasciae, organ capsules, adventitial layer of blood vessels, epineurium, etc.) and shows up in the form of thick, brilliant (strongly birefringent), yellow or red fibers which shine against a dark background when studied with the aid of the Picrosirius-polarization method. On the other hand collagen type III has been disclosed in all structures to which reticulin fibers have been localized (such as endoneurium, arterial tunica media, reticular network of parenchymatous organs, etc.) and displays a weak birefringence of a greenish color when viewed by this method. Collagen type II which usually does not stand out from the amorphous ground substance in routine histologic preparations, is identified with hyaline and elastic cartilages and displays a weak birefringence of a varying color (both of which depend on its interaction with the ground substance) when studied with the aid of the Picrosirius-polarization method. Collagen type IV was shown to display molecular orientation in all basal laminae studied by the Picrosirius-polarization method, since it was disclosed as continuous weakly birefringent sheets in tissue sections.

A striking correlation between the tissue distribution of the genetically distinct types of interstitial collagen and the occurrence of the different GAGs (which argues strongly in favor of a specific interaction) could be demonstrated by means of the methods used. Thus, tissues composed of collagen type I possess small amounts of proteoglycans which contain almost exclusively dermatan sulfate; while tissues containing only collagen type II have high amounts of chondroitin 4/6-sulfates. Collagen type III is the major fibrillary constituent of tissues that possess intermediate levels of proteoglycans, which contain great amounts of heparan sulfate.

The biochemical, light microscopical, and ultrastructural equivalents of these histochemical findings have been emphasized in order to permit the interpretation of the morphologic aspects that can contribute to distinguish these macromolecular components when studying tissue sections either under the light microscope or with the aid of electron microscopy.

ACKNOWLEDGMENTS

Studies cited herein from the authors' laboratory have been supported in part by grant no. 43.83.0610/00 from Financiadora de Estudos e Projetos (FINEP-FNDCT). G. S. Montes is Career Investigator of the Brazilian National Research Council (CNPq). The authors are grateful to Prof. P. A. S. Mourão for helpful criticism of the manuscript.

REFERENCES

1. **Bancroft, J. D. and Stevens, A., Eds.,** *Theory and Practice of Histological Techniques*, Churchill Livingstone, London, 1977, chap. 8 and 11.
2. **Luna, L. G.,** *Manual of Histologic Staining Methods of the Armed Forces Institute of Pathology*, McGraw-Hill, New York, 1968.
3. **Bradbury, P. and Gordon, K.,** Connective tissues and stains, in *Theory and Practice of Histological Techniques*, Bancroft, J. D. and Stevens, A., Eds., Churchill Livingstone, London, 1977, 95.
4. **Horton, W. A.,** Histochemistry — a valuable tool in connective tissue research, *Collagen Rel. Res.*, 4, 231, 1984.
5. **Junqueira, L. C. U. and Montes, G. S.,** Biology of collagen — proteoglycan interaction, *Arch. Histol. Jpn.*, 46, 589, 1983.
6. **Junqueira, L. C. U., Assis Figueiredo, M. T., Torloni, H., and Montes, G. S.,** Differential histologic diagnosis of osteoid — a study on human osteosarcoma collagen by the histochemical Picrosirius-polarization method, *J. Pathol.*, 148, 189, 1986.
7. **Montes, G. S., Krisztán, R. M., Shigihara, K. M., Tokoro, R., Mourão, P. A. S., and Junqueira, L. C. U.,** Histochemical and morphological characterization of reticular fibers, *Histochemistry*, 65, 131, 1980.
8. **Junqueira, L. C. U., Bignolas, G., and Brentani, R. R.,** Picrosirius staining plus polarization microscopy — a specific method for collagen detection in tissue sections, *Histochem. J.*, 11, 447, 1979.
9. **Montes, G. S., Bezerra, M. S. F., and Junqueira, L. C. U.,** Collagen distribution in tissues, in *Ultrastructure of the Connective Tissue Matrix*, Ruggeri, A. and Motta, P. M., Eds., Martinus Nijhoff, The Hague, 1984, chap. 3.
10. **López-de León, A. and Rojkind, M.,** A simple micromethod for collagen and total protein determination in formalin-fixed paraffin-embedded sections, *J. Histochem. Cytochem.*, 33, 737, 1985.
11. **Pérez-Tamayo, R.,** Pathology of collagen degradation, *Am. J. Pathol.*, 92, 509, 1978.
12. **Junqueira, L. C. U., Cossermelli, W., and Brentani, R.,** Differential staining of collagens type I, II and III by Sirius Red and polarization microscopy, *Arch. Histol. Jpn.*, 41, 267, 1978.
13. **Zambrano, N. Z., Montes, G. S., Shigihara, K. M., Sanchez, E. M., and Junqueira, L. C. U.,** Collagen arrangement in cartilages, *Acta Anat.*, 113, 26, 1982.
14. **Kefalides, N. A., Alper, R., and Clark, C. C.,** Biochemistry and metabolism of basement membranes, *Int. Rev. Cytol.*, 61, 167, 1979.
15. **Junqueira, L. C. U., Montes, G. S., Toledo, O. M. S., Bexiga, S. R. R., Gordilho, M. A., and Brentani, R. R.,** Evidence for collagen molecular orientation in basement membranes, *Histochem. J.*, 15, 785, 1983.
16. **Montes, G. S., and Junqueira, L. C. U.,** Biology of collagen, *Rev. Can. Biol. Exp.*, 41, 143, 1982.
17. **Lapiere, Ch. M., Nusgens, B., and Piérard, G. E.,** Interaction between collagen type I and type III in conditioning bundles organization, *Connect. Tissue Res.*, 5, 21, 1977.
18. **Bornstein, P. and Sage, H.,** Structurally distinct collagen types, *Annu. Rev. Biochem.*, 49, 957, 1980.
19. **Eyre, D. R.,** Collagen: molecular diversity in the body's protein scaffold, *Science*, 207, 1315, 1980.
20. **Minor, R. R.,** Collagen metabolism, *Am. J. Pathol.*, 98, 225, 1980.
21. **Pérez-Tamayo, R. and Montfort, I.,** The susceptibility of hepatic collagen to homologous collagenase in human and experimental cirrhosis of the liver, *Am. J. Pathol.*, 100, 427, 1980.
22. **Junqueira, L. C. U., Montes, G. S., and Krisztán, R. M.,** The collagen of the vertebrate peripheral nervous system, *Cell Tissue Res.*, 202, 453, 1979.
23. **Carrasco, F. H., Montes, G. S., Krisztán, R. M., Shigihara, K. M., Carneiro, J., and Junqueira, L. C. U.,** Comparative morphologic and histochemical studies on the collagen of vertebrate arteries, *Blood Vessels*, 18, 296, 1981.

24. **Montes, G. S., Cotta-Pereira, G., and Junqueira, L. C. U.,** The connective tissue matrix of the vertebrate peripheral nervous system, in *Advances in Cellular Neurobiology,* Vol. 5, Fedoroff, S., Ed., Academic Press, Orlando, Fla., 1984, 177.
25. **Junqueira, L. C. U., Montes, G. S., and Sanchez, E. M.,** The influence of tissue section thickness on the study of collagen by the Picrosirius-polarization method, *Histochemistry,* 74, 153, 1982.
26. **Junqueira, L. C. U., Montes, G. S., Almeida Neto, E., Barros, C., and Tedesco-Marchese, A. J.,** The collagen of permanently damaged nerves in human leprosy, *Int. J. Leprosy,* 48, 291, 1980.
27. **Junqueira, L. C. U., Montes, G. S., Kaupert, D., Shigihara, K. M., Bolonhani, T. M., and Krisztán, R. M.,** Morphological and histochemical studies on the collagen in neurinomas, neurofibromas, and fibromas, *J. Neuropathol. Exp. Neurol.,* 40, 123, 1981.
28. **Junqueira, L. C. U., Montes, G. S., Toledo, O. M. S., and Joazeiro, P. P.,** Morphological, histochemical and biochemical observations on the connective tissue matrix of *in situ* and isolated hepatic granulomas in experimental murine schistosomiasis, *Ann. Trop. Med. Parasitol.,* 80, 27, 1986.
29. **Vidal, B. C., Mello, M. L. S., and Pimentel, E. R.,** Polarization microscopy and microspectrophotometry of Sirius Red, Picrosirius and Chlorantine Fast Red aggregates and of their complexes with collagen, *Histochem. J.,* 14, 857, 1982.
30. **Szendroi, M., Vajta, G., Kovács, L., Schaff, Z., and Lapis, K.,** Polarization colours of collagen fibres: a sign of collagen production activity in fibrotic processes, *Acta Morphol. Hung.,* 32, 47, 1984.
31. **Junqueira, L. C. U., Zugaib, M., Montes, G. S., Toledo, O. M. S., Krisztán, R. M., and Shigihara, K. M.,** Morphologic and histochemical evidence for the occurrence of collagenolysis and for the role of neutrophilic polymorphonuclear leukocytes during cervical dilation, *Am. J. Obstet. Gynecol.,* 138, 273, 1980.
32. **Kuttan, R. and Di Ferrante, N.,** Sirius Red-collagen interaction: a method for the measurement of collagen and bacterial collagenase activity, *Biochem. Int.,* 1, 455, 1980.
33. **Luque, E. H., Angulo, E., and Montes, G. S.,** A histochemical and electron microscopic study on the collagen of nerves in the domestic fowl, *J. Anat.,* 137, 171, 1983.
34. **Junqueira, L. C. U., Montes, G. S., Martins, J. E. C., and Joazeiro, P. P.,** Dermal collagen distribution — a histochemical and ultrastructural study, *Histochemistry,* 79, 397, 1983.
35. **Shellswell, G. B., Restall, D. J., Duance, V. C., and Bailey, A. J.,** Identification and differential distribution of collagen types in the central and peripheral nervous system, *FEBS Lett.,* 106, 305, 1979.
36. **Gay, S., Balleisen, L., Remberger, K., Fietzek, P. P., Adelmann, B. C., and Kühn, K.,** Immunohistochemical evidence for the presence of collagen type III in human arterial walls, arterial thrombi, and in leukocytes incubated with collagen in vitro, *Klin. Wochenschr.,* 53, 899, 1975.
37. **Meigel, W. N., Gay, S., and Weber, L.,** Dermal architecture and collagen type distribution, *Arch Dermatol. Res.,* 259, 1, 1977.
38. **Seyer, J. M., Kang, A. H., and Whitaker, J. N.,** The characterization of type I and type III collagens from human peripheral nerve, *Biochem. Biophys. Acta,* 492, 415, 1977.
39. **Trelstad, R. L.,** Human aorta collagens — evidence for three distinct species, *Biochem. Biophys. Res. Commun.,* 57, 717, 1974.
40. **Tajima, S. and Nagai, Y.,** Distribution of macromolecular components in calf dermal connective tissue, *Connect. Tissue Res.,* 7, 65, 1980.
41. **von der Mark, K.,** Localization of collagen types in tissues, *Int. Rev. Connect. Tissue Res.,* 9, 265, 1981.
42. **Gay, S. and Miller, E. J.,** *Collagen in the Physiology and Pathology of Connective Tissue,* Gustav Fischer, Stuttgart, 1978.
43. **Junqueira, L. C. U., Bignolas, G., Mourão, P. A. S., and Bonetti, S. S.,** Quantitation of collagen-proteoglycan interaction in tissue sections, *Connect. Tissue Res.,* 7, 91, 1980.
44. **Junqueira, L. C. U., Toledo, O. M. S., and Montes, G. S.,** Correlation of specific sulfated glycosaminoglycans with collagen types I, II, and III, *Cell Tissue Res.,* 217, 171, 1981.
45. **Biempica, L., Morecki, R., Wu, C. H., Giambrone, M. A., and Rojkind, M.,** Immunocytochemical localization of type B collagen, *Am. J. Pathol.,* 98, 591, 1980.
46. **Konomi, H., Sano, J., and Nagai, Y.,** Immunohistochemical localization of type I, III and IV (basement membrane) collagens in the liver, *Acta Pathol. Jpn.,* 31, 973, 1981.
47. **Brandt, K. D.,** Glycosaminoglycans, in *Textbook of Rheumatology,* Vol. 1, Kelley, W. N., Harris, Jr., E. D., Ruddy, S., and Sledge, C. B., Eds., W. B. Saunders, Philadelphia, 1981, 239.
48. **Yamada, K. M.,** Fibronectin and other structural proteins, in *Cell Biology of Extracellular Matrix,* Hay, E. D., Ed., Plenum Press, New York, 1981, 95.
49. **Hascall, V. C. and Hascall, G. K.,** Proteoglycans, in *Cell Biology of Extracelullar Matrix,* Hay, E. D., Ed., Plenum Press, New York, 1981, 39.
50. **Scott, J. E., Orford, C. R., and Hughes, E. W.,** Proteoglycan-collagen arrangements in developing rat tail tendon — an electron-microscopical and biochemical investigation, *Biochem. J.,* 195, 573, 1981.
51. **Scott, J. E. and Orford, C. R.,** Dermatan sulfate-rich proteoglycan associates with rat tail-tendon collagen at the d band in the gap region, *Biochem. J.,* 197, 213, 1981.

52. **Lindahl, U. and Hook, M.,** Glycosaminoglycans and their binding to biological macromolecules, *Annu. Rev. Biochem.*, 47, 385, 1978.
53. **Rosenberg, L., Hellmann, W., and Kleinschmidt, A. K.,** Electron microscopic studies of proteoglycan aggregates from bovine articular cartilage, *J. Biol. Chem.*, 250, 1877, 1975.
54. **Reale, E., Luciano, L., and Kühn, K.,** Cationic dyes reveal proteoglycans on the surface of epithelial and endothelial kidney cells, *Histochemistry*, 82, 513, 1985.
55. **Shepard, N. and Mitchell, N.,** The localization of proteoglycan by light and electron microscopy using Safranin O, *J. Ultrastruct. Res.*, 54, 451, 1976.
56. **Thyberg, J., Lohmander, S., and Friberg, U.,** Electron microscopic demonstration of proteoglycans in guinea pig epiphyseal cartilage, *J. Ultrastruct. Res.*, 45, 407, 1973.
57. **Cook, H. C.,** Carbohydrates, in *Theory and Practice of Histological Techniques*, Bancroft, J. D. and Stevens, A., Eds., Churchill Livingstone, London, 1977, 141.
58. **Kiviranta, I., Jurvelin, J., Tammi, M., Saamanen, A. -M., and Helminen, H. J.,** Microspectrophotometric quantitation of glycosaminoglycans in articular cartilage sections stained with Safranin O, *Histochemistry*, 82, 249, 1985.
59. **Montes, G. S., Becerra, J., Toledo, O. M. S., Gordilho, M. A., and Junqueira, L. C. U.,** Fine structure and histochemistry of the tail fin ray in teleosts, *Histochemistry*, 75, 363, 1982.
60. **Morales, C. R., Pereyra, L. A., Toledo, O. M. S., and Montes, G. S.,** Histochemical and morphological characterization of migrating mast cells in the bovine gallbladder epithelium, *Histochemistry*, 68, 159, 1980.
61. **Kiviranta, I., Tammi, M., Jurvelin, J., Saamanen, A. -M., and Helminen, H. J.,** Demonstration of chondroitin sulphate and glycoproteins in articular cartilage matrix using periodic acid-Schiff (PAS) method, *Histochemistry*, 83, 303, 1985.
62. **Glant, T. and Lévai, G.,** Localization of antigenic components in proteoglycan aggregate of bovine nasal cartilage, *Histochemistry*, 77, 217, 1983.
63. **Scott, J. E. and Dorling, J.,** Periodate oxidation of acid polysaccharides III. A PAS method for chondroitin sulphates and other glycosamino-glycuronans, *Histochemie*, 19, 295, 1969.
64. **Nimni, M. E.,** Collagen: its structure and function in normal and pathological connective tissues, *Semin. Arthritis Rheum.*, 4, 95, 1974.
65. **Thomas, L.,** Reversible collapse of rabbit ears after intravenous papain, and prevention of recovery by cortisone, *J. Exp. Med.*, 104, 245, 1956.
66. **Martin, G. R., Kleinman, H. K., Terranova, V. P., Ledbetter, S., and Hassell, J. R.,** The regulation of basement membrane formation and cell-matrix interactions by defined supramolecular complexes, in *Basement membranes and cell movement*, Ciba Foundation Symp. 108, Pitman, London, 1984, 197.
67. **Junqueira, L. C. and Carneiro, J.,** *Basic Histology*, 4th ed., Lange Medical, Los Altos, Calif., 1983, 129.
68. **Junqueira, L. C. U., Montes G. S., Mourão, P. A. S., Carneiro, J., Salles, L. M. M., and Bonetti, Ş. S.,** Collagen-proteoglycans interaction during autotomy in the sea cucumber *Stichopus badionotus*, *Rev. Can. Biol.*, 39, 157, 1980.

Chapter 4

STRUCTURE AND ORGANIZATION OF MACROMOLECULES IN BASEMENT MEMBRANES

N. A. Kefalides and R. Alper

TABLE OF CONTENTS

I. Introduction 74

II. The Type IV Collagen Component 74
A. General Aspects 74
B. Domain Structure and Organization of Type IV Collagen 76
1. The Triple Helix 76
2. The Noncollagenous Domains 76
3. The 7-S Domain 78
C. Molecular Models of Type IV Collagen 80

III. Noncollagen Components 81
A. Laminin 81
1. Biochemical Structure 82
2. Biosynthesis 83
3. Specific Binding Properties 85
B. Entactin 85
C. Proteoglycans 86
1. Biosynthesis of HSPG 87
2. Molecular Organization of Proteoglycans in Basement Membranes 87

IV. Supramolecular Organization 88

V. Concluding Remarks 89

Acknowledgments 90

References 91

I. INTRODUCTION

Our knowledge of the molecular composition of basement membranes has grown in the past 6 years as has our understanding of the structure of the component molecules, type IV collagen, laminin, entactin, and heparan sulfate proteoglycan. The use of tumor matrices allowed the extraction of intact macromolecules without having to resort to degradative procedures.[1] This fortunate event combined with the use of rotary shadowing electron microscopy permitted for the first time visualization of laminin molecules in extracts of basement membrane.[2] Intact basement membrane collagen molecules were first isolated by us and Olsen et al. in 1973[3] and shown by transmission electron microscopy to consist of a long fibrillar domain and a globular end. Rotary shadowing electron microscopy of type IV collagen extracted from a mouse tumor demonstrated similar monomer structures.[1] In addition, aggregates of these monomers were present, including dimers and tetramers, which formed by the overlap of monomers at their amino terminus. Additional studies produced the primary amino acid sequence of the α1 and α2 chains of type IV or basement membrane collagen. More recently, cDNA clones have been produced for the carboxyl terminus of the mouse and human α1(IV) and α2(IV) collagen chains.[4-8,8a]

Despite all of this progress on the structure of individual collagen and laminin molecules, our understanding of the in vivo macromolecular organization of basement membranes, i.e., the way these distinct types of molecules interact with themselves and with one another, remains incomplete. The picture of how the collagen and laminin molecules interact with entactin and the proteoglycans in and around basement membranes is even more obscure.

In the present review, we summarize some of the evidence on the macromolecular organization of basement membranes based on the most recent studies. We will first describe structural aspects of type IV collagen and then of the noncollagenous components laminin, entactin, and proteoglycans. An attempt will be made to consider the various supramolecular models proposed in the literature and offer arguments for their validity or invalidity. Finally, we will report on the isolation and characterization of mouse and human cDNA clones for the α1 and α2 chains of type IV collagen as well as for the B-1 chain of mouse tumor laminin.

II. THE TYPE IV COLLAGEN COMPONENT

A. General Aspects

The collagen component of basement membrane has been isolated and characterized chemically, immunologically, and ultrastructurally.[9-11] It has been designated type IV and it is chemically and immunologically distinct from the well-known interstitial collagen types I, II, and III and the so-called "pericellular" collagen type V.[12] Early reports from our laboratory[3,13-15] indicated that the collagen molecule in basement membranes exists in a procollagen-like form, i.e., with its carboxyl and amino termini intact. This concept is consistent with more recent ultrastructural studies[11] which demonstrate that the type IV collagen molecule is composed of long strands, 380 to 390 nm long, containing a globule of 12 nm in diameter at one end. A fraction of anterior lens capsule (ALC) is soluble in 0.5 *M* acetic acid.[3,10] Examination of such an extract by rotary shadowing electron microscopy revealed type IV collagen molecules in different states of aggregation. Monomers, dimers, tetramers, and aggregates of tetramers were observed (Figure 1). Dimers and tetramers appear to form by the overlap of monomers at the amino terminus. Duncan et al.[16] studied the association of procollagen type IV into tetramers in cultures of an endodermal cell line PF-HR9. Initially the triple helical procollagen IV molecules associated at their amino ends by noncovalent interactions which are reversible. This was followed by disulfide bond formation between the molecules which stabilized the tetramers. This association was

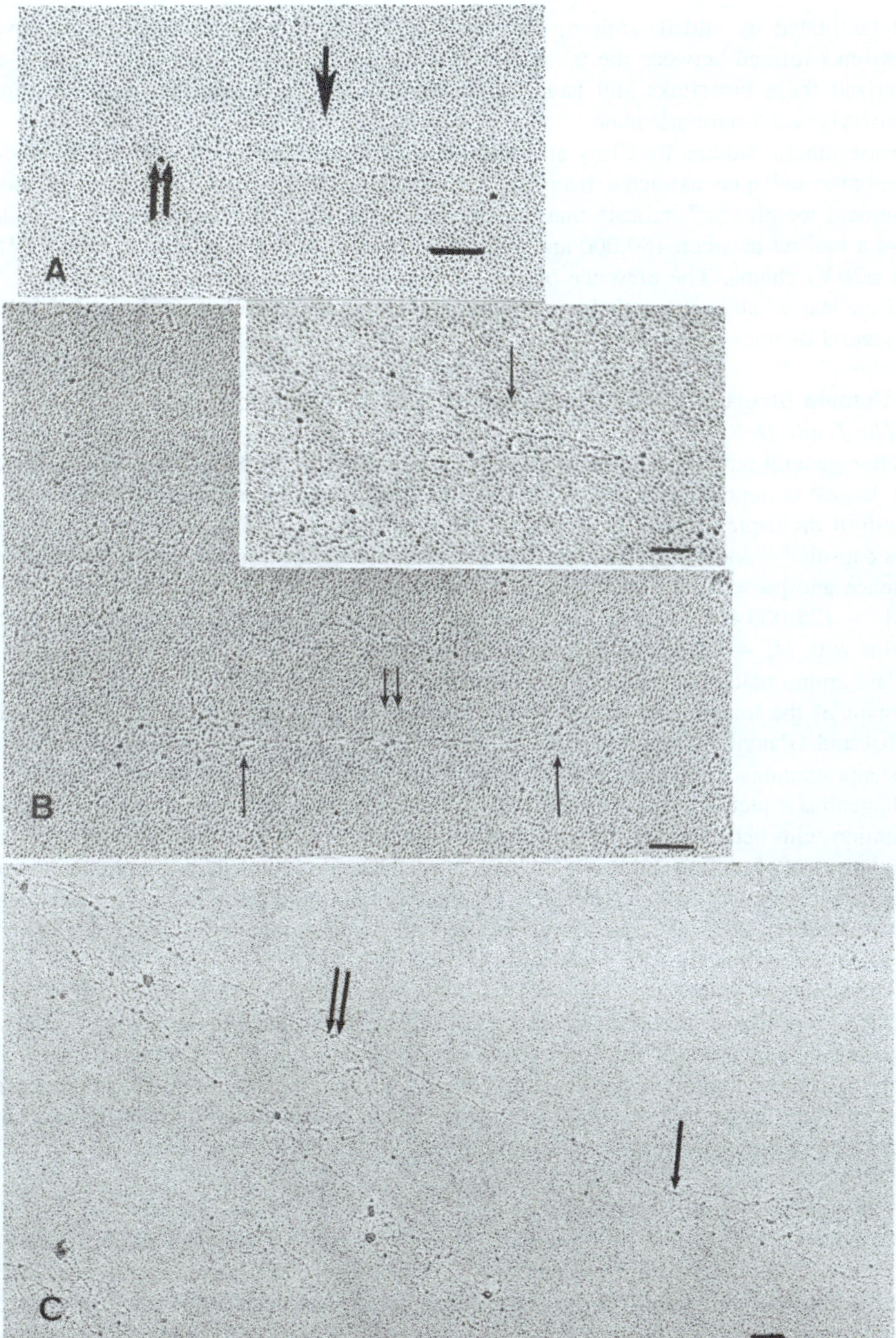

FIGURE 1. Rotary shadowing electron microscopy of 0.5 *M* acetic acid extract of ALC. (A) Dimer formed by the overlap of two type IV collagen molecules at the amino terminus. (B) Inset shows a tetramer formed by the overlap of four type IV collagen molecules at their amino termini. Section below inset shows an interaction between two tetramers via their carboxyl domains. Single arrows point to the amino terminus (7-S domain). Double arrows point to carboxyl termini (NC1 domain). (C) Linear aggregates of several tetramers. Bar = 100 nm.

not facilitated by added laminin, Ca^{2+} or Mg^{2+} ions. Additional nondisulfide covalent crosslinks formed between the molecules at the amino terminus. Bailey et al.[17] have characterized these crosslinks and have shown them to be exclusively the stable oxoimine, hydroxylysino-5-oxonorleucine.

Biosynthetic studies by Clark and Kefalides[18] and electrophoretic analysis of basement membrane collagen extracted from the mouse tumor (EHS) matrix[11] or from glomerular basement membrane[19] indicate that the constituent polypeptide chains of type IV collagen have a mol wt between 180,000 and 170,000. These have been designated pro α1(IV) and pro α2(IV) chains. The presence of additional chains cannot be ruled out at present. It is still unclear whether these chains aggregate to form a single molecule of type IV collagen or several distinct molecules.

B. Domain Structure and Organization of Type IV Collagen

1. The Triple Helix

Four general structural domains have been identified in the type IV collagen molecule. The largest domain is occupied by the triple helix which is interrupted in several areas. The length of the triple helix is about 330 to 350 nm. Collagen chains have been isolated from lens capsule,[9,20] kidney glomerulus,[19,21] and placenta[22] after limited pepsin digestion in the absence and presence of reducing agents. In the absence of reducing agent, α chains with a M_r = 125,000 and 115,000 were obtained, whereas in the presence of reducing agent, α chains with M_r = 95,000 and 75,000 were isolated.

The amino acid sequence of large segments of the α1(IV) chain from the triple helical domain of the human and mouse basement membrane collagen has been determined.[23,24] Babel and Glanville[24] reported that the pepsin-resistant α1(IV) chain isolated from human placenta contains 914 amino acid residues. The sequence contains 12 interruptions of the collagenous triplet Gly-Xaa-Yaa which varied in length from 1 to 11 residues. The distribution of amino acids between Xaa and Yaa position was similar to that found in interstitial collagens but the extent of proline and lysine hydroxylation differed. Although no significant homology was detected between the α1(IV), 95,000 mol wt chain and those of interstitial collagens, comparison with partial sequences of mouse tumor and bovine lens capsule α1(IV) chains revealed approximately 85% homology.

The studies of Babel and Glanville[24] also showed that of the 12 interruptions in the Gly-Xaa-Yaa repeating triplet sequence 6 are caused by the absence of an Xaa or Yaa position resulting in sequences of the form Gly-Xaa-Gly or Gly-Yaa-Gly. The significance of the triple helix interruptions is not absolutely clear but it can be assumed with some degree of certainty that they confer flexibility to the molecule and facilitate interaction with similar molecules or other macromolecular components of basement membranes.

2. The Noncollagenous Domains

Based on chemical and ultrastructural analysis of type IV collagen solubilized by acid extraction or limited pepsin treatment at 4°C[3,8,9,11] followed by bacterial collagenase digestion, a four-domain model for the complete molecule has been proposed.[11] In addition to the long triple helical domain discussed above, there is a short (30 nm) collagenous triple helical domain, the so called 7-S domain, and two noncollagenous segments, the NC1 and NC2 domains. The NC2 domain is interposed between the 7-S and the long helical domains; the NC1 domain occupies the carboxyl terminus of the molecule and has a globular configuration (Figure 1). In this section we will discuss mainly the globular NC1 domain.

The NC1 domain can be obtained by digesting basement membranes with bacterial collagenase and separating the fragments by gel filtration.[11,25] This procedure gives rise to a fraction rich in the NC1 domain monomer (about 27,000 mol wt), dimer (about 54,000 mol wt), and trimer (about 84,000 mol wt). In addition, a fraction rich in the 7-S domain is

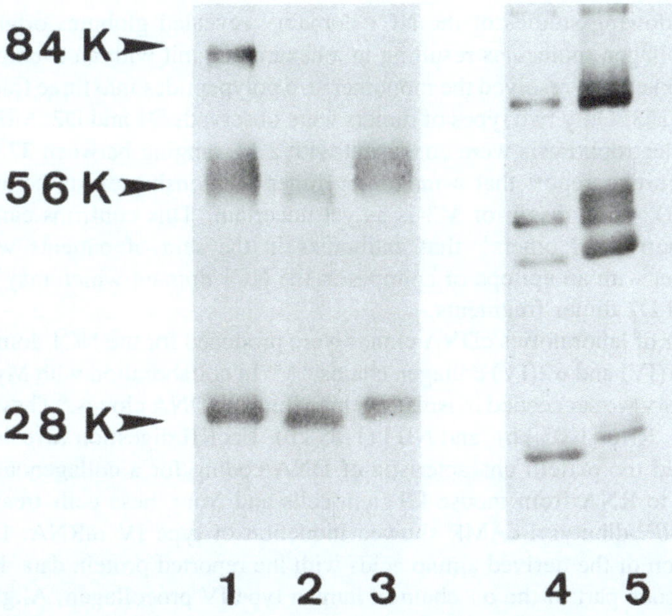

FIGURE 2. Polyacrylamide gel electrophoresis of a bacterial collagenase digest of bovine ALC and human GBM. Samples were run in a 10% SDS-polyacrylamide gel under reducing (lanes 1, 2, and 3) and nonreducing (lanes 4 and 5) conditions. Lanes 1 and 5 represent material obtained after fractionation of the collagenase digest on an Agarose A-5 m column. The fraction analyzed contained the NC1 domain. Unfractionated collagenase digests are shown for ALC (lanes 2 and 4) and for GBM (lane 3).

obtained, which by amino acid analysis is completely collagenous. The NC2 domain is contained in a smaller fraction eluting just after the NC1 containing peak. Gel filtration on agarose A-5M of the collagenase-resistant digest produced additional fractions; one eluting in the void volume and one just after. Amino acid analysis revealed a noncollagenous composition for both fractions. SDS-gel electrophoresis of the peak containing the NC1 domain under nonreducing conditions revealed bands corresponding to monomers, dimers, and trimers migrating faster than the same components electrophoresed under reducing conditions (Figure 2). This phenomenon is observed with the collagenase digest of anterior lens capsule (Figure 2) but not with that of the mouse EHS tumor.[11] This anomaly in mobility following reduction suggests that intramolecular disulfide bonds must occur within each of the three NC1 chains. It also suggests that structural differences must exist between the type IV collagens isolated from the mouse tumor and the normal bovine lens capsule basement membrane. Studies by Weber et al.[26] reveal that in the mouse EHS tumor matrix, the NC1 domains of two type IV collagen molecules combine to form a hexameric globule with an M_r of 170,000. The authors performed amino acid sequence studies of the N-terminal regions of the NC1 domain and found that they start with short triple helical sequences which overlap with the C-terminal triple helix of the $\alpha1(IV)$ and $\alpha2(IV)$ chains, demonstrating that the globule originates from the C-terminus of collagen type IV. Mixtures of the subunits (monomers and dimers) or purified dimers reassemble in neutral buffer into hexamers. Circular dichroic spectra indicate that the structure of the globule is more completely refolded from acid-treated than from urea-treated material. Their data suggest that globules originating from the assembly of a triple helical molecule are stabilized by the adjacent triple helix as well as by covalent intrachain crosslinks. Similar structural studies have been reported for the NC1 domain isolated from an authentic basement membrane, the glomerular basement membrane.[27]

Rotary shadowing studies of the NC1 domain[11] revealed globules originating from two connecting collagen molecules resulting in a hexameric unit with an average diameter of 12 nm. Butkowski et al.[28] resolved the monomer size polypeptides into three fractions designated M1, M2, and M3. Only two types of dimers were observed, D1 and D2. Monomer mobilities on SDS-gel electrophoresis were consistent with a M_r ranging between 32,000 and 28,000. The authors further report that a monomer-dimer relationship exists between M1 and D1 and M2 and D2. The origin of M3 is as yet uncertain. This confirms earlier reports from the same group[29] and others[30] that antibodies in the sera of patients with Goodpasture syndrome react with an epitope or epitopes in the NC1 domain which may reside in the M2 monomer and D2 dimer fragments.

In a number of laboratories cDNA clones were produced for the NC1 domain of the mouse and human α1(IV) and α2(IV) collagen chains.[4-8,8a] In collaboration with Myers and Brinker, in our laboratory we succeeded in isolating three human cDNA clones.[6] They were designated KK4 (1.7 kb), RR6 (1.05 kb), and NB3 (1.15 kb). EcoRII digestion revealed that the larger clone exhibited the pattern characteristic of DNA coding for a collagenous sequence. Blot hybridization to RNA from mouse F9 stem cells and from these cells treated with retinoic acid and N^6, $0^{2'}$-dibutyryl-cAMP showed induction of type IV mRNA. DNA sequencing and comparison of the derived amino acids with the reported protein data demonstrated that the clones encode part of the α1 chain of human type IV procollagen. Alignment of α1(IV) with other human procollagens showed minimal but detectable homology. A small cluster of charged residues in the α chain is partially shared by type IV. Analysis of the carboxyl-terminal α1(IV) peptide showed a repeat symmetry possibly dictated by the six cysteines in each half of the structure. The position of five cysteines in addition to four tyrosine/tryptophan groups allowed a correlation to be drawn between the 3′ noncollagenous type IV region and the larger, highly conserved carboxyl propeptides of other human procollagens. Such similarities in the different chains may define functional domains conserved throughout evolution. Additional studies were carried out comparing the DNA sequence corresponding to the 1/3 kb 3′ untranslated region of the 6.5 kb human procollagen α1(IV) mRNA with the mouse sequence obtained from 3′ cDNA and genomic clones overlapping the 5′ half. A greater degree of interspecies homology was noted in the region occupied by 90 nucleotides 5′ to the poly(A^+)-tail in the major α1(IV) mRNAs.

More recently, Myers et al.[8a] have determined the DNA sequence of a 1.7-kb cDNA corresponding to the complete carboxyl terminal noncollagenous region (NC1) of the human α2(IV) collagen chain. This domain is composed of 227 residues and can be divided into two equal-sized sequential repeats as had been observed for the corresponding 229-residue domain of the α1(IV) chain. The α1(IV) NC1 and the α2(IV) NC1 domains show a 63% homology in their nucleotide and amino acid sequences.

3. The 7-S Domain

The 7-S domain of type IV collagen, a rod like structure, was first identified by Risteli et al.[31] and further characterized by Kühn et al.,[32] and Timpl et al.[11] This domain was initially isolated from the matrix of the mouse EHS tumor after digestion with bacterial collagenase.[11] Similar rod-like structures were isolated after bacterial collagenase digestion of bovine anterior lens capsule and reported in 1973 by Olsen et al.[3] At the time, it was suggested that the short strands "represent a region in the collagen molecule which is resistant to further digestion with collagenase". There is little doubt that the rod-like structure, as revealed by transmission electron microscopy, in that study was similar to the 7-S domain identified later in a variety of basement membranes. Depending on the digestion conditions, a long form (digestion at 20°C) and a short form (digestion at 37°C) of the domain could be identified.[32] Rotary shadowing studies of the short form revealed a compact, rectangular, rod-like structure 30 nm long. The 7-S domain forms from the overlap of four type IV collagen molecules at their amino termini. Preparations of the long form of the 7-S domain showed the same rectangular structure, but in addition, four thinner strands extended outwards from the ends of the rod in a symmetrical fashion. This led to the suggestion that the 7-S domain consists of four triple helical segments which are aligned in an antiparallel or parallel arrangement. The 7-S domain within the type IV collagen tetramer constitutes an important and unique structural feature of the organizational subunit of basement membranes. The high content of cysteine (5 per chain, 15 per triple helix)[33] assures for a potentially rich disulfide bond region (about 30 such bonds per tetramer) and explains its resistance to bacterial collagenase digestion.

Glanville et al.[33] and Siebold et al.[33a] succeeded in determining the complete amino acid sequence of the 7-S domain of the α1(IV) and α2(IV) chains of human placental basement membrane collagen. The α1(IV) region contains 216 amino acids and the N-terminus starts with a nontriple helical region which is 15 residues long, containing four cysteine and two lysine residues as putative crosslink sites. This segment is followed by a 120-residue-long triple helical region, which contains the unusual occurrence of a cysteine residue in the Xaa position of a Gly-Xaa-Yaa triplet. The same authors suggest that since molecules align in an antiparallel fashion in the 7-S domain, this cysteine probably aligns with one of the four cysteines in the amino-terminal end of an adjacent molecule to form an intermolecular disulfide bond. The length of the overlap of two adjacent molecules is about 110 residues. There is a 10-residue-long nonhelical area (positions 136 to 145) which seems to be responsible for the flexible region of the molecules adjacent to the 7-S overlap zone. In the pepsin-derived α1(IV) chain, which has a mol wt of 95,000, almost all the lysine residues in position Yaa of the triplet Gly-Xaa-Yaa are completely hydroxylated and glycosylated.[23] This is also the case for the 7-S region of the α1(IV) chain. Position 99 or 102 is occupied by asparagine with an *N*-glycosidic-bound heteropolysaccharide of the complex type.[33]

In a recent study, Siebold et al.[33a] reported on the complete amino acid sequence of the 7-S domain of the human α2(IV) chain. As in the α1(IV) chain, this region of the α2(IV) chain has a non-triple-helical N-terminal sequence comprising residues 1 to 18 and containing four cysteine residues at identical positions. The triple-helical region, positions 19 to 135, is responsible for the overlapping aggregation of the molecules; it also contains a crosslinking region with a cysteine residue in position 102. This is followed by a second non-triple-helical region, positions 136 to 148, which presumably introduces a flexible region into the type IV collagen molecule. The last segment of the 7-S domain is occupied by a triple helical region, positions 149 to 210. It is estimated that position 102, which is occupied by asparagine, is a carbohydrate attachment site.

Bailey et al.[17] analyzed the long form of 7-S derived from placenta for lysyl-derived crosslinks. In both the N-terminal and C-terminal regions of the long form of 7-S significant proportions of lysyl-derived crosslinks were of the stable oxoimine hydroxylysino-5-oxo-norleucine. The authors suggest that the ketoimine crosslinks in type IV collagen have been stabilized in part by conversion to an unknown nonreducible form, although a sensitive immunoassay failed to show the presence of any pyridinoline. Comparison with interstitial collagen from placenta suggested that the rate of this conversion is much greater in basement membrane collagen.

The intermolecular and intramolecular crosslinks at both ends of the overlap zone are responsible for the high stability of this area of the 7-S domain. This is reflected in the high denaturation temperature of this domain (70°C)[33] compared to interstitial collagens which have a denaturation temperature of 40°C. Antibodies prepared against the 7-S domain from bovine glomerular basement membrane (GBM)[34] reacted with the α1(IV) 130,000-mol wt pepsin-derived fragment isolated from bovine renal cortex indicating common structural antigenic determinants between these two fragments of basement membrane collagen. It should be pointed out that the 130,000-mol wt fragment is generated under nonreducing conditions. When limited pepsin digestion occurs after reduction and alkylation of the molecules the 7-S segment is lost and the 95,000 mol wt α1(IV) chain is obtained.

Antibodies against the whole type IV collagen molecule and the 7-S domain have been detected in the sera of patients with acute poststreptococcal glomerulonephritis.[25] The reactivity of the sera from this type of patient differs from that of patients with Goodpasture syndrome where the reaction is predominantly against the NC1 domain.[27-29]

C. Molecular Models of Type IV Collagen

Transmission[3] and rotary shadowing[10,11] electron microscopy were used to investigate the ultrastructural morphology of type IV collagen isolated from either ALC or mouse EHS tumor. Transmission electron microscopy studies on type IV collagen extracted with weak acetic acid or sodium citrate from anterior lens capsule revealed monomers composed of a thin filament and a globule at one end. Addition of ATP to this extract resulted in the lateral aggregation of the strands and globular ends. It was the use of rotary shadowing electron microscopy that permitted the visualization of the earlier aggregates of such monomers of type IV collagen as extracted from basement membranes. Timpl et al.[11] and Kühn et al.[32] examined acetic acid extracted type IV collagen from mouse EHS tumor by rotary shadowing electron microscopy and observed three major species of aggregates. A dimer formed by the interaction of two monomers at their globular ends. Single strands or monomers were not frequently seen. When present, single strands had a length of about 358 nm and a diameter of 2.5 to 3 nm. The globular end of a triple helical monomer had an average diameter of 8.6 nm whereas in a dimer, the diameter of the globule was 12.9 nm. The third and more complex structure was that of a tetramer, formed by the overlap of monomer molecules at their amino termini to form the 7-S domain of the tetramer. These structures had four strands extending outward from either side of the 7-S domain in a symmetrical fashion. Similar structures were observed in acetic acid extracts of the bovine ALC. These extracts differed from those of the mouse tumor in that monomers and dimers were rarely seen (Figure 1). The few dimers observed formed by overlap at the amino terminus and not at the globular end. The most abundant structures were tetramers and aggregates of tetramers (Figure 1). Aggregates of tetramers appeared in two states. In one state they formed a loose arrangement by interaction of one globule from one tetramer with that of another, leaving the remaining arms of the tetramer free. In another state of aggregation tetramers formed linear polymers through a head-to-head and lateral aggregation (Figures 3 to 5).

The greater proportion of ALC is insoluble in dilute acetic acid. This material can be dispersed with 8 *M* urea giving rise to large, soluble aggregates composed of several tetramers arranged linearly and in parallel. Within these aggregates there are structures resembling laminin (Figures 3 to 5). Laminin molecules occur at distances varying between 400 to 500 nm.[35] The acetic acid insoluble material of ALC can also be dispersed by limited pepsin digestion at 4°C giving rise to α chain size molecules.[9] Within this insoluble fraction in addition to the α chains, one encounters, upon rotary shadowing electron microscopy, a variety of pepsin resistant fragments which still retain the helical filaments and the globules. Some fragments appear as two parallel filaments with a globule at one end. The filaments are about 4.5 nm thick and the globular ends have an average diameter of 12 nm. When the globules appeared to have arisen from the interaction of two tetramers then the average diameter is 23 nm. Other pepsin-resistant fragments appeared to have arisen from two tetramers interacting in parallel. Here the globule from two tetramers would lie side-by-side with the filamentous portion of another tetramer or the 7-S domain of one tetramer would lie side-by-side with the globule of two tetramers.[35]

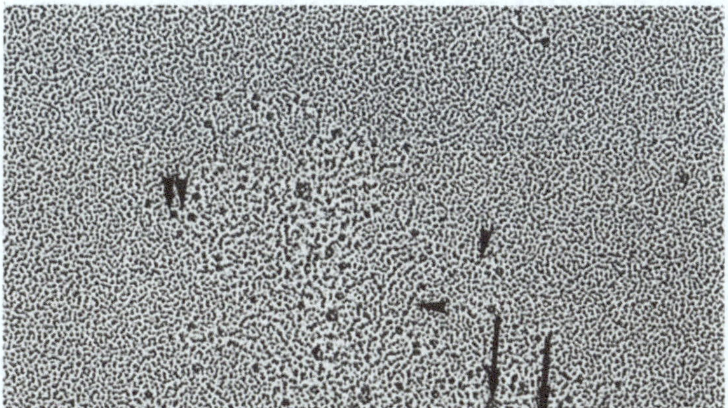

FIGURE 3. Rotary shadowing electron microscopy of 8 *M* urea solubilized ALC. Note the filamentous (single arrowheads) and globular (double arrowheads) domains of type IV collagen. Note the large globular structures within the collagen aggregate (arrows) which may be part of a laminin structure. Bar = 100 nm.

III. NONCOLLAGEN COMPONENTS

In addition to type IV collagen, basement membranes have been shown to have several noncollageneous components most notable of which are laminin and heparan sulfate proteoglycan. In addition, a third protein, entactin, has been described as part of the basement membrane supramolecular complex.

A. Laminin

Laminin is a major noncollagenous glycoprotein component of basement membranes.[2] In addition to its association with the other components of basement membranes, it also plays some important roles in cell attachment, cell growth, tissue development, and differentiation.[36] Laminin has been difficult to isolate from normal basement membranes and tissues due to its poor solubility and most of the knowledge that we have concerning laminin has been derived from studies on extracts of neoplastic tissue systems. However, recent studies on laminin isolated from intact tissues have shown interesting differences from the tumor-derived materials.[37,38]

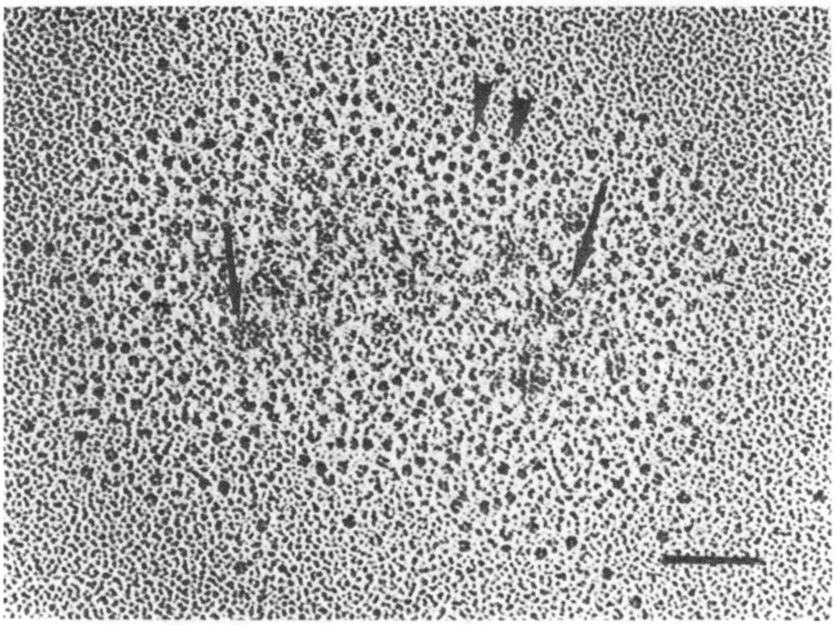

FIGURE 4. Rotary shadowing electron microscopy of an 8-*M* urea extract of ALC. The cluster of tetramers is larger compared to Figure 3. Note the organization of the NC1-like structures at the upper portion of the aggregate (arrowheads). The laminin-like structures are again evident (arrows). (From Kefalides, N. A., Howard, P., and Ohno, N., in *Basement Membranes*, Shibata, S., Ed., Elsevier Science Publishers, Amsterdam, 1985, 73. With permission.)

1. Biochemical Structure

Laminin is a very large molecular complex composed of at least three protein chains associated by disulfide linkages.[2,39] The larger of these subunits, the A chain, has a M_r of about 440,000 whereas the smaller subunits, the B_1 and B_2 chains, have M_r of about 225,000 to 200,000. A third laminin chain, termed M chain, has been shown to be present in preparations from normal tissues but is absent in preparations of neoplastic origin.[37,38] The M chain has an M_r intermediate between the A and B chains. It is not yet clear if the M chain is related to the A or B chain or is an independent gene product.

When examined by rotary shadowing electron microscopy, laminin isolated from tumor as well as normal tissue has been shown to have a twisted cruciform structure (Figure 6) consisting of three short arms and a single long arm with globular domains at the extremities of each arm.[40,41] The subunit composition of tumor laminin appears to involve three chains, one A chain and two B chains. Peptide analysis indicated that each of the B chains (B_1 and B_2) were unique.[39] A second model has been proposed[42] in which it is proposed that the laminin molecule is comprised of three B chains and one A chain. This model was based upon the observation that an epitope reactive with a monoclonal antibody against the laminin A chain was preserved in a proteolytic digest of laminin. Rotary shadowing electron microscopy of the proteolytic fragment revealed the loss of the globular units at the end of the arms along with the disappearance of one of the four arms. Since the monoclonal antibody reacted with the long arm of intact laminin, they concluded that the three short arms derived from B chains, the long arm from the A chain, and that the missing arm in the proteolytic fraction was one of the B chains. However, they concede that their data would also be consistent with the three-chain model if portions of the A and the two B chains occupied a mutually overlapping domain.

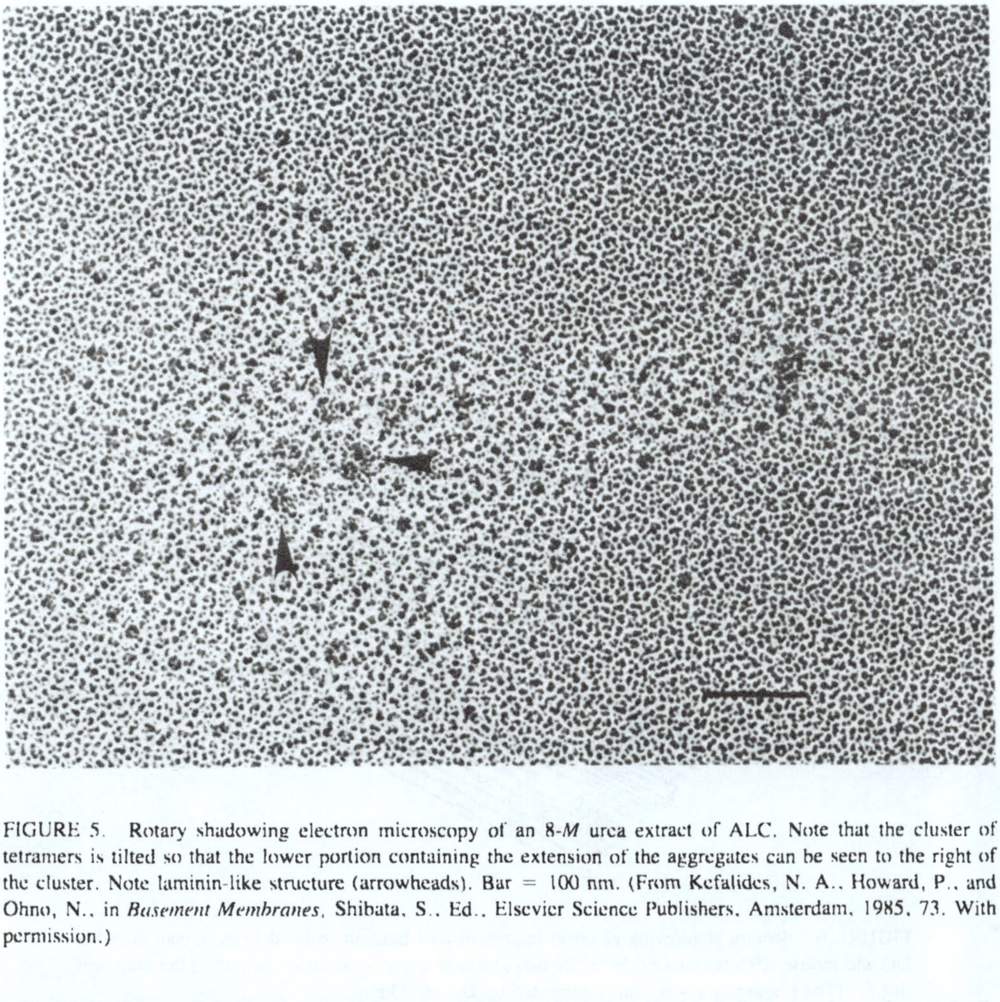

FIGURE 5. Rotary shadowing electron microscopy of an 8-*M* urea extract of ALC. Note that the cluster of tetramers is tilted so that the lower portion containing the extension of the aggregates can be seen to the right of the cluster. Note laminin-like structure (arrowheads). Bar = 100 nm. (From Kefalides, N. A., Howard, P., and Ohno, N., in *Basement Membranes*, Shibata, S., Ed., Elsevier Science Publishers, Amsterdam, 1985, 73. With permission.)

Laminin isolated from tumor sources appears to have a single globular unit at the end of the long arm whereas in laminin from normal tissues containing the M chain, there appear to be additional globular units[41] (Figure 6). As will be seen, this ultrastructure has considerable significance in the specificity of binding to cell surfaces and to type IV collagen.

Two cDNAs for laminin B chains have been prepared from a parietal endoderm cDNA library.[30] The two cDNAs were identical over much of their sequence and together provided 495 residues of the C terminal amino acid sequence. The latter revealed internal repeating sequences having a high probability of a coiled-coil α helix structure.

The laminin molecule has been shown to undergo a two-stage self-association process.[44] The first stage was temperature sensitive and reversible and did not involve divalent cations. The second stage yielded very large polymers and was dependent upon the presence of divalent cations.

Each of the laminin subunits contains asparagine-linked oligosaccharide units containing sialic acid. Analysis of [^{3}H]-glucosamine-labeled laminin subunits isolated from teratocarcinoma-derived parietal endodermal cells[39] suggested that each subunit may have as many as 11 to 14 glycosylation sites.

2. *Biosynthesis*

As indicated previously, most of what is known about the biosynthesis of laminin is derived from studies using tumor-derived cell lines. The synthesis of laminin A and B chains

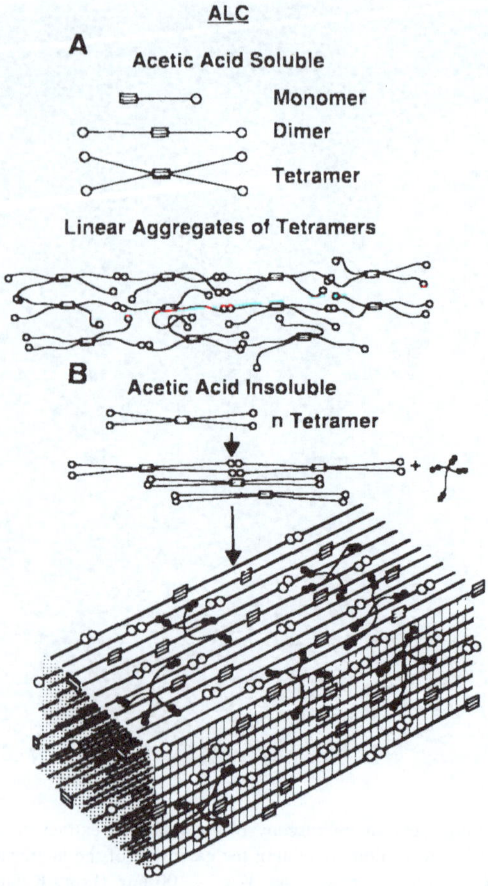

FIGURE 6. Rotary shadowing electron microscopy of laminin isolated from human placenta (A) and mouse EHS tumor (B). Note the two globular domains at the distal end of the long arm in (A). (The photograph was kindly provided by Dr. M. Ohno).

was studied in murine F9 embryonic carcinoma cells.[45] These cells differentiate in response to retinoic acid and cAMP. It was found that the undifferentiated cells incorporate very low amounts of labeled methionine into cytosolic laminin during a 15-min pulse. After 6 days of induction, laminin synthesis was increased over 15- to 20-fold over that seen in undifferentiated cells.

Recent studies have indicated that in F9 cells, only the AB_1B_2 complex goes through the intracellular glycosylation pathway.[46] This glycosylation did not appear to be necessary for subunit assembly. It was suggested that assembly proceeded first by the interaction by the two B chains to form B_1B_2 and finally to the AB_1B_2 structure. In the choriocarcinoma cell line (JAR) it was demonstrated that during the post-translational glycosylation of the A and B chains, the combination of laminin subunits to form the disulfide-linked laminin rapidly occurred at a time when the subunits contained high mannose oligosaccharides but before the final procesing by α-mannosidase I.[47] The production of laminin appeared to be limited by the availability of A chains and any excess B forms accumulated either as uncombined B or as a disulfide-linked B dimer. Pulse-chase experiments established three B forms as intermediate in the assembly of the laminin molecule. Finally, the fully assembled molecule undergoes further oligosaccharide processing and translocation to the cell surface. The mature laminin molecule does not accumulate within the cell but is rapidly externalized upon completion. The uncombined B chains, monomers, or dimers are neither processed nor secreted. Interestingly, in this study it was found that as much as one third of the laminin molecules secreted contained a smaller A subunit that appeared to have been modified by limited proteolytic cleavage. It is not yet known if this is related to the M chain found in normal tissues.[38]

3. Specific Binding Properties

Laminin has been shown to be a multifunctional protein with diverse biological properties.[36] It is capable of influencing processes of differentiation, growth, cell migration, morphology, cell adhesion, and agglutination as well as taking part in the structural assembly of basement membranes. Many of these properties are related to specific interactions of laminin with connective tissue components and cell surface receptors.

Laminin exhibits a preferential binding to type IV collagen as compared to other collagen types[48] suggesting the presence of specific binding sites in both laminin and type IV collagen. Distinct complexes are formed when laminin binds to type IV collagen. Based upon rotary shadowing electron microscopy, it appears that laminin binds to a site located about 140 nm from the NC1 domain of type IV collagen.[49] This binding appears to be mediated through the globular regions of any of the four arms of the laminin molecule. This is reinforced by the observation that laminin fragments lacking the globular domain do not bind to type IV collagen.[50] Precipitation occurs in the presence of equimolar amounts of laminin and type IV collagen[51] and is influenced by variable amounts of heparan sulfate proteoglycan. Fibronectin does not cause the precipitation of type IV collagen. These observations support the concept of specific interactions between laminin and type IV collagen and suggest that such interactions may be involved in the *in situ* deposition of basement membranes.

Cell surface membranes possess a cell substrate attachment antigen (CSAT)[52] that participates in the adhesion of cells to extracellular molecules. The CSAT forms complexes with laminin and fibronectin and binding data strongly implicate this binding as being biologically significant. Laminin seems to primarily affect cells of epithelial origin and appears to function in the attachment of various epithelial cells to basement membranes.[53] It was found that laminin can also function as an agglutinin for fixed erythrocytes. In these studies it was demonstrated that laminin and fibronectin differ markedly in their biologic activity in several specific adhesion assays. A major cell-binding domain was shown to be present near the intersection of the short arms of laminin isolated from human MCF-7 breast carcinoma cells.[54]

Fibroblasts have been shown to attach and spread on laminin substrates in fibronectin-depleted serum and under conditions in which protein synthesis was absent.[55] In this study it was shown that this behavior could be inhibited by antibodies to laminin but not by antibodies to fibronectin. Conversely, cell attachment and spreading on fibronectin substrata could be inhibited by antisera to fibronectin but not to laminin. The observed spreading involved microfilament bundles and focal adhesions. Thus, it appears that cells have distinct receptors for laminin and fibronectin at their cell surfaces.

In studies on the attachment of liver cells to various substrata coated dishes, it was found that laminin promoted the attachment of mouse fetal liver cells but had little influence on adult liver cells.[56] The results using fibronectin were just the opposite. Cells grown on laminin-coated dishes synthesized more α-fetoprotein than cells grown on fibronectin or type IV collagen without any observed differences in the secretions of albumin into the media. These studies suggested an involvement of laminin in cell growth, tissue organization, and cell differentiation during development of liver in vivo.

B. Entactin

Entactin is a novel sulfated glycoprotein that has been identified in the extracellular matrix synthesized by mouse endodermal cells in culture.[57] Polyclonal antibodies against entactin cross react with basement membranes of a variety of tissues. However, in this study, immunoelectron microscopic examination of kidney slices indicated that entactin was localized at the foot processes of epithelial cells facing the underlying basement membrane with lesser amounts in the basement membrane proper. In a later study[58] on appearance of laminin and entactin during development in the mouse embryo, it was shown that although entactin appeared at a later stage of embryonic development, antibodies to entactin and laminin co-localized at all stages once entactin appeared.

The distribution of entactin at the ultrastructural level[59,60] indicated that entactin is an intrinsic compoment of basement membranes but that there are subtle variations in basement membranes from different tissues in both content and/or assembly of the different basement membrane components.

C. Proteoglycans

Much of the earlier work on the structure of basement membranes tended to overlook the possible involvement of proteoglycans, mainly because of the very minute amounts of glycosaminoglycan (GAG)-derived hexuronic acid found in the earlier preparations. However, the studies of Kanwar et al.[61-63] clearly demonstrated the presence within the lamina rara of glomerular basement membrane of a heparan sulfate proteoglycan (HSPG). These observations have been confirmed in studies on GBM[64-68] in vitro and on the synthesis of HSPG by glomeruli.[69] HSPG also has been identified in retinal pigment epithelium,[70] in the external lamina surrounding aortic smooth muscle cells,[71] in the basement membranes surrounding Schwann cells,[72] and in the matrices laid down in culture by several cell lines that synthesize basement membrane components. These include lens epithelium,[73] chick retinal epithelium,[74] the mouse parietal yolk sac,[75] mouse mammary epithelial cells,[76] the tumor-derived cell lines from EHS sarcoma,[77-79] the mouse carcinoma line M1536-BC,[80] and the L_2 rat yolk sac carcinoma.[81]

Various forms of HSPG were isolated from the EHS sarcoma by extraction with saline, collagenase digestion, and guanidine-HCl extraction.[77] These were separated into high and low density fractions. The high density form had a mol wt of about 130,000 and a core protein of about 5,000 to 12,000 and had a considerable degree of poly dispersity. An average molecule consisted of 4 heparan sulfate side chains (M_r 29,000) with a length of 32 nm. The low density form had a M_r of about 400,000 and contained about 50% protein.

In a later study on the EHS sarcoma, two similar fractions were isolated.[78] The core proteins were of different sizes but the heparan sulfate side chains had identical molecular weights. Extraction of the tumor with saline yielded the high density HSPG with a heterodisperse size of 95,000 to 130,000. Subsequent extractions under denaturing conditions yielded the M_r 400,000 HSPG. Antibodies to each preparation cross reacted with both proteoglycans but the larger unit had additional epitopes not present in the smaller.

On the basis of these studies, it would seem that the HSPG of EHS sarcoma consisted of two groups: one, a large, high density proteoglycan and the other a smaller low density unit containing a much smaller core protein, each with four heparan sulfate units per core protein. Biosynthetic studies[79] indicated that a portion of the larger unit is converted to the smaller by a proteolytic step and that each was derived from the same core protein.

Analysis of a proteoglycan isolated from bovine GBM and containing about 75% of the hexuronic acid content of GBM revealed a molecular size of about 200,000 and a composition of about 70% protein and 30% carbohydrate.[68] Alkaline borohydride treatment released heparan sulfate chains with an M_r of about 14,000 indicating the presence of 4 chains per molecule, consistent with the structure of the HSPG from the EHS sarcoma but with a smaller EHS side chain. In a separate study,[69] native and newly synthesized HSPG was isolated from intact glomeruli. The size of the core protein was estimated to be about 18,000, that of the carbohydrate chains about 25,000, and that of the HSPG at 150,000 to 250,000.

It was claimed that the physicochemical characteristics of the renal HSPG differed markedly from that of the EHS sarcoma. However, the properties seemed to be close to that of the high density fraction from the EHS sarcoma.

1. Biosynthesis of HSPG

When rabbit lens epithelial cells were grown on plastic, glass, or dried gelatin or type IV collagen, there was no evidence of basement membrane formation[73] although they secreted the same complement of sulfated molecules. However, if the cells were grown on type I collagen gels, they deposited a discrete basement membrane that followed the contours of the basal surfaces of the cells. The cells secreted basement membrane material that diffused into the gel and associated with the collagen fibrils. Approximately 70% of the HSPG secreted into the medium when grown on the dried substrata became associated with the cell-gel layer. It also was found that 90% of the labeled sulfated material in the gel was in HSPG. Thus, it appears that the major proteoglycan synthesized by lens cells in culture is HSPG.

As indicated previously, the EHS sarcoma synthesizes HSPG.[79] Pulse experiments indicated the synthesis of a M_r 400,000 protein that reacted with antisera against HSPG fractions isolated from the tumor. Pulse-chase experiments indicated the conversion of the M_r 400,000 protein to the large, low-density HSPG which subsequently was degraded to the smaller high-density protein. It was suggested that this represents the chain of events leading to the deposition of the various HSPG fractions into the EHS extracellular matrix.

In another study, primary cultures containing only Schwann cells and sensory nerve cells were shown to secrete a basement membrane which appears to be essential for Schwann cell development.[72] The Schwann cells synthesize two major heparan sulfate-containing proteoglycans of similar densities. After separation of the units by gel filtration, it was found that the layer of the two proteoglycans contained both heparan sulfate (HS) and chondroitin sulfate (CS) chains of M_r 21,000 in a ratio of approximately 3:1. This proteoglycan could be extracted from cultures only under denaturing conditions and was found only when the cells were actively synthesizing the basement membrane. The smaller proteoglycan contained HS and CS chains of M_r 18,000 in a ratio of 4:1. The accumulation of this proteoglycan was independent of basal lamina production. From this it appears that certain basement membranes may contain other GAG components besides HS.

This is consistent with the observations of Hook et al.[75] who reported the presence of a CS proteoglycan produced by mouse parietal yolk sac (PYS) cells in culture. The CS proteoglycan has an M_r of 200,000 to 300,000 and contains 10 to 20 chondroitin sulfate (M_r 16,000 to 14,000) chains attached to a core protein that electrophoresed as a doublet (M_r 34,000 and 27,000). The PYS cells secreted about equal amounts of the two proteoglycans into the medium whereas the predominant proteoglycan in the matrix was HSPG. This appeared to be anchored by interactions involving the core protein rather than the polysaccharide chains.

A hybrid CS-HS proteoglycan was isolated from mouse mammary epithelial cells.[76] CS represented about 15% of the labeled glycosaminoglycans carried by the proteoglycans of the cell surface and basolateral secretions of the cells in culture.

A chondroitin sulfate proteoglycan was the major proteoglycan synthesized by rat 14.5-day parietal yolk sacs in organ culture.[82] This proteoglycan was shown to have a core protein with a M_r of 130,000 to 125,000 with chondroitin-6-sulfate, the predominant GAG, and with a small amount of copolymeric dermatan sulfate, hyaluronic acid, and small amounts of heparan sulfate.

2. Molecular Organization of Proteoglycans in Basement Membranes

Most of the studies to date indicate that the location of the HSPG is within the lamina rara of the basement membrane[64] and the observations of Kanwar and Farquhar[62] and Mynderse et al.[83] suggest a possible role as a charge determinant. Other studies, however, provide evidence of HSPG localization in all basement membrane layers including the lamina densa.[84] To date, little is known on how the proteoglycans interact with the other components of basement membranes. There is some evidence that HSPG can interact in affinity chromatography with laminin, fibronectin, and the NC1 domain of type IV collagen, presumably through the glycosaminoglycan side chains.[77] It is probable that multiple interactive sites are present within the lamina rara for the binding of HSPG and other proteoglycans.

IV. SUPRAMOLECULAR ORGANIZATION

The precise organization of the component molecules of basement membranes into a supramolecular complex in vivo remains elusive despite the numerous attempts to formulate structural models. Practically all proposed models are the product of either a logical arrangement of the various types of molecules based on histochemical and immunolocalization studies or of ultrastructural arrangements seen by electron microscopy of in vitro reconstituted molecules. Although many reasons can be advanced to explain the ephemeral nature of the models, a number of observations are in order. To begin with, the macromolecular organization of one basement membrane, for example GBM, is different from that of ALC or of alveolar basement membrane. The differences in structure are obviously dictated by the functional properties of the tissue in which the basement membrane lies. GBM is primarily involved in the filtration of solutes, alveolar basement membrane in the diffusion of gases, whereas anterior lens capsule (ALC) serves to support the lens and allow light to pass through. Therefore, we cannot speak of a unified, generalized structure of basement membranes. Another factor is the degree of qualitative and quantitative heterogeneity of macromolecules among basement membranes. This would certainly influence interaction between homologous molecules as well as between heterologous components. It can also be argued that the supramolecular organization of basement membranes reflects the state of differentiation of the cell responsible for their synthesis. Undifferentiated and neoplastic cells make imperfect or incomplete basement membranes; it is, therefore, not surprising to see reports of basement membranes containing laminin but no collagen and vice versa.[85] Finally, some concepts for a number of ultrastructural models arose from in vitro studies involving the interaction of similar subunits, e.g., collagen IV molecules or dissimilar subunits, e.g., collagen IV and laminin molecules. The former is exemplified by the work of Yurchenco and Furthmayr[86] and of Duncan et al.[16] who studied the in vitro self-assembly of collagen IV molecules. Yurchenco and Furthmayr,[86] working with the mouse EHS tumor made preparations of dimeric molecules, which formed by head-to-head interaction of monomers. After heat gelation and dialysis against 0.2 *M* ammonium sulfate they examined the preparation in the electron microscope by rotary shadowing. They noted in the majority of fields examined an extensive lattice network in the shape of irregular polygons with extensive side-by-side association of the collagen, two and three triple helices thick. 7-S tetrameric domains could not be identified. When, however, EHS type IV collagen dimers were incubated in tris-buffered saline at room temperature for 24 hr and examined by rotary shadowing, large numbers of tetramers, hexamers, and octamers were revealed. Tetramers, with a 7-S overlap, were also obtained from monomers lacking the globular end which were generated by mild pepsin treatment of EHS collagen. Contrary to the observations of Yurchenco and Furthmayr, Duncan et al.[16] were able to show that 7-S tetrameric collagens can be generated in vitro from monomeric molecules of type IV collagen isolated from the medium of a mouse endodermal cell line PF-HR9. These authors investigated the formation of such tetramers in cell culture and from isolated procollagen IV molecules. After an initial noncovalent association, which is reversible, disulfide bonds form between molecules. This reaction requires an adequate concentration of procollagen and is not facilitated by added laminin, Ca^{2+} or Mg^{2+} ions. The studies of Duncan et al.[16] failed to show the formation of head-to-head dimers as precursors of any larger aggregate.

The studies of Charonis et al.[49] and Rao et al.[50] are examples of attempts to examine the in vitro interaction of dissimilar molecules, namely collagen type IV and laminin. Charonis et al.[49] have shown that laminin forms distinct complexes with type IV collagen in vitro. One binding site appears to occur on the triple helical domain, 140 nm from the carboxyl terminus; another is located within the amino terminal region. The work of Rao et al.[50] showed a similar binding of laminin near the carboxyl terminus of the collagen but no binding was observed near the amino terminus. Ohno,[87] in our laboratory, has examined preparations of normal laminin extracted from human placenta and demonstrated interaction between it and type IV collagen at three sites, one within the amino terminus, one within the carboxyl terminus, and one near the carboxyl terminus. It should be stressed here that the exact mode of interaction between laminin and type IV collagen within the organized basement membrane is still unknown. Our studies, using extracts of anterior lens capsule[10,35] support the observation of Duncan et al.[16] The acetic acid extract of ALC contained predominantly 7-S tetramers with only occasional monomers. Rarely, a dimeric form would be observed but it was formed by an overlap at the amino terminus.[35] Within the acid-extracted fraction of ALC linear aggregates of tetramers were also observed. These linear aggregates formed by interaction of a globular end from one tetramer and a globular end from another (Figure 1C and Figure 7). The remainder of the arms of each tetramer appeared to be free. The ultrastructural studies of in vivo aggregates isolated by dilute acetic acid extraction of ALC would suggest that the structural subunit of this basement membrane is the tetramer which forms by the overlap of monomer type IV collagen molecules at their amino (7-S) terminus. These tetramers form low order linear aggregates which are held together by noncovalent forces as well as disulfide bonds. In this extract, small amounts of laminin, entactin and proteoglycans are present. In the acetic acid-insoluble fraction of the ALC basement membrane, which constitutes more than 85% of the lens capsule, the type IV collagen molecules appear to be held together by strong covalent forces involving both disulfide and lysyl-derived crosslinks. Within this insoluble matrix of ALC the type IV collagen tetramers form compact aggregates by end-to-end and side-to-side association. This is best shown in the rotary shadowing photograph of the 8 *M* of urea extract of the lens capsule (Figures 3 to 5) and in the diagram shown in Figure 7. Within this high order of tetramer aggregation, structures resembling laminin could be demonstrated. We have no documented evidence which would allow us to speculate on the nature of interaction of other components such as entactin and proteoglycans within a basement membrane although imaginary sketches and diagrams of one sort or another have been proposed.

It is hoped that in the future more precise models based on physical and ultrastructural data of intact or fragmented basement membranes will become available.

V. CONCLUDING REMARKS

As can be seen from this discussion, a considerable body of knowledge already exists concerning the structure and synthesis of the individual components that together comprise basement membranes and information is accumulating concerning the nature of the molecular interactions between the collagenous and noncollagenous components and the cells that synthesize and secrete them. These observations have permitted us to draw a conceptual model for at least one basement membrane, the anterior lens capsule of the eye, in which the interactions between the various domains of the type IV procollagen molecules have been shown to exist as aggregates of a tetrameric unit which forms the backbone of the three-dimensional basement membrane through lateral and end-to-end interactions. What remains to be fitted into this structure are the specific interactions of the noncollagen com-

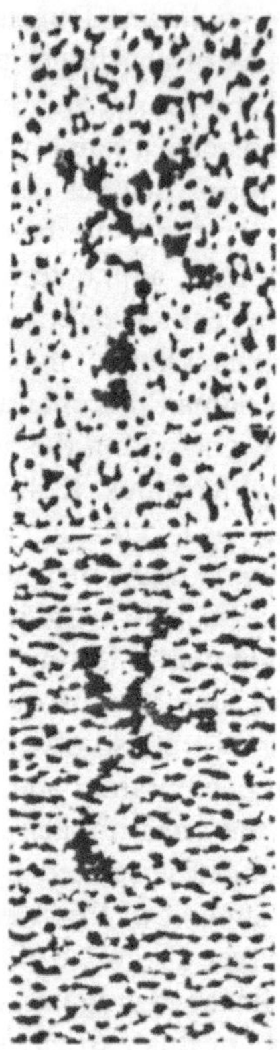

FIGURE 7. Schematic representation of the molecular forms of type IV collagen molecules in the acetic acid soluble (A) and insoluble (B) fractions of ALC. In A are depicted monomers, dimers, and tetramers as well as the formation of linear aggregates of tetramers. Note the loose arrangement of tetramers. In (B) are depicted linear aggregates of tetramers formed by interactions at the NC1 domains. The tetramers aggregate by end-to-end and by three-dimensional lateral interactions. The molecules interact by reducible and nonreducible covalent linkages. Laminin molecules are shown interacting with collagen molecules via their globular domains.

ponents: laminin, entactin, and heparan sulfate, and possibly, other proteoglycans. It is probable that basement membranes from different tissues will have specific supramolecular organizations adapted to their individual functions within each tissue. The supramolecular organizations probably are determined by the relative proportions and types of these individual components and perhaps, to the temporal sequence of their deposition into the extracellular matrix.

Future directions in this area will be aided dramatically by the application of molecular biologic probes to the study of the temporal control of the synthesis of basement membrane components and by advances in immunohistochemical electron microscopy.

ACKNOWLEDGMENTS

We wish to thank Dr. Nobuko Ohno and Ms. Pamela Howard for taking the rotary shadowing pictures. We are grateful to Sherí Johnson for typing the manuscript. This work was supported by NIH grants AM-20553 and HL-29492.

REFERENCES

1. **Timpl, R., Martin, G. R., Bruckner, P., Wick, G., and Wiedemann, H.,** Nature of the collagenous protein in a tumor basement membrane, *Eur. J. Biochem.*, 84, 43, 1978.
2. **Timpl, R., Rohde, H., Robey, P. G., Rennard, S. P., Foidart, J. M., and Martin, G. R.,** Laminin-A glycoprotein from basement membranes, *J. Biol. Chem.*, 254, 9933, 1979.
3. **Olsen, B. R., Alper, R., and Kefalides, N. A.,** Structural characterization of a soluble fraction from lens-capsule basement membrane, *Eur. J. Biochem.*, 38, 220, 1973.
4. **Wang, S.-Y. and Gudas, L. J.,** Isolation of cDNA clones specific for collagen IV and laminin from mouse teratocarcinoma cells, *Proc. Natl. Acad. Sci. U.S.A.*, 80, 5880, 1983.
5. **Kurkinen, M., Barlow, D., Helfman, D. M., Williams, J. G., and Hogan, B. L. M.,** Isolation of cDNA clones for basal lamina components: type IV procollagen, *Nucleic Acids Res.*, 11, 6199, 1983.
6. **Brinker, J. M., Gudas, L. J., Loidl, H. R., Wang, S.-Y., Rosenbloom, J., Kefalides, N. A., and Myers, J. C.,** Restricted homology between human $\alpha 1$ type IV and other procollagen chains, *Proc. Natl. Acad. Sci. U.S.A.*, 82, 3649, 1985.
7. **Pihlajaniemi, T., Tryggvason, K., Myers, J. C., Kurkinen, M., Lebo, R., Cheung, M.-C., Prockop, D. J., and Boyd, C.,** cDNA clones coding for the pro $\alpha 1$(IV) chain of human type IV procollagen revealed unusual homology of amino acid sequences in two halves of the carboxyl-terminal domain, *J. Biol. Chem.*, 260, 7681, 1985.
8. **Oberbaumer, I., Laurent, M., Schwarz, U., Sakurai, Y., Yamada, Y., Vogeli, G., Voss, T., Siebold, B., Glanville, R. W., and Kuhn, K.,** Amino acid sequence of the non-collagenous globular domain (NC1) of the $\alpha 1$(IV) chain of basement membrane collagen as derived from complementary DNA, *Eur. J. Biochem.*, 147, 217, 1985.

8a. **Myers, J. C. Howard, P. S., Jelan, A. M., Dion, A. S., and Macazak, E. J.,** Duplication of type IV collagen COOH-terminal repeats and species-specific expression of $\alpha 1$(IV) and $\alpha 2$(IV) collagen genes, *J. Biol. Chem.*, 262, 1987 (in press).

9. **Dehm, P. and Kefalides, N. A.,** The collagenous component of lens basement membrane. The isolation and characterization of an α-chain size collagenous peptide and its relationship to newly synthesized lens components, *J. Biol. Chem.*, 253, 6680, 1978.
10. **Brinker, J. M., Pegg, M. T., Howard, P. S., and Kefalides, N. A.,** Immunochemical characterization of type IV procollagen from anterior lens capsule, *Collagen Rel. Res.*, 5, 233, 1985.
11. **Timpl, R., Wiedemann, H., Van Delden, V., Furthmayr, H., and Kühn, K.,** A network model for the organization of type IV collagen molecules in basement membranes, *Eur. J. Biochem.*, 120, 203, 1981.
12. **Miller, E. J., Gay, S., Haralson, M. A., Martinez-Hernandez, A., and Rhodes, R. K.,** Chemistry and biology of type V collagen system, in *New Trends in Basement Membrane Research*, Kühn, K., Schoene, H. H., and Timpl, R., Eds., Raven Press, New York, 1982, 99.
13. **Kefalides, N. A.,** Isolation of a collagen from basement membranes containing three identical α-chains, *Biochem. Biophys. Res. Commun.*, 45, 226, 1971.
14. **Kefalides, N. A., Cameron, J. D., Tomichek, E. A., and Yanoff, M.,** Biosynthesis of basement membrane collagen by rabbit corneal endothelium *in vitro*, *J. Biol. Chem.*, 251, 730, 1976.
15. **Minor, R. R., Clark, C. C., Strause, E. L., Koszalka, T. R., Brent, R. L., and Kefalides, N. A.,** Basement membrane procollagen is not converted to collagen in organ cultures of parietal yolk sac endoderm, *J. Biol. Chem.*, 251, 1789, 1976.
16. **Duncan, K. G., Fessler, L. I., Bachinger, H. P., and Fessler, J. H.,** Procollagen IV. Association of tetramers, *J. Biol. Chem.*, 258, 5869, 1983.
17. **Bailey, A. J., Sims, T. J., and Light, N.,** Cross-linking of type IV collagen, *Biochem. J.*, 218, 713, 1984.
18. **Clark, C. C. and Kefalides, N. A.,** Synthesis of basement membrane collagen and non-collagen glycoproteins, in *New Trends in Basement Membrane Research*, Kühn, K., Schoene, H. H., and Timpl, R., Eds., Raven Press, New York, 1982, 155.
19. **Dean, D. C., Barr, J. F., Freytag, J. W., and Hudson, B. G.,** Isolation of type IV procollagen-like polypeptides from glomerular basement membrane. Characterization of pro $\alpha 1$(IV), *J. Biol. Chem.*, 258, 590, 1983.
20. **Grant, M. E. and Heathcote, J. G.,** A comparative study of lens capsule assembly, in *New Trends in Basement Membrane Research*, Kuhn, K, Schoene, H. H., and Timpl, R., Eds., Raven Press, New York, 1982, 195.
21. **Kefalides, N. A.,** Isolation and characterization of the collagen from glomerular basement membrane, *Biochemistry*, 7, 3103, 1968.
22. **Kresina, T. F. and Millers, E. J.,** Isolation and characterization of basement membrane collagen from human placental tissue. Evidence for the presence of two genetically distinct collagen chains, *Biochemistry*, 28, 3089, 1979.

23. **Schuppan, D., Timpl, R., and Glanville, R. W.,** Discontinuities in the triple helical sequence Gly-X-Y of basement membrane (Type IV) collagen, *FEBS Lett.*, 115, 297, 1980.
24. **Babel, W. and Glanville, R. W.,** Structure of human basement membrane (type IV) collagen, *Eur. J. Biochem.*, 143, 545, 1984.
25. **Kefalides, N. A., Pegg, M. T., Ohno, N., Poon-King, T., Zabriskie, J., and Fillit, H.,** Antibodies to basement membrane collagen and to laminin are present in sera from patients with poststreptococcal glomerulonephritis, *J. Exp. Med.*, 163, 588, 1986.
26. **Weber, S., Engel, J., Wiedemann, H., Glanville, R. W., and Timpl, R.,** Subunit structure and assembly of the globular domain of basement membrane collagen type IV, *Eur. J. Biochem.*, 139, 401, 1984.
27. **Wieslander, J., Langeveld, J., Butkowski, R. J., Jodlowski, R., Noelken, M., and Hudson, B. G.,** Physical and immunochemical studies of the globular domain of type IV collagen. Cryptic properties of the Goodpasture antigen, *J. Biol. Chem.*, 260, 8564, 1985.
28. **Butkowski, R. J., Wieslander, J., Wisdom, B. J., Barr, J. F., Noelken, M. E., and Hudson, B. G.,** Properties of the globular domain of type IV collagen and its relationship to the Goodpasture antigen, *J. Biol. Chem.*, 260, 3739, 1985.
29. **Wieslander, J., Barr, J. F., Butkowski, R. J., Edwards, S. J., Bygren, P., Heinegard, D., and Hudson, B. G.,** Goodpasture antigen of the glomerular basement membrane: localization to non-collagenous regions of type IV collagen, *Proc. Natl. Acad. Sci. U.S.A.*, 87, 3838, 1984.
30. **Kefalides, N. A., Ohno, N., and Wilson, C. B.,** Antigenic components of bovine lens capsule that cross-react with serum from Goodpasture's syndrome, *Fed. Proc.*, 43, 779, 1984.
31. **Risteli, J., Bachinger, H. P., Engel, J., Furthmayr, H., and Timpl, R.,** 7-S collagen: characterization of an unusual basement membrane structure, *Eur. J. Biochem.*, 108, 239, 1980.
32. **Kühn, K., Wiedemann, H., Timpl, R., Risteli, J., Dieringer, H., Voss, T., and Glanville, R. W.,** Macromolecular structure of basement membrane collagens. Identification of 7-S collagen as a cross-linking domain of type IV collagen, *FEBS Lett.*, 125, 123, 1981.
33. **Glanville, R. W., Qian, R., Siebold, B., Risteli, J., and Kühn, K.,** Amino acid sequence of the N-terminal aggregation and cross-linking region (7-S domain) of the α1(IV) chain of human basement membrane collagen, *Eur. J. Biochem.*, 152, 213, 1985.
33a. **Siebold, B., Qian, R-Q., Glanville, R. W., Hofmann, H., Deutzmann, R., and Kühn, K.,** Construction of a model for the aggregation and crosslinking 75 domain of type IV collagen based upon an evaluation of the primary structure of the α1 and α2 chains in this region, *Eur. J. Biochem.*, 1987 (in press).
34. **Dixit, S. N., Mainardi, C. L., Beachey, E. H., and Kang, A. H.,** 7-S domain constitutes the amino terminal end of type IV collagen: an immunochemical study, *Collagen Rel. Res.*, 3, 263, 1983.
35. **Kefalides, N. A., Howard, P., and Ohno, N.,** Heterogeneity in basement membranes determines their structure and function, in *Basement Membranes*, Shibata, S., Ed., Elsevier, Amsterdam, 1985, 73.
36. **Kleinman, H. K., Cannon, F. B., Laurie, A. W., Hassell, J. R., Aumailley, M., Terranova, V. P., Martin, G. R., and Dubois-Dalcq, M.,** Biological activities of laminin, *J. Cell. Biochem.*, 27, 317, 1985.
37. **Ohno, M., Martinez-Hernandez, A., Ohno, N., and Kefalides, N. A.,** Isolation of laminin from human placental basement membranes: amnion, chorion and chorionic microvessels, *Biochem. Biophys. Res. Commun.*, 112, 1091, 1983.
38. **Ohno, M., Martinez-Hernandez, A., Ohno, N., and Kefalides, N. A.,** Laminin M is found in placental basement membranes, but not in basement membranes of neoplastic origin, *Connect. Tissue Res.*, 15, 199, 1986.
39. **Howe, C. C. and Dietzchold, B.,** Structural analysis of three subunits of laminin from teratocarcinoma-derived parietal endoderm cells, *Dev. Biol.*, 98, 335, 1983.
40. **Odermatt, E., Furthmayr, H., Timpl, R., Rohde, H., Ott, V., and Engel, J.,** Shape, domain structure and flexibility of laminin, in *New Trends in Basement Membrane Research*, Kuhn, K., Schoene, H., and Timpl, R., Eds., Raven Press, New York, 1982, 79.
41. **Ohno, M., Martinez-Hernandez, A., Ohno, N., and Kefalides, N. A.,** Comparative study of laminin found in normal placental membrane with laminin of neoplastic origin, in *Basement Membranes*, Shibata, S., Ed., Elsevier, Amsterdam, 1985, 3.
42. **Palm, S. L., McCarthy, J. B., and Furcht, L. T.,** Alternate model for the internal structure of laminin, *Biochemistry*, 24, 7753, 1983.
43. **Barlow, D. P., Green, N. M., Kurkinen, M., and Hogan, B. L.,** Sequencing of laminin B chain cDNAs reveals C-terminal regions of coiled-coil alpha helix, *EMBO J.*, 3, 2355, 1984.
44. **Yurchenco, P. D., Tsilibary, E. C., Charonis, A. S., and Furthmayr, H.,** Laminin polymerication *in vitro*. Evidence for a two-step assembly with domain specificity, *J. Biol. Chem.*, 260, 7636, 1985.
45. **Cooper, A. R., Taylor, A., and Hogan, B. L.,** Changes in the rate of laminin and entactin synthesis in F9 embryonal carcinoma cells treated with retinoic acid and cyclic AMP, *Dev. Biol.*, 99, 510, 1983.
46. **Morita, A., Sugimoto, E., and Kitagawa, Y.,** Post-translational assembly and glycosylation of laminin subunits in parietal endoderm-like F9 cells, *Biochem. J.*, 229, 259, 1985.

47. **Peters, B. P., Hartle, R. J., Krzesicki, R. F., Kroll, J. G., Perini, F., Balun, J. E., Goldstein, I. J., and Ruddon, R. W.,** The biosynthesis processing and secretion of laminin by human choriocarcinoma cells, *J. Biol. Chem.*, 260, 14732, 1985.
48. **Woodley, D. T., Rao, C. N., Hassell, S. R., Liotta, L. A., Martin, G. R., and Kleinman, H. K.,** Interactions of basement membrane components, *Biochem. Biophys. Acta*, 761, 278, 1983.
49. **Charonis, A. S., Tsilibary, E. C., Yurchenko, P. D., and Furthmayr, H.,** Binding of laminin to type IV collagen: a morphological study, *J. Cell Biol.*, 100, 1848, 1985.
50. **Rao, C. N., Margulies, I. M. K., and Liotta, L. A.,** Binding domain for laminin on type IV collagen, *Biochem. Biophys. Res. Commun.*, 128, 45, 1985.
51. **Kleinman, H. K., McGarvey, M. L., Hassell, J. R., and Martin, G. R.,** Formation of a supramolecular complex is involved in the reconstitution of basement membrane components, *Biochemistry*, 22, 4969, 1983.
52. **Horwitz, A., Duggan, K., Greggs, R., Decker, C., and Buck, C.,** The cell substrate attachment (CSAT) antigen has properties of a receptor for laminin and fibronectin, *J. Cell Biol.*, 101, 2134, 1985.
53. **Kennedy, D. W., Rohrback, P. H., Martin, G. R., Momoi, T., and Yamada, K. M.,** The adhesive glycoprotein laminin is an agglutinin, *J. Cell Physiol.*, 114, 257, 1983.
54. **Terranova, V. P., Rao, C. N., Kalebic, T., Margulies, I. M., and Liotta, L. A.,** Laminin receptor on human breast carcinoma cells, *Proc. Natl. Acad. Sci. U.S.A.*, 80, 2144, 1983.
55. **Couchman, J. R., Hook, M., Rees, D. A., and Timpl, R.,** Adhesion growth and matrix production by fibroblasts on laminin substrates, *J. Cell Biol.*, 96, 177, 1983.
56. **Hirata, K., Yoshida, Y., Shiramatsu, K., Freeman, A. E., and Hayasaka, H.,** Effects of laminin, fibronectin and type IV collagen on liver cell cultures, *Exp. Cell Biol.*, 51, 121, 1983.
57. **Carlin, B., Jaffe, R., Bender, B., and Chung, A. E.,** Entactin: a novel basal lamina-associated sulfated glycoprotein, *J. Biol. Chem.*, 256, 5209, 1981.
58. **Wu, T.-C., Wan, Y.-J., Chung, A. E., and Damjanov, I.,** Immunohistochemical localization of entactin and laminin in mouse embryos and fetuses, *Dev. Biol.*, 100, 496, 1983.
59. **Martinez-Hernandez, A. and Chung, A. E.,** The ultrastructural localization of two basement membrane components: entactin and laminin in rat tissues, *J. Histochem. Cytochem.*, 32, 289, 1984.
60. **Wan, Y.-J., Wu, T.-C., Chung, A. E., and Damjanov, I.,** Monoclonal antibodies to laminin reveal the heterogeneity of basement membranes in the developing and adult mouse tissues, *J. Cell Biol.*, 98, 971, 1984.
61. **Kanwar, Y. S. and Farquhar, M. G.,** Presence of heparan sulfate in the glomerular basement membrane, *Proc. Natl. Acad. Sci. U.S.A.*, 76, 1303, 1977.
62. **Kanwar, Y. S. and Farquhar, M. G.,** Isolation of glycosaminoglycans (heparan sulfate) from glomerular basement membrane, *Proc. Natl. Acad. Sci. U.S.A.*, 76, 4493, 1979.
63. **Kanwar, Y. S., Hascall, V. C., and Farquhar, M. G.,** Partial investigation of newly synthesized proteoglycans isolated from glomerular basement membrane, *J. Cell Biol.*, 90, 527, 1981.
64. **Stow, J. L., Sawada, H., and Farquhar, M. G.,** Basement membrane heparan sulfate proteoglycans are concentrated in the lamina rarae and in podocytes of the rat renal glomrulus, *Proc. Natl. Acad. Sci. U.S.A.*, 82, 3296, 1985.
65. **Wu, V. Y., Chiu, I. I., and Cohen, M. P.,** Quantitative analysis of glomerular basement membrane glycosaminoglycans and evidence for their binding to adjacent cell membrane lipids, *Anal. Biochem.*, 145, 57, 1985.
66. **Kanwar, Y. S., Veis, A., Kimura, J. H., and Jakubowski, M. L.,** Characterization of heparan sulfate-proteoglycan of glomerular basement membranes, *Proc. Natl. Acad. Sci. U.S.A.*, 81, 762, 1984.
67. **Klein, D. J., Oegema, T. R., Eisenstein, R., Furcht, L., Michael, A. F., and Brown, D. M.,** Renal localization of heparan sulfate proteoglycan by immunohistochemistry, *Am. J. Pathol.*, 111, 323, 1983.
68. **Parthasarathy, N. and Spiro, R. G.,** Isolation and characterization of the heparan sulfate proteoglycan of the bovine glomerular basement membrane, *J. Biol. Chem.*, 259, 12749, 1984.
69. **Kanwar, Y. S., Jakubowski, M. L., Rosenzweig, L. J., and Gibbons, J. T.,** De novo cellular synthesis of sulfated proteoglycans of the developing renal glomerulus *in vivo, Proc. Natl. Acad. Sci. U.S.A.*, 81, 7108, 1984.
70. **Turksen, K., Aubin, J. E., Sodek, J., and Kalnins, V. I.,** Localization of laminin, type IV collagen, fibronectin and heparan sulfate proteoglycan in chick retinal pigment epithelium basement membrane during embryonic development, *J. Histochem. Cytochem.*, 33, 665, 1985.
71. **Clowes, A. W., Clowes, M. M., Gown, A. M., and Wight, T. N.,** Localization of proteoheparan sulfate in rat aorta, *Histochemistry*, 80, 379, 1984.
72. **Mehta, H., Orphe, C., Todd, M. S., Cornbrooks, C. J., and Carey, D. J.,** Synthesis by Schwann cells of basal lamina and membrane associated heparan sulfate proteoglycans, *J. Cell Biol.*, 101, 660, 1985.
73. **Heathcote, J. G., Bruns, R. L., and Orkin, R. W.,** Biosynthesis of sulfated micromolecules by rabbit lens epithelium II. Relationship to basement membrane formation, *J. Cell Biol.*, 99, 861, 1984.

74. **Turksen, K., Aubin, J. E., Sodek, J., and Kalnins, V. I.,** Changes in the distribution of laminin, fibronectin, type IV collagen and heparan sulfate proteoglycan during colony formation by chick retinal pigment epithelial cells *in vitro*, *Collagen Rel. Res.*, 4, 413, 1984.
75. **Hook, M., Couchman, J., Woods, A., Robinson, J., and Christner, J. E.,** Proteoglycans in basement membranes, *Ciba Found. Symp.*, 108, 44, 1984.
76. **David G. and Van den Berghe, H.,** Heparan sulfate — chondroitin-sulfate hybrid proteoglycan of the cell surface and basement membrane of mouse mammary epithelial cells, *J. Biol. Chem.*, 260, 11067, 1985.
77. **Fujiwara, S., and Wiedemann, H., Timpl, R., Lustig, A., and Engel, J.,** Structure and interactions of heparan sulfate proteoglycans from a mouse tumor basement membrane, *Eur. J. Biochem.*, 143, 145, 1984.
78. **Hassell, J. R., Leyshon, W. C., Ledbetter, S. R., Tyree, B., Suzuki, S., Kato, M., Kimata, K., and Kleinman, H. K.,** Isolation of two forms of basement membrane proteoglycans, *J. Biol. Chem.*, 260, 8098, 1985.
79. **Ledbetter, S. R., Tyree, B., Hassell, J. R., and Horigan, E. A.,** Identification of the precursor protein to basement membrane heparan sulfate proteoglycan, *J. Biol. Chem.*, 260, 8106, 1985.
80. **Lowe-Krentz, L. S. and Keller, S. M.,** Multiple heparan sulfate proteoglycans synthesized by basement membrane producing murine embryonal carcinoma cell line, *Biochemistry*, 22, 4412, 1983.
81. **Fenger, M., Wewer, U., and Albrechtsen, R.,** Basement membrane heparan sulfate proteoglycan from the L2 rat yolk sac carcinoma, *FEBS Lett.*, 173, 75, 1984.
82. **Iozzo, R. and Clark, C. C.,** Biosynthesis of proteoglycans by rat embryo parietal yolk sacs in organ culture, *J. Biol. Chem.*, 261, 6658, 1986.
83. **Mynderse, L. A., Hassell, J. R., Kleinman, H. K., Martin, G. R., and Martinez-Hernandez, A.,** Loss of heparan sulfate proteoglycan from glomerular basement membrane of nephrotic rats., *Lab. Invest.*, 48, 292, 1983.
84. **Iozzo, R. V.,** Proteoglycans: structure, function and role in neoplasia, *Lab. Invest.*, 53, 373, 1985.
85. **Birembaut, P., Caron, Y., and Adnet, J.-J.,** Usefulness of basement membrane markers in tumoral pathology, *J. Pathol.*, 145, 283, 1985.
86. **Yurchenco, P. D. and Furthmayr, H.,** Self-assembly of basement membrane collagen, *Biochemistry*, 23, 1839, 1984.
87. **Ohno, M.,** unpublished observations.

Chapter 5

CHANGES IN THE MORPHOLOGY AND CHEMISTRY OF CONNECTIVE TISSUES DURING AGING

H. Bouissou and M. Th. Pieraggi

TABLE OF CONTENTS

I. Introduction 96

II. Aging of Dermal Connective Tissue 96
- A. Dermis in Young Subjects 96
- B. Dermis in Aging Subjects 98
- C. Cultures of Young and Aging Fibroblasts 99
- D. Changes in Fibroblast Appearance During Life 102
- E. Interpretation of the Lesions of Aging 103

III. Aging of the Aorta 104
- A. Age-Related Morphological Changes in Components of the Arterial Wall 104
  - 1. Cells 104
  - 2. Intercellular Components 106
- B. Aging and the Aorta 106
  - 1. In the Child and the Young Adult 106
  - 2. With Aging 106
- C. Interpretation 108
- D. Atheroma, a Complication of Aging 108

IV. Summary 109

References 110

I. INTRODUCTION

Undifferentiated and differentiated (specialized) connective tissues have their origin in the embryonic mesenchyma. Made up of cellular units of synthesis which elaborate an interstitial network composed of collagen, elastin, glycosaminoglycans (GAG), and structural glycoprotiens (SGP), they become individualized according to the their cellular attributes and to the quality and quantity of the four macromolecules which are secreted.

These tissues age in different ways, but for all of them, aging may be considered the result of a biological process which progressively modifies the functioning and the interrelation of their various components.

In this chapter, we have chosen to compare two types of aging: that of the dermis for undifferentiated supportive connective tissue and that of the aortic wall for specialized connective tissue. In these two tissues, the fibroblasts in the one and the smooth muscle cells (SMC) in the other ensure the secretion of the various identical interstitial components, their maintenance by slow turnover,[1,2] and intervene in low-density lipoproteins (LDL) metabolism.[3,4] Also, during life, these two cells become morphologically similar: while the fibroblasts and the SMC, like all mesenchymal cells, can take on macrophagic functions (lipophages), the former can acquire myofilaments and become contractile, while the latter, inversely, can lose their contractile properties and become secretory (SSMC).

II. AGING OF DERMAL CONNECTIVE TISSUE

In aging, major changes within connective components of the dermis are observed: collagen loses its regular and fascicular appearance and the fibroblast cell population becomes "at rest". The resulting dermal aging differs according to the subject and is related to genetic background and exposition to multiple aggressions.

The more important lesions develop in the papillary zone. The deep and mid-dermis do not show important alterations.

The dermal connective tissue in young and old subjects is described before analyzing cultured young and aged fibroblasts and the morphological variations that fibroblasts may present during life.

A. Dermis in Young Subjects

By light microscopy, the papillary dermis shows numerous elastic fibers arranged in a network, with numerous fibrils reaching the epidermis. These are regularly distributed and tend to attach to the basement membrane of the epidermis, according to the schema of Cotta Pereira et al.[5] The oxytalan fibers, composed essentially of SGP, join with the elaunin fibers. In the dermis, the latter run parallel to the stratum germinativum and are composed of SGP and elastin. They continue in the mid- and deep dermis with the elastic fibers composed of elastin and of some SGP. The whole of the elastic network presents spontaneous immunofluorescence, and after reaction with an anti-elastic tissue serum, is clearly stained by Verhoeff's iron hematoxylin[6] and disappears under the action of elastases[6] (Figure 1A, B). Collagen fibers are grouped in thick and compact bundles throughout the thickness of the dermis. In all instances, Sirius Red staining shows yellow, orange, or red fascicles and a few thin green fibers. These observations, in agreement with Montes and Junqueira (this volume, Chapter 3) show the two kinds of cutaneous collagen, type I stained yellow or red and type III in green. This aspect corresponds to normal or "type O" skin.[7,8]

By immunofluorescence, type I collagen is observed throughout the dermis, with coarse collagen bundles in the reticular dermis and thin fibrils in the papillary dermis. Type III collagen is essentially observed in the papillary dermis (dermo-epidermal junction) and surrounding pilosebaceous units. The coarse bundles of the reticular dermis do not stain. A fine meshwork is also observed around dermal capillaries.

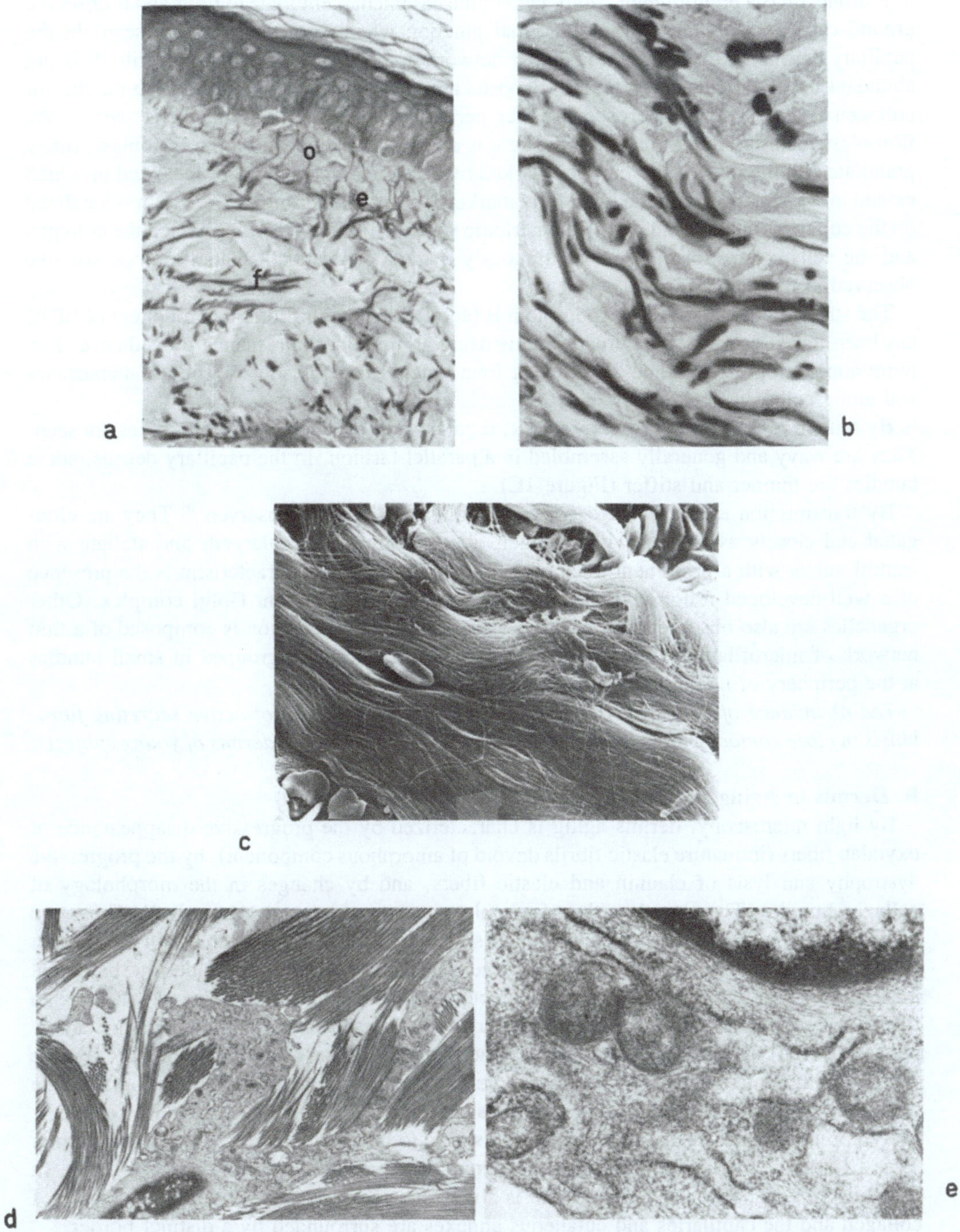

FIGURE 1. Dermis of young subject. (A) Skin type O (20 years old); light microscopy. Numerous oxytalan fibers (O) in papillary dermis; network of elaunic (e) and elastic fibers (f) in reticular dermis. Verhoeff stain, magnification × 40. (B) Skin type O; light microscopy. Elastic fibers in reticular dermis; regular elastic lamellae between collagen bundles. Verhoeff stain, magnification × 100. (C) Skin type O; scanning electron microscopy. Large, regular waved collagen bundles. Magnification × 1500. (D) Skin type O; transmission electron microscopy. Elongated, stellated fibroblast with close connection with collagen fibers. In the cytoplasma, well developed reticulum endoplasmic. Uranyl acetate, lead citrate stain, magnification × 10,000. (e) Skin type O; transmission electron microscopy. The cytoskeleton consists of a fine network throughout the cytoplasm and near of secretory areas. Uranyl acetate, lead citrate stain, magnification × 30,000.

Fibronectin is a predominant component of the basement membrane regions and is observed around capillaries in the dermo-epidermal junction where some clusters are seen. In the papillary dermis, it forms a thin irregular network associated with collagen fibrils. It is not abundant in the reticular dermis; thin deposits are always associated with coarse bundles of collagen I. By electron microscopy, after peroxidase marking, fibronectin is seen in the fibroblasts and the interstitial spaces. There is a continuous deposit on the fibroblast, finely granulated on the plasma membrane, marked by densifications which are localized or which extend in continuation with the interstitial marking. In the interstitium, fibronectin is localized on the collagen bundles and marks the periodic striation of their fibers. Between the collagen and the cells is a felting of fine continuously marked filaments. Isolated clumps are also observed in the interstitium.

The structural localization of apoprotein B (apo B), the major protidic component of LDL, has been investigated in the human dermis using an immunofluorescence procedure and an immunoperoxidase technique.[9] Apo B is found attached to plasmic fibroblast membranes and along collagen fibers.

By scanning electron microscopy, thick, regular, well-oriented collagen bundles are seen. They are wavy and generally assembled in a parallel fashion. In the papillary dermis, some bundles are thinner and stiffer (Figure 1C).

By transmission electron microscopy, fibroblasts are mainly observed.[10] They are elongated and closely associated with collagen fibers (Figure 1D), enlarged, and stellate with central nuclei with a prominent nucleolus. Their most notable characteristic is the presence of a well-developed rough endoplasmic reticulum and a prominent Golgi complex. Other organelles are also observed, but no lipid vacuoles. The cytoskeleton is composed of a thin network of microfilaments throughout the cytoplasm. These are grouped in small bundles at the periphery of nuclei and near secretory areas (Figure 1E).

The abundance of compact collagen bundles and the presence of active secreting fibroblasts in close contact with collagen fascicles is the hallmark of the dermis of young subjects.

B. Dermis in Aging Subjects

By light microscopy, dermis aging is characterized by the progressive disappearance of oxytalan fibers (immature elastic fibrils devoid of amorphous component), by the progressive dystrophy and lysis of elaunin and elastic fibers, and by changes in the morphology of collagen bundles. The latter lose their fascicular pattern and become ''granular''. Thus, two degrees of aging may be described; (1) type I skin, characterized by a decrease in the number of oxytalan fibers; some elastic fibers are seen in places reaching the epidermic basement membrane (Figure 2). (2) type II skin, characterized by complete disappearance of oxytalan fibers, dystrophy and lysis of elaunin (similar to oxytalin fibrils but containing some amorphous material) and elastic fibers, and by the disorganization of collagen bundles (Figure 3A, B) With Sirius Red stain, the intense yellow-orange color of the dermis confirms the predominance of type I collagen and the presence of scanty amounts of type III collagen.

By immunofluorescence,[8] type I collagen is seen to be distributed throughout the dermis. Type III is more predominant in the dermo-epidermal junction, marking the basal layer. In the mid- and deep dermis, fluorescence is slight; it is seen between the bundles of type I collagen and the capillaries and cutaneous annexes are surrounded by a distinct border.

Fibronectin is not abundant. The dermo-epidermal junction is irregularly marked by granular deposits or small clumps. In the papillary dermis, there is slight marking; in the reticular dermis where interstitial marking is not intense, it is however constant around capillaries. By electron microscopy, after peroxidase marking, fibronectin is seen in the cells and the interstitium. Marking on the fibroblast is irregular and discontinuous. The collagen bundles are finely marked along their fibers, but interstitial clumps are rare. There is slight fibrillar felting around the cells.

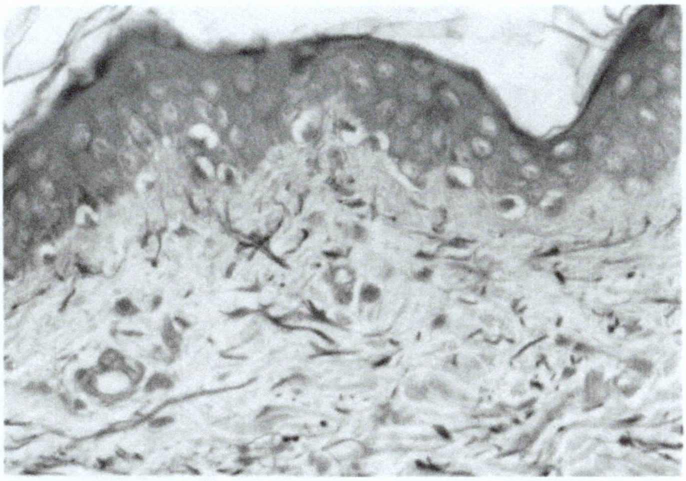

FIGURE 2. Cutaneous aging: skin type I (30 years). In papillary dermis, disappearance of some area of oxytalan fibers. Verhoeff stain, magnification × 40.

Dermal areas containing apo B are more numerous in patients with type II skin and coronary disease than in subjects of the same age with healthy coronary vessels.[9]

By scanning electron microscopy, the superficial dermis appears disordered; collagen bundles are irregular, thin, and often separated by conspicuous interfibrillar spaces (Figure 3C).

Transmission electron microscopy shows short, thin fascicles of collagen separated by empty spaces corresponding to ground substance. With high magnification, occasional numbers of fibers do not show typical periodicity, and extensions appear to separate, forming microfibrils giving a brushed aspect. Elastic fibers are not seen in the papillary dermis and in the reticular dermis they are modified with a smooth moth-eaten aspect. In exposed skin areas, the elastic bands are modified; they become voluminous with abundant fibrils, a sign of elastotic degeneration (Figure 4A, B).[10] Fibroblasts are flattened, and their cytoplasm is poor in rough endoplasmic reticulum and ribosomes; numerous dense bodies of lysosomial type and frequent membrane-bound lipid vacuoles are observed. However, the main modification of aging fibroblasts is the disorganization of the cytoskeleton, characterized by voluminous fascicles or bundles occupying large cytoplasmic areas, often in contact with mitochondrias but without any relation with the plasma membrane.[11] The fibroblast loses its secretory function and becomes a "quiescent" cell, with no contact with collagen. The cell population is isolated within the interstitial matrix. Lipid droplets can be seen in the cytoplasm of aging fibroblasts (Figure 3D, E).

The fragmentation of collagen bundles, the absence of elastic material, and the occurrence of quiescent fibroblasts without any direct contact with collagen represent the morphological changes of aging dermis.

C. Cultures of Young and Aging Fibroblasts

The study of fibroblasts cultured from explants from young and aged subjects, analyzed at identical intervals, shows the same cellular modifications observed in cutaneous biopsy specimens during aging.

In the aged subject fibroblast modifications are as follows: more numerous vacuoles, decrease of the endoplasmic reticulum, and bundling of the cytoskeleton. Fibrillar secretion on the external aspect of the plasma membrane is markedly decreased.

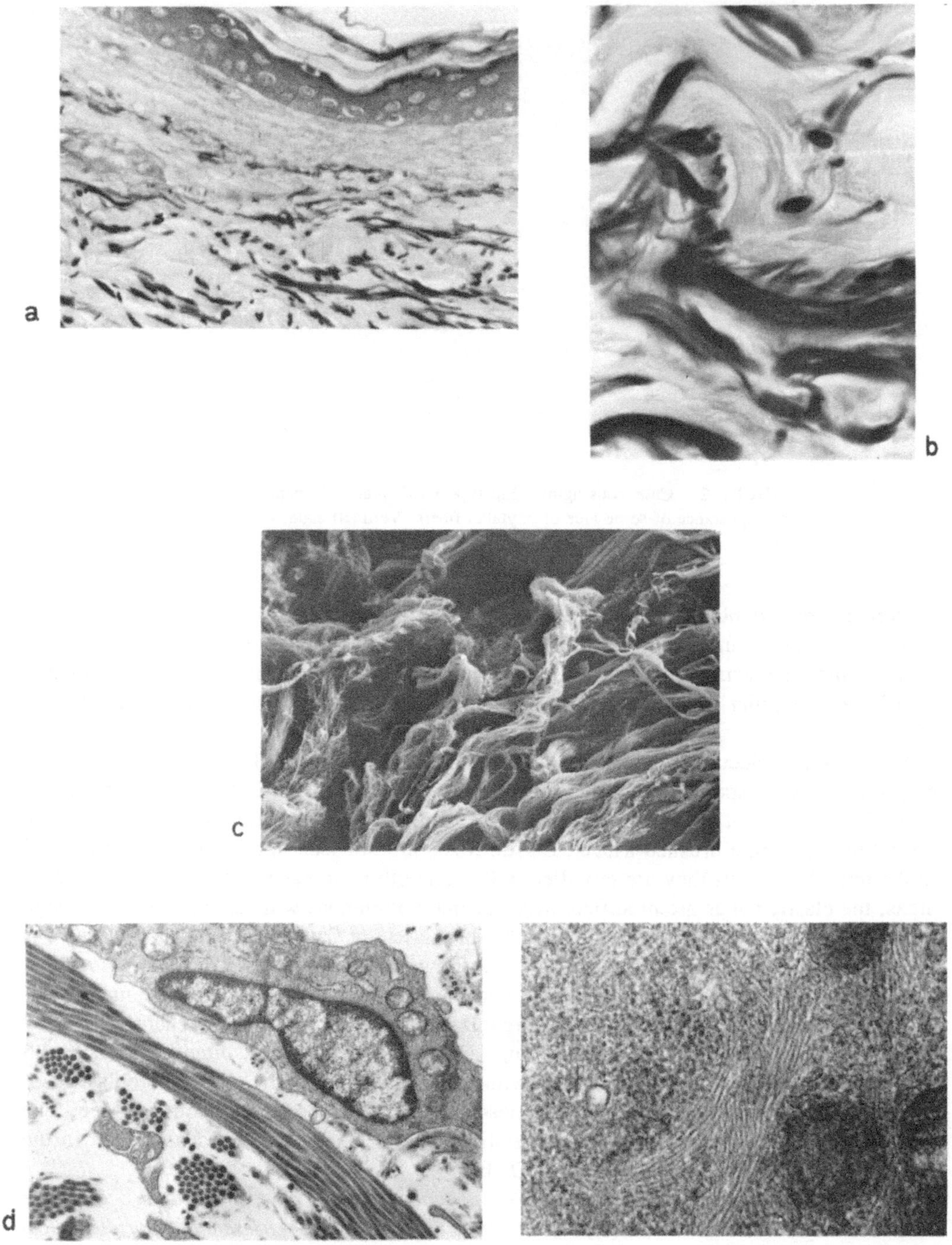

FIGURE 3. Cutaneous aging skin type II. (A) Light microscopy. In papillary dermis disappearance of oxytalan fibers; dystrophic elaunic material. Verhoeff stain, magnification × 40. (B) Light microscopy. Some aspects of dystrophy of elastic fibers in reticular dermis. Verhoeff stain, magnification × 100. (C) Scanning electron microscopy. Collagen bundles are irregular and separated by large spaces. Magnification × 1000. (D) Transmission electron microscopy. Fibroblast is flattened and quiescent, without direct contact with collagen. Uranyl acetate, lead citrate stain, magnification × 10,000. (E) Transmission electron microscopy. The cytoskeleton is characterized by voluminous bundles of filaments. Uranyl acetate, lead citrate stain, magnification × 20,000.

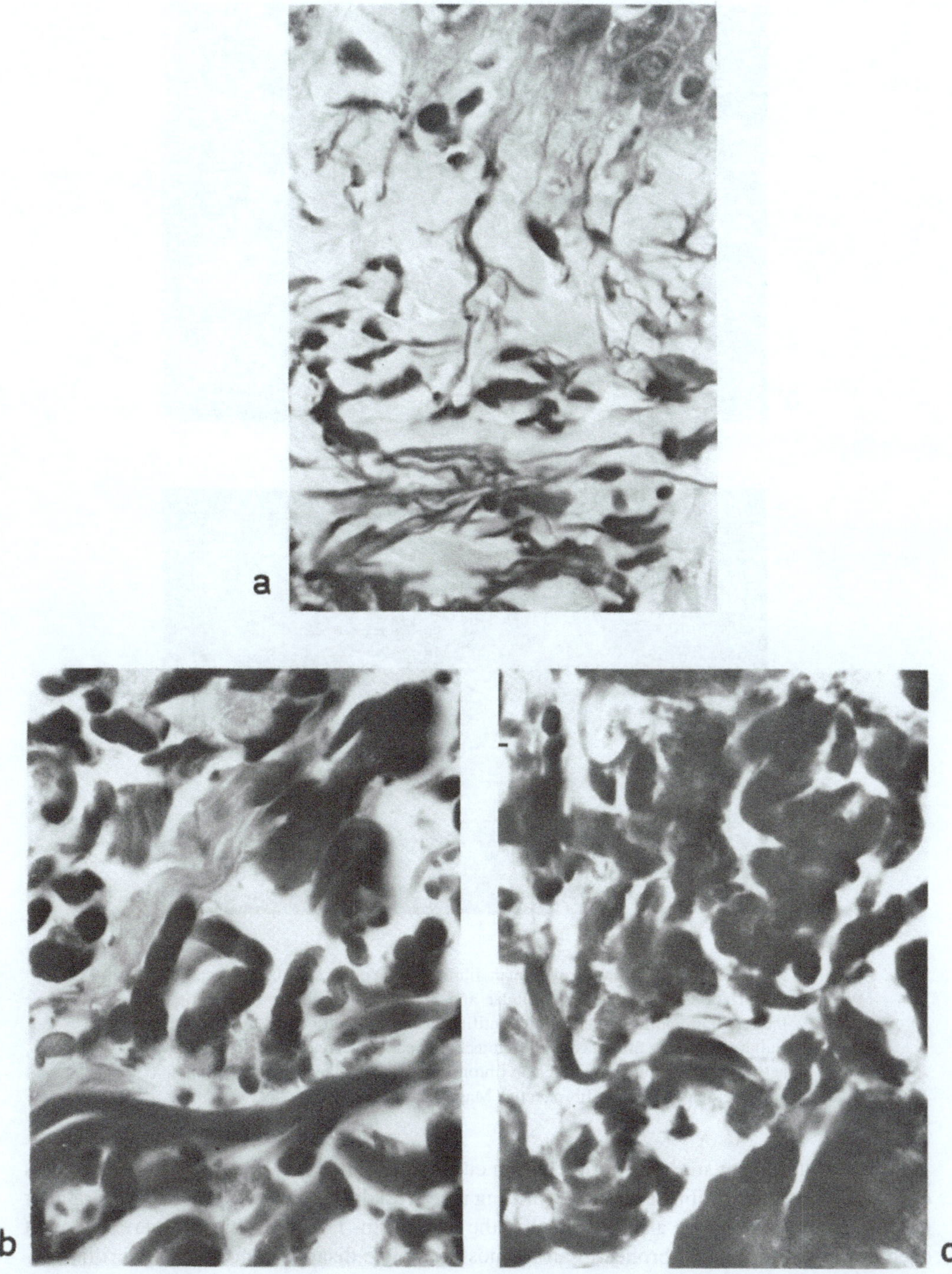

FIGURE 4. Elastotic degeneration. (A) Young subject (19 years). Preauricular skin. Oxytalan, elaunic fibers. Elastic network with hyperplastic and dystrophic changes. Verhoeff stain, magnification × 100. (B) Subject of 49 years. Preauricular skin. Abundant hyperplastic, dystrophic elastic fibers; first phase of elastotic degeneration. Verhoeff stain, magnification × 100. (C) Subject of 50 years. Preauricular skin. Severe elastotic damage in the reticular dermis. Abnormalities of elastic fibers (granular appearance, large dense zones). Verhoeff stain, magnification × 100.

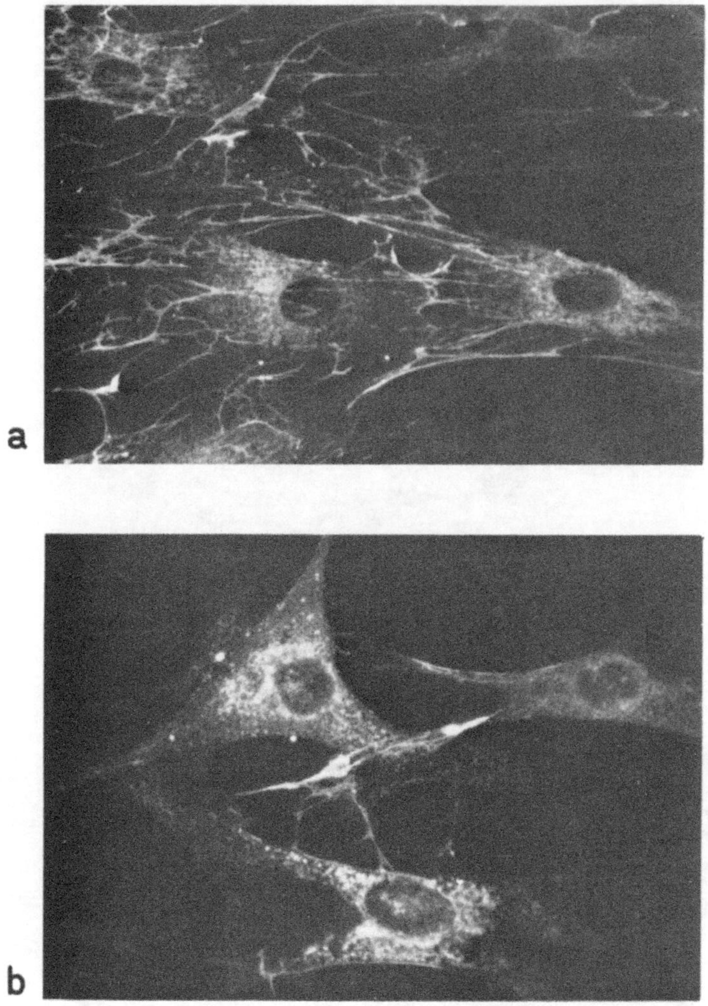

FIGURE 5. (A) Culture of human fibroblasts. Young subject (19 years). Localization of fibrillar extracellular and granular intracellular fibronectin. Indirect immunofluorescence with antifibronectin. Magnification × 400. (B) Culture of human fibroblasts. Old subject (78 years). Decrease of extracellular network. Intracellular perinuclear fibronectin is unchanged. Indirect immunofluorescence with antifibronectin. Magnification × 400.

Immunofluorescent study of fibronectin in cultured fibroblasts from young subjects shows abundant extracellular fibrillar felting, forming a network linking the different cells. Granular perinuclear formations are also observed in the cytoplasm. In aged subjects, there is a great decrease in extracellular fibronectin and almost complete disappearance of the fibrillar network; however, intracytoplasmic granules are observed around the nucleus (Figure 5A, B).

D. Changes in Fibroblast Appearance During Life

The activity of the fibroblast varies according to its situation. Under the effect of various stimuli (enzymatic, chemical etc.), the fibroblast becomes hyperactive, whatever its situation within the organism; this is the case during the healing process, under the influence of numerous stimuli of blood or of cellular origin, such as leucotrienes,[12] fibronectin,[13], C1q[14], etc. Moreover, its aspect and secretory activity vary during the course of the healing process: normal, hypertrophic, or keloid.[15] The fibroblast, even when aging, is thus seen to be a cell which is capable of "reactivation" and recovering efficacious secretory activity.

It is also capable of morphological modification and the acquisition of new properties: it may become a fibroclast capable of phagocytosis, or a myofibroblast capable of contraction.

The fibroclast, described in these terms by Baur et al.[16] and Buntrock et al.[17] but envisaged earlier by several authors, in particular Ten Cate and Freeman,[48] Uzunian,[49] and Svoboda et al.,[50] is an ultramicroscopic entity. It corresponds to a fibroblast containing collagen fibers or fibrils within a vacuole or an invagination of the plasma membrane. In normal dermis, fibroclasts represent 1% of the fibroblast population.[16]

During the healing process, its double function of collagen production and resorption makes the fibroblast the regulator of the quantitative and qualitative collagen equilibrium which is necessary for satisfactory healing. The conditions necessary for this balance are not yet clear, and the intervening factors are unknown.

In 1971, Gabbiani et al.[18] described the presence during healing of a cell "gifted" with contractile properties, which they called a "myofibroblast". They went on to describe it and to study its functions in detail. It is characterized by an abundant ergastoplasm and a cytoskeleton rich in actin and myosin, parallel to the main axis and attached at intervals to the internal aspect of the plasma membrane. It is a macrophagic,[16,17] secretory, and contractile[19] cell.

In 1984, Kocher et al.,[20] studying the three fractions of actin, demonstrated that it is derived from the fibroblast, and like the fibroblast possesses the α and β components of actin, and not the three α, β, γ components like the muscle cell.

B. Interpretation of the Lesions of Aging

Dermal aging is expressed morphologically by disappearance of oxytalan fibers, dystrophy and lysis of elaunic and elastic fibers, changes in collagen architecture, and quiescent fibroblasts.

Collagen modifications are not associated with qualitative and quantitative variations of types I and III. Biochemical studies demonstrate only variations in crosslinks; the dimensions of fibers and fibrils do not change.[21]

The disappearance of oxytalan fibers and the changes in elaunic and elastic fibers are certainly linked with the age-related increase in dermal elastases, which act on the microfibrillar structure of these fibers.[22]

It seems that dermal aging is the result of loss of cohesion between collagens and the fibroblast.[10] This cohesion is dependent on fibronectin, which plays a role in adhesion between cells, between collagen and the cell, and which is in relation with the cytoskeleton. With aging, a decrease in dermal fibronectin is observed in vitro and in vivo, and the cytoskeleton is modified. Thus, aging could be interpreted as a dysfunction of the cell-intercellular matrix unit, in which fibronectin plays a preponderant role. The resulting progressive morphological changes correlate with the alterations in fibroblastic metabolism, in particular in the protein moiety.

The observation of lipid vacuoles in the aged fibroblast may represent the morphological expression of its dysfunction in lipid metabolism. It has already been demonstrated that the fibroblast plays a role in the metabolism of lipoproteins.[4] With aging, this metabolism is modified and lipoproteins (particularly LDL) are stocked within the dermis. Biochemical determination of cutaneous cholesterol and apo B[23] and immunofluorescent procedures with anti-apo B provide evidence of this lipoprotein accumulation.[9]

III. AGING OF THE AORTA

Some useful points on the morphology of the arterial components and the changes brought about in them by aging may be noted here, before describing the young and the aging aorta and its principal complication: atherosclerosis.

A. Age-Related Morphological Changes in Components of the Arterial Wall

1. Cells

The endothelial cell of the young subject has an elongated, voluminous nucleus, and a flat, spreading, thin cytoplasm containing few organelles, actin, and intermediary filaments scattered or in small clumps. Numerous vesicles on the luminal and on the external side, often intercommunicating, give evidence of transcellular exchanges. The basal membrane (collagen IV and V) which they elaborate in vitro[16] is laminated and of uneven thickness. In vitro, this cell secretes collagens I, II, and IV[24] and still in vitro, produces mediators able to stimulate or to inhibit SMC proliferation.[25] With aging, this cell becomes cubic and its microfilaments more numerous[26] resembling those described by Gabbiani et al.[27] as being actin filaments. Its vesicles and cytoplasmic invaginations are more numerous and pronounced. The junctions become irregular and enlarged. Some vesicles may contain lipids. The basal membrane becomes thicker.

The endothelium and its basal membrane form a continuous barrier to the passage of blood macromolecules. Their functioning as such depends on the physical continuity of the endothelium, the closeness of the junctions, the transport capacity of the pinocytotic vesicles, the capacity to break down lipoproteins and lipids, and lastly on the integrity of membranal collagens. This barrier constantly undergoes multiple aggressions: aging, quantity and quality of blood lipids, hypoxia, antigen-antibody complexes, hypertension, blood turbulence, trauma, toxins, etc. All these cause increased permeability with futher entry of blood substances, lipoproteins, and GAG in the subendothelium and the inner layers of the media. These processes account for the formation of the intima in the early years of life and its progressive thickening with age.[28]

During life, especially in the aorta, these injuries result in white or pearly-white fibrous plaques, of varying extent and thickness and often cause destruction of the endothelium and formation of platelet thrombi. The latter are destroyed or incorporated into the intima after their endothelialization and become fibrous. They damage the arterial wall by the liberation of enzymes (elastases in particular), mediators increasing vascular permeability, and a mitogenic factor against mitocytes migrating towards the intima.

The SMC (Figure 6A) oriented paralled to the endothelium in the young subject has numerous myofilaments with irregular densifications,[29] many pinocytotic vesicles, continuous basal membrane, and desmosomes. The nucleus is indented. These cells can mobilize, proliferate, and acquire phagocytic properties. They secrete collagens I and III, SGP, fibronectin, and elastin. They also intervene in the catabolism of LDL[30] which are taken up by specific A receptors. The LDL cholesterol is used in part by the cells themselves and the remainder is transmitted by the B membrane receptors to the HDL, which transport it to the liver via the vasa vasorum. If the quantity of LDL is too great or if there is a catabolic deficiency, the cholesterol may accumulate in the SMC, which take on a foamy appearance (foam cells). This overload may be transitory; if it is severe, it results in asphyxia and the breakdown of the cell, releasing the lipid content into the interstitium where it causes fibrosis.[31]

With aging, during the lysis of one or several elastic layers, the SMC become perpendicular to the arterial lumen; their macrophagic properties towards lipids or cell debris[32] and their secretion of SGP[33] increase. However, the secretion of proelastin slows down;[33] whereas a total collagen remains at a constant level. There may be some change in the percentages of collagens I and III,[34] but this is not yet clearly established.

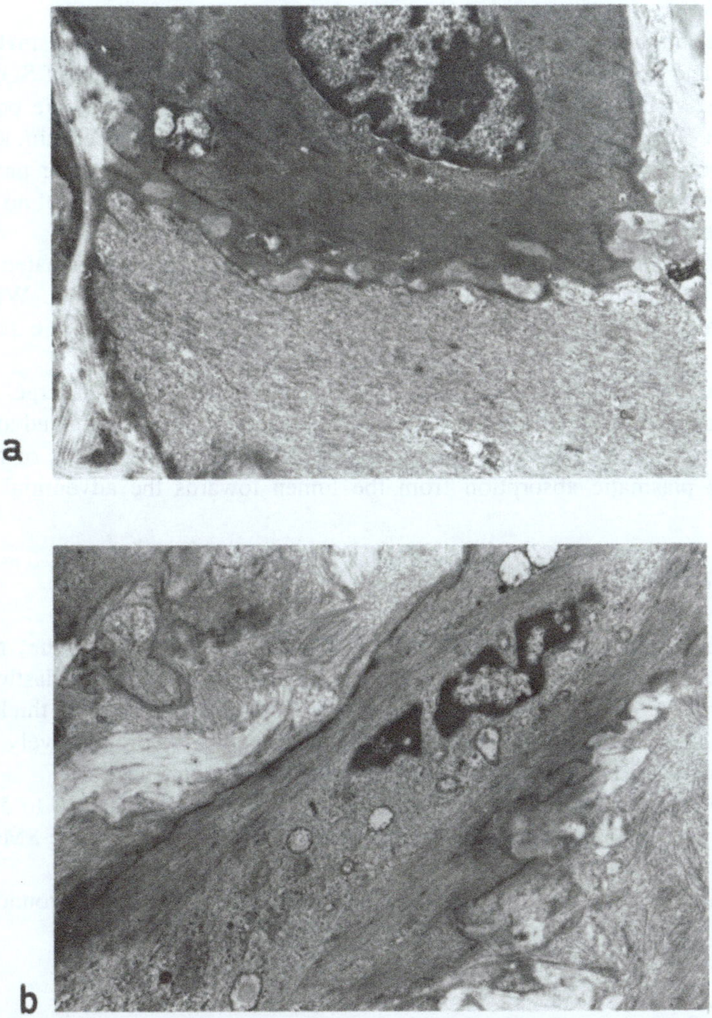

FIGURE 6. (A) Smooth muscle cells in young aorta. Numerous myofilaments in all cytoplasm. Contractile aspect. Uranyl acetate-lead citrate stain, magnification × 10,000. (B) Smooth muscle cell; secretory aspect. Rare myofilaments near plasma membrane. The cytoplasm contains numerous organelles. Uranyl acetate-lead citrate stain, magnification × 10,000.

With aging also, more and more SMC become highly secretory and take on certain aspects of the fibroblast.[35] These SSMC (the work of Kocher[36] showed that they cannot be myofibroblasts[35]) are fusiform, and have a well-developed ergastoplasm (a sign of high secretory activity (Figure 6B). They have a contractile fibrillar system and their basal membrane is often discontinuous. These cells secrete collagens I and III but only produce elastic fibrils or short-lived elastic bands which are soon destroyed. This explains the small elastic filaments frequently noted in the intima and the media of the aorta.

In some aged arteries, in areas which are very poor in collagen and in SMC and very rich in GAG, the cells present all the intermediary aspects between SMC and undifferentiated mesenchymal cells.[35]

A few monocyte-derived histiocytes cross the endothelium and populate the intima. They become macrophages and phagocytize the interstitial blood lipids and cell debris.[37]

Fibroblasts are present only in the arterial adventitia and not in the other two layers.

2. Intercellular Components

The elastic laminae of the media of the aorta are composed of elastic material secreted by the SMC and in part by the endothelial cells.[38,39] They are made up of SGP and elastin. SGP are most visible in immature and in aging fibers, both of which are poor in elastin. Mature elastic fibers, formed once and for all, poor in SGP and rich in elastin, are maintained by slow turnover with the SMC. Their lysis during life may be total or partial, rapid or slow, early or late. They are often replaced by elastic fibrils or minifibers of no physiological value, secreted by the SMC or SSMC.

Fibrillar collagens I and II are visible as small bundles. They are situated in the media between the cells and are mainly secreted by the SMC and the SSMC. With aging, the intima, colonized by the SMC, becomes progressively more fibrous. The same is true of the media.

The arterial and venous vasa vasorum are limited to the adventitia of large- and medium-caliber arteries. The outer third of the media of the large arteries is irrigated and drained by the vasa vasorum, and the endothelium, intima, and internal two thirds of the media are nourished by plasmatic absorption from the lumen towards the adventitial venous vasa vasorum.

B. Aging and the Aorta

1. In the Child and the Young Adult[26,28]

The intima is thin, made up of a fine layer of loose connective tissue, rich in GAG, containing few or no cells, separating the endothelium from the first elastic layer of the media (Figure 7A). During the first years of life, this layer becomes thicker and more organized. GAG, elastic fibrils, and small collagen bundles are progressively observed and the cell population increases at the same time (SMC, macrophages).

The media is made up of parallel elastic layers, which at birth are 40 to 50 in number, and increase to 70 to 80 in the young adult. In these elastic layers are the SMC surrounded by GAG and a small quantity of collagen.

The adventitia consists of fibroblasts in loose connective tissue surrounding the vasa vasorum.

2. With Aging[28,40]

During normal aging, or ''physiological arteriosclerosis'', the aorta becomes thick and sclerotic to a varying degree. The first sign of aging, which occurs towards the age of 15, is thickening of the intima (Figure 7B) due to an accumulation of GAGs and lysis of one or several elastic layers. The corresponding layers of the media thus become part of the intima. This ''intimalization''[28] (Figure 7C) which varies in surface area and depth along the vessel, is always most pronounced in the subdiaphragmatic portion of the aorta and continues throughout life. At the same time, the remaining elastic layers of the media become irregular and fragmented and are increasingly separated by accumulations of GAG and collagen fibers; SSMC appear in the two layers of the arterial wall; the innermost penetrate the intima. The artery becomes fibro-elastic (Figure 7D). Sometimes the lysis of elastic material is accompanied by proliferation of SMC in the internal part of the media and the artery becomes progressively musculo-elastic. When there is a large accumulation of GAG, pseudo-cysts of muciparous material may appear in the damaged media. The artery becomes muco-elastic (Figure 7E). The culmination of this process has been described by Cellina[41] as ''medianecrosis disseminata aorta''.

Between the ages of 15 and 30, fibrous plaques appear in the intima and the internal part of the media. Their number increases with aging. In cross section, they are of more or less homogeneous fibrous consistency, with no yellow center. Microscopically, they consist of parallel collagen fibers interspersed with a few elastic fibers. Within these fibrous plaques are GAG and SMC at various stages of fibroblastic differentiation and therefore secretory.

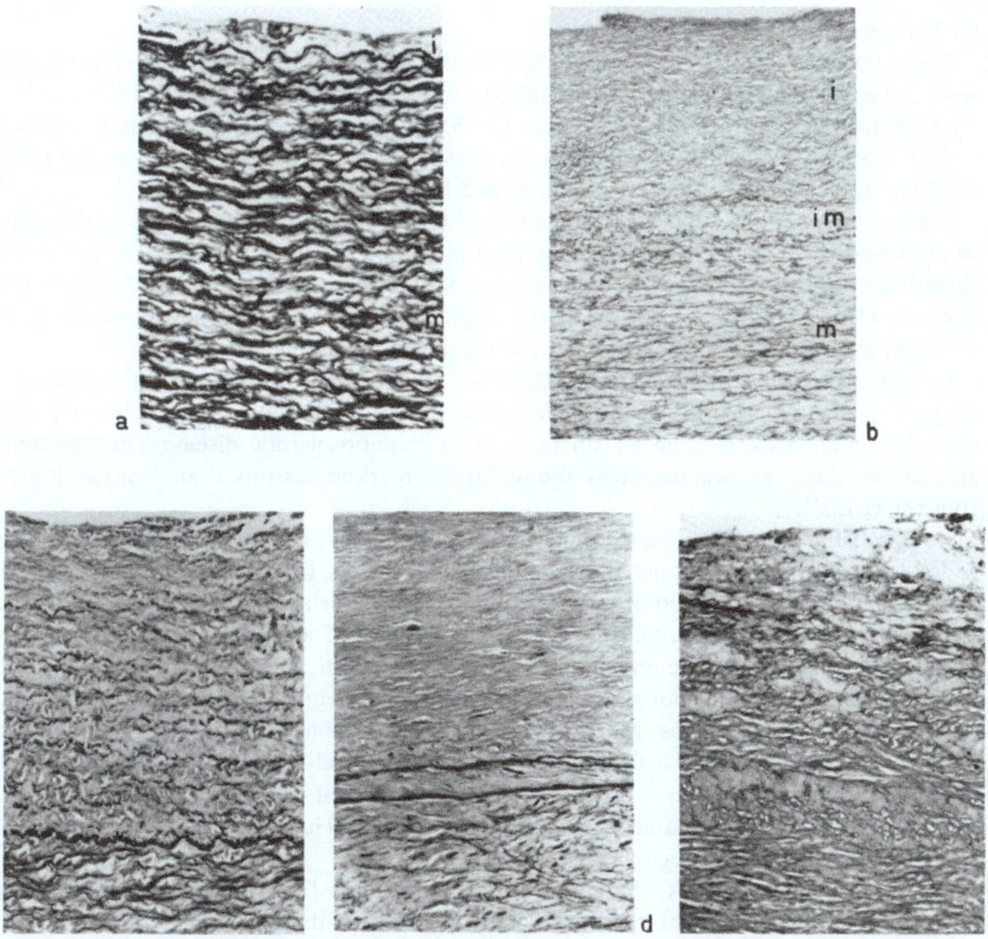

FIGURE 7. Aging of thoracic aorta. (A) Young subject (18 years). Individualization of thin intima layer (i). First elastic lamella marked. In the media (m) thick parallel elastic lamellae. Verhoeff's stain, magnification × 25. (B) Normal aging (65 years). Thick and fibrous intima (i). Intimal-medial junction (im). In the media (m) thinning of elastic lamellae. Verhoeff stain, magnification × 10. (C) Important intimalization process. Intimal thickening results from incorporation in the intima of the medial interlamellar spaces corresponding to the lysed lamellae. New elastic lamella more marked between intima and media. Verhoeff stain, magnification × 25. (D) Fibro-elastic aorta. Fibrous intima. Distinct intimalization process. Thin elastic lamellae in the media and abundant ground substance. Verhoeff stain, magnification × 25. (E) Muco-elastic aorta. Endothelial permeability in enhanced and GAC are present in intima and between elastic lamellae. Blue Alcian P.A.S. stain, magnification × 25.

There may be some foam cells, mainly in the abdominal aorta. These plaques are found in an intima already fibrous and thickened to a varying degree, or in contact with exsiting atheromatous lesions. They always extend beyond the thickened intima into the fragile ''intimalization'' area (see above). We consider them to be hypertrophic fibrous scars, which follow either upon already large and destructive gelatinous plaques formed during periods of marked deposition followed by the arrival of collagen-forming SSMC or upon successive microthrombi which have been integrated and become fibrous. They may also simply result from the deposition or successive depositions of blood lipids in the arterial wall; if these are not metabolized by the SMC, they cause hypertrophic fibrosis of the intimal-medial junction.

C. Interpretation

These changes result from physiological modifications in the components of the arterial wall: the endothelial cells are more permeable, resulting in thickening of the intima and the interlamellary spaces nearest the lumen; the SMC, disoriented, differentiated, secretory (SSMC), are no longer able to maintain the elastic tissue and secrete collagens and GAG, resulting in lysis of elastic layers, fibrosis, and larger interlamellary spaces.

Theoretically, aging in elastic arteries should result in progressive sclerosis, substitution of destroyed elastic fibers, and little overload in the arterial wall of GAG and lipids, with sometimes persistence or even hyperplasmia of SMC. These changes, which doubtless take place slowly, are those observed in most autopsies of aged persons. They allow the vessel to adapt, in spite of the presence of diffuse plaques of atheroma.

In fact, this "physiological arteriosclerosis" is rare, and the majority of autopsies of subjects aged more than 45 to 50 years show intense, precocious arteriosclerosis, often associated with marked stage III atheroma. This "arteriosclerotic disease" or "presenile arteriosclerosis" shows identical though more marked lesions than "physiological arteriosclerosis".

It is probable that the biological quality of the vessel and the environment of life in the wider sense of the term are not the same in the two cases. In arteriosclerotic disease, the marked and brutal penetration of GAG and lipids in the arterial wall accounts for the extent of damage caused, the disorganization and impossibility of adaptation or effective repair. Thus, the stagnation and accumulation of GAG, and in particular lipids, in an arterial wall which is incapable of metabolizing them or transporting them to the vasa vasorum account for the intensity of the lesions. More marked in certain portions of the arteries than in others (hemodynamics is certainly a factor in the preferential localizations of aging, such as the lower abdominal aorta), they will extend with the passing of time, and further lesions will appear at points formerly spared or only slightly involved. The elastic artery is prematurely worn and unequally and irregularly aged.

During this aging process, the adventitia becomes more fibrous. In the vasa vasorum, few changes other than banal microangiopathy are noted with age.

D. Atheroma, a Complication of Aging

The lesions of the connective tissue of the aortic intima and media in physiological arteriosclerosis or in arteriosclerotic disease are almost always accompanied by a lipid overload, rich in esterified cholesterol, which constitutes atheromatous disease.

The morphological aspects of atherosclerosis are mainly due to (1) endothelial hyper-permeability, (2) the retention of serum lipoproteins and GAG by and in the arteriosclerotic arterial wall, and (3) deficient SMC catabolism of LDL.

Two of the initial, "potential" lesions[42] of atheroma result from deficient lipid metabolism: the dot and the streak. The "dot", a macroscopic term describing a yellow elevation measuring 0.5 mm, is due to the proliferation[43] and accumulation in the intima of SMC which have become lipophagic. This lesion, which is reversible if the intracellular lipids are metabolized by these cells, extends and becomes a "streak" through the massive or continuous arrival of LDL and the proliferation of SMC and lipophages. This yellow streak, measuring from a few millimeters to 2 cm, with an indistinct outline, parallel to the vessel, may anastomose with other streaks and form a "meshwork". Interstitial lipid deposits rapidly appear in this lesion, generating fibrosis[31] following on the necrosis of lipophages or the presence in the interstitium of nonmetabolized blood lipids. The irreversible streak may become stabilized or progress. The "gelatinous plaque" or "focal myxoid degeneration", the third potential lesion, results from a massive arrival of GAG.[44] It is an edematous area impinging on the intima to a varying degree. A gray bulla, it is translucid or opaque, rounded or oval, its diameter is from 0.8 mm to 1 cm. This lesion may regress or become fibrotic or more extensive. Invaded by lipids and by lipophages, it then becomes yellow and irreversible.

These three lesions form, by their progression, the "pustule", which is intimal at first, then reaches the intimal-medial area and is surrounded by a fibrous cap. Lenticular, smooth, glistening, yellow, 0.5 cm in diameter and 2 mm thick, it consists mainly of lipophages and interstitial lipids. "Trapped", it cannot be metabolized, is hampered by the fibrosis and disorganized structure of the media, and becomes stabilized and extends. It thus forms the "atheromatous plaque", made up of an accumulation of lanceolate cholesterol crystals, interstitial lipids, foam cells, histiocytes, SSMC, and "walled" within fibrosis. Annular from 1 to 3 cm in diameter, whitish-yellow, cartilaginous, thick (3 to 4 mm), it may become calcified, ulcerated, and thus be the site of thrombosis.

Evolution varies according to the type of arteriosclerosis. In "physiological arteriosclerosis", the processes of lipid penetration, of cell wear, and of interstitial deposition are slow, progressive, and long-term. The aortic wall, which has become fibro-elastic, fibro-musculo-elastic, or musculo-elastic, resists lipid penetration.[45] Atheroma is slight (stage I: streaks; stage II: isolated plaques). It only becomes marked (stage III: coalescent plaques) progressively and at a later stage. (Stage O corresponds to the normal aorta [World Health Organization classification 1957]). In "arteriosclerotic disease", the process is the same. But the disorganization, transformation of the wall, and disruption of lipid and protid exchanges between cells and the interstitium are such that dot, streak, and mucoid plaque are very soon overwhelmed by massive deposits of lipids and protids. These extensive, non-drainable, nonmetabolizable factors of sclerosis form a pustule, then a plaque, which soon sets in and extends. Coalescence of these plaque leads to the classic "paved" appearance of the atheromatous aorta, with its calcified areas, its intraparietal hemorrhages, and its ulcerations giving rise to thrombosis or cholesterolinic embolisms.

The location of the lesions (principally in the abdominal portion of the aorta) is linked with the degree of aging of this vessel and with hemodynamic phenomena.

IV. SUMMARY

During aging, the components of the dermis and the aorta, the former elaborated by the fibroblasts and the latter by the endothelial cells and the SMC, progressively degenerate.

Aging of the dermis, nonspecialized connective tissue, is directly linked with changes in the properties of the functional unit fibroblast-interstitial matrix. Morphologically this is expressed by modifications in the fibroblasts, structural changes in collagen and in elastic fibers, and by a loss of cohesion between cells and the interstitium. During life, no matter at what age, under the influence of various stimuli the fibroblast is capable of becoming active, phagocytic (fibroclast), or contractile (myofibroblast). It is probable that during these various stages (aging, dedifferentiation), LDL metabolism disorders occur which account, at least during aging, for the accumulation of cholesterol and apoprotein B in the interstitium.

Aging of the aorta wall, specialized connective tissue, leads to a fibro-muscular, fibro-elastic, or fibromuco-elastic aorta. This arteriosclerosis is complicated by varying degrees of atheroma. These changes result from (1) increased permeability of the endothelium-basal membrane unit, allowing GAG and lipoproteins to penetrate the wall; (2) lysis of elastic fibers, due partly to platelet elastases and to SMC which proliferate and thus lose their contractile properties to become secretory migrating cells; (3) fibrosis of the intima and media.

These progressive, slow, late-developing phenomena are those of "physiological arteriosclerosis" ("physiological aging"); when they appear early, intensely, and rapidly, they are a sign of "premature aging", "arteriosclerotic disease".

The similarities and the simultaneous occurrence of these two types of aging make it possible for arteriosclerosis and aortic (or coronary) atheroma to be detected by morphological and biochemical examination of the dermis.[46,47]

REFERENCES

1. **Glucksmann, A.,** Cell turn over in the dermis in, *Advances in Biology of Skin*, Montagna and Billingham, Eds., Pergamon Press, Oxford, 1964, 79.
2. **Ross, R.,** The fibroblast and wound repair, *Biol. Rev. Cambridge Philos. Soc.*, 43, 51, 1968.
3. **Basu, S. K., Anderson, R. G. W., Goldstein, J. L., and Brown, M. S.,** Metabolism of cationised lipoproteins by human fibroblasts. Biochemical and morphological correlations, *J. Cell. Biol.*, 74, 119, 1977.
4. **Brown, M. S., Kovanen, P. T., and Goldstein, J. L.,** Regulation of plasma cholesterol by lipoprotein receptors, *Science*, 212, 628, 1981.
5. **Cotta Pereira, C., Guerra Rodrigo, F., Bittencour, T., and Sampaio, S.,** Oxytalan, elaunin and elastic fibers in the human skin, *J. Invest. Dermatol.*, 66, 143, 1986.
6. **Pieraggi, M. Th., Nejjar, I., Julian, M., and Bouissou, H.,** Staining of elastic tissue by Verhoeff's iron hematoxylin, *Ann. Pathol.*, 5, 74, 1986.
7. **Bouissou, H., Pieraggi, M. Th., Julian, M., and Douste-Blazy, L.,** Cutaneous ageing. Its relation with atherosclerosis and atheroma, *Front. Matrix Biol.*, 1, 190, 1973.
8. **Pieraggi, M. Th., Julian, M., Bouissou, H., Stocker, S., and Grimaud, J. A.,** Vieillissement dermique. Etude en immunofluorescence des collagènes I et III et de la fibronectine, *Ann. Pathol.*, 3, 185, 1984.
9. **Pieraggi, M. Th., Julian, M., Pascal, M., and Bouissou, H.,** Localisation de l'apoprotéine B dans le derme humain (étude par immunomarquages optique et electronique), *Nouv. Presse Med.*, 11, 3551, 1982.
10. **Braverman, I. and Fonferko, E.,** Studies in cutaneous aging: I. The elastic fibers network, *J. Invest. Dermatol.*, 78, 434, 1982.
11. **Pieraggi, M. Th., Julian, M., and Bouissou, H.,** Fibroblast changes in cutaneous ageing, *Virchow's Arch. Pathol. Anat.*, 402, 275, 1984.
12. **Mensing, H. and Czarnetzki, B. M.,** Leucotrien B4 induces in vivo fibroblast chemotaxis, *J. Invest. Dermatol.*, 82, 9, 1984.
13. **Saison, M., Van Leuven, F., Cassiman, S. J., and Van den Berghe, M.,** Evidence for a role of fibronectin and other cell surface component in cell layer adhesion of human fibroblasts, *Exp. Cell Res.*, 143, 237, 1983.
14. **Rennard, S. I., Yvang-Fang Chen, Robbens, R. A., Gadek, J. E., and Crystal, R. G.,** Fibronectin mediator cell attachment to Clq a mechanism for the localization of fibrosis in inflammatory diseases, *Clin. Exp. Immunol.*, 54, 239, 1983.
15. **Knapp, T. R., Daniels, J. R., and Kaplan, E. N.,** Pathologic scar formation (morphologic and biochemical correlates), *Am. J. Pathol.*, 86, 47, 1977.
16. **Baur, P. S., Baratt, G. F., Brown, G. M., and Parks, D. H.,** Ultrastructural evidence for the presence of "fibroclasts" and "myofibroclasts" in wound healing tissues, *J. Trauma*, 19, 744, 1979.
17. **Buntrock, P., Schnitzler, S., and Uerlings, I.,** Ultrastrukturelle Charakterisierung von Fibroblasten. Myofibroblasten und fibroklasten in Wundheilungsproze B, *Zentrabl. Allg. Pathol. Pathol. Anat.*, 124, 48, 1980.
18. **Gabbiani, G., Ryan, G. B., and Majno, G.,** Presence of modified fibroblasts in granulation tissue and their possible role in wound contraction, *Experientia*, 27, 549, 1971.
19. **Gabbiani, G., Le Lous, M., Bailey, A. J., Bazin, S., and Delaunay, A.,** Collagen and myofibroblasts of granulation tissue. A chemical, ultrastructural, immunologic study, *Virchows Arch. B.*, 21, 133, 1976.
20. **Kocher, O., Skalli, O., Bloom, W. S., and Gabbiani, G.,** Cytoskeleton of rat aortic smooth muscle cells. Normal conditions and experimental intimal thickening, *Lab. Invest.*, 50, 645, 1984.
21. **Bentley, J. P.,** Aging of collagen, *J. Invest. Dermatol.*, 75, 80, 1979.
22. **Szendroi, M., Meimon, G., Bakala, H., Frances, C., Robert, L., Godeau, G., and Hornebeck, W.,** On the presence of a metalloprotease in human skin fibroblasts, *J. Invest. Dermaltol.*, 83, 224, 1984.
23. **Bouissou, H., De Graeve, J., Solera, M. L., Wulfert, E., and Thiers, J. C.,** Cholesterol and skin apoprotein B in atherosclerosis, *Biomed. Pharmacother.*, 36, 159, 1982.
24. **Levene, C. I. and Heslop, J.,** The synthesis of collagen by cultured aortic and endothelium and its possible role in the pathogenesis of the fibrous atherosclerotic plaque, *J. Mol. Med.*, 2, 145, 1977.
25. **Schwartz, J. M., Reidy, A., and Hansson, G.,** Role of endothelial dysformation in vivo in smooth muscle proliferation, in *Atherosclerosis*, VI, Schetiler, G. and Giotto, A. M., Eds., Springer-Verlag, Berlin, 1983, 350.
26. **Pieraggi, M. Th. and Bouissou, H.,** Aorte humaine: endothelium et atherome, *Ann. Pathol.*, 1, 271, 1981.
27. **Gabbiani, G., Badonnel, M. C., and Rona, G.,** Cytoplasmic contractile apparatus in aortic endothelial cells of hypertensive rats, *Lab. Invest.*, 32, 227, 1975.
28. **Sendrail-Pesque, M., Julian, M., Durroux, R., Fabre, M. Th., Rumeau, J. L., and Bouissou, H.,** Le vieillissement aortique, *Pathol. Biol.*, 17, 223, 1969.

29. **Cooke, P.,** A filamentous cytoskeleton in vertebrate smooth muscle fiber, *J. Cell Biol.*, 68, 539, 1976.
30. **Goldstein, J. L. and Brown, M. S.,** The low density lipoproteins pathway and its relation to atherosclerosis, *Ann. Rev. Biochem.*, 46, 897, 1977.
31. **Adams, C. W. M.,** Lipids, lipoproteins and atherosclerotic lesions, *Proc. R. Soc. Med.*, 64, 902, 1971.
32. **Garfiel, R. E., Chacko, S., and Blose, J.,** Phagocytosis by muscle cells, *Lab. Invest.*, 33, 418, 1975.
33. **Moschetto, Y., De Backer, M., Pizieux, O., Bouissou, H., Pieraggi, M. Th., and Julian, M.,** Evolution de l'élastine d'aortes humaines au cours du vieillissement, *Paroi Arterielle/Arterial Wall*, 2, 161, 1976.
34. **Barnes, M. J.,** Collagens in atherosclerosis, *Collagen, Rel. Res.*, 5, 65, 1985.
35. **Bouissou, H., Pieraggi, M. Th., and Julian, M.,** Cellule mesenchymateuse indifférenciée, fibroblaste, myocyte et myofibroblaste, *Arch. Anat. Cytol. Pathol.*, 20, 17, 1980.
36. **Kocher, O., Skalli, O., Bloom, W. S., and Gabbiani, G.,** Cytoskeleton of rat aortic smooth muscle cells, *Lab. Invest.*, 80, 645, 1984.
37. **Klurfeld, D. M.,** Identification of foam cells in human A.T.S. lesions as macrophages using monoclonal antibodies, *Arch. Pathol. Lab. Med.*, 109, 445, 1985.
38. **Durroux, R. and Bouissou, H.,** Endothelium aortique et elastigénèse, *Pathol. Biol.*, 18, 375, 1970.
39. **Carnes, W. H., Abraham, P. A., and Buonassisi, V.,** Biosynthesis of elastin by an endothelial cell culture, *Biochem. Biophys. Res. Commun.*, 90, 1393, 1979.
40. **Bouissou, H., Pieraggi, M. Th., and Julian, M.,** L'athérome (mise au point), *Ann. Pathol.*, 1, 5, 1981.
41. **Cellina, M.,** Medianecrosis disseminata aorta, *Virchows Arch. Pathol. Anat.*, 280, 65, 1931.
42. **Haust, M. O.,** The morphogenesis and fate of potential and early atherosclerotic lesions in man, *Human Pathol.*, 2, 1, 1971.
43. **Ross, R. and Glomset, J.,** Atherosclerosis and the arterial smooth muscle cell, *Science*, 180, 1336, 1973.
44. **Doerr, W.,** *Perfusion theorie der Arteriosklerose*, Verlag Ed. Stuttgart, 1963.
45. **Rumeau, J. L., Julian, M., Coste, E., Rakotondrainibe, A., and Bouissou, H.,** Qualité de la paroi aortique et degré d'intensité de l'athérome, *Arch. Mal Coeur* (Revue de l'Athérosclérose) Supplément 1, 58, 1969.
46. **Bouissou, H., Pieraggi, M. Th., Julian, M., Buscail, I., Douste-Blazy, L., Latorre, E., and Charlet, J. P.,** Identifying arteriosclerosis and aortic atheromatosis by biopsy, *Atherosclerosis*, 19, 449, 1974.
47. **De Graeve, J., Bouissou, H., Thiers, J. C., Fouet, J., and Valdiguie, P.,** Is cutaneous apoprotein B a better discriminator than serum lipoproteins for atherosclerosis?, *Atherosclerosis*, 52, 301, 1984.
48. **Ten Cate, A. R. and Freeman, E.,** Collagen remodelling by fibroblasts in wound repair. Preliminary observations, *Anat. Rec.*, 174, 543, 1974.
49. **Uzunian, A.,** Electron microscopy of collagen resorption by fibroblasts in wound repair of albino-rat skin, *Rev. Bras. de Pesquisas Med. e Biol.*, 12, 347, 1979.
50. **Svoboda, E., Smiga, A., and De Porter, D.,** Astereologic analysis of collagen phagocytosis by fibroblasts in three soft connective tissues with differing rates of collagen turnover, *Anat. Rec.*, 199, 473, 1981.
51. **Doerr, W.,** *Perfusiontheorie der Arteriosklerose in swanglose Abhanlungen aus dem Gebiet der Normalen und Pathologischen Anatomie*, Vol. 13, Bargmann, W. and Doerr, W., Eds., Georg Thieme Verlag, Stuttgart, 1963.

Chapter 6

FIBROGENIC PROCESSES DURING TISSUE REPAIR

R. F. Diegelmann, W. J. Lindblad, and I. K. Cohen

TABLE OF CONTENTS

I. Introduction 114

II. Biochemical and Cellular Events During Dermal Wound Repair 115
- A. Inflammatory Events 115
- B. Re-establishment of the Dermal and Epidermal Collagen Matrix 117
- C. Wound Maturation 121
- D. Abnormal Dermal Repair 123

III. Collagen Matrix Formation in Response to Injury in Specific Tissues and Organs 124
- A. Tendon 124
- B. Peritoneum 124
- C. Lung 125
- D. Liver 126
- E. Intestine 126
- F. Nerves 128

IV. Investigational Systems of Tissue Repair 128
- A. Repair 128

Acknowledgments 130

References 131

I. INTRODUCTION

Most of our basic understanding of "Fibrogenic Processes During Tissue Repair" is derived from information obtained from dermal repair. Therefore, as fibrogenic processes in other tissues and organ systems are examined, it is necessary to compare these findings to the base information from dermis. While many of the same cellular and fibrogenic responses may be observed in systems other than dermis, it must also be recognized that specialized and unique mechanisms have evolved in different tissues to handle the specific requirements of that tissue.

The schema presented in Figure 1 integrates the specific processes involved in the response of tissue injury. Initiation of the sequence of events begins at the moment of injury. Platelets release their complement of potent "wound healing hormones", and in association with cellular mediators, tissue debris, active amines, and possible foreign materials, the process of inflammation is begun. As a blood clot forms at the site of injury, peripheral leukocytes are trapped in the fibrin-erythrocyte complex and contribute to the inflammation by releasing granules ladened with enzymes and trophic substances. At this time, the tissues adjacent to the central site of injury may be affected by the inflammatory processes. The architectural status of the surrounding tissue may be temporarily altered by fluid accumulation and movement of inflammatory cells through the tissue. If the damage proceeds no further and the small vessels in the area regain their normal permeability, then the temporary tissue disruption due to edema is reversed and the tissue is returned to the exact structure with restoration of function. This temporary, nondamaging inflammatory event with rapid return to normal is defined as *resolution* (Figure 1).

The next level of response to injury is the process of *regeneration*. The inflammatory events preceding regeneration are usually very minimal and are associated with phagocytosis and removal of damaged cell, tissue debris, and any foreign bodies. The cellular elements that were lost to injury are then replaced in their exact anatomic organization and complete structure and function are restored. Only the essential components of the extracellular matrix are re-established. One organ which possesses this capability is the liver which has retained the ability to regenerate to some extent. If overabundant matrix is deposited, then the process is no longer regeneration, but has now entered into the sequence of events leading to scar formation.

If a virgorous inflammatory period ensues after injury, then the mesenchymal elements are directed to begin the process of *repair* by scar deposition. This involves removal of all the damaged components followed by an active movement of collagen-producing cells into the area. As fibroblasts in the surrounding tissues move into the area of injury, other fibroblasts distant to the site begin to proliferate and follow along. The net result is an increased population of very active collagen-producing cells. As the new matrix begins to accumulate and fill the void created by the injury, the process enters a period of transitional repair. The collagen-producing cells continue to resorb and redeposit collagen in an effort to regain the structural integrity that existed prior to injury. The organization of the collagen never achieves the degree of quality of the uninjured matrix and thus the tensile strength of the scar is always weaker compared to the normal. If eventually the remodeling process allows the scar to be deposited in an arrangement which is consistent with restored function, then an optimal result is achieved and the tissue has undergone the process of *functional repair*.

An alternative to functional repair is the undesirable situation where the equilibrium between collagen synthesis and collagen remodeling is not re-established and the injury results in either overdeposition of collagen called *fibrosis* or inadequate integrity referred to as *tissue destruction*. As the various topics in this chapter are discussed, it is helpful to refer back to this schema to appreciate the interrelationship of the key processes involved in the organism's response to injury.

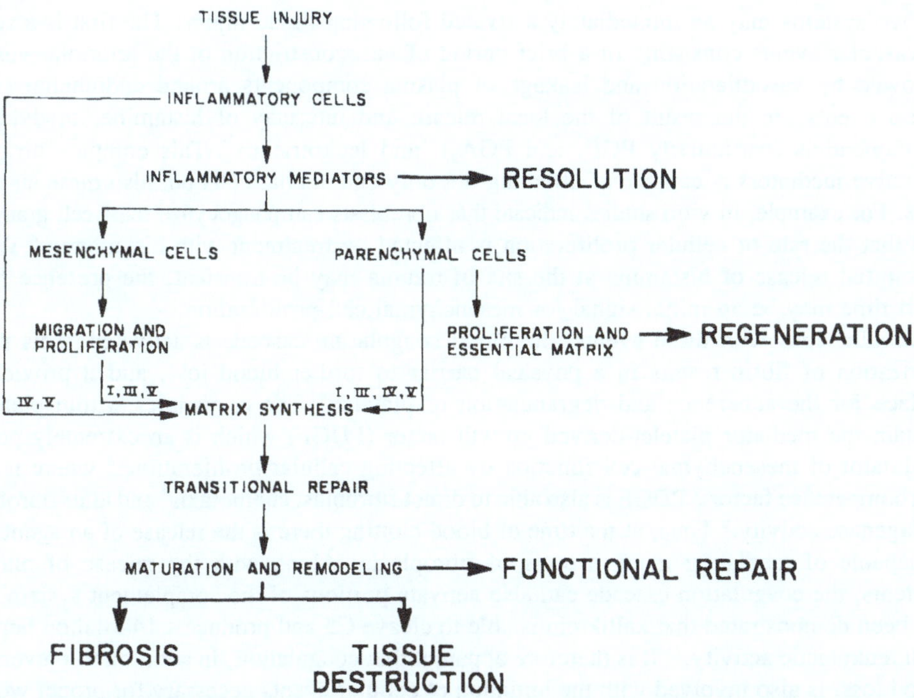

FIGURE 1. Schematic diagram of biological processes and matrix formation following tissue injury.

The critical challenge that now confronts the investigator focusing on tissue injury and tissue response is to design the appropriate model systems to answer the key questions of regulation. Although animal and cell culture systems have been essential for developing the base knowledge, modern techniques must be utilized to focus these questions on the human response to injury. Many of the fibrotic pathologies experienced by humans do not exist in animals. Animals do not develop keloids, hypertrophic scars, adhesions, strictures, cirrhosis, or Crohn's disease as do humans. Now with the availability of increasingly more sensitive techniques such as high performance liquid chromatography (HPLC), nuclear magnetic resonance (NMR), cDNA hybridization analysis, in vitro protein synthesis, high resolution ultrasound, and miniature wound implants, the investigator must be clever enough to apply these powerful technologies to the study of human tissue injury and tissue response.

II. BIOCHEMICAL AND CELLULAR EVENTS DURING DERMAL WOUND REPAIR

A. Inflammatory Events

The inflammatory response to tissue injury is generally considered a protective mechanism which prevents excessive blood loss and development of infections. However, this phase of tissue repair also initiates the long series of events culminating in the replacement of damaged tissue with a mature scar. The events of the inflammatory response are a highly interactive system which will be broken down into its components for the sake of clarity. It must be realized, however, that different processes are invoked concurrently, many times producing a multicellular response which cannot be appreciated unless viewed as a network. This section will deal only with the normal physiological progression of inflammation following injury; it will be left to subsequent sections to examine the effects of aberrant control on fibrogenesis.

Two systems may be immediately activated following tissue injury. The first is a series of vascular events consisting of a brief period of vasoconstriction of the arteriolar vessels followed by vasodilatation and leakage of plasma components around endothelial cells. These events are the result of the local release and interplay of histamine, bradykinin, prostaglandins (particularly PGE_2 and PGA_2), and leukotrienes.[1] This complex array of bioactive mediators is capable of affecting not only endothelial cells but also mesenchymal cells. For example, in vitro studies indicate that fibroblasts can phagocytize mast cell granules[2] and that the rate of cellular proliferation is affected by treatment with histamine.[3,4] While this initial release of histamine at the site of trauma may be transient, the presence for a short time may be an initial signal for mesenchymal cell proliferation.

Simultaneous with these events, the blood coagulation cascade is activated. This polymerization of fibrin results in a physical barrier to further blood loss, and it provides a surface for the adherence and degranulation of platelets. The α granules within platelets contain the mediator platelet-derived growth factor (PDGF) which is an extremely potent modulator of mesenchymal cell function by affecting cellular proliferation[5,6] where it acts as a competence factor.[7] PDGF is also able to direct fibroblast chemotaxis[8] and alter fibroblast collagenase activity.[9] Thus, at the time of blood clotting there is the release of an agent that is capable of mediating many aspects of fibroplasia. Along with the release of platelet contents, the coagulation cascade can also activate portions of the complement system.[10] It has been demonstrated that kallikrein is able to cleave C5 and produce a 14-kdalton peptide with leukotactic activity.[11] It is therefore apparent that coagulation, in addition to preventing blood loss, is also involved with the initiation of cellular events necessary for proper wound healing.

Once hemostasis has been accomplished, a series of cellular migrations commence. Initially, the clot formed at the wound site is populated by inflammatory cells present in the same proportion as that within the circulating blood. However, within a few hours polymorphonuclear leukocytes (PMN) are attracted to the damaged tissue. This leukotaxis results from the local release of a number of compounds that have been demonstrated both in vivo and in vitro to induce directed PMN migration. These mediators include leukotriene B_4,[12] a product of the lipoxygenase pathway of arachidonic acid, and C_{5a}. A potent synthetic PMN chemoattractant, *N*-formylmethionylleucylphenylalanine (fMLP), has recently been demonstrated to be effective in vivo.[13] Presumably, the fMLP moiety is analogous to, or a component of, a naturally occurring material, because this compound is so extremely potent. These mediators produce a chemical gradient through which the PMNs migrate to localize at a site where damaged tissue digestion and bacterial ingestion is required. While the PMN is not required for subsequent cell events of wound healing,[14] it serves a vital defensive role in protection against infection.

The next cell type to actively populate the wound site is the monocyte/macrophage. These cells overcome the PMN for predominance in the wound by 24 to 36 hr postinjury and consist of both tissue macrophages and monocytes recruited from the circulatory system. Peripheral blood monocytes have been shown to chemotax in vitro in response to lymphocyte-derived chemotactic factor,[15] fMLP,[16] and collagen fragments.[17] The possibility that fragments of collagen released during tissue injury and PMN digestion can direct monocytes to the appropriate site is intriguing. However, the high concentration of fragments required ($\mu g/m\ell$) to produce this effect makes it questionable if the cells are responding to this signal in vivo.

The monocyte/macrophage system consists of a complex series of maturation and differentiation events resulting in cells that produce a wide variety of cell products, as well as phagocytizing debris. The macrophage has been shown to secrete over 50 biologically active products which affect multiple cell types. Table 1 presents some macrophage/monocyte mediators and proteins that directly alter fibroblast function, modulate collagen metabolism,

Table 1
MACROPHAGE/MONOCYTE CELL PRODUCTS RELATED TO CONNECTIVE TISSUE METABOLISM

Proteolytic enzymes
Collagenase — type V specific
Elastase — cleaves type III collagen
Enzyme inhibitors
α2-Macroglobulin
Chemotactic factor
MDCF[18]
Factors modulating fibroblasts
PDGF-like factor[19]
Interleukin-1
Collagen synthetic inhibitor[20]
Prostaglandins
Extracellular matrix components
Fibronectin
Type IV collagen[21]

or are extracellular matrix components. It is apparent that this cell is capable of altering virtually every facet of fibroblast function. Recently, several findings have been reported of particular note. A PDGF-like factor was elicited from peripheral blood adherent monocytes activated with either zymosan, immune complexes, or lipopolysaccharide.[19] This material behaved similar to authentic PDGF and may be responsible for several major fibroblast-modulating activities attributed to monocytes. The macrophage also releases a material into cell supernates that specifically inhibits collagen synthesis at the transcriptional level.[20] Further study in this area is required to identify if this inhibition is the result of one or more materials present in supernates of PHA-stimulated macrophages.

The macrophage is also capable of synthesizing extracellular matrix proteins such as fibronectin and collagen.[21,22] The monocytes trapped within the blood clot at the wound site synthesize type IV collagen, thus suggesting that this cell when exposed to a wound environment is capable of laying down a functional basement membrane. This matrix could then be utilized for the migration of subsequent cells to repopulate the wound site.

The final inflammatory cell type to populate the wound bed is the lymphocyte which is composed of two major classes, i.e., B and T cells. The T cell has been examined by several groups for its ability to release lymphokines which can modulate fibroblast function. A number of studies report the presence of lymphokines which may inhibit or stimulate collagen synthesis.[23-25] One may hypothesize that it is the relative concentration of each species that determines the overall state of collagen synthetic modulation, not the absolute level of any one modulator. The fibroblast resides in a very complex milieu at this time since they are being simultaneously exposed to monocyte factors, lymphokines, and environmental conditions, e.g., altered pO_2 and pH. Thus, the fibrogenic outcome is the integration of a number of determinants of the inflammatory response. A summary of the major inflammatory events in tissue repair is presented in Table 2. As noted before, the inflammatory phase is a network of interactions that may have many redundant mechanisms and compensatory systems built in to assure the ordered influx of mesenchymal cells and adequate collagen deposition to fill the wound site. Minor perturbations to the network may be overridden by these compensatory mechanisms. However, if the perturbation is too great, or the affected system is in a pivotal position, then biologically significant abnormalities in wound healing may result.

B. Re-establishment of the Dermal and Epidermal Collagen Matrix

As mentioned in the previous section, there is a host of signals originating from a wound site that function to attract fibroblasts. Based upon information derived from in vitro in-

Table 2
SUMMARY OF MAJOR INFLAMMATORY PHASE EVENTS IN TISSUE REPAIR

Immediate response (0—12 hr)
Vascular changes resulting from release of histamine and other vasoactive mediators
Blood coagulation with resulting PDGF release
Neutrophil chemotaxis occurs
Intermediate response (12—36 hr)
Monocyte/macrophage actively populate wound
Multiple monokines and mediators released
Initial collagen deposition occurs by the monocyte
Late response (>36 hr)
Lymphokines released to modulated fibroblast activity

vestigations, some of the chemoattractants may be degradation products of the damaged connective tissue matrix.[26] Postlethwaite et al.[27] demonstrated that native human collagen types I, II, and III and bacterial collagenase-generated peptides from these collagens were chemotactic for human dermal fibroblasts. More recently, it has been reported that digestion of native collagen with mammalian collagenase releases high molecular weight peptides with approximately threefold greater chemotactic attraction for human dermal fibroblasts.[28] In addition to collagen, tropoelastin, elastin peptides,[29] and synthetic elastin peptides[30] have been shown to attract fibroblasts. In the review by Albini et al.[26] there are at least 20 distinct fibroblast attractants listed. Although each of these substances has been shown to be active for fibroblasts in the concentration range generally regarded as "physiologic", it must be emphasized there is a lack of solid data to suggest any of these substances are involved in attracting fibroblasts in vivo. A recent report by Lawman et al.[31] demonstrated that nerve growth factor at a concentration of 1 n*M* as well as fMLP at 10^{-10} *M* injected into subdermal air sacs in the backs of mice resulted in a typical inflammatory cell response followed by an infiltration of fibroblasts. These types of studies need to be pursued to enhance the base knowledge of in vivo fibroblast chemotaxis. Once the in vivo biochemical signals are identified, then perhaps analogs can be produced to better control the fibroblast response in wound repair.

Several in vitro studies have reported observations linking the macrophage and its soluble products to modulation of fibroblast proliferation.[32] These studies indicate a "bell-shaped", dose-response curve for macrophage-derived factors exists where there is stimulation at low concentrations and inhibition of fibroblast proliferation at higher doses.[33] One would like to speculate that this type of response is also present at the wound site in vivo with the macrophage directing fibroblast activity. But once again, the critical in vivo data just do not exist.

However, studies in the last several years have successfully identified the chemical nature of some of these macrophage "factors". Perhaps the two most important macrophage-elicited factors are interleukin 1[34] and, as mentioned earlier, a platelet-derived growth factor-like activity.[19,35] The PDGF-like mediator apparently functions as a competence factor acting during the G_0-G_1 phase of the cell cycle to prepare cells for entry into the S phase. In order for the cells to go forward in the S phase, progression factors such as insulin and somatomedins A and C are required.[36] Undoubtedly, there are other growth factors such as epidermal growth factor (EGF) and transforming growth factors (TGF alpha and beta) which stimulate fibroblasts to proliferate and could very well be present at the wound healing site.[37,38] As further information on the chemical identity and source of the signals which regulate fibroblast proliferation at the site of the wound is developed, then perhaps one day the pharmacologic means needed to control the process of fibrogenesis will be achieved.[39]

Collagen biosynthesis by a number of cell types involved with dermal wound healing has

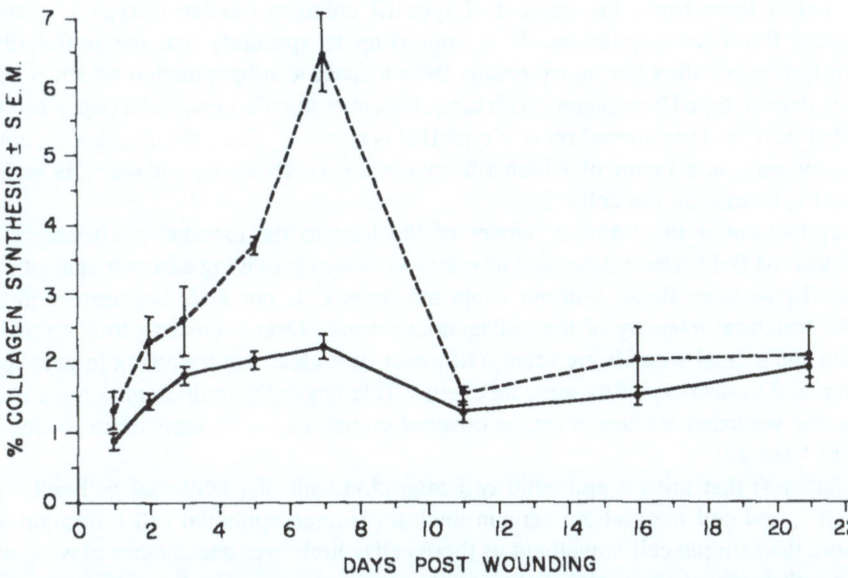

FIGURE 2. Time course of relative collagen synthesis in full-thickness rat skin wounds. Solid line represents normal skin values and dashed line represents wound values. (From Cohen, I. K., Moore, C. D., and Diegelmann, R. F., *Proc. Soc. Exp. Biol. Med.*, 160, 458, 1979. Copyright 1979 by the Society for Experimental Biology and Medicine. With permission.)

been reported. Such cells as epithelial cells,[40,41] endothelial cells,[42] and keratinocytes[43] have all been shown to synthesize different types of collagen. Notably, these cells produce type IV collagen which is an integral part of the basement membrane which underlies the basal surface of the cells. This collagen is required to reconstitute the protective epithelial cell layer and the neovasculature, but it is not the source of collagen responsible for the regain in tissue tensile strength. The fibroblast is the provenance of this latter matrix synthesis.

Early studies from this laboratory indicated that a significant increase in collagen synthesis occurs within healing dermal wounds 48 to 72 hr postinjury (Figure 2).[44] This biosynthetic activity continues to increase, presumably due to increased numbers of fibroblasts at the wound site, until reaching a maximal rate 8 days later. This value, however, represents the summation of all collagen protein synthesized by all of the different cellular sources.

The majority of this newly synthesized collagen was determined to be type I collagen, resulting presumably from fibroblasts.[45] However, tissue from a wound site only 10 hr postinjury was capable of synthesizing elevated levels of type III collagen.[45] These results are consistent with those of Gay et al. who reported that type III collagen was detectable in Cellstics® at 24 to 48 hr following implantation.[46] At that time, no type I collagen was detected using immunohistochemical techniques. There are several possibilities to explain the initial elevation in type III collagen deposition. One possibility is that the total resident fibroblast population in the wound margin is triggered by the wound environment to enhance type III collagen gene expression. While viral transformation of fibroblasts has been shown to increase the phenotypic expression of type III collagen by BALB 3T3 cells,[47] no changes in the fibroblast environment, e.g., pO_2, monokines, lymphokines, have been reported to alter the type III/type I collagen ratio.

Alternatively, the type III collagen may originate from a subpopulation of fibroblasts, triggered by the injury, to rapidly synthesize high levels of type III. It has been recently demonstrated that one can select a subpopulation of fibroblasts which have an elevated phenotypic ratio of type III/type I collagen.[48] In fact, one fibroblast subtype could synthesize approximately three times the amount of type III collagen relative to type I as could the total parent fibroblast population. It is interesting to speculate that the initial fibroblast collagen synthesis following injury results from a specific subpopulation which is triggered to rapidly deposit type III collagen. With time, the entire population of cells deposits collagen, primarily type I, and the normal ratio of type III/I is restored. Thus, the ultimate accumulation of collagen may be a factor of which fibroblasts are synthesizing collagen, as well as the control of synthesis by the cell.

The epithelium is the limiting barrier of the host to the external environment, and it protects against fluid, electrolyte, and heat loss as well as providing a barrier against bacterial invasion. Epithelium alone, without subjacent dermis, is not effective protection since it lacks the structural integrity of the collagenous dermis. Data evaluating total epithelization of human and animal wounds are scanty; however, it is clear that following injury, epithelial migration and mitosis occur as separate events. Teleologically, migration is most important because the wounded surface must be covered completely with epithelium to declare the wound as "healed".

The factor(s) that initiate epithelial cell migration, mitosis, epithelial cell-cell "contact inhibition", and cell maturation, remain unclear. Human epithelial cell migration appears to be more than simple cell movement as these cells climb over one another playing biologic leapfrog as they migrate (epiboly). Serum contains a factor, possibly derived from the platelet, which stimulates epithelial migration in tissue culture,[49] although this may be an indirect effect of stimulation of a dermal component. Collagen appears essential for epithelial migration in tissue culture as epithelial cells produce type V collagen and migration is prevented in the presence of a proline analog which inhibits epithelial collagen production.[41] Therefore, collagen appears to be an essential lattice needed for epithelial as well as fibroblast migration.

Epidermal growth may be regulated by chalones or hormones. Bullough and Laurence first described normal epithelial suppressor substances or chalones[50] which have been espoused by Marks.[51,52] Hormonal regulation has been ascribed to epipolin which is required for epithelial migration and EGF required for mitosis.[53]

Wound contraction is a potent biological force with major clinical implications. For example, contraction of hollow organs such as the common bile duct or esophagus carries a high morbidity and even mortality as luminal impingement inhibits flow of material through these conduits. Similarly, burn scar contraction results in clinical deformity of neck, extremities etc., which requires release and skin replacement for correction of the severe resultant disabilities. Conversely, wound contraction may also be efficacious as recognized by anyone who has suffered a full-thickness open skin wound and observed its "spontaneous healing". Clinically, in any open wound situation, it is wound contraction that closes the open wound and returns continuity to skin. This is the *only* way that skin can be successfully restored without a skin graft if mechanical wound closure with skin sutures is not feasible.

Understanding the mechanism of the active process of contraction still remains somewhat enigmatic, although several observations are clear. Collagen-rich granulation tissue contains fibroblast-like cells with smooth muscle architectual characteristics ("myofibroblasts").[54] When this tissue is studied in vitro, it will contract in the presence of smooth muscle contractile stimulants.[55] These cells have been observed in contracted human tissues such as Dupytyren's nodules,[56] Peyronie's plaques,[57] collagen capsules around breast implants,[58] and immobilized joints to name but a few. Although these cells have been associated with the process of human wound contraction, there is no known correlation between the quantitation of these "myofibroblasts" and the degree of wound contraction. Perhaps the "myofibroblast" is really only seen after the process of contraction has actually occurred which explains the lack of correlation between contraction and the presence of myofibroblasts. Furthermore, it is unclear if "myofibroblasts" actually exist at all as they may simply represent "modified fibroblasts". Ehrlich has raised the interesting hypothesis that it is the components of the particular collagen matrix rather than contracting cells which are of primary importance in contraction.[59] Collagen matrix from "contracting tissues" in the presence of cultured fibroblasts appears to contract more than those collagen matrices derived from "non-contracting tissues". Regardless of the mechanism, pharmacologic control of the contraction process and avoidance of *contractures* will be a major breakthrough in the management of these disabling pathologic processes.

C. Wound Maturation

The events that transpire immediately following tissue injury and during the subsequent 2 to 3 weeks represent an enhancement of systems to deposit extracellular matrix. Mesenchymal cells are stimulated to proliferate and migrate into the wound site, and once there, these cells are stimulated to secrete large quantities of collagen. At some point in time, this process must be dampened to allow the tissue to return to equilibrium where no net loss or gain in extracellular matrix occurs. The process(es) involved with this re-establishment of equilibrium is presently unknown. A number of different mechanisms will be presented dealing with different processes occurring during the wound maturation process.

The collagen fibers deposited in healing wounds are morphologically distinct from the regular collagen fibers of normal dermis[60] (Figure 3). This disordered array may result from changes in the types of crosslinks that occur during the formation of scar tissue. At 2 weeks postinury, the major dermal collagen crosslink in guinea pig is dihydroxylysinonorleucine.[61] In contrast, the normal dermis has hydroxynorleucine as the predominant crosslink. The disparity in collagen crosslink types between the scar and normal dermis diminishes with time until the two tissues are very similar to each other.[62]

The extent of crosslinking has a profound effect on the ability of collagenase to degrade the protein. A two- to threefold increase in the resistance of purified collagen to be degraded by collagenase was obtained by introducing 0.1 Schiff base crosslinks per collagen molecule.[63] While increasing the number of crosslinks may enhance the resistance for protein degradation, changing the type of crosslink may enhance the susceptibility for degradation. This could result from a greater access of enzyme to cleavage sites in the altered collagen fiber. Therefore, understanding the relationship of collagen physicochemical properties and degradation requires information on both cross-link number and type.

The dynamics of collagenase activity during wound healing are not well understood. The presence of collagenolytic activity in healing dermal wounds was demonstrated almost 20 years ago.[64] This study suggested that there is a requisite interaction between epithelium and mesenchyme that stimulates collagenolytic activity. This concept has been the basis for continued studies which have isolated an epithelial-derived cytokine that simulates fibroblasts to secrete collagenase.[65] This may represent yet another controlling mechanism during re-epithelialization and matrix alteration.

Another possible control mechanism for extracellular collagenase is the cellular proliferative rate of fibroblasts. It has been demonstrated that the specific activity of collagenase released into culture medium is cell density dependent.[66] Maximal levels of enzyme were detected only after confluence was reached. However, this has not been our experience using a rat dermal fibroblast line. These cells demonstrate a dramatic decrease ($\sim$ninefold) in collagenase released into the medium following confluence. Our findings support the hypothesis that the mesenchymal cell under a proliferative stimulus will also synthesize high levels of collagenase to reorganize the extracellular matrix and allow the cells space to divide and migrate.

Many factors control extracellular collagenase activity with a few examples presented in Table 3. Of primary importance is the net synthesis of proenzyme under the integrated influence of inflammatory and locally released mediators. Several growth factors have been demonstrated in vitro to stimulate collagenase synthesis. These mediators include PDGF[9]

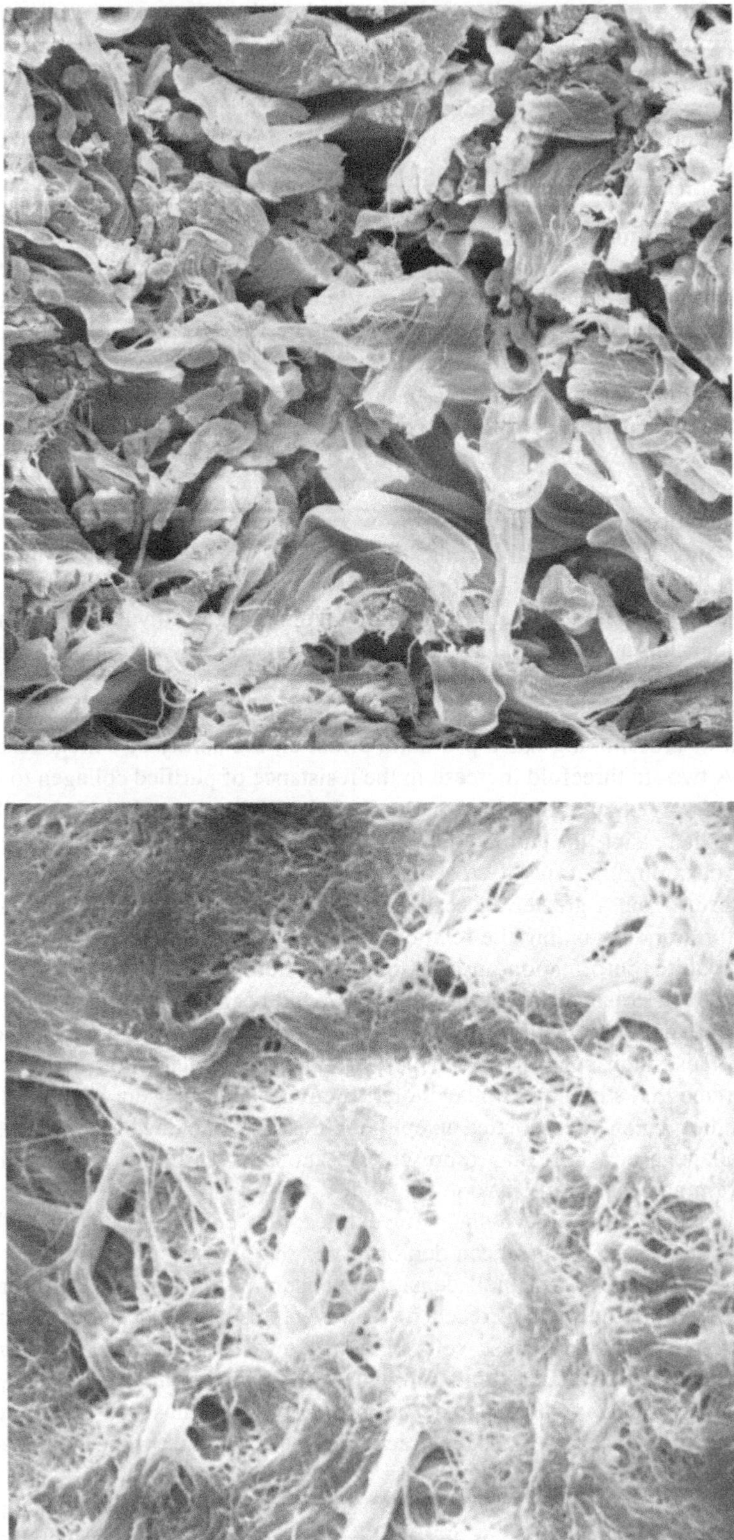

FIGURE 3. Scanning electron micrograph of normal dermis (top panel) and normal scar (lower panel). Magnification × 17,000. (Kindly provided by Dr. C. Ward Kischer, University of Arizona.)

Table 3
FACTORS MODULATING EXTRACELLULAR COLLAGENASE DEGRADATION OF MATRIX

Matrix effects
Collagen type in matrix
Degree of prolyl hydroxylation
Extent and type of collagen crosslinking
Proteoglycan interaction with collagen
Cellular synthesis/secretion
Proliferative state of the fibroblasts
Relative synthesis of procollagenase and procollagenase inhibitor
Environmental
pH
Concentration of serum protease inhibitors
Presence of procollagenase activators

and epidermal growth factor which produce a 3.5-fold increase in enzyme.[67] Many other modulators, including monocyte-derived materials, stimulate procollagenase synthesis and may further mediate the effects of the growth factor.

The re-establishment of a balanced synthesis/degradation equilibrium requires the interaction of many modulators. In this case, the biophysical properties of the matrix along with cellular mediators are probably important for the proper maturation of the scar. In determining how this process occurs, the ability to control matrix deposition during normal and abnormal wound healing will be possible.

D. Abnormal Dermal Repair

Keloid and hypertrophic scars remain an enigma to both the clinical surgeon and wound biologist.[68] Although the two entities are often confused with one another, they represent very different clinical and biological disease processes. Clinically, a keloid is red, raised, and extends beyond the boundaries of the original wound, tending to recur after excision.[69] In contrast, a hypertrophic scar, although red and raised, remains within the boundaries of the original wound and usually regresses in time.[69] Although the incidence of keloid is very low in Caucasians, hypertrophic scar is as common in Caucasions as in darker pigmented races.[70] Many hypotheses have been presented to explain the etiologies of these conditions including thyroid hormone, pigmentation, melanocyte stimulating hormone, estrogens, and sebaceum; however, none have been validated.[71]

Biochemically, both lesions are characterized by overabundant collagen deposition. Prolylhydroxylase activity from biopsies of normal skin, normal scar, hypertrophic scar, and keloid demonstrates that the enzyme has significantly greater activity in keloid than hypertrophic scar and both are significantly greater than other groups.[72] As expected, the rate of collagen synthesis, measured by ^{14}C proline uptake of tissue biopsies, shows that collagen synthesis in keloid is significantly greater than other groups.[73,74] Moreover, keloid fibroblasts grown in culture produce significantly more collagen than normal skin fibroblasts and continue to do so after multiple passages in culture.[75]

Our laboratory has reported that the ratio of types I to III collagen in keloid is similar to normal skin.[76] However, Uitto et al. have observed an overproduction of type I procollagen from keloid fibroblasts with a parallel increase in type I procollagen mRNA levels, while type III procollagen mRNA levels were unchanged.[77] These data suggest that some factor has transformed those dermal fibroblasts to a state where collagen synthesis is aberrantly controlled.

Although it is clear that fibroblasts derived from keloid tend to produce collagen at an excessive rate even after many passages in culture, the etiology of this process remains unclear. It is attractive to hypothesize that keloid formers possess a heterogeneous population of fibroblasts and there is a selection of a subpopulation of high collagen-producing fibroblasts following dermal inury. This selection process may be secondary to changes in tissue pH, pO_2, or lactate. Preliminary data from our laboratory suggest that these factors may indeed be able to select high collagen-producing fibroblast populations.

Other factors may be involved in the selection of a high collagen-producing fibroblast population. For example, immunological factors and the chronic presence of inflammatory cells associated with an immunologic response could stimulate specific fibroblast subpopulations. Several immunologic parameters have been examined in keloid patients including serum immunoglobulins, HLA phenotyping, and Clq levels, with the only positive evidence to date being an excessive deposition of IgG in keloid tissue.[78] The significance of this finding remains unclear. Another inflammatory mediatory found to be elevated in keloids is histamine[79] which suggests that keloid formation may be the result of some type of allergic reaction.

Excessive collagen deposition could also occur because of inhibition in collagenase activity. The evidence that this phenomena occurs is tenable but unsubstantiated. Collagenolytic activity has been reported in keloid tissue utilizing tissue culture medium removed daily from keloid tissue explants.[74,80] These media did not contain enzyme activity for the first 2 days of culture which suggests that there may be collagenase inhibitors ($\alpha 2$ macroglobulin and $\alpha 1$ antitrypsin) present in keloids.[81] These studies require further examination using sophisticated techniques to measure collagenase activity.

III. COLLAGEN MATRIX FORMATION IN RESPONSE TO INJURY IN SPECIFIC TISSUES AND ORGANS

A. Tendon

The tendon is a dense, collagen-rich structure which exhibits remarkable strength. When the tendon is cut or damaged, the subsequent scar that is formed at the site of injury dictates to what extent structure and thus function will be restored. The questions that have been addressed concerning tendon repair focus on the biochemical nature of the collagen deposited in the tendon scar, the source of the collagen, and how these processes are controlled.

The uninjured tendon is composed almost entirely of type I collagen, whereas the tendon scar contains type III collagen in addition to type I.[82] The type III collagen was found to persist at the site of repair for up to 14 months after injury. The continued presence of type III collagen suggests that the small fibrils characteristic of type III are responsible for the greatly reduced tensile strength at the site of the scar.

The other area of interest in tendon repair relates to the cellular origin of the collagen. There is no question that cells in the tendon sheath can very actively invade the site of injury and deposit collagen to form a scar.[83,84] Unfortunately, healing by adhesion formation may restrict tendon motion.

Studies in our laboratory have confirmed the suggestion of Lundborg that tendons can heal without adhesion formation.[85] We have shown that tendons, isolated free of all surrounding synovial elements, have a true intrinsic capacity to form collagen.[86,87] We are now confronted with the challenge to utilize this basic information to design procedures in vivo whereby the tendon fibroblasts within the tendon body can be selectively stimulated to facilitate the repair process and, at the same time, reduce the connective tissue response in the surrounding synovial compartment.

B. Peritoneum

A frequent complication of abdominal surgery is the formation of peritoneal adhesions.[88] These bands of fibrotic collagen can often lead to intestinal obstruction. To date, there is

Table 4
COLLAGEN SYNTHETIC CELLS IN ALVEOLAR STRUCTURES

Cell	Collagen type synthesized
Capillary endothelial	I, III, IV, V[a]
Epithelial	
Type I cells	I, III[b]
Type II cells	IV[c]
Interstitial mesenchymal	I, III

[a] This was determined using endothelial cells from aortic endothelial cells.[42]
[b] The ability of type I cells to produce collagen was examined using a cell line with both type I and type II epithelial cell characteristics.[97]
[c] Reference 98.

very little information concerning the type of collagen, source of collagen, and the exact mechanisms involved in this fibrotic response.

It is very likely that one factor leading to peritoneal adhesion formation is the presence of a fibrin clot which remains after abdominal surgery. Cell culture studies have demonstrated that fibroblasts will migrate along fibers of fibrin and the coarser the fibrin the faster the cells move.[89,90] Other studies have shown that fibrin not only serves as an excellent matrix for fibroblast attachment and elongation, but it also can modulate fibroblast proliferation.[91]

Another very likely stimulus for peritoneal adhesion formation is a foreign body reaction induced by contaminating surgical glove powders.[92] Talc (magnesium silicate) and starch are widely used glove powders and both animal and human studies have shown they can cause chronic inflammatory responses and granuloma formation.[93] Even after washing the surgical gloves with abundant amounts of sterile water or saline, there was still detectable talc[94] or starch.[95] This particular clinical problem represents an excellent example how a small insult such as microscopic foreign body particles can lead to a disastrous surgical complication. This is also an area where additional basic information needs to be developed so the fibrin response and foreign body reactions can be better controlled to avoid the fibrotic complication of peritoneal adhesion formation.

C. Lung

The extracellular matrix of the lung consists of 60 to 65% collagen.[96] This collagen is distributed throughout the organ, thus providing mechanical stability for the lung, and the structural architecture for the alveoli. Impairment of alveolar gaseous exchange following injury may result if a fibrotic response occurs instead of normal repair.

The alveolus contain four cell types which produce collagen (Table 4). Depending on the type of tissue injury, specific populations of cells would be expected to contribute to collagen deposition. For example, exposure of humans or experimental animals to high levels of oxygen can result in lung fibrosis. In animal studies, the type I epithelial cell and the capillary endothelial cells have been found to be particularly susceptible to cell injury.[99] Normally, as the type I cells are killed, the type II epithelial cells proliferate and migrate maintaining an epithelial layer over the interstitium resulting in little scar formation.[99] However, if the re-epithelization is slowed, the mesenchymal cells proliferate and secrete collagen into the extracellular space between the epithelial and endothelial cell layers.[100]

The involvement of inflammatory mediators in modulating the deposition of collagen following lung injury has been extensively studied. One line of pursuit has examined the possible central role of monocyte/macrophages in pulmonary fibrosis. It has been demonstrated that the alveolar macrophage can be stimulated to produce fibronectin and alveolar-macrophage-derived growth factor.[101] These two components are capable of stimulating cells to divide and progress through the cell cycle. If these cells are chronically stimulated, or spontaneously release these mediators, a progressive fibrosis may result.[102] This fibrosis would be secondary to fibroblast proliferation and not necessarily dependent on increased collagen synthesis per cell.

D. Liver

In contrast to many of the other tissues discussed in this chapter, the liver has a relatively low collagen content. Major sites of interstitial collagen (types I, III, and V) deposition have been localized to the portal tracts, central veins, and Glissen's capsule.[103] However, fine fibers of types III and V collagen may also be found along the sinusoids. Further studies have demonstrated the presence of type IV collagen within the different basement membranes and the space of Disse.[104]

Following hepatic injury, there may be a dramatic increase in the quantity of collagen deposited within areas of the lobule not normally containing collagen. This collagen alters the lobular architecture and can lead to further tissue necrosis and poor perfusion of the remaining parenchyma. The cellular origin of this collagen has been studied for numerous years with several cell types and experimental models used. The concept that hepatocytes have the capability to produce collagen in vitro has been demonstrated conclusively by our laboratory.[105,106] These hepatocytes initially synthesize type IV collagen (Figure 4B) followed by a predominance of types I and III collagen. These findings were extended to the in vivo situation where immunochemical techniques detected type IV collagen synthesis by hepatocytes following bile duct ligation (Figure 4A). These latter results have been confirmed using a carbon tetrachloride (CCl_4) model of liver injury.[107] Therefore, within the liver, deposition of collagen following injury may be a function of parenchymal cell collagen synthesis, as well as synthesis by mesenchymal cells.

Another factor that may lead to extracellular matrix accumulation during hepatic fibrosis is the alteration in collagen turnover. It was reported that following chronic CCl_4 administration, the half-life of collagen was greatly extended.[108] An immunochemical study reported that following extended CCl_4 treatment, the presence of collagenase could no longer be detected.[109] However, a subsequent biochemical analysis of hepatic collagenolytic activity using a high-specific activity substrate found progressively increasing levels of enzyme during progression of fibrosis.[110] The enzyme levels plateaued following 12 weeks of treatment, at a point where the fibrosis was irreversible. Although the enzyme levels increased, the hepatic matrix became more resistant to degradation by fibroblast collagenase.[111] This finding could explain the dichotomy of increasing hepatic collagen content in the presence of elevated levels of collagenase.

Thus, the liver may represent another form of healing response. Under normal circumstances the parenchymal cells maintain a very low level of collagen synthesis. This rate of synthesis would serve to maintain the basement membrane structures on the sinusoidal side of the hepatocyte. Once injured, the cell would be stimulated to produce reparative quantities of extracellular matrix which if not controlled could result in overdeposition and hepatic fibrosis.

E. Intestine

Crohn's disease is a debilitating fibrotic process wherein the lumen of the intestine is obstructed by stricture formation. A chronic inflammatory process always precedes the formation of the fibrotic stricture. Until recently, it was assumed that fibroblasts were attracted to the site of inflammation and were responsible for the collagen deposited in the

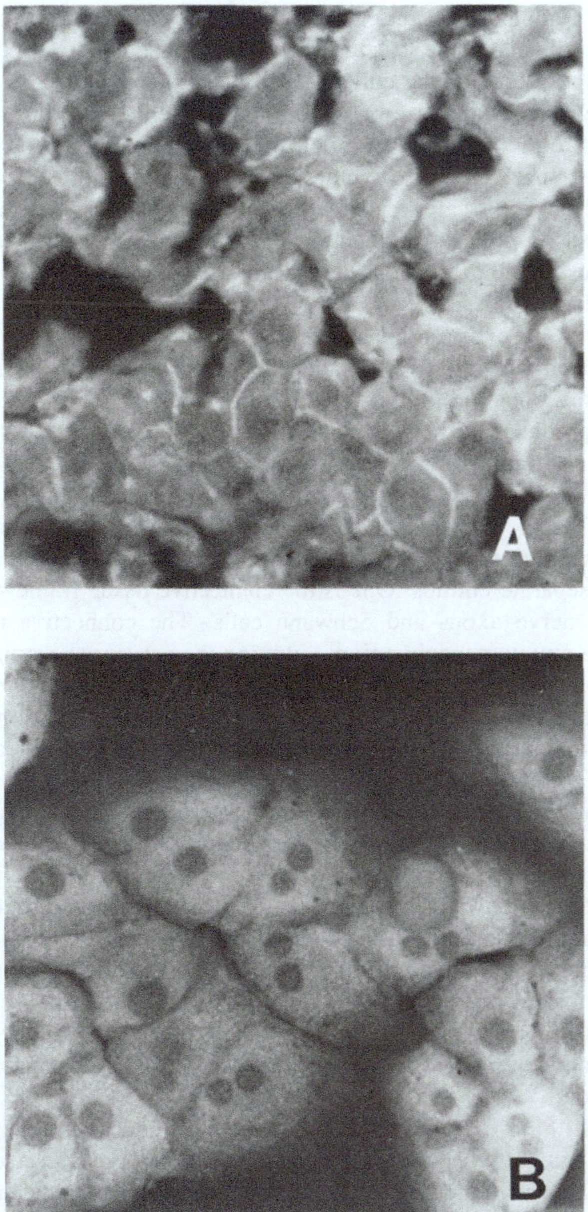

FIGURE 4. (A) A 4-μm section of pathologic rat liver specimen (after experimental bile duct ligation) stained with anitbody to type IV collagen. (B) Four-day-old primary cultures of hepatocytes isolated from regenerated rat liver and stained with antibody to type IV collagen. Positive reactivity was not observed when the pathologic liver slice (A) and the 4-day-old monolayer culture prepared from regenerated liver (B) were exposed to antibodies to collagen types I and III. (From Diegelmann, R. F. et al., *Science*, 219, 1343, 1983. Copyright 1983 by the American Association for the Advancement of Science. With permission.)

stricture. Current findings from this laboratory have now shown there is an accumulation of types IV and V collagen in the stricture[112] and support the hypothesis that intestinal smooth muscle cells are the source of the collagen. Electron microscopic studies have also shown an increase in the number and size of smooth muscle cells in regions of inflamed intestine obtained from a Crohn's patient.[113] Smooth muscle cells have been isolated from human intestine[114] and have been shown to synthesize significant quantities of collagen types I, III, and V.[144] When these cells were treated with glucocorticosteroids (up to 10^{-5} *M*) there was no inhibitory effect on collagen production.[115] Therefore, the resistance of the human intestinal smooth muscle cell to steroids may explain why Crohn's stricture continues to develop in spite of steroid therapy.

Although there is no exact animal model for the study of Crohn's disease, it has been observed that guinea pig intestine responds to physical obstruction by hypertrophy and increased collagen production by the smooth muscle cells.[116] These animal studies support the current data produced in human studies, but the animal model still lacks the crucial element of chronic inflammation.

F. Nerves

There are two distinct types of tissue involved in peripheral nerve healing which should be considered as separate entities. One is the connective tissue framework and the other is composed of the nerve axons and Schwann cells. The connective tissue component is composed of three structures: epineurium investing the nerve; perineurium, a sheath around a bundle of axons; endoneurium, surrounding each axon. These connective tissue layers normally protect the nerve from damage. If the nerve is severed, axonal repair must occur without blocking axonal flow past the site of injury which may be blocked by collagenous scar.

Distal degeneration (Wallerian) of axons may actively involve collagen-producing cells. Within 24 hr of nerve division, these changes begin with fragmentation of myelin. Eight days later, there is chemical degradation of myelin and the axon is actually fragmented by 14 days. At this time, Schwann cell numbers rapidly increase and by the second to fourth week when Schwann cell activity has declined, there is an empty endoneurial tube. Schwann cells and macrophages participate in the degradation of myelin. Unless there is successful nerve repair with axonal regeneration, these endoneurial tubes collapse and are lost among the connective tissue scar which forms. However, in clinical practice, it is of interest to note that sensory nerve repair even 3 years after complete division of the nerve may provide functional sensation.

Perhaps the first thing that must be done is to better understand the biochemical events of the connective tissue involved in nerve repair. Collagen types in nerve tissue have been defined and there appears to be distinct quantitative and qualitative differences in the various nerve connective tissue components.[117] For example, epineurium contains more type I than III collagen. This ratio is reversed in the perineurium, whereas the endoneurium contains mostly type IV, with less amounts of III and I collagens. Schwann cells have the potential to produce significant amounts of collagen composed mainly of types IV and V.[117-119]

If the hypothesis is valid that scar (collagen) prevents axonal flow and, hence, nerve function, then these biochemical characterizations may be of very important clinical significance as one may someday be able to eliminate them from the nerve wound site with pharmacologic agents (e.g., controlled latherism with local beta-aminopropionitrile (BAPN)) or site- and type-specific collagenases.

IV. INVESTIGATIONAL SYSTEMS OF TISSUE REPAIR

A. Repair

Perhaps the most critical factor to be considered when investigating tissue repair is the choice of an appropriate model system. Over the years, many animal models have been used to analyze the various aspects of wound healing. The classic work of Ross and Benditt[120] employed the guinea pig skin wound system. The linear[121] and open skin wounds[122] in the rat models have also proved to be valuable systems to analyze collagen formation following injury. Although the dog,[123] mouse,[124] and rabbit[125,126] models have been studied in the past, they are not generally used to any great extent today. The one animal model which probably provides the closest approximation to human skin is the pig model.[127] The porcine skin closely resembles human skin because of the absence of a well-developed panniculus carnosus, the sparseness of hair, a similar epidermal thickness, and the presence of a subdermal layer of fat. Unfortunately, because of cost, maintenance, and temperament, the pig model does not provide a practical laboratory wound healing model. Therefore, because of purely practical considerations rather than scientific reasons, the rat is the most widely used animal model today. Investigators employing the rat model should be aware that the rat does not require dietary Vitamin C and the distribution of leukocytes is quite different compared to humans. Therefore, caution should be exercised when interpreting data involving the role of inflammation and wound healing dynamics in the rat model.

Analyses of these animal models have provided the basis of our present-day knowledge of the very complex cellular and acellular mechanisms involved in tissue repair. By the use of light and electron microscopy, Ross and Benditt[120] first delineated the early events in guinea pig skin wounds. Animal models have also been used to define the increasing tensile strength of tissue following wounding. The work of Levenson et al.[121] demonstrated that following injury, the rat skin wound regained strength at a rapid and constant rate from day 14 to day 42. By 3 months, the wounds regained their maximal strength of approximately 80% of the strength of unwounded skin. Once again, caution should be exercised in applying data from animal studies to the human situation, since the kinetics of healing in animals are more rapid than humans.

A potential problem which is frequently overlooked when working with animal models is the problem of regional, biological differences within the animal. It has been observed that comparable wounds made at different locations on the backs of mice showed greater ^{3}H-thymidine uptake[128] and greater cell migration[129] in the anterior wounds. Repair measured as wound tensile strength has also been observed to be greater in anterior wounds compared to wounds in more caudal locations.[130] These differences may be due to regional variations in vascular supply as well as overall tissue physiology.[131] The bottom line is that careful experimental design is required when using animal models to ensure that wounds are placed exactly in the same anatomical location to prevent artifacts.

The best model to investigate in order to gain knowledge about human wound repair is the human. Wound healing studies in humans have been very limited because of the non-availability of acceptable testing procedures and sensitive analytical techniques. By the use of histology[132] and gross examination,[133] surgical investigators have obtained some limited clinical impressions concerning healing dynamics in human skin. Although some investigators have succeeded in measuring the rate of collagen synthesis in human skin biopsies by quantitating radioactive proline uptake,[134] these techniques are just not practical to use for any large-scale study. Other investigators have measured tensile strength in human skin wounds,[135] but once again, this type of study is not feasible.

An alternative approach has been to place foreign body material such as viscose cellulose sponge or polyvinyl alcohol sponge in the subcutaneous space in humans.[136,137] The sponges are then removed for hydroxyproline content determination and tensile strength measurements as indexes of collagen deposition and maturation. Unfortunately, the direct placement of these foreign materials causes a very rapid host response leading to encapsulation by fibrous scar tissue. As pointed out by the animal studies of Weiner et al.,[126] it is very difficult to discern actual cellular ingrowth from surrounding adhesions from pre-existing connective tissue. These technical problems can lead to inaccuracies when the material is analyzed for collagen deposition. In addition, because of the vigorous host reaction to the direct implantation of these foreign materials they must be surgically removed.

The latest generation of techniques that has evolved to measure human wound healing is the Cellstic device developed by Viljanto.[138] The Cellstic consists of a silicone wound drain containing a piece of viscose cellulose sponge. The Cellstic drain is placed in the wound closure and traps cells in exudate fluid as it drains from the wound. The sponge can then be removed and analyzed biochemically and histologically to determine the amount of collagen and associated proteins that have accumulated.[46] The problem with this technique is the fact that it does not measure wound healing directly but rather the cells passing from the wound. More recently, Goodson and Hunt[139] have used a small tube of expanded polytetrafluoroethylene (Gore-Tex®, Flagstaff, Arizona) to study human wound healing. The Gore-Tex® tube is placed in the subcutaneous space of the patient and removed after several days to measure accumulation of hydroxyproline. The tube can be slipped easily into place and removed by using local anesthesia and thus has overcome some of the previous problems of patient compliance. Unfortunately, the direct contact of this expanded Gore-Tex® causes a vigorous foreign body reaction with rapid accumulation of a fibrous capsule of collagen. When the tube is withdrawn, most of the newly formed connective tissue remains in the tissue and very little is available for analysis (unpublished observations).

To circumvent these technical problems, our laboratory has developed an implantable device (PVA implant) which can be used to analyze human wound healing.[140,141] The PVA implant contains a small piece of polyvinyl alcohol sponge (approximately 2.5 mg each) encased in a perforated tube of silicone sealed at both ends. The implant can be removed at various times and the material collected and analyzed for hydroxyproline accumulation, using HPLC[142] and percentage neutral salt soluble collagen.[141] Examination by light and electron microscopy to analyze cellular infiltration and collagen organization is also possible. This new procedure incorporates the advantage using a polyvinyl alcohol sponge, yet it overcomes the technical problem of implant encapsulation. The sponge elicits a foreign body reaction but the silicone prevents restricting adhesion formation. The wound insult is standardized by the placement of the same size needle each time and the implantation of very closely matched PVA implants.

In summary, the future of wound healing research depends upon the application of molecular biology techniques. For example, with the use of both in vitro protein synthesis techniques and cDNA probe analysis, it was reported that following fasting, tissues have reduced amounts of procollagen mRNA.[143] In addition, fibroblasts isolated from keloids maintain the capacity to produce increased amounts of collagen in culture,[75] and thus have been reported to be directly related to increased levels of type I procollagen mRNA.[77] The availability and use of these new technologies are essential in order to unfold the complex mechanisms involved in the regulation of connective tissue formation following injury.

ACKNOWLEDGMENTS

The work described in this chapter was supported by NIH grants GM-20298, AM-30856, AM-34151, and GM-25159. The authors are grateful to Mrs. Rae Carole Spivey and Mrs. Fay Akers for their excellent secretarial assistance and Miss Pearl Woolridge for library assistance.

REFERENCES

1. **Sedgwick, A. O. and Willoughby, D. A.,** Initiation of the inflammatory response and its prevention, in *Handbook of Inflammation*, Vol. 5, Bonta, I. L., Bray, M. A., and Parnham, M. J., Eds., Elsevier, Amsterdam, 1985, 27.
2. **Atkins, F. M., Friedman, M. M., Rao, P. V. S., and Metcalfe, D. D.,** Interactions between mast cells, fibroblasts, and connective tissue components, *Int. Arch. Allergy Appl. Immunol.*, 77, 96, 1985.
3. **Russell, J. D., Russell, S. B., and Trupin, K. M.,** The effect of histamine on the growth of cultured fibroblasts isolated from normal and keloid tissue, *J. Cell. Physiol.*, 93, 389, 1977.
4. **Topol, B. M., Lewis, V. L., and Benveniste, K.,** The use of antihistamine to retard the growth of fibroblasts derived from human skin, scar, and keloid, *Plast. Reconstr. Surg.*, 68, 227, 1981.
5. **Ross, R. and Vogel, A.,** The platelet-derived growth factor, *Cell*, 14, 203, 1978.
6. **Scher, C. D., Shepard, R. C., Antoniades, H. N., and Stiles, C. D.,** Platelet-derived growth factor and the regulation of the mammalian fibroblast cell cycle, *Biochim. Biophys. Acta*, 560, 217, 1979.
7. **Deuel, T. F. and Huang, J. S.,** Platelet-derived growth factor. structure, function, and roles in normal and transformed cells, *J. Clin. Invest.*, 74, 669, 1984.
8. **Seppä, H., Grotendorst, G., Seppä, S., Schiffmann, E., and Martin, G. R.,** Platelet-derived growth factor is chemotactic for fibroblasts, *J. Cell Biol.*, 92, 584, 1982.
9. **Bauer, E. A., Cooper, T. W., Hunag, J. S., Altman, J., and Deuel, T. F.,** Stimulation of *in vitro* human skin collagenase expression by platelet-derived growth factor, *Proc. Natl. Acad. Sci. U.S.A.*, 82, 4132, 1985.
10. **Sundsmo, J. S. and Fair, D. S.,** Relationships among the complement, kinin, coagulation, and fibrinolytic systems in the inflammatory reaction, *Clin. Physiol. Biochem.*, 1, 225, 1983.
11. **Wiggins, R. C., Giclas, P. C., and Henson, P. M.,** Chemotactic activity generated from the fifth component of complement by plasma kallikrein of rabbit, *J. Exp. Med.*, 153, 1391, 1981.
12. **Bray, M. A., Ford-Hutchinson, A. W., and Smith, M. J. H.,** Leukotriene B_4. An inflammatory mediator *in vivo*, *Prostaglandins*, 22, 213, 1981.
13. **Ranney, D. F.,** Targeted modulation of acute inflammation, *Science*, 227, 182, 1985.
14. **Simpson, D. and Ross, R.,** The neutrophilic leukocyte in wound repair. A study with anti-neutrophil serum, *J. Clin. Invest.*, 57, 2009, 1972.
15. **Altman, L. C., Snyderman, R., Hausman, M. S., and Mergenhagen, S. E.,** A human mononuclear leukocyte chemotactic factor: characterization, specificity, and kinetics of production by homologous leukocytes, *J. Immunol.*, 110, 801, 1973.
16. **Schiffman, E., Corcoran, B. A., and Wahl, S. M.,** N-formylmethionyl peptides as chemoattractants for leukocytes, *Proc. Natl. Acad. Sci. U.S.A.*, 75, 1059, 1975.
17. **Postlethwaite, A. E. and Kang, A. H.,** Collagen and collagen peptide-induced chemotaxis of human blood monocytes, *J. Exp. Med.*, 143, 1299, 1976.
18. **Diegelmann, R. F., Schuller-Levis, G., Cohen, I. K., and Kaplan, A. M.,** Identification of a low molecular weight, macrophage-derived chemotactic factor for fibroblasts, *Clin. Immunol. Immunopath.*, 41, 331, 1986.
19. **Martinet, Y., Bitterman, P. B., Mornex, J.-F., Grotendorst, G. R., Martin, G. R., and Crystal, R. G.,** Activated human monocytes express the c-sis proto-oncogene and release a mediatory showing PDGF-like activity, *Nature (London)*, 319, 158, 1986.
20. **Narayanan, A. S., Voss, T., and Page, R. C.,** Transcriptional regulation of fibroblast collagen synthesis by activated mononuclear cell supernatants, *Biochem. Biophys. Res. Commun.*, 131, 1028, 1985.
21. **Lindblad, W. J., French, J. A., Redford, K. S., Buenaventura, S. K., and Cohen, I. K.,** Induction of prolyl hydroxylase activity in a nonadherent population of human leukocytes, *Biochem. Biophys. Res. Commun.*, 147, 486, 1987.
22. **Gay, R., Gay, S., Lindblad, W. J., Diegelmann, R. F., and Cohen, I. K.,** Immunohistochemical localization of collagen types in early tissue repair, *Fed. Proc.*, 42, 1888, 1983.
23. **Rosenbloom, J., McArthur, W., Malamud, D., and Jimenez, S.,** Characterization of a lymphokine produced by human T-cells which inhibits collagen synthesis, *Cell Immunol.*, 81, 192, 1983.
24. **Duncan, M. R., Perlish, J. S., and Fleischmajer, R.,** Lymphokine/monokine inhibition of fibroblast proliferation and collagen production: role in progressive systemic sclerosis (PSS), *J. Invest. Dermatol.*, 83, 377, 1984.
25. **Postlethwaite, A. E., Smith, G. N., Mainardi, C. L., Seyer, J. M., and Kang, H. H.,** Lymphocyte modulation of fibroblast function *in vitro*: stimulation and inhibition of collagen production by different effector molecules, *J. Immunol.*, 132, 2470, 1984.
26. **Albini, A., Adelmann-Grill, B. C., and Müller, P. K.,** Fibroblast chemotaxis, *Collagen Rel. Res.*, 5, 283, 1985.
27. **Postlethwaite, A. E., Seyer, J. M., and Kang, A. H.,** Chemotactic attraction of human fibroblasts to type I, II, and III collagens and collagen-derived peptides, *Proc. Natl. Acad. Sci. U.S.A.*, 75, 871, 1978.

28. **Albini, A. and Adelmann-Grill, B. C.,** Collagenolytic cleavage products of collagen type I are chemoattractants for human dermal fibroblasts, *Eur. J. Cell Biol.*, 36, 104, 1985.
29. **Senior, R. M., Griffin, G. L., and Mecham, R. P.,** Chemotactic responses of fibroblasts to tropoelastin and elastin-derived peptides, *J. Clin. Invest.*, 70, 614, 1982.
30. **Senior, R. M., Griffin, G. L., Mecham, R. P., Wrenn, D. S., Prasad, K. U., and Usry, D. W.,** Val-Gly-Val-Ala-Pro-Gly, a repeating peptide in elastin, is chemotactic for fibroblasts and monocytes, *J. Cell Biol.*, 99, 870, 1984.
31. **Lawman, M. J. P., Boyle, M. D. P., Gee, A. P., and Young, M.,** Nerve growth factor accelerates the early cellular events associated with wound healing, *Exp. Mol. Pathol.*, 43, 274, 1985.
32. **Diegelmann, R. F., Cohen, I. K., and Kaplan, A. M.,** The role of macrophages in wound repair: a review, *Plast. Reconstr. Surg.*, 68, 107, 1981.
33. **Diegelmann, R. F., Cohen, I. K., and Kaplan, A. M.,** Effect of macrophages on fibroblast DNA synthesis and proliferation, *Proc. Soc. Exp. Biol. Med.*, 169, 445, 1982.
34. **Schmidt, J. A., Mizel, S. B., Cohen, D., and Green, I.,** Interleukin-1, a potential regulation of fibroblast proliferation, *J. Immunol.*, 128, 2177, 1982.
35. **Shimokado, K., Raines, E. W., Madtes, D. K., Barrett, T. R., Benditt, E. P., and Ross, R.,** A significant part of macrophage-derived growth factor consists of at least two forms of PDGF, *Cell*, 43, 277, 1985.
36. **Stiles, C. D., Pledger, W. J., Tuclah, R. W., Martin, R. G., and Scher, C. D.,** Regulation of the Balb/c 3T3 cell cycle — effects of growth factors, *J. Supramol. Struct.*, 13, 489, 1980.
37. **DeLarco, J. E. and Todaro, G. J.,** Growth factors from murine sarcoma virus-transformed cells, *Proc. Natl. Acad. Sci. U.S.A.*, 75, 4001, 1975.
38. **Sporn, M. B., Roberts, A. B., Shull, J. H., Smith, J. H., Ward, J. M., and Sodek, J.,** Polypeptide transforming growth factors isolated from bovine sources and used for wound healing *in vivo*, *Science*, 219, 1329, 1983.
39. **Grotendorst, G., Pencer, D., Martin, G. R., and Sodek, J.,** Molecular mediators of tissue repair, in *Soft and Hard Tissue Repair*, Hunt, T. K., Heppenstall, R. B., Pines, E., and Rovee, D., Eds., Praeger Scientific, New York, 1984, chap. 2.
40. **Gay, S., Kresina, T. F., Gay, R., Miller, E. J., and Montes, L. F.,** Immunohistochemical demonstration of basement membrane collagen in normal human skin and in psoriasis, *J. Cutan. Pathol.*, 6, 91, 1979.
41. **Stenn, K. S., Madri, J. A., and Roll, F. J.,** Migrating epidermis produces AB_2 collagen and requires continual collagen synthesis for movement, *Nature (London)*, 277, 229, 1979.
42. **Sage, H., Crouck, E., and Bornstein, P.,** Collagen synthesis by bovine aortic endothelial cells in culture, *Biochemistry*, 18, 5433, 1979.
43. **Birkedal-Hansen, H., Munksgaard, E. C., Hansen, I. L., Nellemann, K., and Gay, R.,** Multiple collagen gene expression with Type III predominance in rat mucosal keratinocytes, *Collagen Rel. Res.*, 2, 287, 1982.
44. **Cohen, I. K., Moore, C. D., and Diegelmann, R. F.,** Onset and location of collagen synthesis during wound healing in open rat skin wounds, *Proc. Soc. Exp. Biol. Med.*, 160, 458, 1979.
45. **Clore, J. N., Cohen, I. K., and Diegelmann, R. F.,** Quantitation of collagen Types I and III during wound healing in rat skin, *Proc. Soc. Exp. Biol. Med.*, 161, 337, 1979.
46. **Gay, S., Viljanto, J., Rackallio, J., and Penttinen, R.,** Collagen types in early phases of wound healing in children, *Acta Chir. Scand.*, 144, 205, 1978.
47. **Bateman, J. F. and Peterkofsky, B.,** Mechanisms of Kirsten murine sarcoma virus transformation-induced changes in the collagen phenotype and synthetic rate of BABL 3T3 cells, *Proc. Natl. Acad. Sci. U.S.A.*, 78, 6028, 1981.
48. **Bordin, S., Page, R. C., and Narayanan, A. S.,** Heterogeneity of normal human diploid fibroblasts: isolation and characterization of one phenotype, *Science*, 223, 171, 1984.
49. **Stenn, K. S.,** The role of serum in the epithelial outgrowth of mouse skin explants, *Br. J. Dermatol.*, 98, 411, 1978.
50. **Bullough, W. S. and Laurence, E. B.,** The control of epidermal mitotic activity in the mouse, *Proc. R. Soc. London*, 151, 517, 1960.
51. **Marks, F.,** Isolation of an endogenous inhibitor of epidermal DNA synthesis (G_1 chalone) from pig skin, *Hoppe-Seyler's Z. Physiol. Chem.*, 356, 1989, 1975.
52. **Marks, F.,** Epidermal growth control mechanisms, hyperplasia, and tumor promotion in the skin, *Cancer Res.*, 36, 2636, 1976.
53. **Bertsch, S. and Marks, F.,** Effect of foetal calf serum and epidermal growth factor on DNA synthesis in explants of chick embryo epidermis, *Nature (London)*, 251, 517, 1974.
54. **Gabbiani, G., Ryan, G. B., and Majno, G.,** Presence of modified fibroblasts in granulation tissue and their possible role in wound contraction, *Experientia*, 27, 549, 1971.
55. **Majno, G., Gabbiani, G., Hirschel, B. J., Ryan, G. B., and Statkov, P. R.,** Counteraction of granulation tissue *in vitro:* similarity to smooth muscle, *Science*, 173, 548, 1971.

56. **Gabbiani, G. and Majno, G.,** Dupuytren's contracture: fibroblast contraction? An ultrastructural study. *Am. J. Pathol.*, 66, 131, 1972.
57. **Somers, K. D., Dawson, D. M., Wright, G. L., Jr., Leffell, M. S., Rowe, M. J., Bluemink, G. G., Vande Berg, J. S., Gleischman, S. H., Devine, C. J., Jr., and Horton, C. E.,** Cell culture of Peyronie's disease plaque and normal penile tissue, *J. Urol.*, 127, 585, 1982.
58. **Rudolph, R., Abraham, J., Vecchione, T., Guber, S., and Woodward, M.,** Myofibroblasts and free silicone around breast implants, *Plast. Reconstr. Surg.*, 62, 185, 1978.
59. **Ehrlich, H. P.,** The role of connective tissue matrix in hypertrophic scar contracture, in *Soft and Hard Tissue Repair: Biological and Clinical Aspects*, Hunt, T. K., Heppenstall, R. B., Pines, E., and Rovee, D., Eds., Praeger Press, New York, 1984, chap. 28.
60. **Forrester, J. C., Hunt, T. K., Hayes, T. L., and Pease, R. F. W.,** Scanning electron microscopy of healing wounds, *Nature (London)*, 221, 373, 1969.
61. **Forrest, L., Shuttleworth, A., Jackson, D. S., and Mechanic, G. L.,** A comparison between the reducible intermolecular crosslinks of the collagens from mature dermis and young dermal scar tissue of the guinea pig, *Biochem. Biophys. Res. Commun.*, 46, 1776, 1972.
62. **Barnes, M. J., Morton, L. F., Bailey, A. J., and Bennett, R. C.,** Studies on collagen synthesis in the mature dermal scar in the guinea pig, *Biochem. Soc. Trans.*, 3, 912, 1975.
63. **Vater, C. A., Harris, E. D., and Siegel, R. C.,** Native cross-links in collagen fibrils induce resistance to human synovial collagenase, *Biochem. J.*, 181, 639, 1979.
64. **Grillo, H. C. and Gross, J.,** Collagenolytic activity during mammalian wound repair, *Dev. Biol.*, 15, 300, 1967.
65. **Johnson-Wint, B. and Baver, E. A.,** Stimulation of collagenase synthesis by a 20,000-dalton epithelial cytokine, *J. Biol. Chem.*, 260, 2080, 1985.
66. **Bauer, E. A.,** Cell culture density as a modulator of collagenase expression in normal human fibroblast cultures, *Exp. Cell Res.*, 107, 269, 1977.
67. **Chua, C. C., Geiman, D. E., Keller, G. H., and Ladda, R. L.,** Induction of collagenase secretion in human fibroblast cultures by growth promoting factors, *J. Biol. Chem.*, 260, 5213, 1985.
68. **Murray, J. C., Pollack, S. V., and Pinnell, S. R.,** Keloids: a review, *J. Am. Acad. Dermatol.*, 4, 461, 1981.
69. **Peacock, E. E., Jr., Madden, J. W., and Trier, W. C.,** Biologic basis for the treatment of keloids and hypertrophic scars, *South. Med. J.*, 63, 755, 1970.
70. **Alhady, S. M. A. and Sevanantharajah, K.,** Keloids in various races, *Plast. Reconstr. Surg.*, 44, 564, 1969.
71. **Crikelair, G. F., Ju, D. M. C. and Cosman, B.,** Scars and keloids, in *Reconstructive Plastic Surgery*, Converse, J. M., Ed., W. B. Saunders, Philadelphia, 1977, 413.
72. **Cohen, I. K., Keiser, H. R., and Sjoerdsma, A.,** Collagen synthesis in human keloid and hypertrophic scar, *Surg. Forum*, 22, 488, 1971.
73. **Craig, R. D. P., Schofield, J. D., and Jackson, S. S.,** Collagen biosynthesis in normal human skin, normal and hypertrophic scar and keloid, *Eur. J. Clin. Invest.*, 5, 69, 1975.
74. **Cohen, I. K., Diegelmann, R. F., and Keiser, H. R.,** Collagen metabolism in keloid and hypertrophic scar, in *The Ultrastructure of Collagen*, Longacre, J. J., Ed., Charles C Thomas, Springfield, Ill., 1976, 199.
75. **Diegelmann, R. F., Cohen, I. K., and McCoy, B. J.,** Growth kinetics and collagen synthesis of normal skin, normal scar and keloid fibroblasts *in vitro*, *J. Cell. Physiol.*, 98, 341, 1979.
76. **Clore, J. N., Cohen, I. K., and Diegelmann, R. F.,** Quantitative assay of types I and III collagens synthesized by keloid biopsies and fibroblasts, *Biochim. Biophys. Acta*, 586, 384, 1979.
77. **Uitto, J., Perejda, A. J., Abergel, R. P., Chu, M. L., and Ramirez, F.,** Altered steady-state ratio of type I/III procollagen mRNAs correlates with selectively increased type I procollagen biosynthesis in cultured keloid fibroblasts, *Proc. Natl. Acad. Sci., U.S.A.*, 82, 5935, 1985.
78. **Cohen, I. K., McCoy, B. J., Mohanakumar, T., and Diegelmann, R. F.,** Immunoglobulin, complement, and histocompatibility antigen studies in keloid patients, *Plast. Reconstr. Surg.*, 63, 689, 1979.
79. **Cohen, I. K., Beaven, M. A., Horakova, Z., and Keiser, H. R.,** Histamine and collagen synthesis in keloid and hypertrophic scar, *Surg. Forum*, 23, 509, 1972.
80. **McCoy, B. J. and Cohen, I. K.,** Collagenase in keloid biopsies and fibroblasts, *Connect. Tissue Res.*, 9, 181, 1982.
81. **Diegelmann, R. F., Bryant, C. P., and Cohen, I. K.,** Tissue alpha globulins in keloid formation, *Plast. Reconstr. Surg.*, 59, 418, 1977.
82. **Williams, I. F., McCullagh, K. G., and Silver, I. A.,** The distribution of Types I and III collagen and fibronectin in the healing equine tendon, *Connect. Tissue Res.*, 12, 211, 1984.
83. **Potenza, A. D.,** Tendon healing within the flexor digital sheath in the dog, *J. Bone Joint Surg.*, 44A, 49, 1962.

84. **Peacock, E. E., Jr.,** Biological principles in the healing of long tendons, *Surg. Clin. N. Am.*, 45, 461, 1965.
85. **Lundborg, G.,** Experimental flexor tendon healing without adhesion formation — a new concept of tendon nutrition and intrinsic healing mechanisms, *Hand*, 8, 235, 1976.
86. **Becker, H., Graham, M. F., Cohen, I. K., and Diegelmann, R. F.,** Intrinsic tendon cell proliferation in tissue culture, *J. Hand Surg.*, 6, 616, 1981.
87. **Graham, M. F., Becker, H., Cohen, I. K., Merritt, W., and Diegelmann, R. F.,** Intrinsic tendon fibroplasia: documentation by *in vitro* studies, *J. Orthopaed. Surg.*, 1, 251, 1984.
88. **Ellis, H.,** The cause and prevention of postoperative intraperitoneal adhesions, *Surg. Gynecol. Obstet.*, 133, 497, 1971.
89. **Willmer, E. N., Ed.,** Morphological problems of cell type, shape and identification, in *Cells and Tissues in Culture*, Vol. I, Academic Press, New York, 1965, 150.
90. **Roka, L. and Bleyl, H.,** Fibrinogen and seine umbauprodukte, *Fortschr. Hamatol.*, 3, 195, 1965.
91. **Pohl, J. and Christophers, E.,** Growth characteristics of skin fibroblasts and 3T3 cells entrapped by polymerizing fibrin, *In Vitro*, 15, 624, 1979.
92. **Tolbert, T. W. and Brown, J. L.,** Surface powders on surgical gloves, *Arch. Surg.*, 115, 729, 1980.
93. **Sheikh, K. M. A., Duggal, K., Relfson, M., Gignac, S., and Rowden, G.,** An experimental histopathologic study of surgical glove powders, *Arch. Surg.*, 119, 215, 1984.
94. **Wilson, D. F. and Garach, V.,** Surgical glove starch granuloma, *Oral Surg.*, 51, 342, 1981.
95. **Jagelman, D. G. and Ellis, H.,** Starch intraperitoneal-adhesion formation, *Br. J. Surg.*, 60, 111, 1973.
96. **Bradley, K. H., Breul, S., and Crystal, R. G.,** Collagen in the human lung: composition and quantitation of rates of synthesis, *J. Clin. Invest.*, 55, 543, 1975.
97. **Fulmer, J., Elson, N., Bradley, K., Ferrans, V., and Crystal, R. G.,** Comparison of type-specific collagen synthesized by lung epithelial and mesenchymal cells, *Clin. Res.*, 503A, 1977.
98. **Sage, H., Farin, F. M., Striker, G. E., and Fisher, A. B.,** Granular pneumocytes in primary culture secrete several major components of the extracellular matrix, *Biochemistry*, 22, 2148, 1983.
99. **Bowden, D. H. and Adamson, J. Y. R.,** Reparative changes following pulmonary cell injury: ultrastructural, cytodynamic, and surfactant studies in mice after oxygen exposure, *Arch. Pathol.*, 92, 279, 1971.
100. **Adamson, J. Y. R. and Bowden, D. H.,** Pulmonary injury and repair: organ culture studies of murine lung after oxygen, *Arch. Pathol. Lab. Med.*, 100, 640, 1976.
101. **Rennard, S. I., Bitterman, P. B., and Crystal, R. G.,** Mechanisms of fibrosis, *Am. Rev. Respir. Dis.*, 130, 492, 1984.
102. **Bitterman, P. B., Adelberg, S., and Crystal, R. G.,** Mechanisms of pulmonary fibrosis: spontaneous release of the alveolar macrophage derived growth factor in the interstitial lung disorders, *J. Clin. Invest.*, 72, 1801, 1983.
103. **Biempica, L., Morecki, R., Wu, C. H., Giambrone, M-A., and Rojkind, M.,** Immunocytochemical localization of Type B collagen, *Am. J. Pathol.*, 98, 591, 1980.
104. **Hahn, E., Wick, G., Pencer, D., and Timpl, R.,** Distribution of basement proteins in normal and fibrotic human liver: collagen Type IV, laminin, and fibronectin, *Gut*, 21, 63, 1980.
105. **Guzelian, P. S., Qureshi, G. D., and Diegelmann, R. F.,** Collagen synthesis by the hepatocyte: studies in primary cultures of parenchymal cells from the adult rat liver, *Collagen Rel. Res.*, 1, 83, 1981.
106. **Diegelmann, R. F., Guzelian, P. S., Gay, R., and Gay, S.,** Collagen formation by the hepatocyte in primary monolayer culture *in vivo*, *Science*, 219, 1343, 1983.
107. **Martinez-Hernandez, A.,** The hepatic extracellular matrix. II. Electron immunohistochemical studies in rats with CCl_4-induced cirrhosis, *Lab. Invest.*, 53, 166, 1985.
108. **Hutterer, F., Eisenstadt, M., and Rubin, E.,** Turnover of hepatic collagen in reversible and irreversible fibrosis, *Experientia*, 26, 244, 1970.
109. **Montfort, I. and Perez-Tamayo, R.,** Collagenase in experimental carbon tetrachloride cirrhosis of the liver, *Am. J. Pathol.*, 92, 411, 1978.
110. **Lindblad, W. J. and Fuller, G. C.,** Hepatic collagenase activity during carbon tetrachloride induced fibrosis, *Fundam. Appl. Toxicol.*, 3, 34, 1983.
111. **Lindblad, W. J. and Fuller, G. C.,** An improved method for the quantitation of mammalian collagenase activity and its use to determine hepatic extracellular matrix susceptibility to degradation, *Clin. Chem.*, 28, 2134, 1982.
112. **Graham, M. F., Elson, C. O., Diegelmann, R. F., Gay, S., and Gay, R.,** Abnormal accumulation of basement membrane (Type IV) and cytoskeletal (Type V) collagens in the strictures of Crohn's disease: the probable role of intestinal smooth muscle cells, *Gastroenterology*, 86, 1096, 1984.
113. **Dvorak, A. M., Osage, J. E., Monohan, R. A., and Dickersin, G. R.,** Crohn's disease: transmission electron microscopic studies. III. Target tissues. Proliferation of and injury to smooth muscle and the autonomic nervous system, *Hum. Pathol.*, 11, 620, 1980.
114. **Graham, M. F., Diegelmann, R. F., Elson, C. O., Bitar, K. N., and Ehrlich, H. P.,** Isolation and culture of human intestinal smooth muscle cells, *Proc. Soc. Exp. Biol. Med.*, 1976, 503, 1984.

115. **Drucker, D. E. M., Graham, M. F., Diegelmann, R. F., and Elson, C. O.,** The high rate of collagen synthesis by human intestinal smooth muscle (HISM) cells is resistant to inhibition by corticosteroids, *J. Cell Biol.*, 101, 95A, 1985.
116. **Gabella, G.,** Hypertrophic smooth muscle. V. Collagen and other extracellular materials, vascularization, *Cell Tissue Res.*, 235, 275, 1984.
117. **Gay, R. E., Gay, S., and Jones, R. E., Jr.,** Histological and immunohistological identification of collagen in basement membranes of Schwann cells of neurofibromas, *Am. J. Dermatopathol.*, 5, 317, 1983.
118. **Church, R. L., Tanzer, M. L., and Pfeiffer, S. E.,** Collagen and procollagen production by a clonal line of Schwann cells, *Proc. Natl. Acad. Sci. U.S.A.*, 70, 1943, 1973.
119. **Carey, D. J., Eldridge, C. F., Cornbrooks, C. J., Timpl, R., and Bunge, R. P.,** Biosynthesis of Type IV collagen by cultured rat Schwann cells, *J. Cell Biol.*, 97, 473, 1983.
120. **Ross, R. and Benditt, E. P.,** Wound healing and collagen formation. I. Sequential changes in components of guinea pig skin wounds observed in the electron microscope, *J. Biophys. Biochem.*, 11, 677, 1961.
121. **Levenson, S. M., Geever, E. F., Crowley, L. V., Oates, J. F., III, Bernard, C. W., and Rosen, H.,** The healing of rat skin wounds, *Ann. Surg.*, 161, 293, 1965.
122. **Stein, H. D. and Keiser, H. R.,** Collagen metabolism in granulating wounds, *J. Surg. Res.*, 11, 277, 1971.
123. **Botsford, T. W.,** The tensile strength of sutured skin wounds during healing, *Surg. Gynecol. Obstet.*, 72, 690, 1941.
124. **Benveniste, K. and Thut, P.,** The effect of chronic alcoholism on wound healing, *Proc. Soc. Exp. Biol. Med.*, 166, 568, 1981.
125. **Sandblom, P. and Muren, M. K.,** Differences between the rate of healing of wounds inflicted with short time interval, *Ann. Surg.*, 110, 449, 1951.
126. **Wiener, S. L., Wiener, R., Platt, N., Urivetzky, M., and Meilman, E.,** The need for modification of the polyvinyl sponge model of connective tissue growth: histologic and biochemical studies in the rabbit, *Connect. Tissue Res.*, 3, 213, 1975.
127. **Wachtel, T. L., Shuck, J. M., Eaton, R. P., Schade, D., and Shuck, L. W.,** Glucagon, insulin and glucose relationships in a porcine experimental burn model, *J. Surg. Res.*, 24, 70, 1978.
128. **Kobayashi, K.,** Regional differences in mitotic activity due to injury in mouse skin, *Cell Tissue Res.*, 175, 319, 1976.
129. **Thevenet, A. and Sengel, P.,** Modalitiés de la cicatrisation der flanc d'embryons de poulet de 5 jours, *Acad. Sci. (Paris)*, 276, 3379, 1973.
130. **Kullander, S. and Olsson, A.,** On the tensile strength of healing cutaneous wounds in pregnant rats, *Acta Endocrinol.*, 41, 314, 1962.
131. **Arey, L. B.,** Wound healing, *Physiol. Rev.*, 16, 327, 1936.
132. **Ross, R. and Odland, G.,** Human wound repair. II. Inflammatory cells, epithelial-mesenchymal interrelations, and fibrogenesis, *J. Cell Biol.*, 39, 152, 1968.
133. **Pories, W. J., Henzel, J. H., Rob, C. G., and Strain, W. H.,** Acceleration of wound healing in man with zinc sulphate given by mouth, *Lancet*, 1, 121, 1967.
134. **Uitto, J.,** A method for studying collagen biosynthesis in human skin biopsies in vitro, *Biochim. Biophys. Acta*, 201, 438, 1970.
135. **Lindstedt, E. and Sandblom, P.,** Wound healing in man: tensile strength of healing wounds in some patient groups, *Ann. Surg.*, 181, 842, 1975.
136. **Viljanto, J.,** A sponge implantation method for testing connective tissue regeneration in surgical patients, *Acta Chir. Scand.*, 135, 297, 1969.
137. **Peacock, E. E., Jr. and Madden, J. W.,** Administration of beta-aminopropionitrile to human beings with urethral strictures: a preliminary report, *Am. J. Surg.*, 136, 600, 1978.
138. **Viljanto, J.,** Cellstic: a device for wound healing studies in man. Description of the method, *J. Surg. Res.*, 20, 115, 1976.
139. **Goodson, W. H., III and Hunt, T. K.,** Development of a new miniature method for the study of wound healing in human subjects, *J. Surg. Res.*, 33, 394, 1982.
140. **Diegelmann, R. F., Lindblad, W. J., and Cohen, I. K.,** Measurement of collagen deposition following standardized subcutaneous injury, *J. N.Y. Acad. Sci.*, 460, 419, 1986.
141. **Diegelmann, R. F., Lindblad, W. J., and Cohen, I. K.,** A subcutaneous implant for wound healing studies in humans, *J. Surg. Res.*, 40, 229, 1986.
142. **Lindblad, W. J. and Diegelmann, R. F.,** Quantitation of hydroxyproline isomers in acid hydrolysates by high-performance liquid chromatography, *Anal. Biochem.*, 138, 390, 1984.
143. **Spanheimer, R. G. and Peterkofsky, B.,** A specific decrease in collagen synthesis in acutely fasted, vitamin C-supplemented, guinea pigs, *J. Biol. Chem.*, 260, 3955, 1985.
144. **Graham, M. F., Drucker, D. E. M., Diegelmann, R. F., and Elson, C. O.,** Collagen synthesis by human intestinal smooth muscle cells, *Gastroenterology*, 92, 400, 1987.

Chapter 7

THE LIVER AS A BIOECOLOGICAL SYSTEM: MODIFICATIONS DURING REGENERATION AND REPAIR

Marcos Rojkind and Marisabel Mourelle

TABLE OF CONTENTS

I. Introduction 138

II. Liver Composition 138
- A. Extracellular Matrix 138
  - 1. Collagens 138
    - a. Type I Collagen 141
    - b. Type III Collagen 141
    - c. Type IV Collagen 142
    - d. Type V Collagen 142
    - e. Type VI Collagen 143
  - 2. Noncollagenous Proteins 143
    - a. Laminin 143
    - b. Fibronectin 143
    - c. Other Glycoproteins 143
  - 3. Glycosaminoglycans and Proteoglycans 143

III. Metabolism of Connective Tissue Components 144
- A. Collagen Synthesis 145
- B. Collagen Degradation 146
- C. Collagen Half-Life 146
- D. Biosynthesis of Noncollagenous Components 146

IV. Modifications of the Extracellular Matrix During Regeneration and Repair 146
- A. Collagen Half-Life 148
- B. Collagenolytic Activity 148

V. Cell-Cell and Cell Matrix Interactions 148

VI. Diagnostic Procedures 151

VII. Treatment of Liver Cirrhosis 151

VIII. The Liver as a Bioecosystem 151
- A. Data in Support of the Ecosystem Hypothesis 152
- B. What are the Mechanisms Involved in Maintaining Homeostasis? 152

References 153

I. INTRODUCTION

With the development of simple techniques for the isolation and culture of liver cells (all types), it has been possible to study the role of individual components of the connective tissue in maintaining the phenotype of cultured hepatocytes. Furthermore, the interactions between the different cells that compose the liver can be explored in order to determine the normal cell-cell and cell-matrix interactions which may be essential for maintaining the normal liver architecture and function.

In this chapter we shall describe the composition of cells and matrix of normal rat liver as well as the changes that occur during regeneration and fibrosis. Finally, a speculative model of the liver as a bioecosystem will be presented, based on our current knowledge of cell-cell and cell-matrix interation and their alterations in chronic liver diseases leading to fibrosis and cirrhosis.[1-12]

II. LIVER COMPOSITION

The liver is composed of several cell types which are embedded in a complex extracellular matrix. In general, the bulk of the tissue is made up of cells, and only a relatively small amount is represented by the extracellular matrix. Approximately 80% of the liver volume is occupied by cells and 20% by the elements of the extracellular space.[13-15] The hepatocytes account for 92% of total cell volume, whereas, the sum of endothelial, Kupffer, and Ito cells correspond to 8%. The extracellular space contains all the components of the extracellular matrix and includes the different spaces that connect the liver with other systems (blood vessels, lymphatics, and bile ducts) (Table 1).

The cells within the liver have unique features. The endothelial cells are fenestrated and there is no basement membrane between the sinusoids and the plasma membrane of the hepatocytes.[16] The hepatocytes are not homogeneous with respect to their metabolic activities, and different cells within the hepatic acinus (zone 3 vs. zone 1) are performing different functions.[17,18] The Kupffer cells sit inside the sinusoids but extend projections that allow them to interact and anchor on the surface of the endothelial cells. In some regions the projections appear to go through the fenestrations.[19] The Ito cells, are small cells that store Vitamin A and are located inside the Disse space in the junctions formed between the endothelial cells and the hepatocytes.[20]

A. Extracellular Matrix

Approximately 97% of the extracellular space corresponds to the space of Disse and the sinusoidal lumen. Both are interconnected with each other through the fenestrations of the endothelial cells. Because of the free flow of macromolecules between sinusoids and Disse space, there is no permeability barrier, and except for the cells (under normal conditions), all the elements present in plasma reach the sinusoidal surface of the hepatocytes.

The extracellular matrix is composed of a variety of macromolecules which include collagens, noncollagenous glycoproteins, glycosaminoglycans (GAGs), and proteoglycans. Water, electrolytes, and export proteins are flow-through elements and could be present in trace amounts.

1. Collagens

A complete analysis of liver collagens has been performed in human, rat, and mouse liver.[21-28] Although there are important qualitative and quantitative differences, the four normal genetic types of collagen identified in human liver, mainly types I, III, IV, and V have been detected also in rat and mouse livers. In human liver, the concentration of collagen is 5.5 ± 1.6 mg/g of fresh tissue[23] and in the rat the concentration is 0.91 ± 0.15[29] (Table 2). In mouse liver, collagen concentration is similar to that found in human.[25,30]

Table 1
PERCENT VOLUME COMPOSITION OF THE LIVER[a]

	Percent composition	Approximate number of cells/g fresh tissue
Cells	84[b]—78[c]	
Parenchymal (hepatocytes)	78—72	120 × 10[6] [e]
Nonparenchymal	6.3—5.8	
Endothelial	2.8—2.6	42 × 10[6] [f]
Kupffer	2.1—2.0	12 × 10[6] [f]
Fat-storing	1.4—1.3	
Extracellular Space	16—22	
Disse	4.9—N.D.[g]	
Sinusoidal lumen	10.6—N.D.	
Biliari canaliculi	0.4—N.D.	
Collagens	0.1[d]	

[a] Taken from Ref. 4.
[b] Taken from Ref. 14.
[c] Taken from Ref. 15.
[d] Calculated from hydroxyproline values.[29]
[e] Calculated from the relative volumes as determined by Blouin et al.[14]
[f] Taken from Ref. 183.
[g] N.D., not determined.

Table 2
COLLAGEN TYPES IN NORMAL AND CIRRHOTIC LIVER[a]

	Human[b] (mg/g)				Rat[c] (mg/g)			
	Normal	% of total	Cirrhotic	% of total	Normal	% of total	Cirrhotic	% of total
Total collagen	5.5 ± 1.6	—	30.0	—	0.91 ± 0.15	—	8.0	
Type I	2.0	36.0	15.0	50.0	0.40	43.9	3.2	40.0
Type I-trimer	0.05	0.9	0.9	3.0	0.03	3.2	0.3	3.7
Type III	2.0	36.0	8.0	26.0	0.40	43.9	3.4	42.5
Type IV	0.5	9.0	7.0	23.0	0.06	6.5	{ 0.8	10.0
Type V	0.9	16.3	—	—	0.05	5.5		
Ratio III/I + III	47.3[d] 54.2[f]		37.5[d] 39.6[f]		41.3[e] 51.1[g]		37.5[e] 30.4[g]	

[a] Modified from Rojkind and Ponce-Noyola.[4]
[b] Values are average and were obtained from data by Seyer et al.[22] and Rojkind et al.[23] and were calculated assuming that a normal liver contains 5.5 mg of collagen/g and a cirrhotic liver contains 30 mg of collagen/g.
[c] Values are average and were obtained from data by Seyer[24] and Rojkind and Rojkind,[27] calculated assuming that a normal liver contains 0.91 mg of collagen/g and a cirrhotic liver contains 8 mg of collagen/g.
[d] Taken from Seyer et al.[22] Ratios were determined by the CNBr method.
[e] Taken from Seyer.[24] Ratios were determined by the CNBr method.
[f] Taken from Rojkind et al.[23] Ratios were determined after solubilization of collagens with pepsin.
[g] Taken from Rojkind and Rojkind.[27] Values were obtained after solubilization of collagens with pepsin.

The quantitative analysis of the different genetic types of collagen in the liver and other parenchymatous tissues has been hampered, in part, by the small concentration of collagen, its great insolubility, the methods used for isolation and quantitation, and the polymorphism and heterogeneity of its molecules. The first problem has been partially solved with the development of a method that permits the preparation of intact connective tissue fibers

(Biomatrix) from many parenchymatous tissues by a relatively simple procedure.[31] The fibers obtained contain 90 to 95% of tissue collagen, small amounts of noncollagenous glycoproteins, and glycosaminoglycans. Of the total protein content of liver biomatrix, approximately 60% corresponds to collagen and 40% to the noncollagenous proteins.

Two methods have been used for solubilization and quantitation of liver collagens: pepsin digestion or cyanogen bromide (CNBr) solibilization of the connective tissue remaining after removing part of the noncollagenous proteins.[21-28] Of the two methods used, CNBr digestion of the liver renders the highest yields of solubilized collagen (90 to 95%).[22] The mixture of CNBr-peptides can be analyzed by carboxymethyl (CM)-cellulose and agarose chromatography. Although the recovery of peptides from CM-cellulose is only of the order of 70 to 80%, when calibration mixtures of type I and type III collagens are digested with CNBr, and the specific CB-8 peptides derived from each collagen are estimated, the correct ratios are obtained.[22] The disadvantage of the CNBr method is that it cleaves all the collagen types into peptides, and collagens other than I and III cannot be analyzed quantitatively. Furthermore, the presence of the trimer of [α1(I)] which is found in rat and human liver in 3% concentration could bias the determination of type I collagen.[23] Recently, a method has been developed for the analysis of collagen types in samples containing less than 100 mg of tissue. Collagen is quantitatively solubilized with CNBr after removing all the noncollagenous proteins with a detergent. The peptide mixutre is then resolved by SDS-polyacrylamide gel electrophoresis, and the ratio of type I/III collagens is calculated from the densitometric tracings of Coomassie blue stained gels.[32]

Solubilization of collagen with pepsin also has its problems. The types and yields of collagens obtained depend on the amount and time of pepsin digestion. It has been suggested that types IV and V are rendered soluble after a single incubation with pepsin. Aggregates of type I collagen may remain in the insoluble residue even after several incubations with pepsin.[33] The yield of collagen solubilized from normal liver (60%) differs greatly from that obtained from cirrhotic liver (95%).[22,23] Finally, salt fractionation does not always yield pure components and ionic exchange chromatography for chain separation is required. In spite of the limitations of both methods, the ratio of type I/III collagens determined in normal and cirrhotic liver by either method gave very similar results. The slightly higher percentage of type III collagen obtained with pepsin, could be due in part, to the presence of highly crosslinked type I collagen in the insoluble pepsin residue.[33] For biochemical studies of collagen, pepsin digestion is still the best method available and has been applied successfully for the isolation of large quantities of type III collagen that have been used for the determination of the complete amino acid sequence of the helical region of this genetic type of collagen.[34,35]

Recently, a simple quantitative procedure was developed for determining the distinct genetic types of collagen present in the liver.[28] After pepsin solubilization, aliquots are subjected to SDS-acrylamide gels (single or bidimensional gels) and then stained with Sirius Red, a dye that binds with high selectivity and high affinity to collagen.[36,37] The amount of dye bound is proportional to the concentration of collagen, and within certain limits, the amount bound to all types of collagen is very similar. The concentration of collagen can be estimated by two procedures: densitometric tracing of cylindrical gels or colorimetric reading of the dye eluted after trypsin digestion of the denatured collagen chains. Both procedures yield identical results.[28]

One of the problems in dealing with biological materials obtained from patients is the small amount of tissue available for analysis. Therefore, very sensitive and reliable methods are still needed to circumvent this problem. Recently, a simple procedure was developed that allows the determination of microgram quantities of collagen in 10-μm thick sections of tissues fixed in formalin and embedded in paraffin. The method combines two dyes (Sirius Red and Fast Green) that under the conditions of the assay bind selectively to collagen and

Table 3
DISTRIBUTION OF EXTRACELLULAR MATRIX IN NORMAL LIVER[6,9]

Collagen type	Molecular formula	Liver localization
I	$[\alpha 1(I)]_2\alpha 2$	Thick collagen bundles in portal triads and vascular regions; occasionally seen in centrolobular areas; forms the dense connective tissue matrix
I-trimer	$[\alpha 1(I)]_3$	Present in very small concentrations in normal liver; increases in cirrhotic livers; its distribution is not known
III	$[\alpha 1(III)]_3$	It represents some, but not all, the reticulin fibers; it is associated with fibronectin and forms the loose connective tissue matrix. It is present in portal triads, terminal venule zones, and centrolobular regions
IV	$[\alpha 1(IV)]_3$ αC αD	It is localized around smooth muscle cells in vessels, around bile ducts, ductules, and nerve axons, and in the endothelial lining of hepatic arteries and lymphatic vessels; in the parenchyma it is localized around sinusoids between endothelial lining and liver plates; it is associated with laminin in most sites
V	$(\alpha B)_3 = [\alpha 1(V)]_3$ $(\alpha A)_3 = [\alpha 2(V)]_3$ $(\alpha A)\alpha B_2 = [\alpha 1(V)]_2[\alpha 2V]$	It is present in portal and terminal venule areas; in the centrolobular region it shows a perisinusoidal distribution and may be in close contact with plasma membrane of hepatocytes
Fibronectin		It is codistributed with type IV collagen; it is present in the perisinusoidal region, perhaps associated with type III collagen; it is absent from the interstitial matrix of portal triads; it is increased in liver cirrhosis, but its concentration is not known
Laminin		Codistributed with type IV collagen except for the perisinusoidal region; it is increased in cirrhotic liver; its concentration is not known

noncollagenous proteins respectively. The dyes absorb light at distinct wavelengths, they can be eluted from the tissue sections, and their absorbance can be determined in a colorimeter. The concentration of collagen and protein is then estimated from the absorbance values obtained, after determining the color equivalence for each dye.[38] The method has been successfully applied to human liver biopsies obtained from cirrhotic patients.[39]

a. Type I Collagen

Type I collagen is present in large concentration in the liver and it accounts for 33 to 45% of total rat and human liver[21-23,26,28] (Table 2). In normal mouse liver type I collagen solubilized after pepsin digestion accounts for up to 65% of the total.[25] It corresponds to the thick collagen bundles which stain blue with Mallory stains. It is present in the liver capsule, in portal triads, large vessels, and around terminal venules.[40-47] It has been found inside the Disse space, often in direct contact with the hepatocyte plasmalemma and with endothelial cells. It is more abundant at points of branching or inflection. It has been suggested that it may form the scaffold of the hepatic lobule (Tables 3 and 4).[47]

b. Type III Collagen

Type III collagen is also present in large concentration and the ratio of type I/type III is 1.[22-24,27] Type III collagen stains black with silver but does not correspond to all the reticulum fibers of the liver. It is associated with type I collagen in portal triads and in terminal venule zones. It forms the fine framework of the liver and it is present in the interlobular regions inside the Disse space[40,42-45] in close association with the processes of Ito cells.[48,49] Type

Table 4
AMINO ACID COMPOSITION OF RAT LIVER COLLAGENS

Amino acid	Residues/1000 amino acid residues					
	α1(I)	α2	[α1(I)]$_3$	α1(III)	α2(V)	α1(V)
3-Hydroxyproline	0	0	0	0	0	0
4-Hydroxyproline	90	102	80	118	112	84
Aspartic acid	46	46	51	48	43	58
Threonine[a]	18	16	21	13	15	19
Serine[a]	24	16	43	38	28	29
Glutamic acid	77	77	86	78	96	108
Proline	128	121	101	109	116	121
Glycine	332	337	320	336	334	311
Alanine	123	101	115	85	77	58
Cystine	0	0	0	2	0	0
Valine	23	22	23	17	26	24
Methionine[b]	5	6	7	7	7	7
Isoleucine	9	11	11	14	15	21
Leucine	21	23	23	19	30	40
Tyrosine[a]	1	1	2	2	2	2
Phenylalanine	14	13	14	11	21	13
Hydroxylysine	12	17	18	16	23	35
Lysine	28	24	25	25	20	22
Histidine	3	8	4	13	8	10
Arginine	46	49	56	46	36	36

[a] Not corrected for losses after acid hydrolysis.
[b] Determined as the sum of methioine and methionine sulfoxides.

From Rojkind, M. and Kershenobich, D., *Liver Annual*, Vol. 1, Arias, I. M., Frenkel, M., and Wilson, J. H. P., Eds., Excerpta Medica, Amsterdam, 1981, 126. With permission.

III collagen fibrils are of smaller diameter than type I collagen fibrils perhaps due to their deposition in the tissue as procollagen molecules[50-52] (Table 4).

c. Type IV Collagen

Type IV collagen may represent up to 10% of total liver collagen.[23] The concentration of type IV collagen in human liver is approximately 500 μg/g of fresh tissue and in the rat, it is 50 to 60 μg/g of fresh tissue[27] (Table 2). Type IV collagen is present in sites containing a basement membrane. It has been found around arteries and lymphatics, bile ducts and ductules, and nerve endings. It forms a discontinuous sheet along the total length of the sinusoid. It is localized between the endothelial lining of the sinusoids and the liver plates, being closer to the endothelial cells than to the hepatocyte plasmalemma.[47,51,53] It is important to stress, that in the sinusoidal region there is no morphologically detectable basement membrane, and that laminin, is absent.[47,53]

d. Type V Collagen

Type V collagen accounts for 7 to 10% of total liver collagen[23] (Table 2). In normal liver, it is approximately 800 to 900 μg/g of fresh tissue, and in the rat, it is 40 to 50 μg/g.[4,6,27] Its exact chain composition is unknown, although from the stoichiometry of the chains isolated on CM-cellulose, liver type V collagen may be $[a1(V)2]_2a2(V)$.[23] This collagen is present in portal triads and central venule areas. However, it is more conspicuous at the sinusoidal site of the hepatocytes. It follows a basket-like pattern and in some instances, appears to involve the hepatocyte[41] (Tables 3 and 4).

e. Type VI Collagen

Several years ago, a new collagen chain (E collagen) was isolated from human liver and its amino acid composition determined.[23] This collagen has been named type VI collagen and has been shown to be an important nonfibrillar component of the interstitial matrix of the portal tracts[54] (Table 4).

2. *Noncollagenous Proteins*

The liver extracellular matrix contains several noncollagenous proteins, some of which have been studied by immunocytochemical techniques. The proteins are very insoluble and therefore, we lack quantitative information regarding their actual concentration in the tissue, or the changes occurring in disease. The best-characterized components are laminin and fibronectin.[47,53,55]

a. Laminin

Laminin is localized in basement membranes. It is present beneath the endothelial lining of hepatic arteries, portal veins, and lymphatic vessels. It is also found around smooth muscle cells of vessel walls, bile ducts and ductules, and nerve axons.[47,53,55] Laminin is absent from the sinusoidal lining (Table 3).

b. Fibronectin

Fibronectin has a spotty distribution in the basement membranes of portal capillaries, but is absent from the basement membranes of large arteries, veins, and the biliary tree.[47] In other locations, such as portal triads and terminal venule areas, fibronectin is closely associated with collagen fibrils. However, it is also found as free fibrils.[47,53] In the space of Disse, it forms irregular clusters of amorphous material in close contact with the nonvascular surface of the sinusoidal lining cells. It is also found attached to the plasma membrane of the hepatocytes[47] (Table 3).

c. Other Glycoproteins

During the course of our studies performed to determine the factors present in liver biomatrix that participate in hepatocyte attachment, a group of proteins was extracted with dilute acetic acid.[56] The glycoprotein extract is very heterogeneous and it contains several components ranging in molecular weight from 40 to 300 kdaltons. The glycoproteins are rich in acidic and hydrophobic amino acids and in general, they resemble the composition of the structural glycoproteins of the connective tissues isolated by others from soft tissues[57-59] (Table 5). The amount of glycoprotein solubilized from biomatrix is about 10 mg/g of fibers and it corresponds to 100 μg of proteins extracted per gram of fresh liver.

3. *Glycosaminoglycans and Proteoglycans*

The organization of the proteoglycan monomer in the liver has not been fully investigated, although there is some information regarding its GAGs. Two of the best studies have shown conflicting results. One group[61] isolated four types of GAGs from rat liver. The total amount reported per liver (10 g) as uronic acid was 310 μg of glucuronic acid equivalents. Approximately 172 μg corresponded to the nonsulfated hyaluronic acid (54% of total GAGs) and 146 μg to a mixture of sulfated GAGs (30% chondroitin sulfate and 16% of heparin and heparan sulfate). Another group[62] used two-dimensional electrophoresis for the separation and quantitation of GAGs of normal and lathyritic rat liver. They found that only 10% corresponded to hyaluronic acid, 7% to dermatan sulfate, 7% to chondroitin sulfates, and 75% to heparan sulfate. The very high concentration of heparan sulfate found is not surprising, since this is the major GAG of liver plasma membrane.[63-66] Approximately 50% of the heparan sulfate molecules are weakly associated with the membrane and are easily

Table 5
AMINO ACID COMPOSITION OF COLLAGENASE-RESISTANT PROTEINS OBTAINED FROM CONNECTIVE TISSUE FIBERS (BIOMATRIX) OF DIFFERENT ORGANS

	Residues/1000 amino acid residues				
Amino acid	**Liver**[a]	**Spleen**	**Kidney**	**Lung**	**Heart**
4-Hydroxyproline	0	11	29	13	0
Aspartic acid	77	78	99	119	110
Threonine	60	30	41	48	53
Serine	69	14	32	39	46
Glutamic acid	104	64	90	90	61
Proline	68	67	71	45	57
Glycine	122	220	169	130	149
Alanine	97	129	71	80	94
1/2 Cystine	*[b]	*	*	*	*
Valine	50	64	59	65	62
Methionine	12	*	*	*	*
Isoleucine	35	44	44	50	46
Leucine	88	70	70	69	80
Tyrosine	30	15	14	31	33
Phenylalanine	37	30	35	42	36
Hydroxylysine	0	0	0	0	0
Lysine	74	39	52	68	68
Histidine	22	14	20	14	17
Arginine	58	36	38	34	38

[a] Modified from Rojkind et al.[185]
[b] Asterisk indicates present but not determined.

displaced by other GAGs. However, the remaining 50% is strongly associated with the membrane lipid bilayer through a very hydrophobic protein. This species can be extracted from the membrane with detergents or after cleavage of the core protein with proteolytic enzymes.[64-66]

The proteoglycan monomer complex has a mol wt of approximately 75 kdaltons. The protein core has a mol wt of 17 kdaltons and the four GAG chains have a mol wt of 14 kdaltons each.[63-67] Recently, an antibody raised against a heparan sulfate proteoglycan isolated from liver microsomes was used to study the localization of the GAG. The results clearly showed that heparan sulfate proteoglycans "are largely confined to the sinusoidal plasmalemmal domain and to biosynthetic and endocytic compartments of hepatocytes".[66]

III. METABOLISM OF CONNECTIVE TISSUE COMPONENTS

A large variety of cells possess the capacity to produce variable amounts of collagen in culture and also contain the enzyme prolylhydroxylase.[68,69] Therefore, it is important to establish which cell or cells is involved in collagen synthesis under normal physiological conditions (growth and regeneration) or pathological states (fibrosis and cirrhosis).

Hepatocytes in primary culture or cell lines derived from the liver that maintain the capacity to synthesize albumin, also produce collagen, GAGs, laminin, and fibronectin.[62,70-80] Liver-derived endothelial cells,[81] lipocytes,[82-84] and myofibroblasts[85,86] produce collagen. In addition, lipocytes produce laminin.[83] From the data obtained from these cells, it has been estimated that lipocytes are the main collagen-producing cells of normal liver; they produce 10-fold more collagen than that synthesized by hepatocytes and 20-fold more than that

Table 6
SOME METABOLIC DIFFERENCES BETWEEN NORMAL AND CIRRHOTIC LIVER

	Normal	Cirrhotic
Liver weight	7—10 g	7—10 g
Collagen		
Content	1 mg/g	5 mg/g
Type I	40%	60%
Ratio I/III	1.0	>2
Half-life		
Type I	30—35 days	17—20 days
Type III	15—17 days	10—12 days
Synthesis	0.32 ± 0.04 μmol Hypor/6 hr/g	5.20 ± 1.33 μmol Hypor/6 hr/g
Proline dependence	±	+ + + +, saturation occurs at 0.48 m*M*
Effect of lactate	None	Stimulation
Actin concentration	1.4 mg/g	2.1 mg/g
β:γ ratio	60:40	40:60
Nonparenchymal cells	35 × 10^6	70 × 10^6
Myofibroblasts	Few and grow slowly	Many and grow fast
Total tRNA	0.34 mg/g	0.33 mg/g
Functional tRNA	21.6 ± 3.9%	76.0 ± 12.5%
Proline pool	0.2 m*M*	0.4 m*M*
Prolyl-tRNA acylated		
In vivo	1.22 nmol/mg tRNA	1.34 nmol/mg tRNA
In vitro	1.50 nmol/mg tRNA	2.08 nmol/mg tRNA
Prolyl actylating enzymes		
V_{max}	11 nmol/mg/10 min	25 nmol/mg/10 min
Km_{app}	4.0 × 10^{-4} *M*	4.0 × 10^{-4} *M*
Proline made from arginine	0.93 ± 0.20 μmol/g/6 hr	1.82 ± 0.44 μmol/g/6 hr
Prolyl-oxidase activity	4.7°C 0.72 nmol Δ'P5C/min/mg protein	1.7 ± 0.47 nmol Δ'P5C/min/mg protein
cAMP levels	600 pmol/g	1500 pmol/g

From Rojkind, M. et al. *Pharmacological, Morphological, and Physiological Aspects of Liver Aging*, Van Bezooijen, C. F. A., Ed., Eurage, Rijswijk, The Netherlands, 1983, 59. With permission.

manufactured by endothelial cells.[83] However, it is important to stress that cell cultures may resemble more closely a process of wound healing than a normal physiological state. Cells that have been separated from their matrix by mechanical or enzymatic methods will have to synthesize part or all the components of the matrix in order to warrant their survival in culture. In vivo, the amounts and types of protein produced by the cells, are regulated in part, by the microenvironment; i.e., cells and matrix.

A. Collagen Synthesis

In the normal liver the amount of collagen produced is small. In vitro studies using the mouse liver have shown that approximately 0.32 ± 0.04 μmol of radioactive hydroxyproline are produced from precursor radioactive proline per 6 hr of incubation at 37°C per gram of wet tissue.[11,30] The synthesis is independent of the concentration of the pool of free proline[29] (Table 6). The main collagens produced in vivo are I and III although small amounts of V and perhaps IV collagens are also made (Table 7). The ratio of collagens synthesized bears a direct relationship to the collagen composition of the tissue, thus suggesting, that collagen deposition in the normal liver reflects the biosynthetic capacity of the organ.[27]

B. Collagen Degradation

Collagenolytic activity in the liver was first demonstrated in Kupffer cells.[87] It has been shown also that hepatocytes[88,89] and fibroblasts[90] produce collagenase. However, the production of collagenase by the latter only occurs in co-cultures containing 75% liver fibroblasts and 25% hepatocytes and if the culture medium contains phorbol esters.[90] The collagenolytic activity is detected after the cells have been maintained for 6 days in culture. At this time the hepatocytes start to disintegrate. The mechanism by which one cell stimulates the production of collagenase by the other, is unknown. However, it could be related to the production of hormone-like materials (paracrine and autocrine systems).[91]

With the use of immunocytochemical methods, collagenase has been detected in close association with almost every single collagen fibril of the normal tissue.[92,93] Thus, collagenase is produced and secreted, and then, in the extracellular matrix it binds to collagen. It is thus possible, that collagenolytic activity measured in vitro using tissue homogenates and radioactive collagen reflects the amount of collagenase that was initially bound to the collagen fibrils and was released either during homogenization or after digestion of the collagen fibril to which it was attached.[94] Furthermore, collagenases can be produced as precursor or can be bound to inhibitors, thus making its quantitation difficult. Recently, an immunoassay to measure collagenase has been developed. The method is simple and provides reliable results with regard to enzyme protein concentration. However, it does not provide information pertaining to the number of active molecules.[95]

C. Collagen Half-Life

The half-life of liver collagen has been measured in vivo after a pulse of radioactive proline followed by the isolation and purification of individual collagen chains of type I and type III collagen. The results showed that the half-life of the radioactive type I collagen is 30 to 35 days and that of radioactive type III is 15 to 17 days (Table 6). Data to be shown later (*vide infra*) indicate that this may not reflect the actual half-life of all the collagen molecules (I or III) and that at least two different metabolic pools are present in the liver.[96]

D. Biosyntheis of Noncollagenous Components

Different laboratories have demonstrated that hepatocytes,[97,98] Kupffer cells,[99] endothelial cells,[100-102] and liver myofibroblasts[81] produce fibronectin. In addition, hepatocytes[97], Ito cells,[83,84] and liver myofibroblasts[86] synthesize laminin in culture.

IV. MODIFICATIONS OF THE EXTRACELLULAR MATRIX DURING REGENERATION AND REPAIR

One of the constant features of many organs is the relationship between its connective tissue and its cellular components. Several years ago, the ratio between the two was defined as the parenchyma/stroma ratio.[103,105] It was further shown that this constant is maintained during physiological but lost during pathological situations.[105] During liver regeneration, several changes in the organization of the tissue are observed. First, there is swelling of the cells, and therefore, the amount of connective tissue left in proportion to liver weight is decreased. Then, (24 to 48 hr after hepatectomy) the cells start to proliferate and continue to multiply for several days. The connective tissue increases slowly, and although it does not follow precisely the increase in tissue mass, at the end of 2 to 3 weeks, a normal parenchyma/stroma is re-established.[106,107]

In liver cirrhosis, the parenchyma/stroma ratio is altered. There is destruction of liver cells, arrival of circulating monocytes,[108] proliferation of fibroblasts (myofibroblasts),[109,110] and excess collagen deposition.[1-3,5-7,9-13] Although all the collagens are increased, the predominant collagen is type I. In addition, the connective tissue is deposited in abnormal

Table 7
RELATIVE PERCENT OF COLLAGEN TYPES I, III, AND V SYNTHESIZED IN VIVO AND DEPOSITED INTO LIVER BIOMATRIX OF NORMAL, FIBROTIC, AND HEPATECTOMIZED RATS 24 HR AFTER A PULSE WITH 2,3-[^{3}H]-PROLINE

Collagen type	Normal	Fibrotic	Hepatectomized 5 days	Hepatectomized 11 days
I	44.0	52.4	48.0	32.0
III	41.7	41.8	21.3	57.2
V	14.1	5.7	29.7	10.6

Note: Animals received a single pulse of 100 μCi of 2,3-[^{3}H]-proline. Pools of 5 to 6 livers were prepared for the analysis. The relative percentages were estimated from the radioactivity recovered in collagens I, III, and V.

From Rojkind, M. et al., *Collagen Rel. Res.*, 3, 335, 1983. With permission.

locations and the fibrous septa that surround the nodules of apparently regenerating hepatocytes contain all the collagen types mixed together.[42,46,53,111] Moreover, the cells produce laminin, a component normally absent from the sinusoids and real basement membranes are formed that isolate the liver cells from the sinusoidal components.[111,112] The permeability barrier formed could further modify liver cell function.

In studies performed with cirrhotic livers of different etiologies, it was established that in early lesions, the ratio of type I/type III collagen is normal (livers containing less than 15 to 20 mg of collagen per gram of fresh liver). However, in livers with advanced fibrosis (over 15 to 20 mg of collagen per gram of tissue) the ratio increased above 1.[23] Similarly, in cirrhotic animals, the ratio is slightly greater than 1.[27] As shown above, the changes in the ratio of type I to type III collagen are independent of the nature of the etiologic agent. Thus suggesting, that with respect to collagen, all the scars are the same. This implies, that independently of the mechanism by which tissue damage is produced (each group of chemical and biological agents can act by different mechanisms), there is a final common pathway leading to cirrhosis.

During growth and regeneration the deposition of a matrix is essential for the establishment of normal cell-cell and cell-matrix interactions. During pathological conditions, such as it occurs in liver cirrhosis, the matrix produced is abnormal, and in addition, it is not deposited in the correct sites. This is accompanied by abnormal cell-cell and cell-matrix interactions that could further contribute to the alterations observed in liver function.

It has been suggested that collagens produced during regeneration are those needed to reform sinusoids and reestablish the normal liver architecture. Therefore, the different cells within the liver that have the capacity to produce either collagen and/or collagenase could contribute to tissue remodeling. In vivo experiments have shown that collagen synthesis is increased 5 days after partial hepatectomy (Table 7) but returns to normal values 10 to 14 days posthepatectomy.[27] The types of collagen synthesized are I, III, and V. However, in contrast to the results obtained with normal or fibrotic liver, there is a relative increase in the production of type V collagen[27] (Table 7). Although it has been suggested that the production of type V collagen could be considered as a sign of sinusoid formation[27] or as an indicator of rapid cell proliferation,[113] several investigators have shown that type V collagen is increased in pathological states such as inflammatory and fibroproliferative diseases.[114-116]

Eleven days posthepatectomy, when collagen synthesis decreases to almost normal values, the production of type V collagen is also normal. However, at this time a relative increase in the production of type III collagen is observed.[27] These results differ greatly from those obtained in animals treated with CCl_4. In the latter, the predominant collagen produced is type I collagen. The data obtained with the regenerating liver clearly suggest, that during growth, collagen deposition is a highly coordinated process and that the correct amounts and types of collagen are manufactured. It is also possible to suggest, that the cells involved in collagen synthesis during regeneration are not the same producing collagen during repair.

A. Collagen Half-Life

There is no information pertaining to the half-life of collagen produced during regeneration. There is only information obtained in vivo with livers of cirrhotic animals that had been treated with CCl_4 for 7 weeks. The results obtained demonstrated that radioactive collagen produced after a single pulse with radioactive proline is active metabolically. Types I and III collagen have a 50% shorter half-life in the cirrhotic liver as compared with collagens obtained from normal liver[96] (Table 6). In spite of that, the liver remains cirrhotic 2 months after discontinuation of CCl_4 administration, thus suggesting that all the collagen molecules are not degraded equally.

From these and other experiments, mainly the changes in distribution of collagenase observed in cirrhotic animals (the collagenase disappears from the regeneration nodules and is only present in the most outer portion of the fibrous septa),[92] it has been postulated that at least, two different metabolic pools are present: one consisting of small diameter fibrils or newly synthesized collagen molecules (actively degraded), and the other made of the old collagen molecules which are present in the inner portion of the septa (not actively degraded).[96] Accordingly, collagen degradation is a complex process that depends of many factors (cells and mediators), among which the three-dimensional organization of the fibers is also important. Collagen of normal liver does not differ qualitatively from cirrhotic liver collagen in its susceptibility to digestion by homologous or heterologous collagenase,[117] thus suggesting that once collagenase reaches the fibers, they are digested.

B. Collagenolytic Activity

Collagenolytic activity has been determined in animal and human liver. In murine schistosomiasis it follows the pattern of collagen deposition.[118,119] In rat, baboon, and human liver, the activity has been found normal or increased. However, in advanced cirrhosis, the collagenolytic activity is decreased.[120-123]

V. CELL-CELL AND CELL-MATRIX INTERACTIONS

In a highly organized society such as the liver, each of its components is present in a fixed concentration (within certain limits) and in a unique distribution. Accordingly, the interactions to be established with cells and matrix will depend on its three-dimensional organization. When cell-cell and cell-matrix interactions are changed due to destruction of hepatocytes, or the arrival of circulating monocytes, etc., the normal interactions are lost, and new contacts are established. Under these new sets of interactions, liver cell function could be impaired.

The normal interactions between cells and matrix have been studied in vitro. It has been shown that freshly isolated hepatocytes attach to a variety of proteins such as fibronectin, laminin, and collagen type I and type IV.[55,123-127] However, in some substrates, the cells do not show spreading, and in general, when maintained in culture in the aforementioned substrates, they show a modification in their phenotype and die within a few weeks.[128-130] Therefore, it appears that proteins other than those tested may be necessary for the cells to express their function in culture.

Studies from our laboratory have suggested that tissue-specific proteins may be required for cells to maintain function. Accordingly, these proteins can be found preferentially in the tissue from which the cells are obtained. We therefore extracted the connective tissue of the liver (liver biomatrix) as a fibrous structure free of cells and used it as a substratum to culture cells.[131-133] The results obtained suggest that adult rat hepatocytes maintained on rat liver biomatrix survive for several months in culture and express a variety of normal functions.

The protein or proteins responsible for the long-term survival of the hepatocytes is unknown. However, it was demonstrated that collagenase digestion of the matrix did not alter its properties. However, proteolytic digestion destroyed its capacity to facilitate the attachment and survival of the hepatocytes.[131] It is thus suggested that a noncollagenous component(s) is (are) responsible for sustaining liver function in vitro.

Attempts have been made to isolate and characterize the noncollagenous components of liver biomatrix involved in cell attachment. Initially, an acetic acid extract that contained proteins of high molecular weight was prepared. This extract was shown to facilitate the attachment of freshly isolated hepatocytes. There was a high specificity since similar extracts obtained from other parenchymatous organs such as lung, heart, spleen, and kidney, failed to provide high yields of attachment.[56]

Tissue specificity in maintaining differentiated functions of cells has been confirmed by others. Epithelial cells obtained from mammary gland attach and survive on liver biomatrix. However, they do not express a complete normal phenotype and eventually they lose their function. However, the survival and expression of specific functions are greatly improved when the cells are plated on mammary gland biomatrix.[134] Similarly, kidney-derived epithelial cells (MDCK) attach preferentially to kidney biomatrix[135] and Sertoli cells express their function when cultured on a specific matrix.[136]

The selective attachment to a specific matrix has to be associated to the presence of specific membrane proteins that interact with the matrix. Therefore, since the membrane receptors to some hormones[137,138] and the nature of the matrix produced during regeneration differ from normal,[27,55] it is important to analyze the changes in cell-matrix interactions that take place during active replication of hepatocytes.

It has been shown by others that during cell regeneration post hepatectomy, laminin is produced and deposited in the sinusoidal region.[55] In addition, the attachment of freshly isolated hepatocytes to fibronectin, which is normally high, decreases to very low values. Simultaneously, the attachment to laminin, which is low in normal cells, increases. When regeneration has been completed, the behavior with regard to attachment to fibronectin and laminin returns to normal.[55]

Malignant cells have the capacity to become loose, travel, attach to basement membranes, and proliferate in other organs (matastasis). Their capacity to metastasize is associated with a change in matrix recognition by the plasma membrane of the tumor cells.[139-141] We have investigated the behavior of a variety of Morris hepatomas that show different degrees of malignancy (growth rate and differentiation). We tested their capacity to attach to glycoproteins extracted from a variety of biomatrices. The results obtained clearly indicate that although the malignant cells maintained their capacity to attach to liver-derived biomatrix proteins, they now attached with much greater efficiency to heterologous biomatrices.[135] When comparing their attachment efficiency to kidney biomatrix with their proliferative capacity, the latter determined by the decrease in glucagon receptors, a direct correlation between both parameters was obtained.[135] The capacity to recognize proteins derived from an heterologous matrix is not an intrinsic property of the tumor. Rapidly proliferating cells obtained 24 hr after a partial hepatectomy, show exactly the same behavior (Table 8 Figure 1). It is possible that the high attachment efficiency to kidney is associated with the presence of laminin in the extract. This component is present in large quantities in the kidney.

The capacity of rapidly growing cells to attach to kidney-derived glycoproteins has been used to selectively isolate "preneoplastic cells". We have recently shown that the isolated

Table 8
RELATIVE ATTACHMENT EFFICIENCIES OF ADULT RAT HEPATOCYTES AND HEPATOMAS ON BIOMATRIX DERIVED GLYCOPROTEINS

	Glycoprotein-aggregate (% attachment)					
Cell type	Liver	Kidney	Heart	Lung	Spleen	Empty
Normal	100	22 ± 6	27 ± 12	30 ± 6	50 ± 6	50
Regenerating	100	147 ± 21	ND[a]	ND	ND	56 ± 2
Hepatomas						
9618-A	100	81	47	82	28	46
5123-TC	100	106 ± 8	81 ± 9	79 ± 6	79 ± 11	ND
7777	100	129 ± 10	76	ND	ND	72
8999	100	135	71	107	106	53
5123-D	100	187 ± 35	70 ± 17	138 ± 15	151 ± 14	42
Kidney						
MDCK	100	172	134	112	112	48

Note: Values are expressed as % attachment relative to liver binding. Results derived from three or more experiments show the standard deviation. All others are means of duplicate analysis of two experiments.

[a] ND, not determined.

From Mourelle, M. et al., *Hepatology*, 3, 303, 1983. With permission.

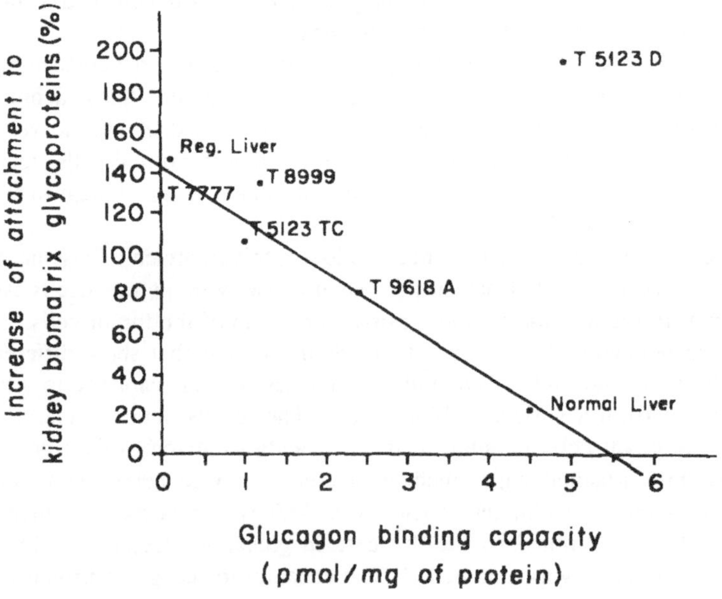

FIGURE 1. Relationship between attachment efficiency of normal, regenerating, and tumoral hepatocytes to kidney biomatrix glycoproteins (expressed as percent attachment relative to liver biomatrix binding) and glucagon binding capacity expressed at pmol/mg of protein. (From Mourelle, M. et al., *Hepatology*, 3, 303, 1983. With permission.)

hepatocytes obtained from carcinogen-treated rats that attach to kidney-derived glycoproteins are highly enriched in gamma-glutamyltranspeptidase activity and incorporate thymidine in culture 20-fold above control values.[142]

VI. DIAGNOSTIC PROCEDURES

Although the liver biopsy is a safe procedure that allows the establishment of a correct diagnosis of liver cirrhosis, because of ethical reasons and risk factors, it cannot be performed frequently. Therefore, efforts have been made to develop simple techniques that will allow the follow-up of the patient, especially when under a given therapeutic scheme. Assays include the measurement of every single enzyme involved in post-translational modification of collagen.[143] The most useful test developed so far is the immunoassay for the procollagen peptide derived from the amino-terminal end of type III collagen.[144-146] Although several groups have used the assay and have reported their values, it has been difficult to interpret the results correctly.[147-159] In order to give a correct interpretation to the serum values reported in the literature, critical experiments have yet to be performed. It is important to establish if the release of the propeptide occurs normally during active synthesis and deposition of type III collagen in the tissue, or if the cleavage of the propeptide is a late event that takes place prior to the catabolism of type III procollagen. Several groups have already indicated that liver type III collagen is indeed type III procollagen.[40,52,145]

VII. TREATMENT OF LIVER CIRRHOSIS

It has been postulated that a better understanding of the pathophysiology of liver cirrhosis will allow the development of a rational approach to therapy. Although for many years it was strongly believed that liver cirrhosis in man was an irreversible process that lead to patient death, several reports in the literature[160,161] and our own work using colchicine as an antifibrogenic agent, have shown that cirrhosis is indeed reversible. However, much work is still needed to determine the actual mechanism of action of colchicine in modifying the course of the disease. Beneficial effects other than those expected simply from its antimicrotubular activity have been observed,[162,163] including the inhibition of the arrival of monocytes into the injured liver,[108] the inhibition in the production of interleukin-1 (IL-1) by circulating monocytes,[167] and the modification in several plasma membrane enzymes,[168,169] perhaps due to changes in the cholesterol/phospholipid ratio.[169]

VIII. THE LIVER AS A BIOECOSYSTEM

It was already described that the liver is composed of cells, extracellular spaces, and extracellular matrix, all distributed in a highly ordered and specific manner that gives the liver its characteristic morphologic appearance. Each component, within certain limits, is constant (number of cells, volume occupied, etc.) and maintains a constant proportion with regard to the other components (cells and matrix). The function is highly dependent on the structural organization, and therefore, alterations in liver structure are also accompanied by disturbances in its function.

Although many systemic factors (metabolites, hormones, neurological stimulation, etc.) are important for cell differentiation and regulation of growth and intermediary metabolism, local factors produced by parenchymal and nonparenchymal cells (autocrine and paracrine-like systems), play an important role in maintaining homeostasis; when these mechanisms are altered, the organization of the liver is modified and the function is impaired. One of the important features of chronic liver disease, such as cirrhosis, is the distortion of the architecture. This results from the destruction of parenchymal cells, the arrival and proliferation of nonparenchymal cells, and from the deposition and contraction of an abnormal connective tissue matrix. In the new structural organization of the tissue, abnormal cell-cell and cell-matrix interactions are established. Because of the complexity of the interactions and their role in regulating liver cell function and structure, a new approach to the study of chronic liver disease is needed. Therefore, the liver should be considered as a small ecological unit within a greater universe composed of many ecosystems. Accordingly, when one ecosystem fails, the others suffer functional modifications.

A. Data in Support of the Ecosystem Hypothesis

Simple techniques have been developed for the isolation and culture of parenchymal cells. Results from several laboratories have clearly shown that plastic dishes or dishes coated with type I collagen, fibronectin, or laminin are not sufficient to maintain a normal phenotypic expression of cultured hepatocytes.[170,171] The cells lose, within hours, important metabolic activities that they perform in vivo.

Two methods have provided encouraging results with regard to the phenotypic expression of liver cells and have allowed the preparation of long term cultures of rat and human hepatocytes. One of the methods has been described above and provides the cells with a complex matrix that contains most of the connective tissue components present in the intact liver.[131-133,172] The second method utilizes a co-culture of liver-derived epithelial cells with hepatocytes. Under these culture conditions, the epithelial cells provide the matrix (and perhaps metaoblites) that is needed for the hepatocytes to maintain their phenotype.[173-175] The epithelial cells cannot be replaced by other cells or by conditioned medium. These results clearly suggest, as described above, that liver function results from the multiple interactions established between cells and between cells and matrix.

In end-stage liver disease the main alterations are as follows:

1. Decrease of liver parenchymal cells due to cell inury and cell death.
2. Increase in nonparenchymal cells, mainly due to arrival of circulating mononuclear cells.
3. Increased deposition of connective tissue components with formation of septa that surround nodules of hepatocytes.
4. Changes in the distribution of connective tissue components with formation basement membrane-like structures that isolate the hepatocytes from the nutrients flowing through the sinusoids.
5. Changes in the interactions between cells and between cells and matrix with modifications in the phenotypic expression of the hepatocytes.

The above-mentioned biochemical and anatomical modifications are accompanied by functional and hemodynamic alterations.

B. What are the Mechanisms Involved in Maintaining Homeostasis?

The mechanisms involved in maintaining homeostasis are unknown. However, we could speculate that each cell within the liver controls either growth and/or metabolism of other cells.[176,177] In addition, each cell plays an important role in maintaining the amount, type, and distribution of connective tissue components.

After acute injury, the decrease in functional capacity of the liver stimulates tissue regeneration. If a single lesion is produced and the animal survives, the proliferation of liver cells is either accompanied or followed by an increase in nonparenchymal cells. In addition, collagen synthesis and degradation will be coordinated processes that will result in the production of normal collagen types, the right concentration of each component, and their correct distribution.[27] At the end of the process, when regeneration ceases, collagen synthesis returns to normal.[27] In addition, the normal interactions and the normal functions are reestablished.

During chronic liver injury, homeostasis is not maintained. Chemotactic substances are produced that bring into the injured liver circulating mononuclear cells.[108] These cells will now produce factors that will attract fibroblasts and will stimulate their proliferation.[178-180] In addition, factors that stimulate collagen biosynthesis will be produced.[181-182] Although some of these abnormalites may also occur in the regenerating liver, in chronic liver injury, the stimulus is sustained. Moreover, it is possible that the recruitment of fibroblasts (Ito cells, myofibroblasts, or regular fibroblasts) is a selection process, and that the newly recruited cells are clones that differ from the normal cells in their doubling time, collagen synthesizing capacity, and response to stimulation by mononuclear cell factors. Therefore, these cells could remain in the tissue, temporarily or permanently, even when the noxious agent has been removed.

Although the above suggestions are highly speculative, nonetheless, they open new perspectives to study the liver as a unit, and should allow a better understanding of the role of each cell and each component of the connective tissue in maintaining homeostasis.

REFERENCES

1. **Rojkind, M. and Kershenobich, D.,** Hepatic fibrosis, *Clin. Gastroenterol.*, 10, 737, 1981.
2. **Rojkind, M. and Kershenobich, D.,** Fibrosis and cirrhosis, in *Liver Annual*, Vol. 1, Arias, I. M., Frenkel, M., and Wilson, J. H. P., Eds., Excerpta Medica, Amsterdam, 1981, 126.
3. **Rojkind, M. and Kershenobich, D.,** Liver extracellular matrix: fibrosis and cirrhosis, in *Liver Annual*, Vol. 2, Arias, I. M., Frenkel, M., and Wilson, J. H. P., Eds., Excerpta Medica, Amsterdam, 1982, 180.
4. **Rojkind, M. and Ponce-Noyola, P.,** The extracellular matrix of the liver, *Collagen Rel. Res.*, 2, 151, 1982.
5. **Rojkind, M. and Kershenobich, D.,** The extracellular matrix: fibrosis and cirrhosis, in *Liver Annual*, Vol. 3, Arias, I. M., Frenkel, M., and Wilson, J. H. P., Eds., Excerpta Medica, Amsterdam, 1983, 189.
6. **Rojkind, M. and Perez-Tamayo, R.,** Liver fibrosis, *Int. Rev. Connect. Tissue Res.*, 10, 333, 1983.
7. **Rojkind, M. and Kershenobich, D.,** The extracellular matrix: fibrosis and cirrhosis, in *Liver Annual*, Vol. 4, Arias, I. M., Frenkel, M., and Wilson, J. H. P., Eds., Excerpta Medica, Amsterdam, 1984, 239.
8. **Rojkind, M. and Kershenobich, D.,** Diagnostic procedures for liver fibrosis, in *Diagnostic Procedures in the Evaluation of Hepatic Disease*, Beker, S., Ed., Alan, R. Liss, New York, 1983, 73.
9. **Rojkind, M. and Kershenobich, D.,** Liver fibrosis: a dynamic process?, in *Clinical Hepatology*, Csomos, G. and Thaler, H., Eds., Springer-Verlag, Berlin, 1983, 126.
10. **Rojkind, M.,** Collagen metabolism in the liver, in *Alcoholic Liver Disease*, Hall, P., Ed., Edward Arnold, London, 1985, 90.
11. **Rojkind, M.,** Liver cirrhosis, in *Current Hepatology*, Vol. 5, Gitnick, G., Ed., Year Book Medical, Chicago, 1985, 465.
12. **Rojkind, M.,** Liver fibrosis, in *Current Hepatology*, Vol. 6, Gitnick, G., Ed., Year Book Medical, Chicago, 1986, 111.
13. **Rojkind, M. and Kershenobich, D.,** Extracellular matrix: fibrosis and cirrhosis, in *Liver Annual*, Vol. 5, Arias, I. M., Frenkel, M., and Wilson, J. H. P., Eds., Excerpta Medica, Amsterdam, 1986, 210.
14. **Blouin, A., Bolender, R. T., and Weibel, E. R.,** Distribution of organelles and membranes between hepatocytes and non-hepatocytes in the rat liver parenchyma, *J. Cell Biol.*, 72, 441, 1977.
15. **Van de Werve, G.,** Isolation and characteristics of hepatocytes, *Toxicology*, 18, 179, 1980.
16. **Wisse, E., De Zanger, R. B., Jacobs, R. et al.,** Scanning electron microscope observations on the structure of portal veins, sinusoids and central veins in rat liver, *Scan. Electron Microsc.*, 3, 1441, 1983.
17. **Jungermann, K. and Katz, N.,** Functional hepatocellular heterogeneity, *Hepatology*, 2, 385, 1982.
18. **Gumucio, J. J. and Miller, D. L.,** Liver cell heterogeneity, in *The Liver Biology and Pathobiology*, Arias, I. M., Popper, H., Schachter, D. et al., Eds., Raven Press, New York, 1982, 647.
19. **Jones, E. A. and Summerfield, J. A.,** Kupffer cells, in *The Liver Biology and Pathobiology*, Arias, I. M., Popper, H., Schachter, D. et al., Eds., Raven Press, New York, 1982, 507.
20. **Bioulac-Sage, P. and Balabaud, C.,** La Cellule perisinusoidale, *Gastroenterol. Clin. Biol.*, 9, 312, 1985.

21. **Rojkind, M. and Martinez-Palomo, A.,** Increase in type I and type III collagens in human alcoholic liver cirrhosis, *Proc. Natl. Acad. Sci. U.S.A.*, 73, 539, 1976.
22. **Seyer, J. M., Hutcheson, E. T., and Kang, A. H.,** Collagen polymorphism in normal and cirrhotic human liver, *J. Clin. Invest.*, 59, 241, 1977.
23. **Rojkind, M., Giambrone, M. A., and Biempica, L.,** Collagen types in normal and cirrhotic liver, *Gastroenterology*, 76, 710, 1979.
24. **Seyer, J. M.,** Interstitial collagen polymorphism in rat liver with CCl-induced cirrhosis, *Biochim. Biophys. Acta*, 629, 490, 1980.
25. **Wu, H. C., Giambrone, M. A., Howard, D. J. et al.,** The nature of the collagen in hepatic fibrosis in advanced murine schistosomiasis, *Hepatology*, 2, 366, 1982.
26. **Hassanein, H., Herbage, D., Chevalier, O. et al.,** Solubilization and characterization of human liver collagens in *Schistosomiasis mansoni*, *Cell. Mol. Biol.*, 29, 139, 1983.
27. **Rojkind, M., Rojkind, M. H., and Cordero-Hernandez, J.,** In vivo collagen synthesis and deposition in fibrotic and regenerating rat livers, *Collagen Rel. Res.*, 3, 335, 1983.
28. **Armendariz-Borunda, J. and Rojkind, M.,** A simple quantitative method for collagen typing in tissue samples: its application to human liver with schistosomiasis, *Collagen Rel. Res.*, 4, 35, 1984.
29. **Ehrinpreis, M. N., Giambrone, M. A., and Rojkind, M.,** Liver proline oxidase activity and collagen synthesis in rats with cirrhosis induced by CCl, *Biochim. Biophys. Acta*, 629, 184, 1980.
30. **Dunn, M. A., Rojkind, M., Warren, K. S. et al.,** Liver collagen synthesis in murine schistosomiasis, *J. Clin. Invest.*, 59, 666, 1977.
31. **Rojkind, M., Gatmaitan, Z., Mackensen, S. et al.,** Connective tissue biomatrix: its isolation and utilization for long-term cultures of normal rat hepatocytes, *J. Cell. Biol.*, 87, 255, 1980.
32. **Laurent, G. J., Cockerill, P., Mac Anulty, R. J. et al.,** A simplified method for quantitation of the relative amounts of type I and type III collagen in small tissue samples, *Anal. Biochem.*, 113, 301, 1981.
33. **Rauterberg, J., Schlief, E., and Pott, G.,** The collagen of the normal and fibrotic liver. Characterization of collagen derived peptides solubilized by CNBr cleavage and by low temperature pepsin treatment, in *Proceedings of Connective Tissue of the Normal and Fibrotic Human Liver*, Gerlach, U., Pott, G., Rauterberg, J., and Voss, B., Eds., Georg Thieme Verlag, Stuttgart, 1982, 36.
34. **Seyer, J. M., Mainardi, C., and Kang, A. H.,** Covalent structure of collagen: amino acid sequence of l(III)-CB5 from type III collagen of human liver, *Biochemistry*, 19, 1583, 1980.
35. **Seyer, J. M. and Kang, A. H.,** Covalent structure of collagen: amino acid sequence of alpha-1(III)-CB9 from type-III collagen of human-liver, *Biochemistry*, 20, 2621, 1981.
36. **Junqueira, L. C. U., Bignolas, G., and Brentani, R. R.,** Picrosirius staining plus polarization microscopy, a specific method for collagen detection in tissue sections, *Histochem. J.*, 11, 447, 1979.
37. **Junqueira, L. C. U., Bignolas, G., and Brentani, R. R.,** A simple and sensitive method for the quantitative estimation of collagen, *Anal. Biochem.*, 94, 96, 1979.
38. **López-De-León, A. and Rojkind, M.,** A simple micromethod for collagen and total protein determination in formalin-fixed paraffin-embedded sections, *J. Histochem. Cytochem.*, 33, 737, 1985.
39. **Jímenez, W., Pares, A., Caballería, J. et al.,** Measurement of fibrosis in needle liver biopsies: evaluation of a colorimetric method, *Hepatology*, 5, 815, 1985.
40. **Gay, S., Fietzek, P. P., Remberger, P., et al.,** Liver cirrhosis: immunofluorescence and biochemical studies demonstrate 2 types of collagen, *Klin. Woschenshr.*, 53, 205, 1975.
41. **Biempica, L., Morecki, R., Wu, C. H. et al.,** Immunocytochemical localization of type B collagen. A component of basement membrane in human liver, *Am. J. Pathol.*, 98, 591, 1980.
42. **Grimaud, J. A., Druguet, M., Peyrol, S. et al.,** Collagen immunotyping in human liver. Light and electron microscope study, *J. Histochem. Cytochem.*, 28, 1145, 1980.
43. **Voss, B., Rauterberg, J., Allam, S. et al.,** Distribution of collagen type I and III and of two collagenous components of basement membrane in human liver, *Pathol. Res. Pract.*, 170, 50, 1980.
44. **Konomi, H., Sano, J., and Nagai, Y.,** Immunohistochemical localization of type I, type III and type IV collagens in the liver, *Acta Pathol. Jpn.*, 31, 973, 1981.
45. **Sano, J., Sato, S., Ishizaki, M. et al.,** Types I, III, and IV (basement membrane) collagens in the bovine liver parenchyma: electron microscopic localization by the peroxidase-labeled antibody method, *Biomed. Res.*, 2, 546, 1981.
46. **Biempica, L., Dunn, M. A., Kamel, I. A. et al.,** Liver collagen type characterization in human schistosomiasis. A histological, ultrastructural, and immunocytochemical correlation, *Am. J. Trop. Med. Hyg.*, 32, 316, 1983.
47. **Martinez-Hernández, A.,** The hepatic extracellular matrix. I. Electron immunohistochemical studies in normal rat liver, *Lab. Invest.*, 51, 57, 1984.
48. **Kent, G., Gay, S., Inouye, T. et al.,** Vitamin A containing lipocytes and formation of type III collagen in liver injury, *Proc. Natl. Acad. Sci. U.S.A.*, 73, 3719, 1976.
49. **Oikawa, K.,** A scanning electron microscopic study of the liver of mice, *Tohoku J. Exp. Med.*, 129, 373, 1979.

50. **Wick, G., Brunner, H., Penner, E. et al.,** The diagnostic application of specific antiprocollagen sera. II. Analysis of liver biopsies, *Int. Arch. Allergy Appl. Immunol.*, 56, 316, 1978.
51. **Gay, S.,** Collagen in liver fibrosis, *Ital. J. Gastroenterol.*, 12, 30, 1980.
52. **Sato, S., Leo, M. A., and Lieber, C. S.,** Ultrastructural localization of type III procollagen in baboon liver, *Am. J. Pathol.*, 122, 212, 1986.
53. **Hahn, E., Wick, G., Pencev, D. et al.,** Distribution of basement membrane proteins in normal and fibrotic human liver: collagen type IV, laminin and fibronectin, *Gut*, 21, 63, 1980.
54. **Hahn, E. G., Schuppan, D., and Hahn, U.,** Expression of new collagen types by portal fibroblasts in obstructive bile duct disease (abstr.), *Hepatology*, 3, 861, 1983.
55. **Carlsson, R., Engvall, E., Freeman, A. et al.,** Laminin and fibronectin in cell adhesion: enhanced adhesion of cells from regenerating liver to laminin, *Proc. Natl. Acad. Sci. U.S.A.*, 78, 2403, 1981.
56. **Ponce, P., Cordero, J., and Rojkind, M.,** A noncollagenous matrix for attachment of rat hepatocytes in culture, *Hepatology*, 1, 204, 1981.
57. **Timpl, R., Wolff, I., and Weiser, M.,** A new class of structural proteins from connective tissue, *Biochim. Biophys. Acta*, 168, 168, 1968.
58. **Timpl, R., Wolff, I., and Weiser, M.,** Acidic structural proteins of connective tissue. I. Solubilization and preliminary chemical characterization, *Biochim. Biophys. Acta*, 194, 112, 1969.
59. **Furthmayr, H. and Timpl, R.,** Acidic structural proteins of connective tissue. II. Evidence for disulfide cross-links, *Biochim. Biophys. Acta*, 220, 396, 1970.
60. **Robert, L., Junqua, S., and Moczar, M.,** Structural glycoproteins of the intercellular matrix, *Front. Matrix Biol.*, 3, 113, 1976.
61. **Koizumi, P., Nakamaura, N., and Abe, H.,** Changes in acid mucopolysaccharides in the liver in hepatic fibrosis, *Biochim. Biophys. Acta*, 148, 749, 1967.
62. **Ninomiya, Y., Hata, R., and Nagai, Y.,** Glycosaminoglycan synthesis by liver parenchymal cells clones in culture and its change with transformation, *Biochim. Biophys. Acta*, 629, 349, 1980.
63. **Oldberg, A., Kjellen, L., and Hook, M.,** Cell surface heparan sulfate. Isolation and characterization of a proteoglycan from rat liver membranes, *J. Biol. Chem.*, 254, 8505, 1979.
64. **Kjellen, L., Pettersson, I., and Hook, M.,** Cell surface heparan sulfate: an intercalated membrane proteoglycan, *Proc. Natl. Acad. Sci. U.S.A.*, 78, 5371, 1981.
65. **Norling, B., Glimelus, B., and Wasteson, A.,** Heparan sulfate proteoglycan of cultured cells: demonstration of a lipid and a matrix associated form, *Biochem. Biophys. Res. Commun.*, 103, 1265, 1981.
66. **Lea Stow, J., Kjellen, L., Unger, E. et al.,** Hepatic heparan sulfate proteoglycans are concentrated on the sinusoidal plasmalemmal domain and in intracellular organelles of hepatocytes, *J. Cell Biol.*, 100, 975, 1985.
67. **Dietrich, C. P., Sampaio, L. O., and Toledo, O. M. S.,** Characteristic distribution of sulfated mucopolysaccharides in different tissue and in their respective mitochondria, *Biochem. Biophys. Res. Commun.*, 71, 1, 1976.
68. **Goldberg, B. and Green, H.,** The synthesis of collagen and protocollagen hydroxylase by fibroblastic and nonfibroblastic cell lines, *Proc. Natl. Acad. Sci. U.S.A.*, 59, 1110, 1968.
69. **Langness, U. and Udenfriend, S.,** Collagen biosynthesis in non-fibroblastic cell lines, *Proc. Natl. Acad. Sci. U.S.A.*, 71, 50, 1974.
70. **Prince, R., Klein, V., Sudhakaran, P. R. et al.,** Metabolism of sulfated glycosaminoglycans in rat hepatocytes. Synthesis of heparan sulfate and distribution into cellular and extracellular pools, *Biochim. Biophys. Acta*, 630, 402, 1980.
71. **Guzelian, P. S. and Diegelmann, R. F.,** Localization of collagen prolylhydroxylase to the hepatocytes, *Exp. Cell Res.*, 123, 269, 1979.
72. **Guzelian, P. S., Diegelman, R. F., Qureshi, G. D. et al.,** Immunohistochemical evidence for synthesis of collagen by the hepatocyte, *Gastroenterology*, 79, 1024, 1980.
73. **Diegelmann, R. F., Cohen, I. K., and Guzelian, P. S.,** Rapid degradation of newly synthesized collagen by primary cultures of adult rat hepatocytes, *Biochem. Biophys. Res. Commun.*, 97, 819, 1980.
74. **Hata, R., Ninomiya, Y., Nagai, Y. et al.,** Biosynthesis of interstitial types of collagen by albumin producing rat liver parenchymal cell (hepatocyte) clone in culture, *Biochemistry*, 19, 169, 1980.
75. **Guzelian, P. S., Quereshi, G. D., and Diegelmann, R. F.,** Collagen synthesis by the hepatocytes: studies in primary cultures of parenchymal cells from adult rat liver, *Collagen Rel. Res.*, 1, 83, 1981.
76. **Sakakibara, K., Suzuki, T., Tsukada, Y. et al.,** Distribution of type I collagen and basement membrane proteins in the cultures of cloned epithelial liver cells, *Cell Struct. Funct.*, 7, 213, 1982.
77. **Tseng, S. C. G., Lee, P. C., Ells, P. F. et al.,** Collagen production by rat hepatocytes and sinusoidal cells in primary monolayer culture, *Hepatology*, 2, 13, 1982.
78. **Hatahara, T. and Seyer, J. M.,** Procollagen production by rat hepatocytes in primary culture, *Biochim. Biophys. Acta*, 716, 431, 1982.
79. **Diegelman, R. F., Guzelian, P. S., Gay, R. et al.,** Collagen formation by the hepatocytes in primary monolayer culture and in vivo, *Science*, 219, 1343, 1983.

80. **Hata, R. and Ninomiya, Y.,** Hepatocytes (hepatic parenchymal cells) produce the major part of collagen in vivo, *Biochem. Int.*, 8, 181, 1984.
81. **Voss, B., Rauterberg, J., Pott, G. et al.,** Non-parenchymal cells cultivated from explants of fibrotic liver resemble endothelial and smooth muscle cells from blood vessel walls, *Hepatology*, 2, 19, 1982.
82. **Senoo, H., Hata, R. I., Nagai, Y. et al.,** Stellate cells (vitamin A-storing cells) are the primary site of collagen synthesis in non-parenchymal cells in the liver, *Biomed. Res.*, 5, 451, 1984.
83. **Friedman, S. L., Roll, F. J., Boyles, J. et al.,** Hepatic lipocytes: the principal collagen producing cells of normal rat liver, *Proc. Natl. Acad. Sci. U.S.A.*, 82, 8681, 1985.
84. **De Leuw, A. M., Mcarthy, S. P., Geerts, A. et al.,** Purified rat liver fat-storing cells in culture divide and contain collagen, *Hepatology*, 4, 392, 1984.
85. **Holt, K., Bennett, M., and Chojkier, M.,** Acetaldehyde stimulates collagen and non-collagen protein production by human fibroblasts, *Hepatology*, 4, 843, 1984.
86. **Savolainen, E. R., Leo, M. A., Timpl, R. et al.,** Acetaldehyde and lactate stimulate collagen synthesis of cultured baboon liver myofibroblasts, *Gastroenterology*, 87, 777, 1984.
87. **Fujiwara, K., Sakai, T., Oda, T. et al.,** The presence of collagenase in Kupffer cell in rat liver, *Biochem. Biophys. Res. Commun.*, 54, 531, 1973.
88. **Nagai, Y., Hori, H., Hata, I. R. et al.,** Collagenase production by adult rat hepatocytes in primary culture, *Biomed. Res.*, 3, 345, 1982.
89. **Maruyama, K., Okazaki, I., Kobayashi, T. et al.,** Collagenase production by rabbit liver cells in monolayer culture, *J. Lab. Clin. Med.*, 102, 543, 1983.
90. **Okazaki, I., Maruyama, K., Kashiwazaki, K. et al.,** Mechanism of collagenase production by liver cells, in *Pathobiology of Hepatic Fibrosis*, Hirayama, C. and Kivirikko, K. E., Eds., Elsevier, Amsterdam, 1985, 141.
91. **Bradshaw, R. A. and Sporn, M. B.,** Polypeptide growth factors and the regulation of cell growth and differentiation, *Fed. Proc.*, 42, 2590, 1983.
92. **Montfort, I. and Pérez-Tamayo, R.,** Collagenase in experimental carbon tetrachloride cirrhosis of the liver, *Am. J. Pathol.*, 92, 411, 1978.
93. **Biempica, L., Takahashi, S., Biempica, S. et al.,** Immunohistochemical localization of collagenase in hepatic murine schistosomiasis, *J. Histochem. Cytochem.*, 31, 488, 1983.
94. **Pérez-Tamayo, R.,** Pathology of collagen degradation: review, *Am. J. Pathol.*, 92, 509, 1978.
95. **Takahashi, S. and Koda, K.,** Radioimmunoassay of soluble and insoluble collagenase in fibrotic liver, *Biochem. J.*, 220, 157, 1984.
96. **Rojkind, M., Giambrone, M. A., and Takahashi, S.,** Collagen polymorphism in normal and fibrotic liver, *J. U.O.E.H. (Jpn.)*, Suppl. 4, 157, 1982.
97. **Foidart, J. N., Berman, J. J., Paglia, L. et al.,** Synthesis of fibronectin, laminin, and several collagens by a liver-derived epithelial cell line, *Lab. Invest.*, 42, 525, 1980.
98. **Voss, B., Allam, S., Rauterberg, J. et al.,** Primary cultures of rat hepatocytes synthesize fibronectin, *Biochem. Biophys. Res. Commun.*, 90, 1348, 1979.
99. **Wahl, S. M., Tsukamoto, Y., Obriest, R. et al.,** Stimulation of fibroblast activity by soluble mediators, in *Proceedings of Connective Tissue of the Normal and Fibrotic Human Liver*, Gerlash, U., Pott, G., Raterberg, J., and Voss, B., Eds., Georg Thieme Verlag, Stuttgart, 1982, 50.
100. **Birdwell, C. R., Gospodarowicz, D., and Nicolson, G. L.,** Identification, localization, and role of fibronectin in cultured bovine endothelial cells, *Proc. Natl. Acad. Sci. U.S.A.*, 75, 3273, 1978.
101. **Jaffe, E. A. and Mosher, D. F.** Synthesis of fibronectin by cultured human endothelial cells, *J. Exp. Med.*, 147, 1779, 1978.
102. **Macarak, E. J., Kirby, E., Kirk, T. et al.,** Synthesis of cold-insoluble globulin by cultured calf endothelial cells, *Proc. Natl. Acad. Sci. U.S.A.*, 75, 2621, 1978.
103. **Pérez-Tamayo, R.,** Some aspects of connective tissue of the liver, *Prog. Liver Dis.*, 2, 192, 1965.
104. **Montfort, I. and Pérez-Tamayo, R.,** The parenchyma/collagen ratio in normal and hypertrophic rat kidneys, *Proc. Soc. Exp. Biol. Med.*, 110, 731, 1962.
105. **Montfort, I. and Pérez-Tamayo, R.,** The muscle-collagen ratio in normal and hypertrophic human hearts, *Lab. Invest.*, 11, 463, 1962.
106. **Harkness, R. D.,** Collagen in regenerating liver of the rat, *J. Physiol.*, 17, 257, 1952.
107. **Harkness, R. D.,** Regeneration of the liver, *Br. Med. Bull.*, 13, 87, 1957.
108. **Rojkind, M. and Valadez, G.,** Regulation of fibroblast proliferation by Kupffer cells and monocytes, in *Proceedings of the Ciba Foundation,* Symp. 114, Evered, D., and Whelan, J., Eds., Pitman, London, 1985, 208.
109. **Nakano, M. and Lieber, C. S.,** Ultrastructure of initial stages of perivenular fibrosis in alcohol-fed baboons, *Am. J. Pathol.*, 106, 145, 1982.
110. **Lieber, C. S.,** Precursor lesions of cirrhosis, *Alcohol and Alcoholism,* 18, 5, 1983.
111. **Martinez-Hernández, A.,** The hepatic extracellular matrix. II. Electron immunohistochemical studies in rats with CCl-induced cirrhosis, *Lab. Invest.*, 53, 166, 1985.

112. **Bianchi, F. B., Biagini, G., Ballardini, G. et al.,** Basement membrane production by hepatocytes in chronic liver disease, *Hepatology*, 4, 1167, 1984.
113. **Murata, K., Kudo, M., Onuma, F. et al.,** Changes of collagen types at various stages of human liver cirrhosis, *Hepatogastroenterology*, 31, 158, 1984.
114. **Ooshima, A.,** Collagen, B chains: increased proportion in human atherosclerosis, *Science*, 213, 666, 1981.
115. **Morton, L. F. and Barnes, M. J.,** Collagen polymorphism in the normal and diseased blood vessel wall. Investigation of collagen I, III, and V, *Atherosclerosis*, 42, 41, 1982.
116. **Narayanan, A. S. and Page, R. C.,** Synthesis of type V collagen by fibroblasts derived from normal, inflamed and hyperplastic human connective tissues, *Collagen Rel. Res.*, 5, 297, 1985.
117. **Pérez-Tamayo, R. and Montfort, I.,** The susceptibility of hepatic collagen to homologous collagenase in human and experimental cirrhosis of the liver, *Am. J. Pathol.*, 100, 427, 1980.
118. **Takahashi, S., Dunn, M. A., And Seifter, S.,** Liver collagenase in murine schistosomiasis, *Gastroenterology*, 78, 1425, 1980.
119. **Takahashi, S. and Kobayashi, M.,** Concanavalin A reduces liver collagen accumulation in murine schistosomiasis, *Hepatology*, 2, 249, 1982.
120. **Okazaki, I. and Maruyama, K.,** Collagenase activity in experimental hepatic fibrosis, *Nature (London)*, 252, 49, 1974.
121. **Okasaki, I. and Maruyama, K.,** Mammalian collagenase in the process of hepatic fibrosis, *J. U.O.E.H. (Jpn.)*, 2, 401, 1980.
122. **Maruyama, K., Feinman, L., Okazaki, I. et al.,** Direct measurement of neutral collagenase activity in homogenates from baboon and human liver, *Biochim. Biophys. Acta*, 658, 124, 1981.
123. **Kleinman, H. K., Klebe, R. J., and Martin, G. R.,** Role of collagenous matrices in the adhesion and growth of cells, *J. Cell Biol.*, 88, 473, 1981.
124. **Johansson, S., Kjellen, L., Hook, M. et al.,** Substrate adhesion of rat hepatocytes: a comparison of laminin and fibronectin as attachment proteins, *J. Cell Biol.*, 90, 260, 1981.
125. **Rubin, K., Hook, M., Obrink, B. et al.,** Substrate adhesion of rat hepatocytes: mechanism of attachment to collagen substrates, *Cell*, 24, 463, 1981.
126. **Obrink, B.,** Hepatocyte collagen adhesion, *Meth. Enzymol.*, 82, 513, 1982.
127. **Strom, S. C. and Michalopoulos, G.,** Collagen as a substrate for cell growth and differentiation, *Meth. Enzymol.*, 82, 544, 1982.
128. **Sirica, A. E., Richards, W., Tsukada, Y. et al.,** Fetal phenotypic expression by adult rat hepatocytes on collagen gel/nylon meshes, *Proc. Natl. Acad. Sci. U.S.A.*, 76, 283, 1979.
129. **Mak, W. W. N., Sattler, C. A., and Pitot, H. C.,** Accumulation of actin microfilaments in adult rat hepatocytes cultured on collagen gel/nylon mesh, *Cancer Res.*, 40, 4552, 1980.
130. **Mak, W. W. N. and Pitot, H. C.,** Microfilament accumulation and the transport of amino acids and glucose in adult rat hepatocytes cultured on collagen gel/nylon mesh, *Biochem. Biophys. Res. Commun.*, 98, 203, 1981.
131. **Rojkind, M., Gatmaitan, Z., Mackensen, S. et al.,** Connective tissue biomatrix. Its isolation and utilization for long-term culture of normal rat hepatocytes, *J. Cell Biol.*, 87, 255, 1980.
132. **Reid, L. M., Gatmaitan, Z., Arias, I. et al.,** Long-term cultures of normal rat hepatocytes on liver biomatrix, *Ann. N.Y. Acad. Sci.*, 349, 70, 1980.
133. **Reid, L. and Jefferson, D. M.,** Culturing hepatocytes and other differentiated cells, *Hepatology*, 3, 548, 1984.
134. **Wicha, M. S., Lowrie, G., Kohn, E. et al.,** Extracellular matrix promotes mammary epithelial growth and differentiation in vitro, *Proc. Natl. Acad. Sci. U.S.A.*, 79, 3213, 1982.
135. **Mourelle, M., Cordero-Hernandez, J., Ponce-Noyola, P. et al.,** Abnormal matrix recognition by Morris hepatomas correlates with low glucagon binding capacity, *Hepatology*, 3, 303, 1983.
136. **Tung, P. S. and Fritz, I. B.,** Extracellular matrix promotes rat Sertoli cell histotypic expression in vitro, *Biol. Reprod.*, 30, 213, 1984.
137. **Mourelle, M. and Rubalcava, B.,** Changes in the insulin and glucagon receptors in the regenerating liver following intoxication with carbon tetrachloride, *Biochem. Biophys. Res. Commun.*, 88, 189, 1979.
138. **Mourelle, M. and Rubalcava, B.,** Regeneration of the liver after carbon tetrachloride, *J. Biol. Chem.*, 256, 1656, 1981.
139. **Kramer, R. H. and Nicolson, G. L.,** Interactions of tumor cells with vascular endothelial cell monolayers: a model for metastatic invasion, *Proc. Natl. Acad. Sci. U.S.A.*, 76, 5704, 1979.
140. **Kramer, R. H., González, R., and Nicolson, G. L.,** Metastatic tumor cells adhere preferentially to the extracellular matrix underlying vascular endothelial cells, *Int. J. Cancer*, 26, 639, 1980.
141. **Terranova, V. P., Williams, J. E., Liotta, L. A. et al.,** Modulation of the metastatic activity of melanoma cells by laminin and fibronectin, *Science*, 226, 982, 1984.
142. **Tóres, M. E. and Rojkind, M.,** Selection of liver preneoplastic cells on kidney biomatrix (abstr.), *Hepatology*, 5, 1037, 1985.

143. **Rojkind, M. and Kershenobich, D.,** Diagnostic procedures for liver fibrosis, in *Diagnostic Procedures in the Evaluation of Hepatic Disease*, Beker, S., Ed., Alan R. Liss, New York, 1983, 73.
144. **Rohde, H., Hahn, E., and Timpl, R.,** Radioimmunoassay for aminoterminal procollagen peptides in liver disease, *Fresenius Z. Analyt. Chem.*, 290, 151, 1978.
145. **Wick, G., Brunner, H., Penner, E. et al.,** The diagnostic application of specific antiprocollagen sera. II. Analysis of liver biopsies, *Int. Arch. Allergy Appl. Immunol.*, 56, 316, 1978.
146. **Rohde, H., Vargas, M., Hahn, E. et al.,** Radioimmunoassay for type III procollagen peptide and its application to human liver disease, *Eur. J. Clin. Invest.*, 9, 451, 1979.
147. **Timpl, R., and Glanville, R. W.,** The aminopropeptide of collagen, *Clin. Orthopaed.*, 158, 224, 1981.
148. **Readsch, R., Stiehl, A., Waldherr, R. et al.,** Procollagen type III peptide serum concentrations in chronic persistent and chronic active hepatitis and in cirrhosis of the liver and their diagnostic value, *Gastroenterologie*, 20, 738, 1982.
149. **Niemela, O., Risteli, L., Sotaniemi, E. A. et al.,** Heterogeneity of the antigens related to the aminoterminal propeptide of type III procollagen in human serum, *Clin. Chim. Acta*, 124, 39, 1982.
150. **Bolarin, D. M., Savolainen, E. R., and Kivirikko, K. I.,** Enzymes of collagen synthesis and type III procollagen amino-propeptide in serum from Nigerians with hepatocellular carcinoma and other malignant diseases, *Int. J. Cancer*, 29, 401, 1982.
151. **Savolainen, E. R., Miettinen, T. A., and Pikkarainen, P.,** Enzymes of collagen synthesis and type III procollagen aminopropeptide in the evaluation of D-penicillamine and medroxyprogesterone treatments of primary biliary cirrhosis, *Gut*, 24, 136, 1983.
152. **Rohde, H., Langer, I., Krieg, T. et al.,** Serum and urine analysis of the aminoterminal procollagen peptide type III by radioimmunoassay with antibody Fab fragments, *Collagen Rel. Res.*, 3, 371, 1983.
153. **Niemela, O., Risteli, L., Sotianemi, E. A. et al.,** Amino terminal propeptide of type III procollagen in serum in alcoholic liver disease, *Gastroenterology*, 85, 254, 1983.
154. **Frei, A., Zimmermann, A., and Weigand, K.,** The N-terminal propeptide of collagen type III in serum reflects activity and degree of fibrosis in patients with chronic liver disaese, *Hepatology*, 4, 830, 1984.
155. **Weigand, K., Zaugg, P. Y., Frei, A. et al.,** Long-term follow up of serum N-terminal propeptide of collagen type III levels in patients with chronic liver disease, *Hepatology*, 4, 835, 1984.
156. **Savolainen, E. R., Goldberg, B., Leo, M. A. et al.,** Diagnostic value of serum procollagen peptide measurements in alcoholic liver disease, *Alcoholism Clin. Exp. Res.*, 8, 384, 1984.
157. **Hahn, E. G.,** Blood analysis for liver fibrosis, *J. Hepatol.*, 1, 67, 1984.
158. **Rojkind, M.,** The blue glass and the predictive value of serum amino terminal propeptide of type III procollagen as a marker of liver fibrosis, *Hepatology*, 4, 977, 1984.
159. **Galambos, M. R., Collins, D. C., and Galambos, J. T.,** A radioimmunoassay procedure for type III procollagen: its use in the detection of hepatic fibrosis, *Hepatology*, 5, 38, 1985.
160. **Pérez-Tamayo, R.,** Cirrhosis of the liver: a reversible disease?, in *Pathology Annual*, Vol. 2, Sommers, S. C. and Rosen, P. P., Eds., Appleton-Century-Crofts, New York, 1979, 183.
161. **Rojkind, M. and Dunn, M. A.,** Hepatic fibrosis, *Gastroenterology*, 76, 849, 1979.
162. **Rojkind, M., Uribe, M., and Kershenobich, D.,** Colchicine and the treatment of liver cirrhosis, *Lancet*, 7793, 38, 1973.
163. **Rojkind, M. and Kershenobich, D.,** Effect of colchicine on collagen, albumin and trasferin synthesis by cirrhotic rat liver slices, *Biochim. Biophys. Acta*, 378, 415, 1975.
164. **Kershenobich, D., Uribe, M., Suarez, G. I. et al.,** Colchicine, serum albumin and alkaline phosphatase, *Ann. Intern. Med.*, 89, 144, 1978.
165. **Kershenobich, D., Uribe, M., Suarez, G. I. et al.,** Treatment of liver cirrhosis with colchicine: a double blind randomized trial, *Gastroenterology*, 77, 532, 1979.
166. **Rojkind, M.,** Antifibrotic therapy in liver disease, *Ital. J. Gastroenterol.*, 12, 97, 1980.
167. **Kershenobich, D., Alcocer, J., Quiroga, A. et al.,** Effect of colchicine on immunoregulartory T-lymphocytes and monocytes in patients with primary biliary cirrhosis (abstr.), *Clin. Sci.*, 32, 490, 1984.
168. **Mourelle, M., Rojkind, M., and Rubalcava, B.,** Colchicine improves alterations in the liver adenylate cyclase system of cirrhotic rats, *Toxicology*, 21, 213, 1981.
169. **Yahuaca, P., Amaya, A., Rojkind, M. et al.,** Cryptic ATPase activities in plasma membranes of CCl — cirrhotic rats. Its modulation by changes in cholesterol-phospholipid ratios, *Lab. Invest.*, 53, 541, 1985.
170. **Borek, C. and Williams, G. M., Eds.,** Differentiation and carcinogenesis in liver cell cultures, *Ann. New York Acad. Sci.*, 349, 1980.
171. **Harris, R. A. and Cornell, N. W., Eds.,** *Isolation characterization and use of hepatocytes*, Elsevier Biomedical, New York, Amsterdam, 1983.
172. **Reid, L. M. and Rojkind, M.,** New techniques for culturing differentiated cells: reconstituted basement membrane rafts, *Meth. Enzymol.*, 58, 263, 1979.
173. **Guguen-Guillouzo, Ch., Clement, B., Lescoat, G. et al.,** Modulation of human fetal hepatocyte survival and differentiation by interactions with a rat liver epithelial cell line, *Dev. Biol.*, 105, 211, 1984.

174. **Guillouzo, A., Delers, F., Clement, B. et al.,** Long-term production of acute-phase proteins by adult rat hepatocytes co-cultured with another liver cell type in serum-free medium, *Biochem. Biophys. Res. Commun.*, 120, 311, 1984.
175. **Clement, B., Guguen-Guillouso, Ch., Campio, J. P. et al.,** Long-term co-cultures of adult human hepatocytes with rat liver epithelial cells: modulation of albumin secretion and accumulation of extracellular material, *Hepatology*, 4, 373, 1984.
176. **Thomas, K. A. and Gimenez-Gallego, G.,** Fibroblast growth factors: broad spectrum mitogens with potent angiogenic activity, *TIBS*, 11, 81, 1986.
177. **Wang, J. L. and Ming Hsu, Y.,** Negative regulators of cell growth, *TIBS*, 11, 24, 1986.
178. **Wyler, D. J., Wahl, S. M., and Wahl, L. M.,** Hepatic fibrosis in schistosomiasis: egg granulomas secrete fibroblast stimulating factor in vitro, *Science*, 202, 438, 1978.
179. **Wyler, D. J., Stadecker, M. J., Dinarello, Ch. A. et al.,** Fibroblast stimulation in schistosomiasis. V. Egg granuloma macrophages spontaneously secrete a fibroblast stimulating factor, *J. Immunol.*, 132, 3142, 1984.
180. **Wyler, D. J.,** Regulation of fibroblast functions by products of schistosomal egg granulomas: protential role in the pathogenesis of hepatic fibrosis, in *Cytopathology of Parasitic Disease*, Pittman, London, 1983, 190.
181. **Hatahara, T. and Seyer, J. M.,** Isolation and characterization of a fibrogenic factor from CCl-damaged rat liver, *Biochim. Biophys. Acta*, 716, 377, 1982.
182. **Raghow, R., Gossage, D., Seyer, J. M. et al.,** Transcriptional regulation of type I collagen genes in cultured fibroblasts by a factor isolated from thioacetamide-induced fibrotic rat liver, *J. Biol. Chem.*, 259, 12718, 1984.
183. **Knook, D. L. and Leeuw, A. M.,** Cell types of the normal liver and their possible role in fibrogenesis, in *Proceeding of Connective Tissue of the Normal and Fibrotic Human Liver*, Gerlach, U., Pott, G., Rauterberg, J., and Voss, B., Eds., Georg Thieme Verlag, Stuttgart, 1982, 67.
184. **Rojkind, M., Reyes-Romero, M., Cerbón-Ambriz, J. et al.,** Cell-cell and cell-matrix interactions in liver cirrhosis, in *Pharmacological, Morphological and Physiological Aspects of Liver Aging*, Van Bezooijen, C. F. A., Ed., Eurage, The Netherlands, 1983, 59.
185. **Rojkind, M., Cordero-Hernandez, J., and Ponce, P.,** Non-collagenous glycoproteins of the connective tissue and biomatrix, in *Progress in Clinical and Biological Research*, Vol. 154, Berk, P. D., Castro-Malaspina, H., and Wasserman, L. R., Eds., Alan R. Liss, New York, 1984, 103.

Chapter 8

METHOD OF TREATMENT OF FIBROTIC LESIONS BY TOPICAL ADMINISTRATION OF LATHYROGENIC DRUGS

Milos Chvapil

TABLE OF CONTENTS

I. Introduction 162

II. Dynamics of Scar Formation 162

III. Pharmacology of Fibrosis 163

IV. Maturation (Polymerization) of Collagen 163

V. Toxicity of Systemically Administered Lathyrogens 164

VI. Indications for Topical Treatment 165
  A. Peritendinous Adhesions 165
  B. Perineural Adhesions 166
  C. Joint Stiffness 166
  D. Burn Scar Contractures 167

VII. Evidence for the Effectiveness and Safety of Topical βAPN 167
  A. Percutaneous Resorption of βAPN 167
  B. Safety Information on Topical βAPN 170
  C. Effectiveness of Topical βAPN — Animal Studies 170

VIII. Preliminary Information on Ongoing Human Trials 171

IX. Final Comments 173

References 173

I. INTRODUCTION

In human and in veterinary medicine there has always been an urgent need for an efficient method of inhibiting abnormal accumulation of collagen in the form of fibrotic nodules, excessive scars, peritendinous adhesions, stiff joints, and other pathologies. Only in the last decade has the knowledge on various aspects of collagen metabolism and collagen synthesis regulating mechanisms made it possible to apply basic knowledge to more integrated and complex systems as represented by various pathological situations related to abnormal collagen deposition and stabilization of its structure. The goal of various treatments aimed at the inhibition of collagen metabolism is to *selectively* inhibit the synthesis and deposition of this fibrous protein *specifically* in the fibrotic lesion, but not in intact tissues. However, an efficient method of interfering with the fibrotic lesions has not yet been developed. It has been my experience that inhibition of the maturation (polymerization, stabilization of the structure) of collagen has so far been the most successful and powerful method of interfering with the physical properties of collagenous tissues manifested as clinical pathologies of scar contractures, fibrotic strictures, joint stiffness etc. Apparently, it is not the total content of collagenous structures within a certain tissue, but rather the physical properties, controlled by the degree of polymerization of the collagenous matrix that represents the real danger to tissue or organ function. When collagen is crosslinked, possibly by stable covalent cross-links involving the function of the enzyme lysyl oxidase, it is less degradable by mammalian collagenases and forms a compact rigid scar that binds less water. As indicated below, the formation of a scar, which by maturation of its structural macromolecules ultimately shrinks, contracts, and forms deformities and strictures, represents the real danger for the patient. It is the object of this review to present recent findings on topical interference with fibrotic tissue deposits, peritendinous adhesions, stiff joints, and other fibrotic lesions by a group of drugs called lathyrogens. Lathyrogens are defined as a group of chemical substances that inhibit, by any method, the formation of intermolecular and intramolecular covalent crosslinks in collagen or elastin structures. Systemically administered lathyrogens have been very thoroughly studied from both a basic and a clinical point of view for the last 2 decades.[1-4] Although my concept of using lathyrogens in topical application is based on previous studies and impressive clinical results mainly with β-aminopropionitrile (βAPN), this review will analyze the reasons why such an effective and needed treatment never became a clinical modality.

In order to justify the reasons for administering certain lathyrogenic drugs topically into the site of the fibrotic lesion, it is necessary to briefly review several topics related to this therapeutic method.

II. DYNAMICS OF SCAR FORMATION

Scar is composed mainly of collagen synthesized by fibroblasts. Once the collagen molecules are secreted from the fibroblasts into the extracellular environment, they undergo intramolecular and intermolecular covalent crosslinking to form collagen fibers that become progressively less soluble, are more resistant to enzymatic degradation, and acquire mechanical strength. This process of stabilization of the collagen superstructure is also called maturation or polymerization of collagen.

Formation of a scar is the result of a fibroproliferative inflammation. Scar is an imperfect method of repairing tissue defects: its mechanical and functional characteristics are inferior to the original-substituted tissue.[5,6] Scar continuously undergoes a remodeling, which is reflected in the higher metabolic activity of a repair tissue.[7] It disrupts the structural-functional integrity of the organ. Still, with the inability of higher organisms to regenerate the tissue defects *ad integrum*, it is the only way of repairing deep wounds. Several steps could be identified in the course of fibrotic reaction. The injury activates fibroblasts to produce more glycosaminoglycans (GAGs) and collagen. A common part of the injury is bleeding by ruptured vessels. Blood is a known factor of inducing fibrosis either through chemotactic activity of fibrin degradation products or fibrin serving as a guideline for invasion of the clot by fibroblasts. In ruptured or injured tendons, the presence of blood and blood proteins, mainly of fibrin, forms bridges between the tendon and its sheath to form fibrotic peritendinous adhesions that immobilize the joint by impairing the gliding function of the tendon.[8,9] Once collagen is formed and accumulates at the site of the injury, it matures and the abnormal functions of the tissue then manifest themselves.

In the dynamics of the fibroproliferative inflammation, various reactions reach their maximum and then decline with time. Activity of lysyl oxidase, the enzyme that is involved in the polymerization of collagen, is highest at much later stages after collagen has been accumulated in the intercellular space. The highest activity of lysyl oxidase in the wound coincides with formation of a stable collagen within the injured tissue.[10,11] Consequently, the optimal time of "pharmacological" interference with a certain step of the inflammation is at the time of maximum incidence and activity of this specific process.[12,13]

The fibrotic, collagenous tissue is in a continuous metabolic turnover. Both synthesis and degradation of collagen are increased and their relative activities change during healing. It could be assumed that by decreasing the maturation of the scar collagen by the administration of a lathyrogenic agent, the pool of labile collagen is increased.[14,15] These forms would be more readily available to digestion by the collagenase system and by other proteolytic enzymes in the later stage. This should result in a reduction of the size and volume of the fibrotic mass. However, this assumption has not been experimentally verified, as many studies indicate that lathyrogens interfere with collagen cross-linking without changing the total mass of collagen.

III. PHARMACOLOGY OF FIBROSIS

There exist several stages unique for collagen synthesis that could be used as a target for so-called "specific interference" with collagen deposition and maturation. These stages are

1. Hydroxylation of prolyl and lysyl residues by appropriate hydroxylases
2. Glycosylation of some ϵ-NH_2 hydroxylysyl groups
3. Subsequent secretion of the molecule out of the cell
4. In the extracellular space, the molecule further polymerizes or matures under the effect of lysyl oxidase, which forms the basis for development of stable covalent cross-links
5. At different stages of collagen synthesis the molecule is degradable; the "younger" less stable molecules are degraded faster by tissue collagenases

Several sophisticated methods were developed and studied that aimed at reducing collagen deposition or polymerization in the fibrotic lesion.[12,13,16-19] In all these situations, the medication was administered either perorally or parenterally, e.g., by systemic route. Thus, the drug was distributed among all body tissue components and body fluids, causing its dilution and the need to administer a high dose of the drug. This may explain why after systemic administration of many of these drugs at the effective dose proposed for the treatment of fibrotic lesions, (proline analogs, lathyrogens, chelating agents, colchicine, etc.) their therapeutic effect is close to the toxic effect. Thus, there is a permanent risk of general toxicity of the drug used.[20,21]

IV. MATURATION (POLYMERIZATION) OF COLLAGEN

The enzyme forming the basis for the polymerization of collagen is lysyl oxidase. Lysyl

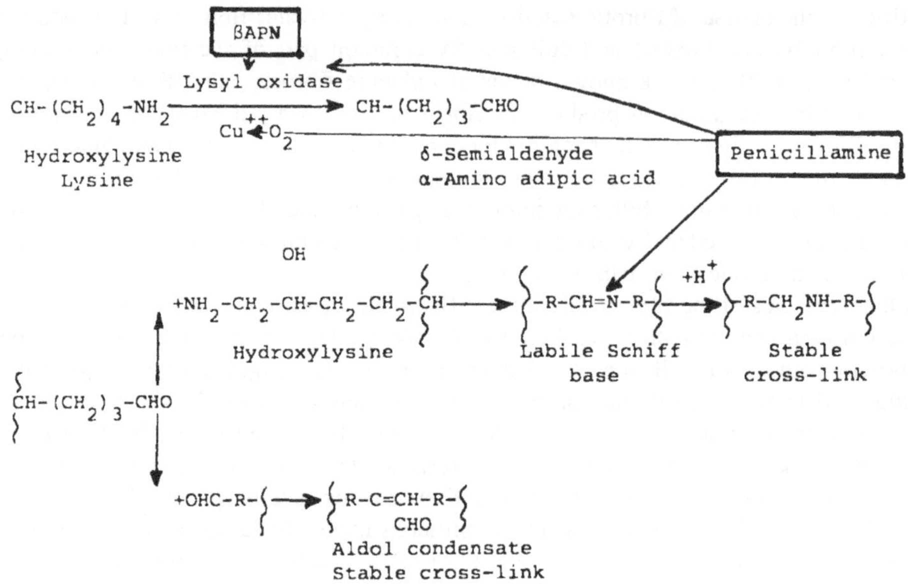

SCHEME 1. Suggested methods of the effect of lathyrogens on the covalent cross-links in the collagen structure.

oxidase oxidatively deaminates specific ε-amino groups of peptidyllysine and hydroxylysine residues of collagen and elastin. The aldehyde product then forms, nonenzymatically, either Schiff base adducts with other specific ε-amino peptidyllysine or hydroxylysine residues or aldol condensates with other preformed aldehydic components. These crosslinking reactions lend structural integrity to collagenous and elastinous connective tissue.

In principle, we may interfere directly with the function of lysyl oxidase by various lathyrogenic agents, βAPN and aminoacetonitril being the most effective. In closed systems, micromolar concentrations of the above drugs are inhibitory to lysyl oxidase. Inhibition of lysyl oxidase by βAPN is irreversible. As lysyl oxidase has a very high turnover and its activity in a granuloma tissue after a single systemic moderate dose of βAPN recovers within 6 to 12 hours, a continuous, effective therapeutic level of a lathyrogen should be kept. Another possibility for inhibition of collagen polymerization is to block the formed aldehydes, as shown for D-penicillamine, thus preventing condensation of aldehydes or formation of Schiff bases as shown in Scheme 1.[23,24]

Since βAPN and D-penicillamine affect different sites in the formation of covalent cross-links, when used together their effect is additive. The lathyrogenic properties of penicillamine are derived from its ability to chelate aldehydes formed by the action of lysyl oxidase on the ε-amino group of tropocollagen lysine, and to cleave the nonreduced Schiff base crosslink. In addition, direct inhibition of lysyl oxidase by D-penicillamine was demonstrated.[23,24] The overall effectiveness of systemically administered D-penicillamine in decreasing the structural stability of collagen has been well-documented.[25] Several clinical studies using the treatment with D-penicillamine have indicated a high incidence of acute hypersensitivity reactions.[26] Striking changes in the metal content of various tissues and fluids after D-penicillamine treatment were also reported.[27]

V. TOXICITY OF SYSTEMICALLY ADMINISTERED LATHYROGENS

The systemic toxicity of βAPN has been the major obstacle in using this drug on a large scale in clinical medicine. There is no doubt that at lathyrogenic dosages of βAPN, the animals stop growing, lose body weight, and lower their food intake. Also, the animals' behavior is changed — they do not clean their fur and are irritated if they are touched.[28] All of this indicates general toxicity. Good evidence exists, however, that various lathyrogenic drugs, mainly βAPN were very effective in blocking the polymerization of collagen in various animal models of fibrotic lesion.

Because of success in animal studies, Keiser and Sjoerdsma[29] studied the effect of βAPN in patients with scleroderma. In short-term treatment with a dose of 2 g/day, no toxic side effects were noted. However, longer courses of treatment were associated with prohibitive reactions, such as allergic skin rash and hemolytic anemia. Peacock and Madden[30] employed βAPN in clinical trials with patients undergoing flexor tendon surgery. Although a significant number of these patients exhibited toxicity or hypersensitivity to the agent, decreased covalent bonding of collagen with significant clinical benefit was demonstrated. The same authors[31] reported no signs of toxicity of βAPN administered at a daily dose of 1 g to five patients after mechanical dilation of urethral strictures. The authors suggested that the lack of non-toxicity of βAPN resulted from using highly purified βAPN-fumarate.

Based on the common incidence of systemic toxicity of βAPN in both animal and human trials, I arrived at the conclusion that the only method of preventing the toxic problems of this clinically important and needed drug is to administer βAPN, and other lathyrogens, topically, either by local injection(s) of the drug(s) into the site of fibrotic lesion or percutaneously by applying the drug onto the intact skin overlying the fibrotic injury. It was assumed that in both methods of application the drug reaches the target tissue in the concentrated form and dilutes minimally so that a several times lower dose of the drug than in systemic application can be used to achieve the desired effect.

VI. INDICATIONS FOR TOPICAL TREATMENT

There exist several clinical symptoms related to abnormal collagen deposition and maturation that could be beneficially treated by topically administered lathyrogens. These include scar contractures, such as after deep burns, hypertrophic scars, keloids, peritendinous adhesions after injury to the tendon, perineural adhesions, stiffness of periarticular tissue of immobilized and surgically treated joints, Dupuytren's contracture, and other fibrotic lesions common in human and veterinary medicine.

A. Peritendinous Adhesions

These adhesions are a major problem in hand surgery. For many years it was believed that the tendon was an inert, nearly avascular structure with highly specialized cells incapable of contributing to the healing process. Healing was then thought to take place by the ingrowth of fibroblasts and capillaries from the peritendinous structures (one scar concept of healing).[32] As a consequence of this philosophy the belief was that tenden adhesions are necessary for tendon healing. These same tendon adhesions are the cause of the marked limitation of gliding function that occurs following many tendon injuries, particularly those of flexor tendons of the hand. The pathogenesis of the adhesion relates to the invasion of fibrogenic cells into the fibrin clot formed due to injury-related bleeding between the tendon and its sheath. Following collagenization and subsequent crosslinking (resulting in the maturation of the deposited collagen), the fragile fibrin adhesion changes into rigid, strong structures limiting or preventing the gliding of the tendon in its sheath. It was shown experimentally in rats and dogs that inhibition of collagen crosslinking in and around tenolyzed tissue by systemically administered βAPN could significantly improve the tendon gliding fraction or reduce stiffness.[13,33,34] In both species, however, the systemic administration of the lathyrogen resulted in undesirable side effects.

Recent research clearly documented that intrinsic repair takes place in the tendon and that tendon cells have potential for repair in a tissue that has a sufficient vessel network.[35,36] Thus, the adhesions can be viewed as only a nondesirable phenomenon.

B. Perineural Adhesions

One of the major complications to axon regeneration of severed nerve ends is excessive scar tissue formation. In their opinions on repair of peripheral nerves, Peacock and Madden[39] indicate that, contrary to tendons, connective tissue within peripheral nerves is highly reactive. Ingrowth of excessive collagenous structures forms a barrier reducing the possibility of re-establishing nerve axons' continuity. Other clinical evidence suggests that external compression of peripheral nerves by dense scar tissue can be responsible for failure of transmission of nerve impulses or axonal regeneration.

Both perineural and intraneural adhesions may limit the essential elasticity of a nerve. This elasticity is essential to avoid a stretch injury to a normal nerve or disruption of a repair in a transected-repaired nerve.

Peacock and Madden[37] were primarily concerned with the formation of scar blocking the proliferation of axons rather than the biomechanical problem. There are a number of papers, chiefly those of Millesi,[38] that address the biomechanical problem. Briefly, if a peripheral nerve is cut in two, it will retract several centimeters: it is therefore always under tension. The clinical problem with repair is that any suture line in the nerve is always under tension and the repair tends to disrupt for that reason alone in spite of posturing of the adjacent joints. When "scar contracture" from peri- and intraneural fibrosis occurs in addition, the problem is potentially worsened.

C. Joint Stiffness

Inflammatory changes in or around the joint, whether from injury, disease, or plaster immobilization in the treatment of fractures and soft tissue injuries have a profound effect on the composition and structure of connective tissue components. Several studies, mainly by Akeson and his group,[39-42] demonstrated changes in the biochemical composition of periaticular fibrous connective tissue after immobilization of synovial joints: water and GAG composition and content were reduced. Collagen crosslinking in periarticular tissue was altered by increasing several identified intermolecular crosslinks.[43] There was no change in collagen mass or collagen type. Therefore, increased joint stiffness after immobilization, documented by arthrographic torque-angular deformation is ascribed to the formation of intra- and intermolecular crosslinks among collagen fibers. Complete rehabilitation of a patient may be prolonged or prevented by joint stiffness.

Immobilization affects four major tissue areas:

- Muscles — atrophy due to the loss of myofibrillar structure and increased muscle contraction, both phenomena reversible after restoration of the motility.
- Cartilage — changes can be characterized as degenerative, including thinning, irregularities, and fibrillation of the hyalin cartilage surfaces, followed by encroachment of the joint by fibro-fatty connective tissue with eventual obliteration of the joint.
- Ligamentous changes are not important to our topic of localized use of lathyrogens and were mentioned above.
- Joint contractures — are the most important and devastating effect of joint immobilization. Loss of motion caused by shrinkage in the capsular structures are followed by ingrowth of connective tissue from the periphery. In the late 1960s Peacock and Madden[34] first noted changes in the collagen extractibility following immobilization of joints. Akeson's work detailed this finding by several excellent studies.

Furlow and Peacock[44] showed in rats that inhibition of collagen cross-linking with βAPN in and around tenolyzed tissue or around the immobilized joint significantly improved the tendon gliding function and reduced joint stiffness. Riley and Peacock[45] produced these findings in dogs.

D. Burn Scar Contractures

"Contractures and hypertrophic scars are two of the most frustrating sequelae of thermal injury . . . and are so common, especially in children, that they are frequently accepted by many physicians as the natural course of events following thermal injury."[49] Contractures of hypertrophic scar is a condition found almost exclusively in human beings.[50] There are surprisingly few studies on the mechanisms of scar contractures. The various hypotheses were presented by Brody et al.[51] as related to:

1. Action of myofibroblasts[52]
2. Buckling of immature connective tissue in an adhesive milieu[51]
3. Locking the buckled collagen in the contracted portion by the macropolysaccharide "glue"[51]

In my opinion, none of the above fits the present knowledge on the dynamics of scarring. In principle, early wound contraction starts a few days after skin excision and, in some species, amounts to 80% reduction of the size of the lesion within 7 to 10 days of the lesion. This phenomenon is related mainly to the contractive effect of myofibroblasts, although other factors cannot be forgotten.[53] Wound contracture is a late phenomenon in the wound healing process. It occurs at times when both the population of fibroblasts, even more myofibroblasts, and the content of glycosaminoglycans are reduced. At the time of contracture, the scar is mainly a collagen deposit, which in addition progressively matures by the formation of covalent crosslinks. Once collagen mass crosslinks, its capacity to bind fluids, to swell and be pliable, is reduced.

Thus, contracture reflects stabilization of collagen structures of the scar by covalent crosslinks, initiated by the effect of lysyl oxidase.[5]

VII. EVIDENCE FOR THE EFFECTIVENESS AND SAFETY OF TOPICAL βAPN

A. Percutaneous Resorption of βAPN

Inhibition of collagen maturation in specific diseases can be achieved either by topical application of βAPN or other lathyrogens onto the skin or by direct injection into the lesion. The major limitation of topical drug administration onto the skin surface is the imposing impermeability of the skin which forms a chemical and waterproofing seal. It was shown by several studies that the rate of percutaneous resorption is limited mainly by the stratum corneum, formed by cornified cells. Substantial knowledge has been accumulated on the permeability of the skin.

It was shown that the rate of transport of some solutes through the skin depends on:

1. Polar-nonpolar nature of the substance
2. Hydration of the skin
3. Blood supply
4. Modification of the stratum corneum by chemicals

Accordingly, noncharged molecules penetrate faster. Fleischer et al.[46] demonstrated this by showing four times faster penetration rate of βAPN-base over that of βAPN-fumarate. Hoffman et al.[47] documented a dose-effect relation by administering βAPN on intact or incised skin at a dose of 5 to 200 μℓ. Breaking strength of the skin wound was inhibited

Table 1
EFFECT OF VEHICLE AND SKIN PRETREATMENT ON THE PERCUTANEOUS ABSORPTION OF βAPN-FUMARATE IN THE RAT[46]

Group	Drug recovery in the urine (% recovered in 6 hr)
1. βAPN-F	
In phosphate buffer on intact skin	0.97
2. βAPN-F	
In phosphate buffer	
Skin washed with 2% Ivory® soap	2.17
3. βAPN-F	
In phosphate buffer	
Skin washed with 2 % Ivory® soap, occlusive bandage	10.56
4. βAPN-F	
In phosphate buffer in 60% DMSO	11.66
5. βAPN-F	
In phosphate buffer in 60% DMSO	
Skin washed in 2% Ivory® soap	10.10
6. βAPN-F[a]	
In phosphate buffer	
Skin washed with 2% Ivory® soap	
Applied in hydrophilic container, occlusive bandage	35.00

Note: ^{14}C-βAPN-fumarate, 2 μCi in 30 mg βAPN-fumarate/mℓ, 0.1 mℓ applied/4 cm^2 skin area. DMSO is dimethyl sufoxide.

[a] % recovered in 24 hr.

in relation to the dose used. An occlusive bandage with the solutes considerably enhances penetration. UV-irradiation of the skin, causing erythema, is supporting percutaneous resorption. Detergents, dimethylsulfoxide (DMSO), and several other solvents have been shown to substantially increase the process (see Table 1).

In the early days of our research program we were very enthusiastic about using βAPN as a free base, due to its faster penetration through the skin and also because it exists as a fluid with easy dispensing. We found, however, that in light and at room temperature free base βAPN decomposes forming toxic acrylonitrile and ammonia. Thus, the instability of the βAPN-free base is the major obstacle in using this drug.

The data in Table 1 indicate the importance of cleaning the skin from contaminating lipids and keratinous debris with Ivory® soap (group 2). It also shows the significant role of occlusive dressings on the magnitude of percutaneous transport (group 3). The enhancement by DMSO is documented by the results of groups 4 and 5. The major accomplishment was the continuous delivery of the drug through skin covered by an occlusive dressing of surgical adhesive tape. By this method (group 6) the delivery was increased almost 40 times when compared with group 1.

Continuous βAPN delivery is essential as 6 hr after a short time exposure of the skin to ^{14}C-βAPN less than 1% of the dose was found within the skin sections, suggesting rapid drug clearance. Topical application of βAPN on the intact skin and over the sponge-induced granuloma tissue inhibited the activity of lysyl oxidase in the granuloma and significantly increased the extractibility of collagen.[46] The slow and continuous transport of βAPN across the skin barrier with appearance in the urine is shown in Table 2.

In a series of experiments, Arem[68] clearly documented the percutaneous absorption of βAPN. In both the rabbit and the dog, island pedicle flaps were raised on a single vascular

Table 2
RECOVERY OF β-APN FROM THE URINE AFTER ITS ADMINISTRATION ON THE SKIN OF RATS THROUGH A HYDROPHILIC BAG CONTAINER

Time (hr)	Recovery in urine (% of the total dose)
4	0.50
8	1.01
12	2.37
24	5.40
72	9.90
96	15.2
120	22.5

Note: Time indicates the urine sampling period. Recovery is given in percent of the dose administered into the hydrophilic bag.

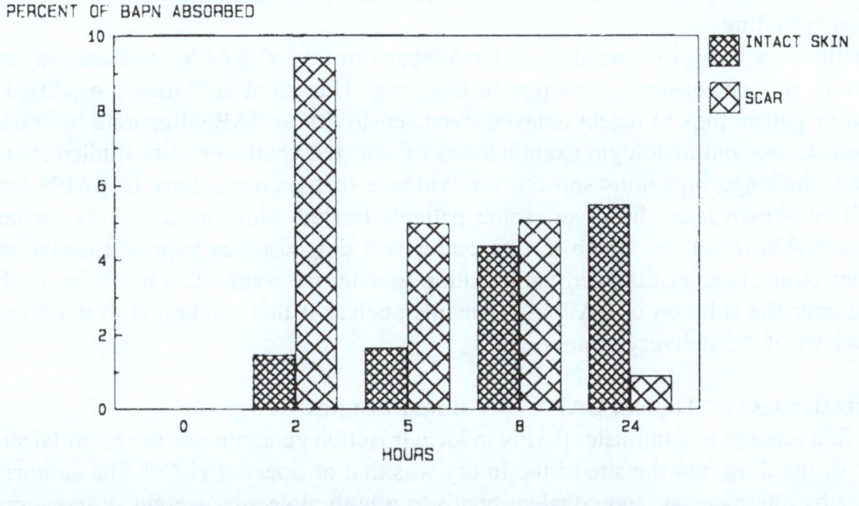

FIGURE 1. Transport of ^{14}C-β-APN-fumarate across intact skin and the scar in rats. In a total of 17 rats a full thickness wound was excised on the back, splinted to prevent wound contraction, and left to heal for 21 days. At this time the same dose of ^{14}C-β-APN-fumarate was topically applied on a 4-cm^2 area of the scarred wound and on intact skin of the same area covered with impermeable adhesive polyurethane tape. At given time intervals, the thoroughly washed skin was excised, digested, and counted for radioactivity, according to Cravens and Chvapil.[48]

pedicle and βAPN applied to the skin surface. Not only did the radioactive βAPN appear in the venous outflow, but the kinetics of absorption through the skin followed the typical pattern of percutaneous absorption of other drugs in these experimental animals.

Although the percutaneous transport of βAPN across intact skin in an effective dose has been documented by clinical effectiveness and direct chemical assays,[46,47] the drug will be applied often over the surgical wound and repair tissue. In order to ascertain the kinetics of βAPN-fumarate transport across skin and scar, Cravens and Chvapil[48] inflicted full thickness skin wounds in rats and prevented the wound contraction by splinting the skin wound edges. In 14 days the wound was healed by a scar. ^{14}C-βAPN-fumarate at the same dose was applied over the scar or intact skin and covered by an occlusive adhesive tape. The data (Figure 1) show the fast and effective absorption of the drug by the scar tissue, which acts almost as a sponge. This contrasted with a continuous and slow penetration of βAPN-F across intact skin.

B. Safety Information on Topical βAPN

Hoffman et al.[47] studied cytotoxicity of βAPN as a free base or as a fumarate salt after intraperitoneal (i.p.) and topical skin application. They found that βAPN-fumarate has a low i.p. toxicity of 3 g/kg and free base is approximately 2.6 times more toxic. The authors were not able to determine the actual LD_{50} for either form of βAPN in cutaneous administration as even the highest technically possible dose used (12.8 g/kg) did not show any signs of systemic toxicity.

Signs of skin toxicity were found only with free base βAPN, never with fumarate salt applied onto the intact skin of various species (mice, rat, pig, rabbit).[47,48]

No adverse effects of topically applied βAPN free base were observed on the skin of the leg of the chicken studied for peritendinous adhesions.

The toxicity of βAPN-fumarate in local injections into peritendinous space,[55] or onto the sciatic nerve[54] and to the partially exposed region of the spinal cord in rats at a 1-m*M* βAPN dose showed the reduced evoked potentials, increased latency, and decreased amplitude of cortical somatosensory-evoked potential as criteria for local toxicity of βAPN. At lower concentrations of βAPN no signs of toxic effects on the nervous system structures were recorded.[69] This study was conducted with the aim to test the possible sustained delivery of βAPN in cases of spinal cord injuries where the ingrowth of collagenous scar interferes with the axon sprouting.

Repetitive exposures of the skin on the subcutis to topical βAPN-fumarate may result in developing a hypersensitivity reaction to this drug. Ulreich et al.[56] used a modified Draize method in guinea pigs to detect delayed hypersensitivity to βAPN-fumarate by intradermal injection. Gross and histologic examinations of selected challenge sites studied 24 to 72 hr after the challenge injections showed no evidence of hypersensitivity to βAPN-fumarate. As will be shown later, however, some patients treated with topical βAPN formulations developed skin irritation, which in one center was diagnosed as hypersensitivity reaction; the other clinical center assigned the irritation to skin maceration due to occlusive dressing applied over the solution of βAPN. It remains unclear if this reaction is to βAPN or other components of the delivery system.

C. Effectiveness of Topical βAPN — Animal Studies

The first attempt to administer βAPN in local injection guaranteeing the essential sustained release of the drug into the site of the injury was that of Speer et al.[57,58] The authors linked the drug by alternate salt and covalent bonds to a high molecular weight, viscous polymer. The complex was then injected locally around the injured tendon. The drug remained at the site of the injury for a sufficient period of time to reduce the peritendinous adhesions. It was found, however, that the high molecular weight polymers themselves had a fibrogenic effect and several efforts to overcome this complication failed.

Biochemistry of subcutaneous granuloma induced by polyvinylalcohol sponge implantation was studied after painting the intact skin over the lesion with βAPN free base or fumarate.[59] Fleischer et al.[46] found that significant inhibition of lysyl oxidase activity in reactive granuloma tissue by a single dose of 20 μℓ βAPN free base also inhibited significantly lysyl oxidase activity of the granuloma tissue underlying the area of topical application as well as that on the opposite site not receiving the treatment. This indicates the resorption and diffusion of the drug. However, the extractibility of collagen into 0.45 *N* NaCl (4°C) was significantly increased only in the granuloma tissue of the βAPN treated area.

In a thorough study, Lata[59] documented that topical βAPN-fumarate applied in a solution with 30% DMSO on the intact skin over the reactive granuloma in rats had no effect on the total content of collagen, noncollagenous proteins, or DNA. Significant changes were observed in the extractibility of collagen of the granuloma tissue into saline, urea, and thiocyanate. The amount of aldehydes in extracted collagen was significantly reduced. Lata also showed that the drug administered with DMSO was more effective than the drug in saline only. These data indicate that the treatment did not induce general toxic effect as only specific interference with collagen stability was recorded. Rakin et al.[54] arrived at the same conclusions: local application of βAPN-fumarate alone or in combination with DMSO significantly inhibited the activity of lysyl oxidase in the reactive granuloma without changing the total content of collagen.

Karpman et al.[60] induced peritendinous adhesions in a chicken foot sublimis tendon. Three weeks after injury a tenolysis was performed and topcial βAPN was applied for 5 days as infusion underneath an impermeable membrane wrapping the tenolyzed area. The work of chicken foot tendon flexion of sham control vs. βAPN treated, showed significant gliding function after topical βAPN application (34.3 ± 8.9 vs. 15.1 ± 3.7 g/degree flexion, $p < 0.01$).

Speer et al.[55] used the same chicken foot tendon model and a similar experimental design on a larger group of animals and arrived at identical results: βAPN reduced by one third the force required to effect tendon gliding and flexion of the joints of the involved digit. While this study still used now abandoned βAPN free base, we repeated the experiment with βAPN-fumarate[61] and reproduced the above reported effects.

Karpman et al.[60] also documented the effectiveness of topical βAPN free base on the joint stiffness model of immobilized rat knee. The knee was immobilized for 15 days, during which time the medication was applied over the whole knee joint. The calculated work of extension of the control vs. treated stiff joint showed significant improvement of joint extensibility after topical βAPN (11.8 ± 0.93 vs. 6.3 ± 0.3 g/degree flexion, $p < 0.0001$).

Tendovaginitis or tendinitis is a common injury mainly in racing horses, causing long-lasting disability, lameness. Once the injury proceeds to the chronic stage, permanent peritendinous adhesions limit the horse's work. Sanders and Chvapil[62] used direct multiple injections of βAPN-fumarate solution into the site of the injured tendon and peritendinous adhesions. Through a G-22 needle they injected approximately 0.3 mℓ at various sites, using a total of 5 to 10 mℓ of the 5 m*M* βAPN fumarate for one treatment. The treatment was repeated every other day for a total of 10 days. The clinical picture of the injury was evaluated both by clinical semiquantitative and objective criteria measuring edema circumference of the leg at the injury site, length of the stride, local skin temperature, and force the horse applied on the leg. The semiquantitative criteria included scoring the pain, degree of lameness, swelling, and induction of the lesion. The results (Figure 2) based on the clinical evaluation of 43 horses showed significant and lasting improvements in 37 horses. It was of interest that after the first local injection treatment, the overall clinical picture of the "bowed tendon" worsened; later a continuous improvement was registered. The authors found it essential that during the treatment which started no sooner than 8 to 10 days after the injury, that the pharmacological treatment had to be combined with controlled physical activity. The injections of saline (control) were ineffective.

Recently, the study was repeated in another group of 65 horses, treated with saline (control) and various doses of βAPN, ranging from 1.75 m*M* to 5.5 m*M*. This time, however, the authors used quantitative criteria of the tendon injury, including measurements of skin temperature, circumference of the injured tendon, length of the stride, and force the horse put on the injured leg. The change in these paramenters was compared with the noninjured leg and followed up for 30 to 60 days after the original 10 days of treatment. The data showed that the maximum effect was obtained with the highest dose of βAPN. Dose of 3.0 m*M* and lower dosages were not effective.

VIII. PRELIMINARY INFORMATION ON ONGOING HUMAN TRIALS

Since 1984, after obtaining FDA approval for new drug investigation, three clinical centers

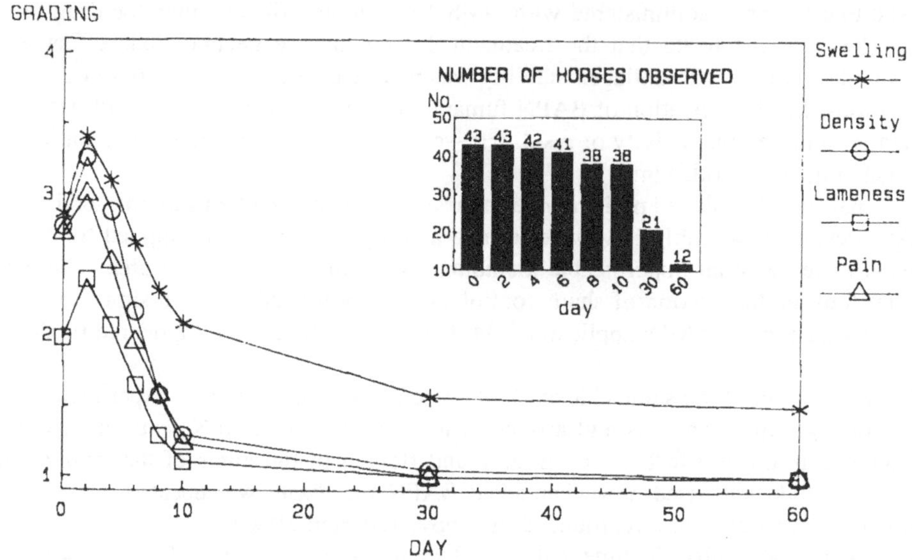

FIGURE 2. Effect of local injection of β-APN-fumarate into tendons and peritendinous sheath of horses with tendovaginitis. Dose was 2.5 or 5.0 m*M* βAPN (as no significant difference was found between treatments, they were combined). A total of 43 horses were treated with topical injections of the lesions with several injections 2 cm apart, totaling 5 to 7 mℓ per treatment. The treatment started 8 to 10 days postinjury and was repeated every second day, five times. The swelling, density of the injured area, degree of lameness, and pain were scored on a scale from 1 to 4, 1 being normal, 4 being extensive change. The horses were observed up to 60 days after starting the treatment, according to Sanders and Chvapil.[62]

have been involved in testing βAPN-fumarate in clinical situations. The clinical pathologies included patients with surgery on the temporomandibular joint, hand surgery patients, and patients with excised keloids and hypertrophic scars. In addition, a new protocol was approved using topical lathyrogenic drugs in patients with scar contractures after deep burns.

Although at this time it would be premature to talk about the effectiveness of the ongoing treatment, some features of the testing protocols may contribute to better understanding of the potentials of this new pharmacological procedure:

In principle, the drug βAPN-fumarate was used only in three different formulations:

- Solution containing 5 mg βAPN per milliliter of 0.5% Azone (Nelson Labs, Irvine, California) solution with 0.001% Triton X-100 to allow emulsification of water-insoluble Azone.
- Ointment containing 5% βAPN in Aquafore base.
- Powder of βAPN-fumarate trapped on gauze bandages, containing 1 mg βAPN per cm^2 of the dressing. After applying the gauze over the area it was soaked with Azone solution to enhance the percutaneous absorption.

In all three applications, an occlusive adhesive impermeable tape was placed over the medication to further enhance the transport.

The treatment protocol calls for initiation of the treatment 7 to 10 days after the surgery to allow for closing of the surgical wound, which otherwise will be hindered by surgical βAPN. The treatment should last at least 6 weeks, corresponding to enhanced activity of lysyl oxidase in the repair tissue.

Originally we preferred continuous delivery of the drug into the site of the injury. With the recent knowledge of fast penetration of βAPN through the scar-repair tissue, we now favor repetitive, short-lasting applications preventing the irritation of the exposed skin.

IX. FINAL COMMENTS

Although this chapter reviews the experimental data mainly of one research group at the University of Arizona, the results of others using topical lathyrogens in various animal models of fibrotic lesions should be mentioned. This mainly refers to several studies by Moorhead who used topical βAPN to limit the degree of post-traumatic fibrotic proliferation in rabbits[63] and to change the curvature and compliance of corneal tissue to radical keratotomy[64] in the rabbit. A similar effect of topical βAPN (33% ointment) was reported by Kogan and Leeds,[65] who significantly reduced the initial flattening of the rabbit cornea after radical keratotomy.

Moorhead[66] also reported on lesser conjunctival scarring in rabbits when the injury was treated consecutively for 21 days with topical βAPN. All these studies as well as the one testing the effect of βAPN in controlling regression of radical keratotomy in humans[67] showed no toxic adverse effect of βAPN.

As with every new therapeutic method, proper indications, optimal formulations, and adequate therapeutic regimens should be established for every clinical pathology. More studies are needed to establish the optimal use of these drugs in definite pathologies involving the maturation of collagenous structures.

REFERENCES

1. **Barrow, M. V., Simpson, C. F., and Miller, E. J.,** Lathyrism: a review, *Q. Rev. Biol.*, 46, 101, 1974.
2. **Levene, C. I.,** Lathyrism, in *Molecular Pathology of Connective Tissues*, Perez-Tamayo, R. and Rojkind, M., Eds., Marcel Dekker, New York, 1973, chap. 4.
3. **Chvapil, M.,** Experimental modifications of collagen synthesis and degradation and their therapeutic applications, in *Collagen in Health and Disease*, Weiss, J. B. and Jayson, M. I. V., Eds., Churchill Livingston, New York, 1982, chap. 12.
4. **Rosmus, J., Trnavsky, K., and Deyl, S.,** Editorial, experimental lathyrism — a molecular disease, *Biochem. Pharmacol.* 15, 1405, 1966.
5. **Chvapil, M. and Koopmann, C. F., Jr.,** Scar formation: physiology and pathological states, *Otolarynogol. Clin. North Am.*, 17, 265, 1984.
6. **Ketchum, L. D., Cohen, I. K., and Master, F. W.,** Hypertrophic scars and keloids, a collective review, *Plast. Reconstruct. Surg.*, 53, 140, 1974.
7. **Peacock, E. E., Jr., Madden, J. W., and Trier, W. C.,** Biologic basis for the treatment of keloids and hypertrophic scars, *South. Med. J.*, 63, 755, 1970.
8. **Peacock, E. E., Jr. and van Winkle, W., Jr.,** *Wound Repair*, W. B. Saunders, Philadelphia, 1976, 19.
9. **Ellis, H.,** The causes and prevention of intestinal adhesions, *Br. J. Surg.*, 69, 241, 1982.
10. **Pinto, J. S. and Bentley, J. P.,** The timecourse of collagen cross-linking, *Biochim. Biophys. Acta*, 354, 254, 1974.
11. **Chvapil, M., McCarthy, D. W., Misiorowski, R. L., Madden, J. W., and Peacock, E. E., Jr.,** Activity and extractability of lysyl oxidase and collagen proteins in developing granuloma tissue, *Proc. Sco. Exp. Biol. Med.*, 146, 688, 1974.
12. **Chavpil, M.,** Minireview, pharmacology of fibrosis: definitions, limits and perspectives, *Life Sci.*, 16, 1345, 1975.
13. **Peacock, E. E., Jr.,** Pharamcological control of surgical scar tissue, *Am. Surg.*, 44, 693, 1978.
14. **Peacock, E. E., Jr. and Madden, J. W.,** Some studies on the effect of β-aminopropionitrile on collagen in healing wounds, *Surgery*, 60, 7, 1966.
15. **Peacock, E. E., Jr.,** Collagenolysis: the other side of the equation, *World J. Surg.*, 4, 297, 1980.
16. **Peacock, E. E., Jr.,** Biologic frontiers in the control of healing, *Am. J. Surg.*, 126, 708, 1973.
17. **Fuller, G. C.,** Perspectives for the use of collagen synthesis inhibitors as antifibrotic agents, *J. Med. Chem.*, 24, 651, 1981.

18. **Uitto, J., Ryhänen, L., Tan, E. M., Oikarinen, A. I., and Zaragoza, E. J.,** Pharmacological inhibition of excessive collagen deposition in fibrotic diseases, *Fed. Proc.*, 43, 2815, 1984.
19. **Chvapil, M., Peacock, E. E., Jr., Carlson, E. C., Blau, S., Steinbronn, K., and Morton, D.,** Colchicine and wound healing, *J. Surg. Res.*, 28, 49, 1980.
20. **Chvapil, M., Misiorowski, R., and Eskelson, C.,** On the mechanisms of β-aminopropionitrile toxicity, *J. Surg. Res.*, 31, 151, 1981.
21. **Madden, J. W., Chvapil, M., Carlson, E. C., and Ryan, J. N.,** Toxicity and metabolic effects of 3,4-dehydroproline in mice, *Toxicol. Appl. Pharmacol.*, 26, 426, 1973.
22. **Arem, A. J., Misiorowski, R., and Chvapil, M.,** Effects of low-dose BAPN on wound healing, *J. Surg. Res.*, 27, 228, 1979.
23. **Nimni, M. E. and Bavetta, L. A.,** Collagen defect induced by penicillamine, *Science*, 150, 905, 1966.
24. **Nimni, M. E.,** A defect in the intramolecular and intermolecular crosslinking of collagen caused by penicillamine. I. metabolic and functional abnormalities in soft tissues, *J. Biol. Chem.*, 243, 1457, 1968.
25. **Nimni, M. E., Deshmukh, K., and Gerth, N.,** Collagen defect induced by penicillamine, *Nature (New Biol.)*, 240, 220, 1972.
26. **Irion, M., Yasuhira, K., and Takeda, T.,** D-penicillamine toxicity in mice. II. Concentrations of Cu, Zn, and Fe related to development of toxicity, *Toxicol. Appl. Pharmacol.*, 63, 1, 1982.
27. **Kuchinskas, E. J. and Rosen, Y.,** Metal chelates of DL-penicillamine, *Arch. Biochem. Biophys.*, 97, 370, 1962.
28. **Naber, E. C., Scott, K., and Johnson, R. M.,** Relationships of divalaent cations to experimental lathyrism and collagen formation, *Fed. Proc.*, 26, 121, 1967.
29. **Keiser, H. R. and Sjoerdsma, A.,** Studies on beta-aminopropionitrile in patients with scleroderma, *Clin. Pharmacol. Ther.*, 8, 593, 1967.
30. **Peacock, E. E., Jr. and Madden, J. W.,** Some studies on the effects of β-aminopropionitrile in patients with injured flexor tendons, *Surgery*, 71, 922, 1972.
31. **Peacock, E. E., Jr. and Madden, J. W.,** Administration of beta-aminopropionitrile to human beings with urethral strictures: a preliminary report, *Am. J. Surg.*, 136, 600, 1978.
32. **Matthews, P.,** The pathology of flexor tendon repair, *Hand*, 11, 233, 1979.
33. **Peacock, E. E., Jr.,** Fundamental aspects of wound healing relating to the restoration of gliding functions after tendon repair, *Surgery*, 119, 241, 1964.
34. **Peacock, E. E., Jr. and Madden, J. W.,** Some studies on the effects of β-aminopropionitrile in patients with injured flexor tendons, *Surgery*, 66, 215, 1969.
35. **Lundborg, G.,** Experimental flexor tendon healing without adhesion formation — a new concept of tendon nutrition and intrinsic healing mechanisms, a preliminary report, *Hand*, 8, 235, 1976.
36. **Lundborg, G. and Rank, F.,** Experimental studies on cellular mechanisms involved in healing of animal and human flexor tendon in synovial environment, *Hand*, 12, 3, 1980.
37. **Peacock, E. E., Jr. and Madden, J. W.,** Some studies on the effects of β-aminopropionitrile in patients with injured flexor tendons, *Surgery*, 66, 215, 1969.
38. **Millesi, H.,** Clinical aspects of nerve healing, in *Wound Healing, International Symposium on Wound Healing Rotterdam 1974*, Gibson, T. and van der Meulen, J. C., Eds., Found. Int. Coop. Med. Sci., Rotterdam, 1974, 232.
39. **Akeson, W. H.,** An experimental study of joint stiffness, *J. Bone Joint Surg.*, 43-A, 1022, 1961.
40. **Akeson, W. H., Amiel, D., and LaViolette, D.,** The connective-tissue response to immobility: a study of chondroitin-4 and 6-sulfate and dermatan sulfate changes in periarticular connective tissue of control and immobilized knees of dogs, *Clin. Orthop.*, 51, 183, 1967.
41. **Akeson, W. H., Woo, S. L-Y., Amiel, D., and Matthews, J. V.,** Biomechanical and biochemical changes in the periarticular connective tissue during contracture development in the immobilized rabbit knee, *Connect. Tissue Res.*, 2, 315, 1974.
42. **Akeson, W. H., Woo, S. L-Y., Amiel, D., Coutts, R. D., and Daniel, D.,** The connective tissue response to immobility: biochemical changes in periarticular connective tissue of the immobilized rabbit knee, *Clin. Orthop.*, 93, 356, 1973.
43. **Akeson, W. H., Amiel, D., Mechanic, G. L., Woo, S. L-Y., Harwood, F. L., and Hamer, M. L.,** Collagen cross-linking alterations in joint contractures: changes in the reducible cross-links in periarticular connective tissue collagen after nine weeks of immobilization, *Connect. Tissue Res.*, 5, 15, 1977.
44. **Furlow, L. and Peacock, E. E., Jr.,** Effect of beta-aminopropionitrile on joint stiffness in rats, *Ann. Surg.*, 165, 442, 1967.
45. **Riley, W. B., Jr. and Peacock, E. E., Jr.,** The identification, distribution and significance of a collagenolytic enzyme in human tissue, *Proc. Soc. Exp. Biol. Med.*, 124, 207, 1967.
46. **Fleisher, J. H., Misiorowski, R., Owen, J. A., and Chvapil, M.,** Topical application of β-aminopropionitrile, *Life Sci.*, 29, 2553, 1981.
47. **Hoffman, D. L., Owen, J. A., and Chvapil, M.,** Healing of skin incision wounds treated with topically applied BAPN free base in the rat, *Exp. Mol. Pathol.*, 39, 154, 1983.

48. **Cravens, R. B., Jr. and Chvapil, M.,** unpublished observation.
49. **Larson, D. L., Abston, S., Dobrkovsky, M., Huang, T., Evans, E. B., and Lewis, S. R.,** The prevention and correction of burn scar contracture and hypertrophy, monograph, Shriners Burns Institute, University of Texas, Medical Branch, Galveston, 1974.
50. **Peacock, E. E., Jr.,** Pharmacologic control of surface scarring in human beings, *Ann. Surg.*, 193, 592, 1982.
51. **Brody, G. S., Peng, S. T. J., and Landel, R. F.,** The etiology of hypertrophic scar contracture: another view, *Plast. Reconstr. Surg.*, 67, 673, 1981.
52. **Majno, G., Gabbiani, G. M., Hirschel, B. J., Ryan, G. B., and Statkov, P. R.,** Contraction of granulation tissue in vitro: similarity to smooth muscle, *Science*, 173, 548, 1971.
53. **van Winkle, W., Jr.,** Wound contraction, *Surg. Gynecol. Obstet.*, 125, 131, 1967.
54. **Rankin, L. L., Chvapil, M., Misiorowski, R., Johnson, P., and Weinstein, P. R.,** Diffusion characteristics of β-aminopropionitrile in peripheral nerve, *Exp. Neurol.*, 79, 97, 1983.
55. **Speer, D. P., Feldman, S., and Chvapil, M.,** The control of peritendinous adhesions using topical β-aminopropionitrile base, *J. Surg. Res.*, 38, 252, 1985.
56. **Ulreich, J. B., Holubec, H., and Chvapil, M.,** BAPN hypersensitivity in the guinea pig test, FDA submission, Food and Drug Administration, Washington, D.C., 1985.
57. **Speer, D. P., Chvapil, M., Brendel, K., and Peacock, E. E., Jr.,** Use of large-molecular-weight compounds to produce local lathyrism in healing wounds, *Surg. Forum*, 24, 37, 1973.
58. **Speer, D. P., Peacock, E. E., Jr., and Chvapil, M.,** The use of large molecular weight compounds to produce local lathyrism in healing wounds, *J. Surg. Res.*, 19, 169, 1975.
59. **Lata, S.,** The Chemistry of Topical Effect of Beta-Aminopropionitrile on Reactive Granuloma Tissue, Ph.D. thesis, University of Madras, Madras, India, 1986.
60. **Karpman, R., Speer, D. P., and Chvapil, M.,** Control of peritendinous adhesions and joint stiffness by a topical lathyrogen, *Trans. Orthop. Res. Soc.*, 6, 202, 1981.
61. **Chvapil, M.,** Method of Treatment of Fibrotic Lesions by Topical Administration of Lathrogenic Drugs, U.S. Patent 4,485,088, November 27, 1984.
62. **Sanders, W. and Chvapil, M.,** Treatment of tendinitis and tendovaginitis in horses by local injection of β-aminopropionitrile, *Vet. Rec.*, in press.
63. **Moorhead, L. C.,** Effects of beta-aminopropionitrile after posterior penetrating injury in the rabbit, *Am. J. Ophthalmol.*, 95, 97, 1983.
64. **Moorhead, L. C., Carroll, J., Constance, G., Jenkins, D. E. III., and Armeniades, C. D.,** Effects of Topical treatment with β-aminopropionitrile after radial keratotomy in the rabbit, *Arch. Ophthalmol.*, 102, 304, 1984.
65. **Kogan, L. L. and Leeds, K.,** Enhancement of radial keratotomy by chemical inhibition of collagen cross-linkages: a preliminary report, *Ann Ophthalmol.*, 15, 842, 1983.
66. **Moorhead, L. C.,** Inhibition of collagen cross-linking: a new approach to ocular scarring, *Curr. Eye Res.*, 1, 77, 1981.
67. **Moorhead, L. C. and Hoffman, R.,** Effects of beta-aminopropionitrile in controlling regression after radial keratomy in humans (abstr.), *Association for Research in Vision and Ophthalmology, Annual Meeting*, 1985, 150.
68. **Arem, A.,** unpublished observation.
69. **Gunasekaran, S. and Chvapil, M.,** unpublished observation.

Chapter 9

HIERARCHICAL STRUCTURE OF COLLAGEN AND ITS RELATIONSHIP TO THE PHYSICAL PROPERTIES OF TENDON

E. Baer, J. J. Cassidy, and A. Hiltner

TABLE OF CONTENTS

I. Introduction 178

II. Hierarchical Structure of Collagen 178
- A. Collagen Fibrils 178
- B. Extrafibrillar Matrix 178

III. Tendon 178
- A. Anatomy 179
- B. Mechanical Properties 182
- C. Effect of Maturity/Aging 185
- D. Effect of Hydration on the Extrafibrillar Matrix 186

IV. Hierarchical Structure in Other Species and Tissues 188
- A. Tendon 188
- B. Intestine 188
- C. Intervertebral Disk 189

V. Summary 191

References 198

I. INTRODUCTION

Collagen in soft connective tissue is a macromolecular multicomposite material organized into complex architectures. In general terms, the tissue consists of collagen fibrils embedded in an amorphous gel-like matrix. We apply the term hierarchical structure to collagen to indicate that a highly organized and well-defined structure exists at all levels from the molecule through the fibril to the tissue. It is this structural hierarchy which permits collagenous tissues to be "manufactured" from identical macromolecular components, with only the macroscopic organization varied to meet the requirements of a specific application.

The role of the hierarchical structure of collagen in determining the physical properties of a tissue will be illustrated using the tendon. This tissue serves as the primary linkage between muscle and bone and functions mainly in uniaxial tension. The effects of maturation and aging on collagen fibrils will also be described including the role of water content in the extrafibrillar matrix. Finally, the universality of hierarchical structure in collagenous tissues will be examined. In particular, a comparison will be made between tendon, intestine, and intervertebral disk. In all these cases, it will be shown that the hierarchical architecture of the collagenous component reflects the mechanical loading modes which are imposed by the particular physiological conditions.

II. HIERARCHICAL STRUCTURE OF COLLAGEN

A. Collagen Fibrils

Collagen is a fibrous protein whose basic chemical constituents are glycine, proline, and hydroxyproline with some other amino acids present in variable proportions. These chains of amino acids are uniformly 290 nm in length and form a triple helical structure called tropocollagen. Five tropocollagen molecules are staggered longitudinally, overlapping by about one quarter of their molecular length, to form a microfibril 3.6 nm in diameter. This quarter stagger, combined with the gap between the ends of successive tropocollagen mol ecules is responsible for the 64 nm periodicity observed by electron microscopy and X-ray diffraction. These microfibrils have both polar and nonpolar segments that are readily accessible for specific inter- and intrafibrillar interactions. Microfibrils are packed in a tetragonal lattice to form subfibrils about 30 nm in diameter. These subfibrils are then assembled into the collagen fibril, the basic unit of connective tissue. These fibrils vary in diameter from 50 to 500 nm depending on the tissue type and age of the animal (Figure 1).[1,2] The fibrils are subsequently surrounded by the extrafibrillar matrix and oriented to form a specific tissue. The physical properties of that tissue are derived from this hierarchical design.

B. Extrafibrillar Matrix

The matrix material which surrounds the collagen fibrils is an amorphous gel-like material composed of proteoglycans and water. The proteoglycans consist of a core protein of high molecular weight hyaluronic acid, with numerous pendant mucopolysaccharide molecules, specifically chondroitin sulfates 4 and 6 and keratan sulfate. The relative proportions of these mucopolysaccharides vary between tissue types, anatomic locations within a single tissue, and with the age of the tissue. When swollen with water, this ultra high molecular weight proteoglycan aggregate forms an extremely efficient network that connects and maintains the hierarchical architecture of the collagen fibrils. The properties of this matrix are strongly dependent on and regulated by the quantity of free and bound water.[2]

III. TENDON

The tendon joins muscle to bone and as such is subjected to predominantly uniaxial tensile

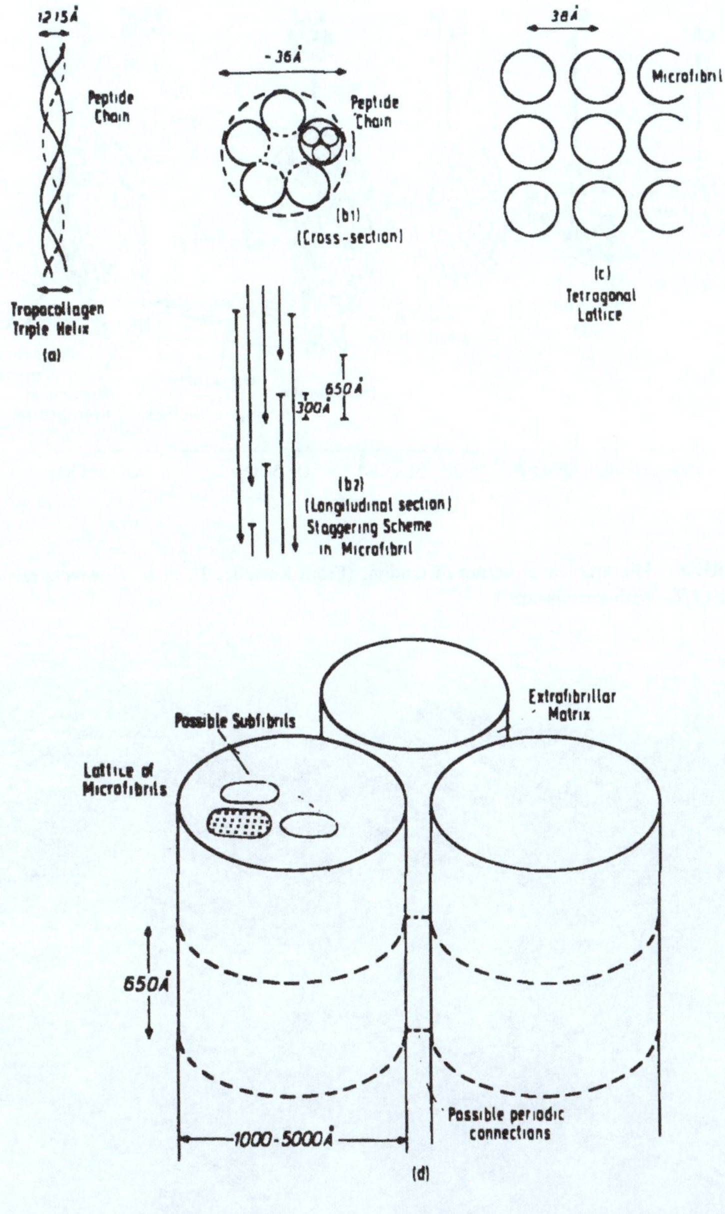

FIGURE 1. Structural hierarchies in collagen from the molecule through the fibril level. (a) Tropocollagen, perspective view. (b) Microfibril: (b_1) cross-sectional view, (b_2) longitudinal view with quarter stagger illustrated. (c) Lattice formed by microfibrils. (d) The collagen fibril, perspective view. (From Baer, E. et al., *Proc. Colston Conf.*, 1974, 189. With permission.)

loads. It represents the simplest loading configuration among the collagenous tissues in the body and will permit extrapolation to more complex loading geometries later in this chapter. Most data presented herein are derived from experiments using the tail tendon of the Sprague-Dawley rat with exceptions as noted.

A. Anatomy

The hierarchical structure of tendon is as described in Figure 2.[3] Collagen fibrils are arranged in parallel and oriented along the longitudinal axis (Figure 3).[4] The fibrils are

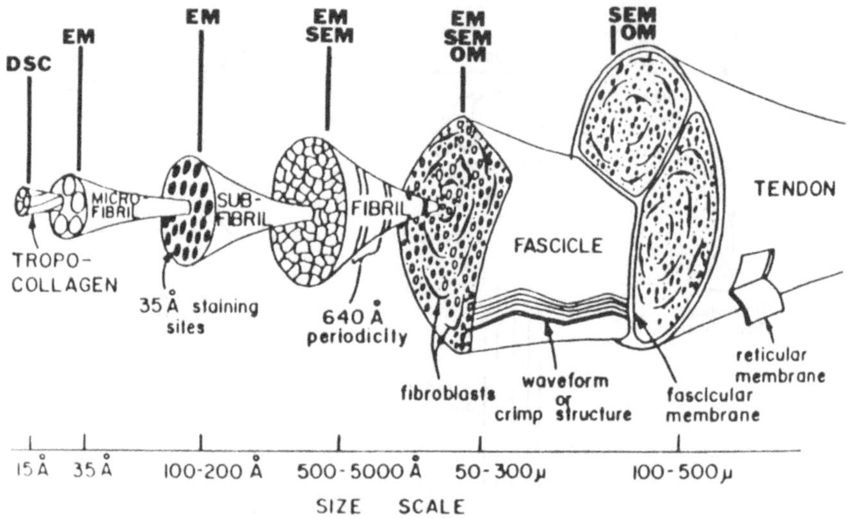

FIGURE 2. Hierarchical structure of tendon. (From Kastelic, J., et al. *Connect. Tissue Res.*, 6, 11, 1978. With permission.)

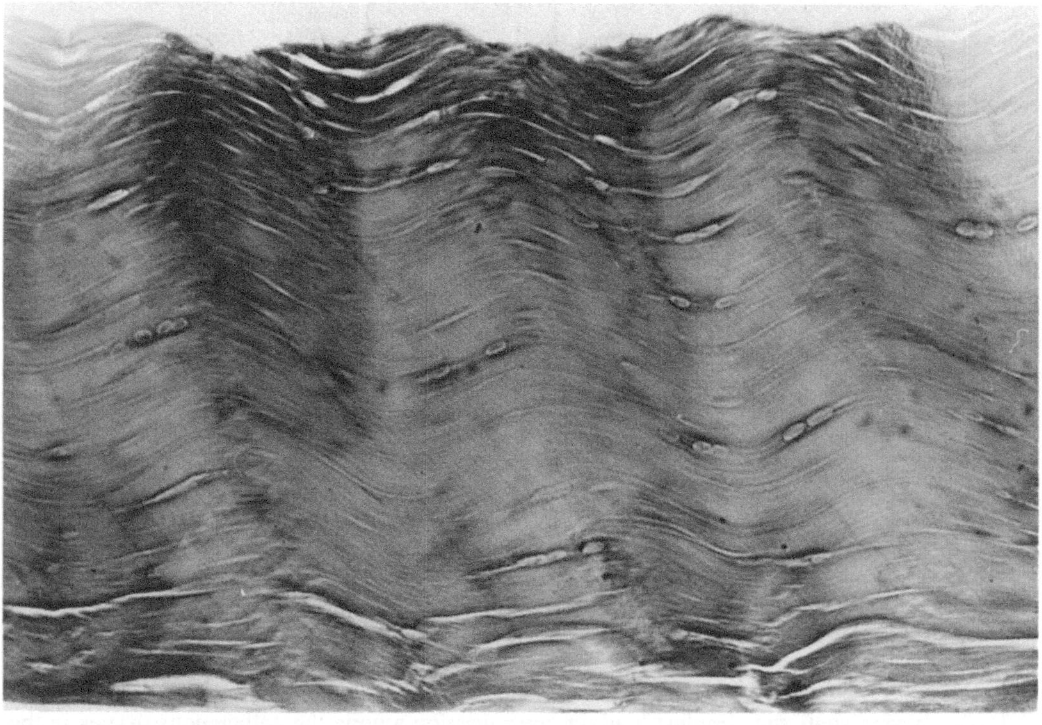

FIGURE 3. Collagen fibril arranged in parallel and crimped in register transversely. (From Niven, H., et al., *Collagen Rel. Res.*, 2, 131, 1982. With permission.)

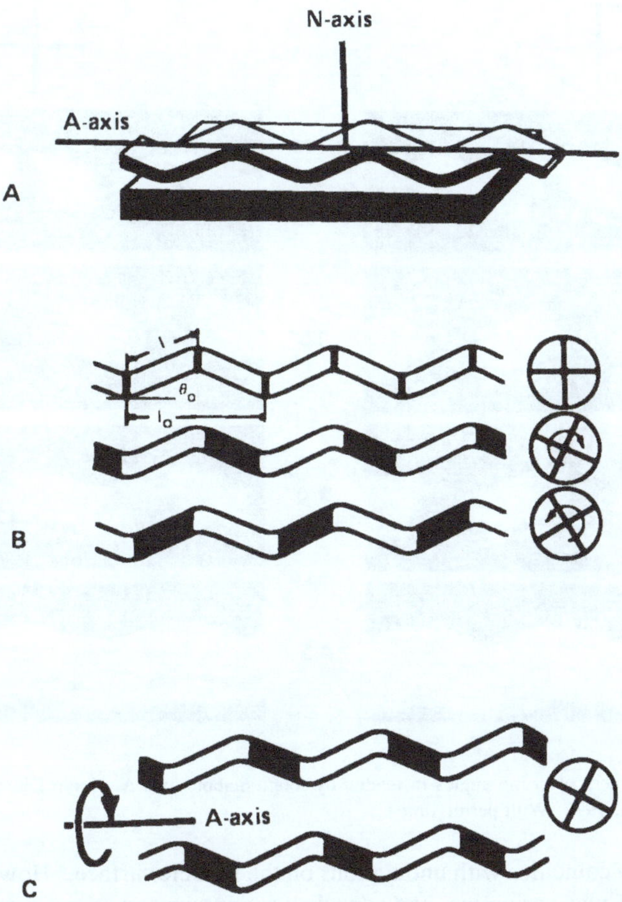

FIGURE 4. Diagrams illustrating the behavior of the planar crimped collagen fibrils in the optical microscope between crossed polarizers. (A) Axis A is the logitudinal axis of the fibril. Axis N is normal to the planar crimp. (B) Effect of fibril rotation about axis N (measuring crimp angles) with parameters of the planar crimp identified: l_o, the crimp period, θ_o, the crimp angle. (C) Effect of rotation about axis A (testing planarity of crimp). (From Diamant, J., et al., *Proc. Royal Soc. Lond.*, B180, 293, 1972. With permission.)

characterized by a smooth waveform with a sharply defined crimp angle. The waveform is a planar zig-zag as evidenced by rotation of the tendon between crossed polarizers in the optical microscope (Figures 4 and 5).[5] Further confirmation that the crimp is planar and not helical is offered by rotation about the long axis of the tendon in polarized light and by small angle X-ray diffraction.[6,7] The period of this waveform is 200 μm and the crimp angles fall in the range of 8 to 20°.[5] X-ray diffraction studies indicate that the fibrils are crimped in register.[6] The origin of this waveform has not been deduced. Possible explanations include specific "crimp sites" in the fibrils, "bimetallic strip" effects, internal tensions, matrix shrinkage, and fibril-matrix interactions.[8] The latter theory is favored at present because it offers the best explanation of the changes in mechanical properties observed as a function of fibril:matrix volume ratio during maturation and aging.

The basic structural unit of the tendon is the fascicle. It has a triangular cross section approximately 150 to 300 μm in diameter (Figure 6).[4] The waveform in the collagen fibrils is parallel to the long axis of the fascicle and persists through its entire thickness.[6] This

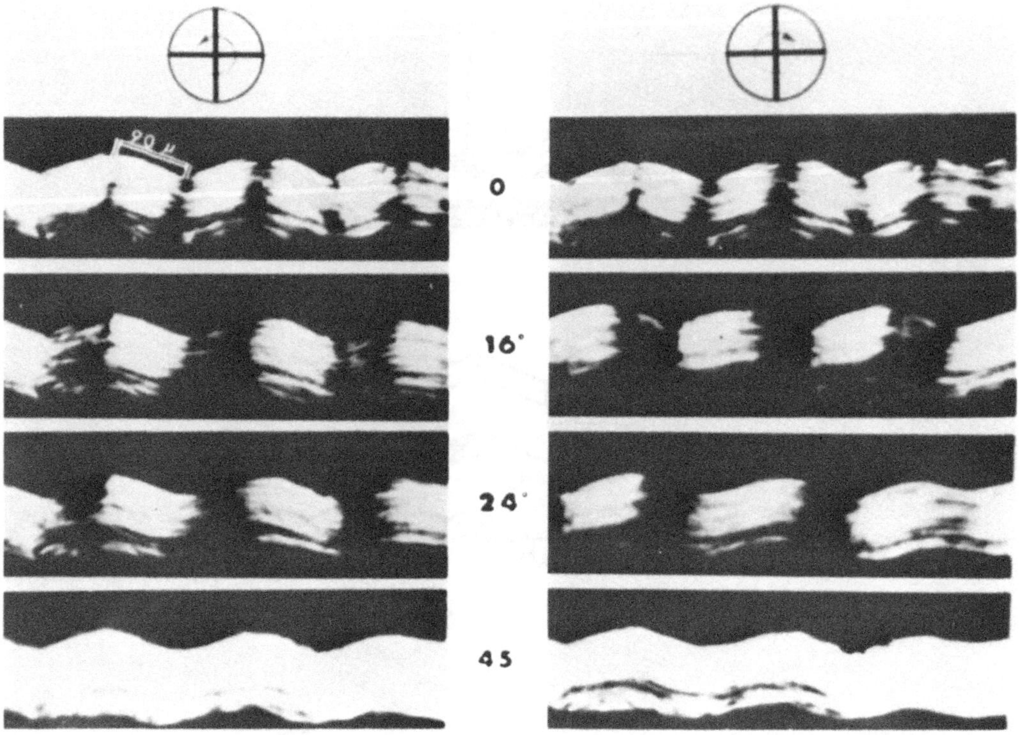

FIGURE 5. Measurement of crimp angles in tendon by rotation about axis A. (From Diamant, J., et al., *Proc. R. Soc. Lond.*, B180, 293, 1972. With permission.)

waveform usually coincides with undulations on the fascicle surface.[4] However, longitudinal slip planes, out-of-plane crimping, and edge dislocations are sometimes present at the surface. Rotation about the long axis of the tendon shows that the direction of the crimp on one side of the fascicle is the reverse of the other (Figure 7).[3] This phase change appears to result from the presence of slip planes and shear effects rather than considerations of cylindrical symmetry.[4] Fascicles are packed two to three together in the rat tail (Figure 8).[4] Neighboring fascicles appear to be crimped in register. While there is no evidence of interfascicle connections, there may be mechanical interlocking of surface undulations.[7]

B. Mechanical Properties

The hierarchical structure of tendon is designed to accommodate uniaxial tensile forces and as such is organized as an anisotropic material. The stress-strain curve of tendon is characterized by four distinct regions (Figure 9).[3] The inital part of the curve is referred to as the toe and is nonlinear. This corresponds anatomically to a gradual straightening of the collagen fibrils and removal of the waveform (Figure 10). Most physiological strains fall within this region. A linear elastic region follows which represents a reversible extension of the collagen. Yield region I is characterized by irreversible elongation and damage by a shear-slip mechanism between collagen microfibrils (Figure 11). The waveform does not reappear following a recovery period. This region is only present in mature tissues. Yield region II is identified by a decrease in modulus from region I and persists to failure. Damage is characterized by slippage of tropocollagen units within the microfibrils and by permanent dissociation between microfibrils by shear/slip (Figure 12). Another phenomenon that is observed is "recrimping", the appearance of smaller crimped waves superimposed on the inital waveform. The extent of the recrimping increases with amount of damage accumulated

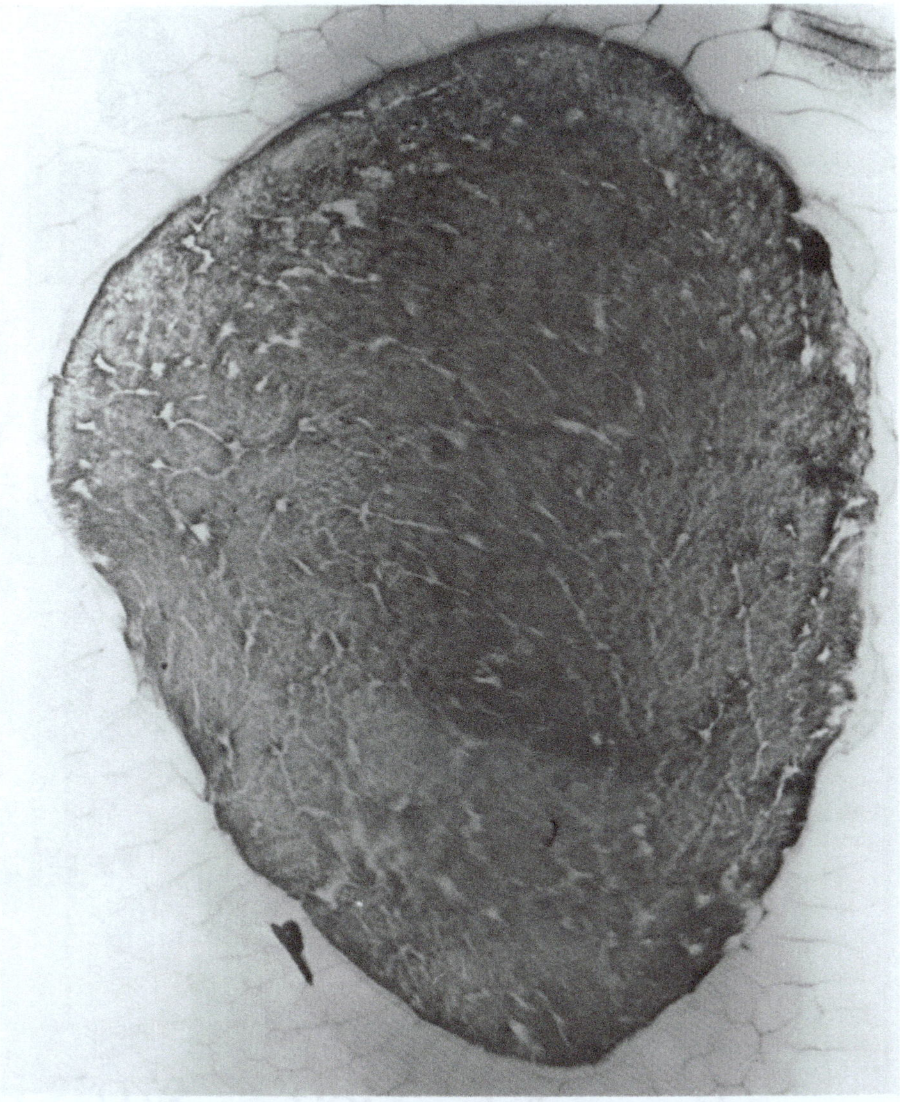

FIGURE 6. Cross-section of a tendon fascicle showing triangular shape. (From Niven, H. et al., *Collagen Rel. Res.*, 2, 131, 1982. With permission.)

until it completely replaces the original waveform (Figure 13). This is believed to occur because of an irregular distribution of damage from the center to the surface of the tendon fascicle. Surface fibrils are deformed until disassociated while some underlying fibrils remain intact. As these fibrils recover their original length they pull the overstretched fibrils back, forcing them to recrimp in a shorter length. At low strain rates, this curve is invariant with regard to strain rate.[9]

A first attempt at modeling the tensile behavior of the tendon used an elastica design with inflexible hinges.[5] This model agreed approximately with the experimental data, especially at higher strains. Later designs used a blunted zig-zag waveform and hypothesized the recovering force to come from elastic shear in the extrafibrillar matrix.[10] The current ''sequential straightening and loading model'' predicts tendon behavior based upon properties arising from the hierarchical structure of the tissue (Figure 14).[22] Resistance to deformation only comes from the elasticity of straightened collagen fibrils; crimped fibrils have negligible

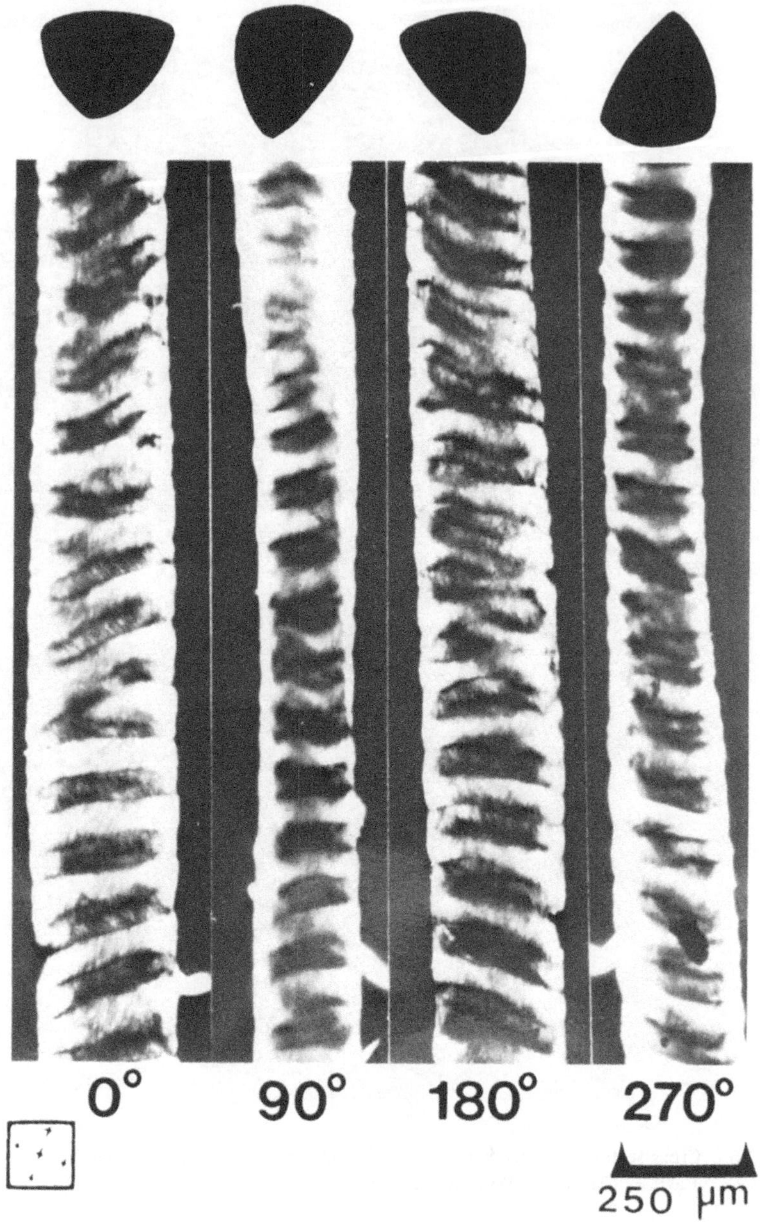

FIGURE 7. Formalin fixed fascicle viewed between crossed polarizers as it is rotated at 90° intervals about its logitudinal axis. (From Niven, H. et al., *Collagen Rel. Res.*, 2, 131, 1982. With permission.)

resistance. A range of crimp angles is assumed in the fibrils of a fascicle. Consequently, slack afforded by the crimps varies between fibrils. The toe region of the stress-strain curve results from sequential straightening and weight bearing. The linear portion of the curve begins when all fibrils have been pulled straight. This model closely fits morphological observations and mechanical test data for a significant range of animal ages.

For cyclic loading in the toe region to constant maximum stress, the modulus increases in each cycle, area of the hysteresis loop decreases, and maximum stress decreases. After a period of rest, the values of maximum stress and hysteresis loop area return to initial values.

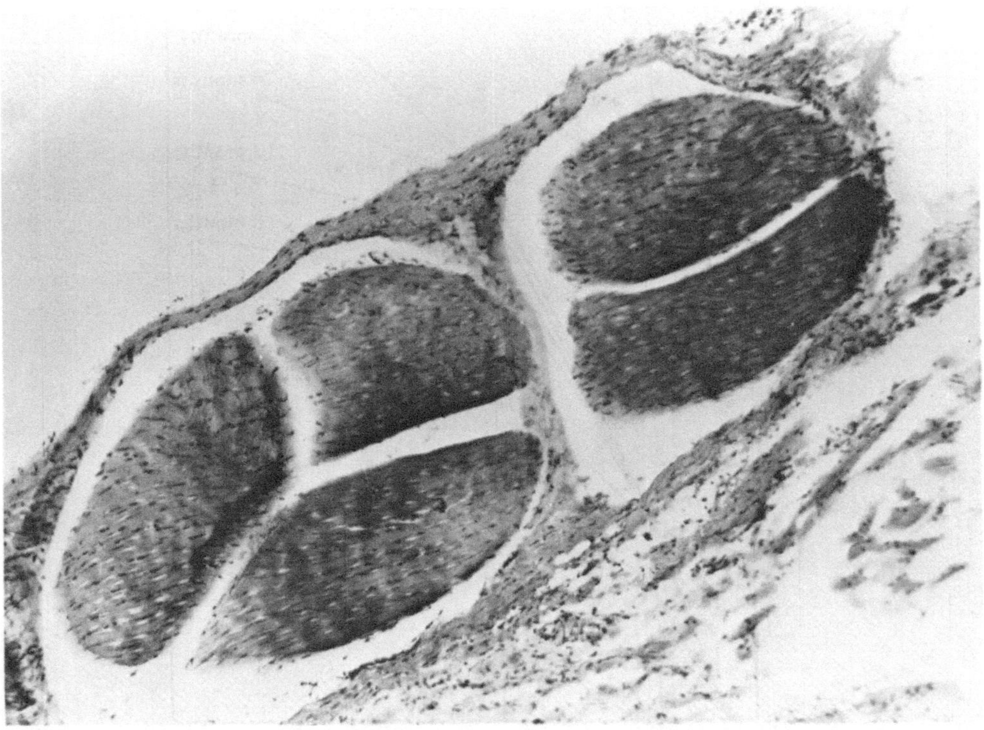

FIGURE 8. Fascicle packing in rat tail tendon. (From Niven, H. et al., *Collagen Rel. Res.*, 2, 131, 1982. With permission.)

These results can be explained in physical terms by improved fibril orientation. Cyclic loading in the linear elastic region results in some transient changes in mechanical properties which are fully recoverable with rest. The waveform parameters do not change (Figure 15). Cyclic deformation into the first yield region results in significant property changes, the most notable of which is a decrease in modulus (Figure 16). These changes result from cumulative and irreversible damage to the hierarchical structure of the tendon. Recrimping is observed that increases with the number of loading cycles.[11]

C. Effect of Maturity/Aging

The effect of aging on the mechanical response of tendon to uniaxial tension has been studied extensively. The changes observed during maturation are manifested in all regions of the stress-strain curve while those associated with aging of a mature organism appear only in the toe and damage regimes.

The diameter of collagen fibrils increases with age from 30 nm for newborn tendon up to 450 nm at maturity. With aging there are no further changes. For all ages beyond newborn, the fibril diameters are broadly distributed with the thinner fibrils between the larger ones.[9] The crimp period of the collagen fibrils increases and the crimp angles decrease with age from birth. This is associated with a decrease in the length of the toe region of the stress-strain curve (Figure 17).[5] The modulus of the tendon also increases up to maturity then remains invariant throughout the life of the animal.[12] Since the modulus is related to the properties of the collagen fibrils themselves, this indicates that there is no direct aging effect on the collagen.

The most pronounced effect of aging in the tendon appears in the yield regime of the mechanical test curve. In mature tendon, two distinct yield regions are found representing

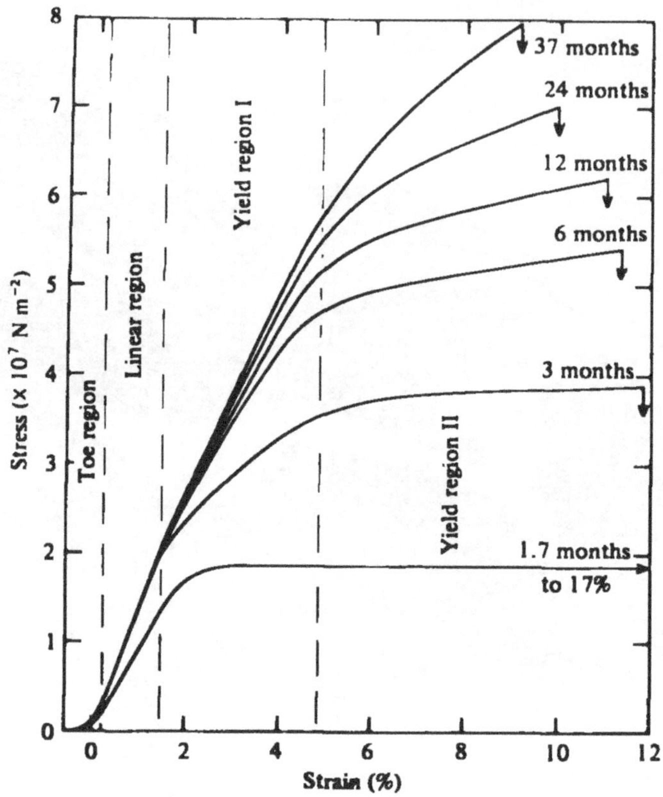

FIGURE 9. Stress-strain behavior of rat tail as a function of age. Tested in neutral saline; 8%/min strain rate. (From Kastelic, J. and Baer, E., *Soc. Exp. Biol. Symp.*, XXXIV, 1980, 397. Academic Press, With permission.)

two damage processes. In immature animals, only a single yield region is seen.[11] With increasing age, the slope of the curve in yield region I approaches that of the linear elastic region and the maximum yield stress increases (Figure 9).[3] These results support the contention that aging effects may be attributed to strengthening of intrafibrillar bonding interactions rather than changes in the structural hierarchy of the collagen itself.[9]

D. Effect of Hydration on the Extrafibrillar Matrix

Investigators have employed a variety of techniques including nuclear magnetic resonance (NMR), infrared (IR), differential scanning calorimetry (DSC), and X-ray to study the effects of water content on the properties of collagen.[13] As a result, four distinct hydration regimes have been identified. The first regime (0 to 7% water by weight) is associated with structural water which is incorporated into the tropocollagen triple helix. The second regime (7 to 25%) is water molecules bound to polar side chains and localized between helices within the microfibril. At high water content (> 45%) free water is sorbed which along with mucopolysaccharides forms the interfibrillar matrix gel. A transitional regime (25 to 45%) exists where both bound and free water are sorbed. The structural and bound water molecules, then, hydrate the collagen while the free water hydrates the matrix.[12]

The results of mechanical testing as a function of percent hydration are illustrated in Figure 18. Native collagen, by comparison, is approximately 63% water by weight. As the tendon is dehydrated through the free and transitional regimes, the shape of the stress-strain curves changes. The curvature of the toe region decreases and the yield stress increases. The modulus increases dramatically from 63 to 43% then remains constant thereafter. At

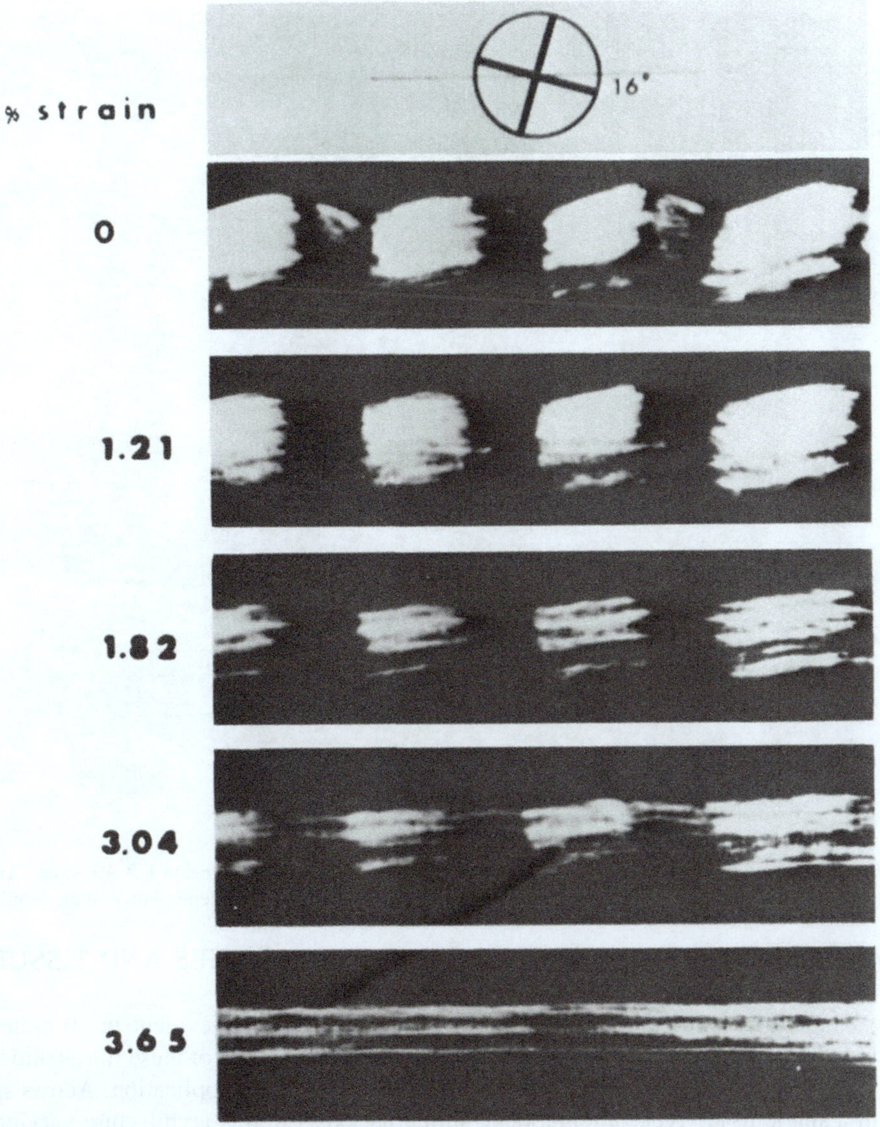

FIGURE 10. Partially dissected tendon showing crimp pullout as a function of applied strain. (From Diamant, J., et al., *Proc. R. Soc. Lond.*, B180, 293, 1972. With permission.)

water contents below 22% the tendon exhibits a sharp increase in initial slope and yield stress. At low strains, a shoulder appears prior to the toe and linear regions. At 7 to 25% water content (the bound water regime), a yield plateau is observed after the linear region. At water contents below 7% (the structural water regime), this plateau is absent.[14]

Observations of tendons undergoing dehydration have associated water loss with a decrease in the 64-nm banding period in the collagen fibril.[15] However, loss of free water does not lead to fibril shrinkage. It may decrease separation between sub- and microfibrils. This may increase electrostatic interactions between subfibrils, giving rise to strain hardening of the tissue similar to that observed with increased cross-link density.[14] The shoulder seen at low strains after removal of free water may result from altered properties of the interfibrillar matrix. It may represent a loss of cohesion between fibrils and matrix or yielding of the matrix material itself.

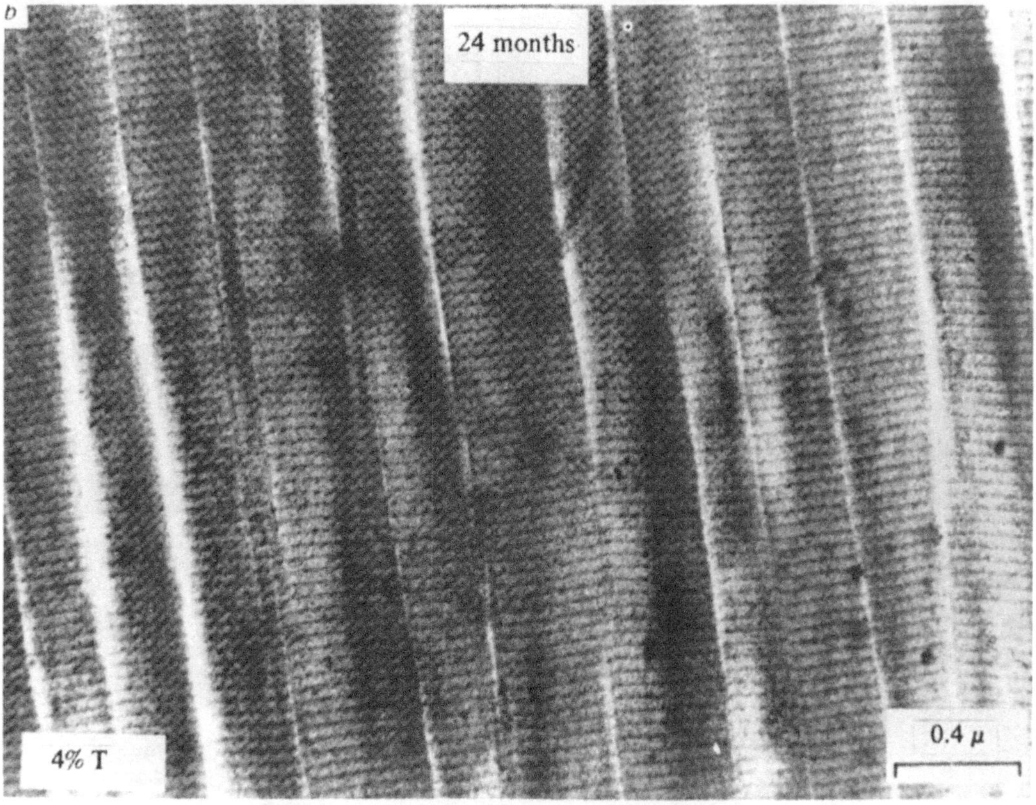

FIGURE 11. Electron micrograph of mature rat tail tendon deformed into yield region I at 4% strain. Taut fixed. (From Kastelic, J. and Baer, E., *Soc. Exp. Biol. Symp.*, XXXIV, 1980, 397. Academic Press. With permission.)

IV. HIERARCHICAL STRUCTURE IN OTHER SPECIES AND TISSUES

The hierarchical structure of collagen up to the fibril level is evident in all connective tissues studied. Variation between tissues results from organization of these fibrils into higher order structures to meet the specific physical requirements of the application. Across species lines in a single tissue type, a remarkable similarity exists in this architecture varying only in the crimp parameters.

A. Tendon

Interspecies variation in tendon appears to occur primarily in the parameters of the collagen fibril waveform, not in the structural organization of the tissue. In observing tendon specimens harvested from the rat, kangaroo, and human, one finds the anatomy to be identical. However, the crimp period and angle vary (Table 1).[16] The implication of this finding is that the collagen waveform and its effects on mechanical properties are universal among species and anatomic location. The hierarchical structure of the tendon has remained essentially constant despite great differences in the magnitude of the loads applied, the impact requirements of the tissue, and the lifespan of the animal. The variation that is seen may simply reflect the fact that while the tissues harvested from each animal are homologous they are not identical in terms of physical size or magnitude of the loads applied.

B. Intestine

The hierarchical structure of collagen in the rat intestine is identical to that of the tendon

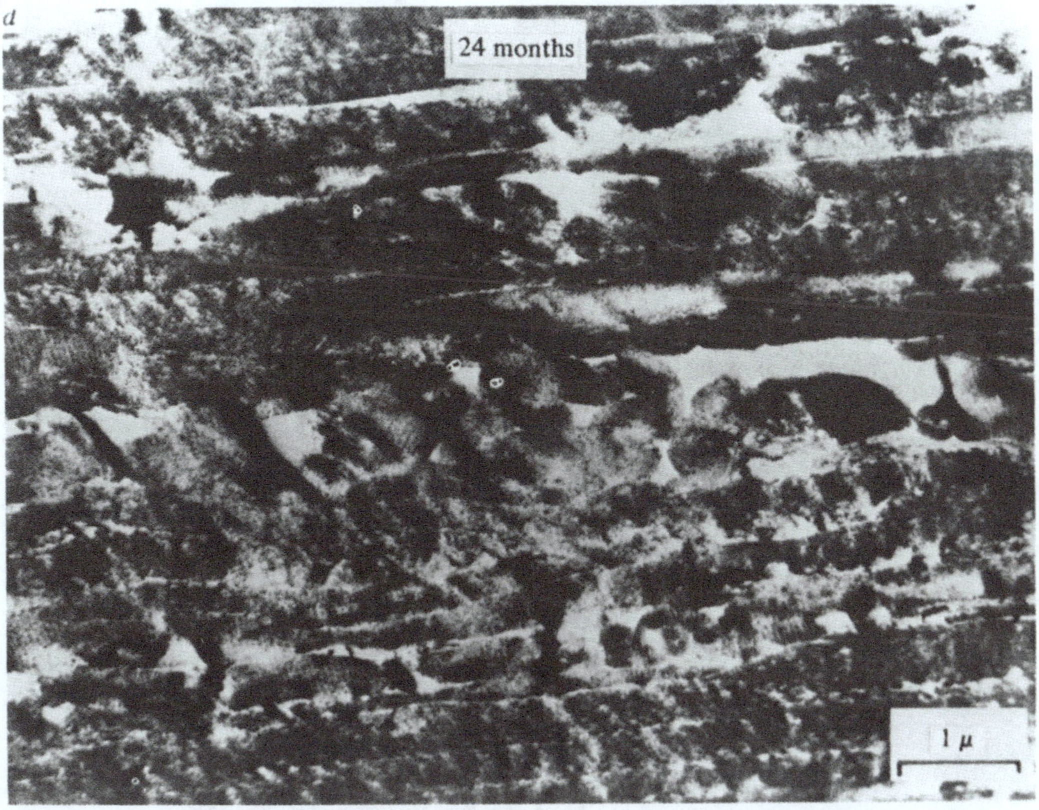

FIGURE 12. Electron micrograph of mature rat tail tendon deformed into yield region II at 8.5% strain. Relaxed fixed. (From Kastelic, J. and Baer, E., *Soc. Exp. Biol. Symp.*, XXXIV, 1980, 397. Academic Press. With permission.)

up through the fibril level (Figure 19).[17] However, there are two types of collagen present in the intestine. Type III collagen is found as "fibrettes" which form a loosely woven meshwork encircling the smooth muscle cells.[20] The type I collagen fibrils are organized in an anisotropic lay-up to accommodate the mechanical requirements of the digestive tract. As a bolus of food enters the intestine, the lumen is deformed radially and longitudinally.

The type I collagen fibrils in the intestine are morphologically similar to those of the tendon and have a crimp period in the range of data from other collagenous tissues. However, the crimp angles are considerably larger and are asymmetric, being 30° ascending and 56° descending. The collagen fibrils are oriented in alternating layers at ±30° to the longitudinal axis (Figure 20).[18] The most notable difference between the tendon and intestine is the absence of the fascicle in the structural hierarchy.[17]

Testing of specimens cut from whole intestine has shown that the mechanical properties of the intestine are derived primarily from the collagenous layers. A characteristic stress-strain curve with toe and linear elastic regions and a single yield region has been identified. As in tendon, the regions of the stress-strain curve correspond to removal of the crimp from the fibrils, stretching of the collagen, and irreversible damage to the intestinal structure, respectively. In addition, as the intestine is stretched, the fibrils reorient themselves within the layers to accommodate the deformation. The high degree of fiber organization produces highly anisotropic mechanical properties (Figure 21).[19]

C. Intervertebral Disk

The intervertebral disk is a composite structure in which a gel-like nucleus pulposus of

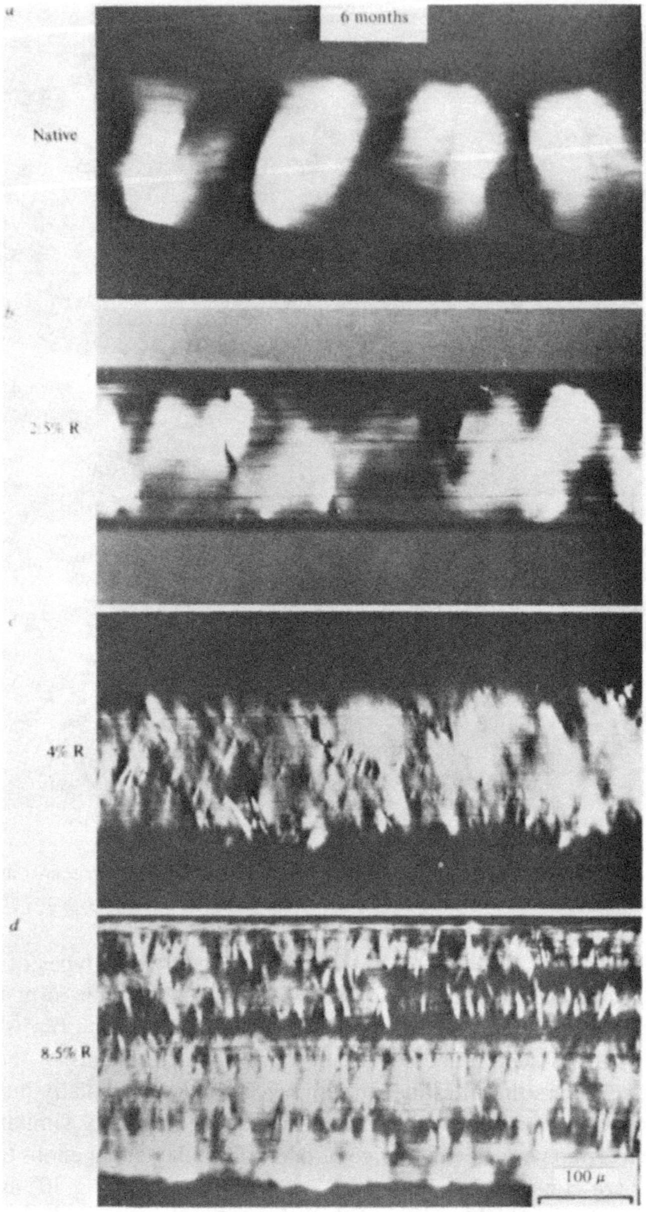

FIGURE 13. Optical micrographs of tendon recrimping in 6-month-old rat. Relaxed fixation at indicated strains. (From Kastelic, J. and Baer, E., *Soc. Exp. Biol. Symp.*, XXXIV, 1980, 397. Academic Press. With permission.)

type II collagen is contained within the concentric fibrous lamellae of the annulus fibrosus. The disk is anchored to the vertebral bodies by cartilage endplates. This composite structure is subjected to axial compression and shear forces. The walls of the annulus are composed of crimped collagen type I fibrils whose morphology is similar to that of the tendon (Figure 22). The crimp period is approximately 20 μm and the crimp angle is 22°. The fibrils are oriented at ±65° to the axis of the spine in alternating layers (Figure 23).

It is hypothesized that the nucleus pulposus transmits axial compressive stresses to hoop

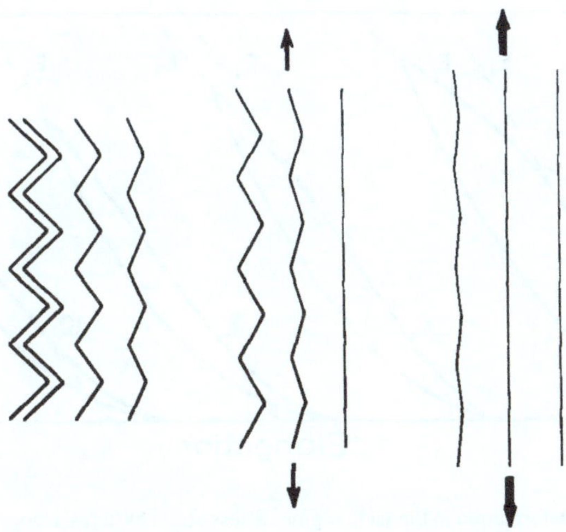

FIGURE 14. Sequential straightening and loading model for deformation of tendon collagen. (From Kastelic, J., et al., *J. Biomech.*, 13, 887, 1980, Pergamon Press, Elmsford, N.Y. With permission.)

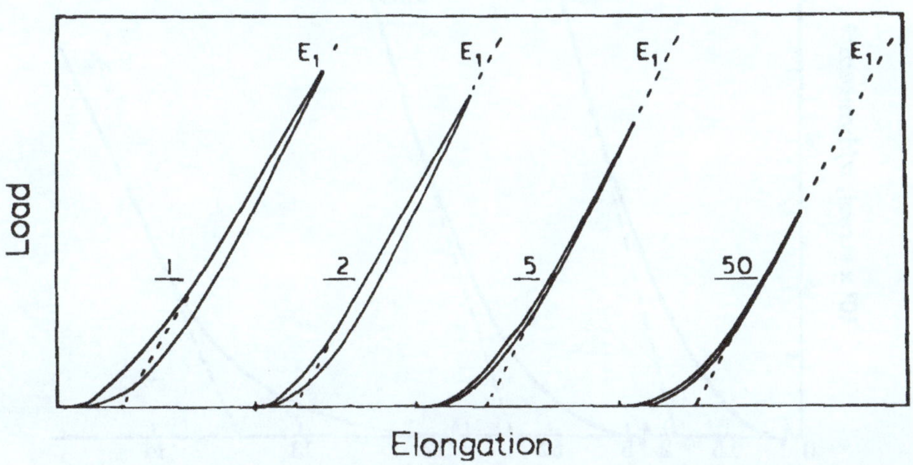

FIGURE 15. Cyclic deformation in the linear region. Stress-strain hysteresis loops at constant maximum strain at 1, 2, 5, and 50 cycles. Tendon is 15 months old. Maximum initial strain is 1.4%. (From Torp, S. et al., *Proc. Colston Conf.*, 1975, 197. With permission.)

tensile stresses in the fibrils of the annulus fibrosus. Mechanical testing of whole disk preparations with viable bone above and below yields a stress-strain curve which is similar to that of other collagenous tissues (Figure 24). Observations on disks fixed in formalin under compressive strains show progressive pullout of the fibril crimp with increased applied load (Table 2). This result appears to support this model of disk behavior.[21]

V. SUMMARY

In summary, the hierarchical structure of collagen appears to be universal across tissue types and species from the molecular through the fibrillar levels. The collagen fibril, then, appears to be the basic structural unit of soft connective tissue. In some tissues a higher

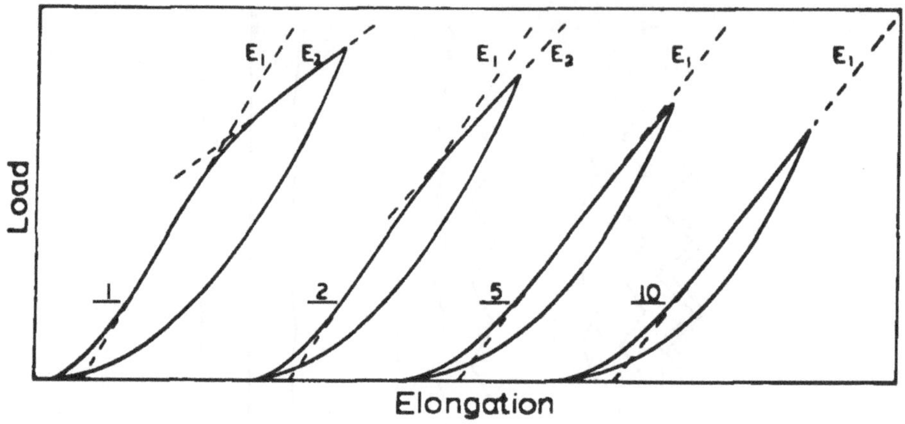

FIGURE 16. Cyclic deformation in the yield region. Stress-strain hysteresis loops at constant maximum strain at 1, 2, 5, and 10 cycles. Tendon is 7 months old. Maximum initial strain is 3.1% (From Torp, S. et al., *Proc. Colston Conf.*, 1975, 197. With permission.)

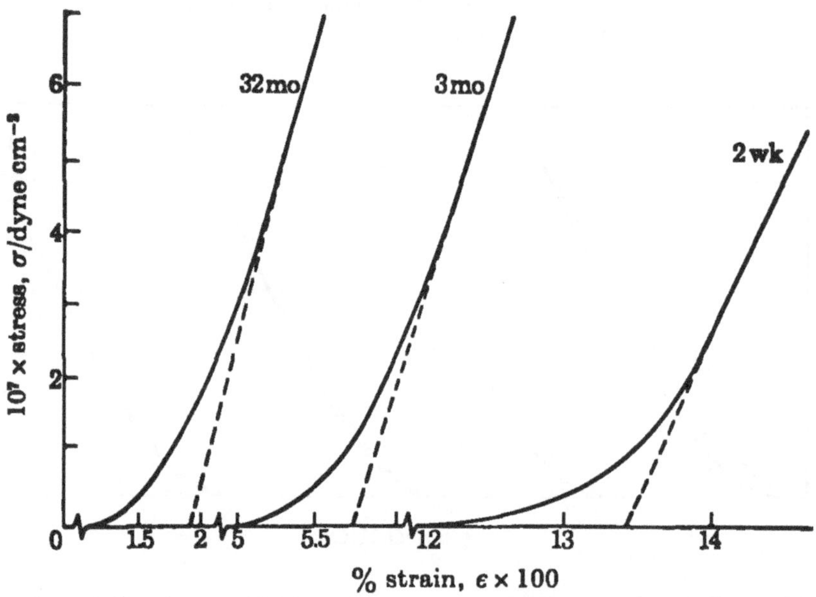

FIGURE 17. Length of the toe region of stress-strain curve as a function of age in the rat tail tendon. (From Diamant, J., et al., *Proc. R. Soc. Lond.*, B180, 293, 1972. With permission.)

order structure, such as the fascicle in tendon, may serve as the primary "building block" of the tissue.

Collagenous tissues are organized in such a manner as to accommodate the physiological stresses placed upon them. A clear relationship has been established between the mechanical properties of the tendon in tension and specific properties of the structural hierarchy of the collagen. This same relationship exists in other tissues such as intestine and intervertebral disk where the loading modes are more complex.

The mechanical properties of all collagenous tissues change with maturation and aging and are also dependent upon the properties of the extrafibrillar matrix, most significantly the degree of hydration.

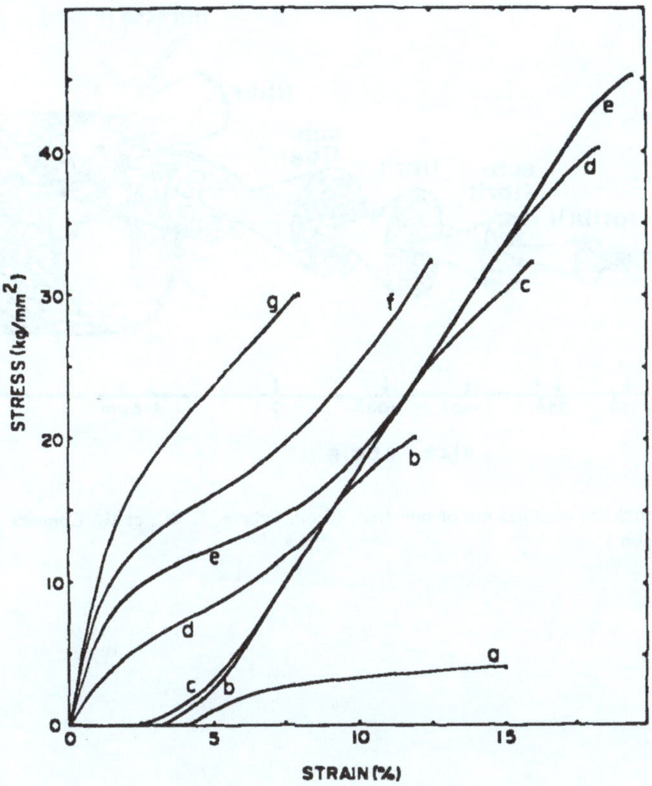

FIGURE 18. Stress-strain behavior of rat tail tendon as a function of water content in 3-month-old rats. Native tendon; percent hydration is (a) 63, (b) 43, (c) 22, (d) 15, (e) 11, (f) 6, and (g) 2. Temperature at 25°C; 8 %/min strain rate. (From Betsch, D. F., M.S. thesis, Case Western Reserve University, Cleveland, Ohio, 1982. With permission.)

Table 1
REPRESENTATIVE CRIMP PARAMETERS OF TENDONS FROM VARIOUS SOURCES

Source and age	Microns	Degrees
Rat Tail (14 months)	200	12
Human diaphragm (51 years)	120	12
Kangaroo tail (11.7 years)	150	8—9
Human achilles (46 years)	40—100	6—8

From Dale, W. C. et al., *Experientia*, 28, 1293, 1972. With permission.

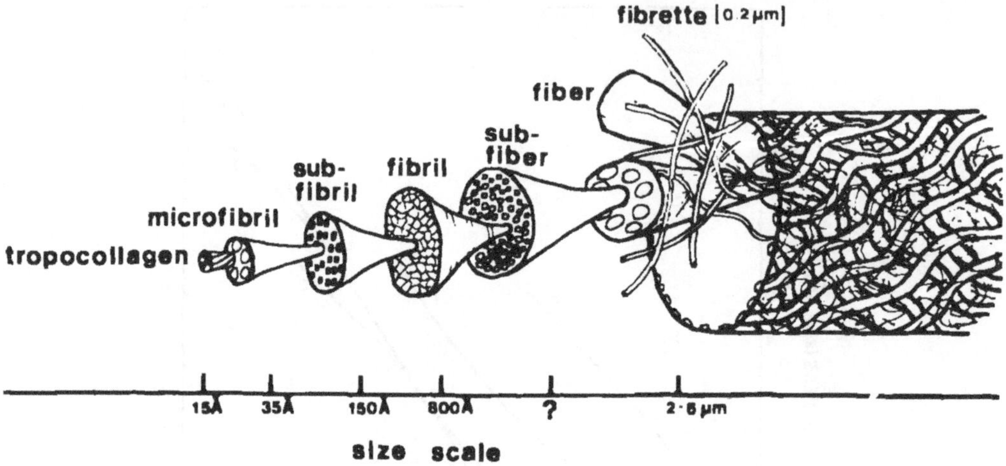

FIGURE 19. Hierarchical organization of intestine. (From Orberg, J. W., et al., *Connect. Tissue Res.*, 9, 187, 1982. With permission.)

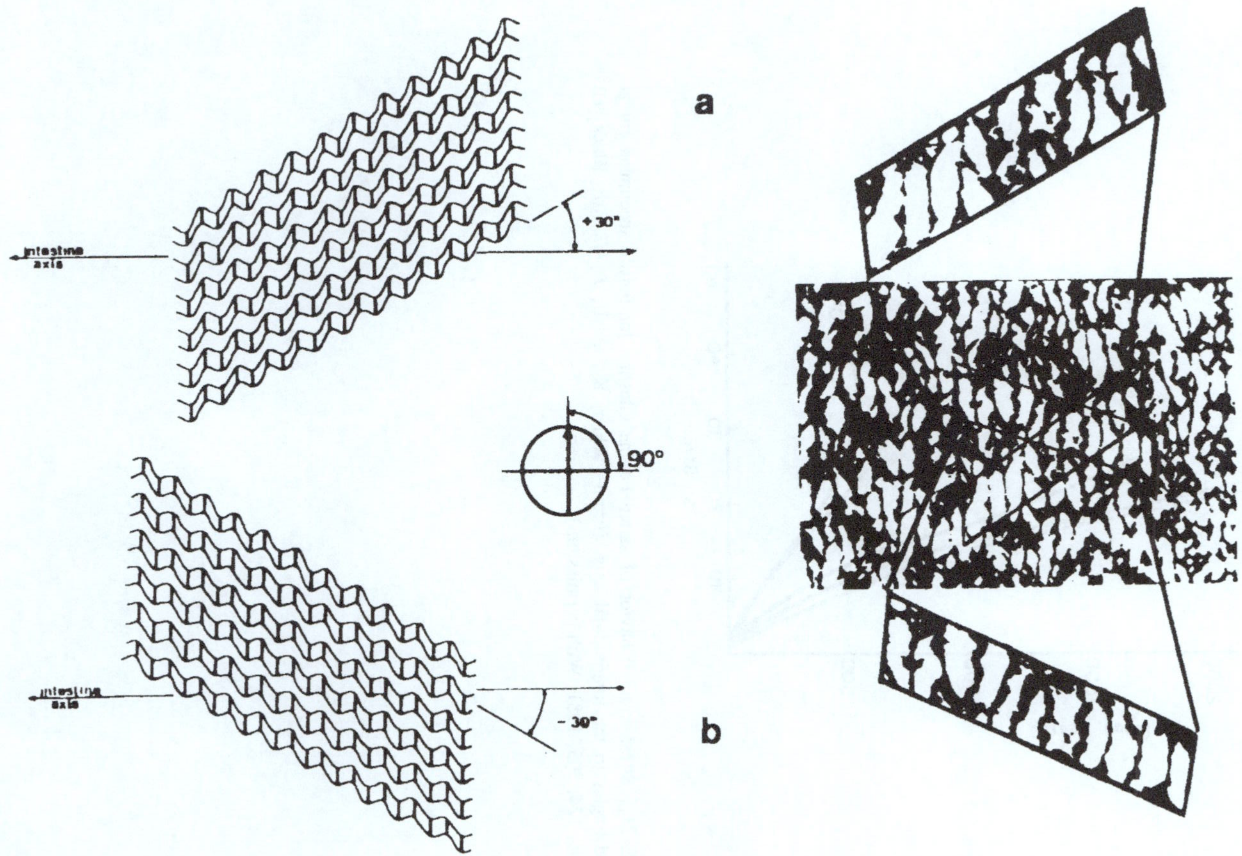

FIGURE 20. Comparison of schematic illustration of extinction patterns with optical micrograph of unstretched intestine between crossed polarizers at 90° to longitudinal axis of intestine. (a) Planar crimped fibrils at +30°. (b) Planar crimped fibrils at −30°. (From Orberg, J. W. et al., *Connect. Tissue Res.*, 11, 285, 1983. With permission.)

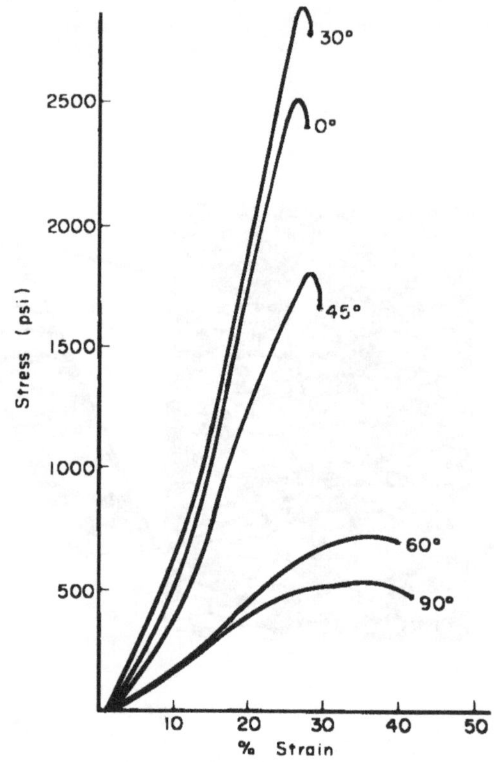

FIGURE 21. Stress-strain behavior of salted bovine intestine for tensile specimens cut at indicated angles to the longitudinal axis. (From Fackler, K., et al., *J. Microsc.* Blackwell Scientific, 124, 305, 1981. With permission.)

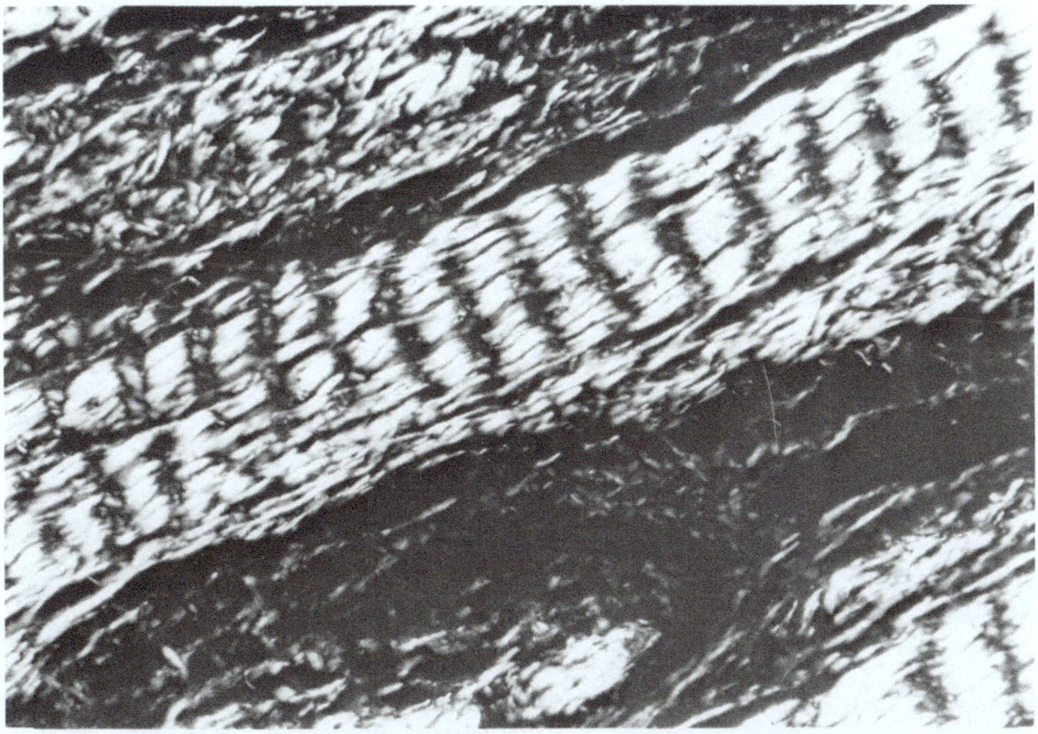

FIGURE 22. Crimped collagen fibrils in the annulus fibrosus of the human intervertebral disk sectioned through transverse plane perpendicular to the spinal axis.[21]

FIGURE 23. Collagen fibrils in the annulus fibrosus oriented $\pm 60°$ to transverse plane in alternating layers.[21]

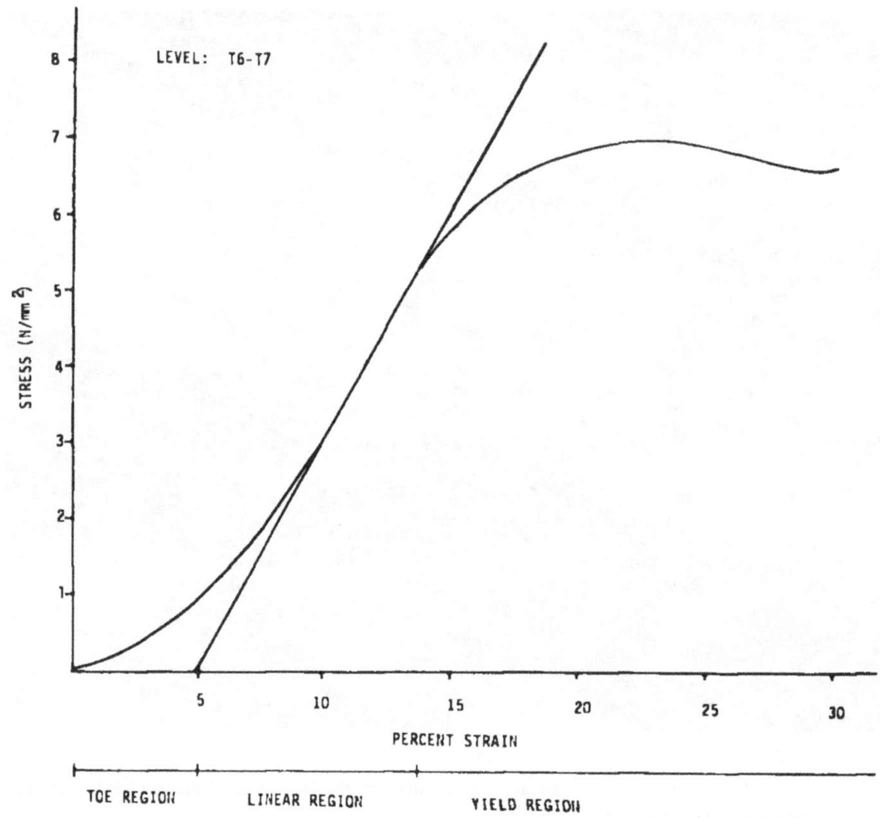

FIGURE 24. Stress-strain behavior of intervertebral disk under axial compression. Disk illustrated is between 6th and 7th thoracic vertebra.[21]

Table 2
CRIMP PARAMETERS OF THE FIBRILS IN THE ANNULUS FIBROSUS AND ANGLE BETWEEN LAMELLAE AS A FUNCTION OF APPLIED AXIAL COMPRESSIVE STRAIN[21]

	Applied strain (%)		
	0	**10**	**20**
Angle between lamellae (deg.)	55 ± 2	48 ± 6	42 ± 3
Crimp angle (deg.)	21 ± 2	14 ± 1	9 ± 2
Crimp period (μm)	21 ± 3	26 ± 6	33 ± 1

REFERENCES

1. **Baer, E., Gathercole, L. J., and Keller, A.,** Structure hierarchies in tendon collagen: an interim summary, *Proc. Colston Conf.*, 1974, 189.
2. **Hiltner, A., Cassidy, J. J., and Baer, E.,** Mechanical properties of biological polymers, *Annu. Rev. Mater. Sci.*, 15, 455, 1985.

3. **Kastelic, J. and Baer, E.,** Deformation in tendon collagen, *Soc. Exp. Biol. Symp.*, XXXIV, 1980, 397.
4. **Niven, H., Baer, E., and Hiltner, A.,** Organization of collagen fibers in rat tail tendon at the optical microscope level, *Collagen Rel. Res.*, 2, 131, 1982.
5. **Diamant, J., Keller, A., Baer, E., Litt, M., and Arridge, R. G. C.,** Collagen; ultrastructure and its relation to mechanical properties as a function of ageing, *Proc. R. Soc. Lond.*, B180, 293, 1972.
6. **Gathercole, L. J. and Keller, A.,** X-ray diffraction effects related to superstructure in rat tail tendon collagen, *Biochim. Biophys. Acta*, 535, 235, 1978.
7. **Kastelic, J., Galeski, A., and Baer, E.,** The multicomposite structure of tendon, *Connect. Tissue Res.*, 6, 11, 1978.
8. **Viidik, A.,** in *Biology of Collagen*, Viidik, A. and Vurst, J., Eds., Academic Press, New York, 1980, 257.
9. **Torp, S., Baer, E., and Friedman, B.,** Effects of age and of mechanical deformation on the ultrastructure of tendon, *Proc. Colston Conf.*, 1974, 223.
10. **Betsch, D. F. and Baer, E.,** Structural and mechanical properties of rat tail tendon, *Biorheology*, 17, 83, 1980.
11. **Torp, S., Arridge, R. G. C., Armeniades, C. D., and Baer, E.,** Structure-property relationships in tendon as a function of age, *Proc. Colston Conf.*, 1975, 197.
12. **Baer, E., Gathercole, L. J., and Keller, A.,** in *Structures of Fibrous Proteins*, Atkins, E. D. T. and Keller, A., Eds., Butterworths, London, 1975, 189.
13. **Nomura, S., Hiltner, A., Lando, J., and Baer, E.,** Interaction of water with native collagen, *Biopolymers*, 16, 231, 1977.
14. **Betsch, D. F.,** The Effect of Water on the Mechanical Properties of Tendon Collagen, M.S. thesis, Case Western Reserve University, Cleveland, Ohio, 1982.
15. **Rougvie, M. A. and Baer, R. S.,** An x-ray diffration investigation of swelling by collagen, *J. Am. Leather Chem. Assoc.*, 48, 735, 1953.
16. **Dale, W. C., Baer, E., Keller, A., and Kohn, R. R.,** On the ultrastructure of mammalian tendon, *Experientia*, 28, 1293, 1972.
17. **Orberg, J. W., Klein, L., and Hiltner, A.,** Scanning electron microscopy of collagen fibers in intestine, *Connect. Tissue Res.*, 9, 187, 1982.
18. **Orberg, J. W., Baer, E., and Hiltner, A.,** Organization of collagen fibers in the intestine, *Connect. Tissue Res.*, 11, 285, 1983.
19. **Fackler, K., Klein, L., and Hiltner, A.,** Polarizing light microscopy of intestine and its relationship to mechanical behavior, *J. Microsc.*, 124, 305, 1981.
20. **Anderson, J. M., Yaffe, A., and Hiltner, A.,** unpublished results.
21. **Cassidy, J. J.,** unpublished results.
22. **Kastelic, J., Palley, I., and Baer, E.,** A structural mechanical model for tendon crimping, *J. Biomech.* 13, 887, 1980.

Chapter 10

CONTRIBUTIONS OF ELASTIN AND COLLAGEN ORGANIZATION TO PASSIVE MECHANICAL PROPERTIES OF ARTERIAL TISSUE

Robert J. Boucek

TABLE OF CONTENTS

I. Introduction 202

II. Elastin Synthesis and Lamellar Formation in Arterial Tissue 202

III. Elastin Organization and the Mechanical Properties of Aortic Tissue 204
A. The Analyses of Elastin Structure, Content, and Organization in Arterial Tissue 207

IV. Collagen Synthesis and Organization into Fibers and Bundles 208

V. Recognition of Defective Collagen Organization in Skin and Aortic Tissues 211
A. Intermolecular Crosslink Profiles 211
B. Thermal Contraction and Bacterial Collagenase Digestion 212

VI. Role of Glycoproteins and Proteoglycans in Elastin and Collagen Organization and in the Mechanical Properties of Arterial Tissue 215

VII. Elastin and Collagen Organization and the Passive Mechanical Properties of Aortic Tissue 216

VIII. Summary and Concluding Remarks 217

References 219

I. INTRODUCTION

In serving as blood conduits, arteries have extraordinarily diverse mechanical requirements. These include a low impedance for forward blood flow, a capability for rapid and externally mediated expansion or contraction of blood transport capabilities, a minimal hysteresis in recoiling from pulse-pressure-volume deformations, and a large tensile strength for supporting high pressure loading. Such a wide mechanical repertoire is met by complementary anatomic and biophysical properties of three fibrous proteins: actin-myosin of smooth muscle cells, elastin of the elastic membranes, and the collagens in fibers and bundles of the medial and adventitial layers. The three fibrous proteins are anatomically and functionally linked into a tissue fabric. Identifying contributions of elastin and collagen organization to the mechanical properties of this fabric is the objective of the following review.

Studies of the passive mechanical properties of arteries in vitro using the elastomeric approach of force-deformation analyses began with the work of Roy,[1] Burton,[2] and Bergel[3] who identified the nonlinear relationship between force and deformation of these tissues. Roach and Burton[4] suggested that the initial response to mechanical loading was due to deformations of the elastin-rich membranes while the collagen-rich membranes limited the changes associated with large loads. Initially, the concentrations of elastin and collagen were considered as the major determinants of the passive mechanical response to loading in vitro, however, subsequent work identified the anatomic distribution,[5,6] the macromolecular organization,[7] and the surrounding proteoglycan material of elastin and collagen[8] as important contributors. But assessing the contributions of elastin and collagen organization to the mechanical properties of arteries requires special biochemical and biophysical indexes. To appreciate these indexes, the synthesis and the stepwise organization of elastin and collagen are reviewed.

II. ELASTIN SYNTHESIS AND LAMELLAR FORMATION IN ARTERIAL TISSUE

Following the transcription from an extensive genome,[9] the messenger RNA ($mRNA_E$) undergoes major splicing processing in forming the signal for elastin synthesis. Synthesis of structural elastin of arteries is largely regulated by the $mRNA_E$ levels.[10] Translation of cytoplasmic $mRNA_E$ proceeds on membrane-bound polysomes[11] which synthesizes a signal peptide-containing tropoelastin monomer into the cisternae of the rough endoplasmic reticulum. Following a series of packaging events involving the Golgi apparatus and a coalescing of coated vesicles,[12] the tropoelastin arrives at the cell surface and is secreted into the extracellular space. Then a series of extracellular events modifies the tropoelastin molecules through a lysyloxidase-catalyzed oxidative deamination of specific lysyl residues to reactive aldehydes (allysines)[13,14] and aligns the tropoelastin by an interaction with a glycoprotein microfibril[15] (Figure 1). These extracellular events are essential for elastin fiber formation.

With the extracellular alignment of the modified tropoelastin, alanine-enriched regions containing lysine or allysine pairs separated by one to three alanine residues interact with neighboring lysyl or allysyl residues.[16] Aldol condensation between two allysines or allysine-lysyl epsilon amine Schiff base reaction leads to intramolecular bifunctional crosslinks, the aldol condensation product (ACP) and dehydrolysinonorleucine (*LNL)[17] (Figure 1). Then the soluble elastin is converted to an insoluble elastin polymer. The initial intramolecular crosslinks are chemically unstable and, with maturation of the elastin polymer into a macropolymer, give way to more stable bifunctional (LNL) and polyfunctional (the desmosines, DES)[18] crosslinks. The crosslinks developing in the alanine-rich region of the elastin molecule organize a compact and possibly an α-helical region, while the remaining portion containing hydrophobic amino acid residues are probably the more extensible elements of the macropolymer.[19]

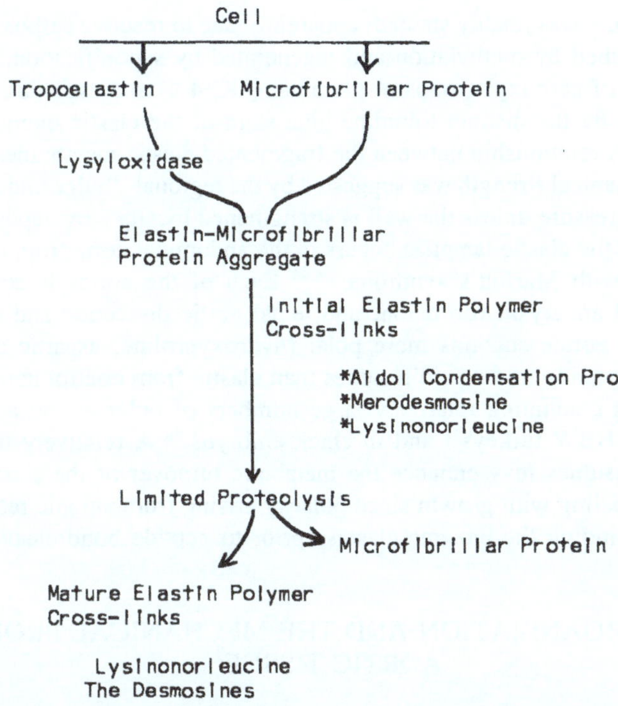

FIGURE 1. Elastin synthesis and fibrillogenesis. Asterisks indicate chemically unstable compounds.

Elastin gene expression and protein synthesis in large arteries (porcine aorta) undergo a multifold increase in the immediate 1 to 3 weeks following birth and "largely subside by 110 days" of postnatal age.[20] Morphologically, the elastin in the elastic membranes is arranged as parallel sheets which begin in the subendothelial region and extend around smooth-muscle cells and fine fibers of collagen in the medial layer. Thickened elastic membranes, seen in resistance-sized arteries as the internal and external elastic lamella, contain microfibrils and mature elastin (79%) in intimate association with collagen fibrils (17%).[21]

According to the careful morphologic studies of resistance arteries from 96 humans ranging in age from newborns and infants (21 cases), 2 to 20 years (11 cases), 20 to 40 years (25 cases), 50 to 60 years (23 cases), and 60 to 84 years (16 cases) by Wolff,[22] the internal elastic membrane commences as a single layer which underlies the thin endothelium. A new elastic membrane appears toward the end of the first year as thin lines of fibrous connective tissue staining as elastica in the subendothelial region and is part of a forming intimal layer. After the first year of life, the "new" elastic-containing layer reaches or even exceeds the thickness of the first layer. With age, the primary internal elastic membrane becomes extremely thin or disappears, while the "new" elastic layer thickens and develops distinctive tinctorial properties.

Thinning and fragmentation of the internal elastic membrane in arteries occur in many animals, including birds. That a loss of integrity in the elastic membranes and the organization of elastin in these mebranes are related to the mechanical properties of the arterial tissue is suggested by studies of the canine carotid artery,[23] the abdominal aorta from the domesticated turkey (Broad-Breasted White — BBW)[24] and from the aorta of patients with Marfan's syndrome.[25] In the canine carotid artery, Lehoczky, et al.[23] found evidence of a histochemical modification of the elastin in the fragmented internal elastic membranes in a small number of animals. Whereas the intact elastic membranes did not stain with toluidine blue, the fragmented membrane was readily stained, apparently due to reactive carboxyl groups, since the stain was abolished by methylation and regenerated by saponification. The possibility of the involvement of carboxyl groups of glutamic (pK_1 4.4) and aspartic (pK_1 3.0 to 4.7) acid was suggested by the distinct toluidine blue stain of the elastic membrane at pH 5.0 but not at pH 4.0. A relationship between the fragmented elastic membranes (lamellae) and the artery wall mechanical strength was suggested by the regional "bulge under the distending force of the blood pressure unless the wall is strengthened locally" by deposits of collagen.

Fragmentation of the elastic lamellae occurs in the abdominal aorta from the BBW turkey and from patients with Marfan's syndrome.[24,25] Each of the aortas is complicated by a dilatation, localized aneurysm formation, and lethal aortic dissection and rupture. Elastin isolated from these aortae contains more polar (hydroxyproline, aspartic acid, threonine, serine, and glutamic acid) amino acid residues than elastin from control aortic tissue (Table 1).[26] Aortic elastins containing relatively large numbers of polar amino acid residues are found in immature BBW turkeys[27] and in chick embryos.[28] A relatively large number of polar amino acid residues may enhance the metabolic turnover of the elastin required for arterial tissue remodeling with growth since polarity arising from anionic residues may play a decisive role in binding the enzyme elastase prior to peptide bond cleavage and elastolysis.[29-31]

III. ELASTIN ORGANIZATION AND THE MECHANICAL PROPERTIES OF AORTIC TISSUE

While an assessment of the elastic lamellae integrity is possible by light microscopy, the ultrastructure organization of elastin cannot be resolved even by the elctron microscope. At the resolution of the electron microscope, the fragmented elastic lamellae of the BAPN-fed BBW turkey or from patients with Marfan's syndrome are irregularly shaped with electron opaque elevations on their surface.[32,33] Identifying defects in molecular or macromolecular organization of elastin in aortic tissue requires indirect biochemical measurements.

Since elastin organization into a mature polymer proceeds in a stepwise manner and involves lysyl-derived crosslinks (Figure 1), the numbers of lysyl residues, the amounts of crosslinks associated with the initial steps relative to those identified with the mature polymer provide biochemical means for assessing elastin organization. The numbers of lysyl residues are higher in elastin isolated from the abdominal as compared to the thoracic aorta of the BBW turkey[27] and are significantly greater in elastin isolated from aortae of Marfan patients when compared to values for the elastin of the control[26] (Table 1); and the amounts of the chemically unstable bifunctional crosslink, LNL, relative to the polyfunctional and chemically stable crosslinks, DES are higher in the abdominal than in the thoracic aorta of the BBW turkey and in the aorta from a Marfan patient as compared to a control (Table 1). Furthermore, the density of DES is significantly lower in aortic elastin from Marfan patients than in the elastin from the control tissue. The large number of lysyl residues in the isolated elastin, the crosslink ratio values, and the density of DES in the elastin suggest proportionately large amounts of immature elastin in aortic tissues of the BBW turkey and from the Marfan patient displaying elastic lamina fragmentation.

Associations between the lysyl crosslinking reactions and the mechanical properties of aortic tissue were extended by observations of β-aminopropionitrile fumarate (BAPN) induced aortic aneurysms in a number of laboratory animals including the BBW turkey.[34] Whereas some 3 to 6% of a given flock of BBW turkeys may be "internal bleeders" and die from ruptured aortic aneurysm,[35] more than 40% develop lethal aneurysm when fed BAPN. BAPN, a rather selective lysyloxidase inhibitor,[36] reduces the density of lysyl-derived aldehydes in chick aortic elastin,[28] increases the length-tension characteristics,[37] and reduces the tensile strength of aortic tissues of the BBW turkey when compared to tissue from control

Table 1
ELASTIN CHEMISTRY AND AORTIC ANEURYSM: POLAR AMINO ACID AND CROSSLINKING RESIDUES

	Polar amino acids (Residue/1000 residues)							DES	
	Hypro	Asx	Thr	Ser	GLx	Lys	Lysyl crosslinks $NaB[^3H]_4$-reducible	Lysine equiv. (Residue/1000 residues)	Elastin (nmol/mg)
BBW turkey aorta[a]							*LNL (cpm/mg)		
Thoracic	14.5	4.0	8.3	4.1	13.7	2.1	1.53	2.00*	
	± 1.0	± 0.2	± 0.1	± 0.2	± 0.3	± 0.1	± 0.03	± 0.02	
Abdominal	15.6	6.3	10.0	6.9	17.8	3.9	2.12	2.30	
	± 0.9	± 0.4	± 0.4	± 0.5	± 0.4	± 0.1	± 0.10	± 0.03	
Human aorta[b]							*LNL (cpm/nmol)		
Control	7.5	3.0	2.5	2.8	13.3	8.8	313[c]	6.8	34[c]
	± 2.0	± 1.3	± 1.6	± 2.6	± 2.7	± 1.9		± 0.2	
Marfan	12.6	5.3	5.6	6.7	14.9	13.8	459[c]	3.7	19[c]
	± 6.0	± 0.6	± 1.3	± 2.4	± 4.2	± 3.3		± 0.3	

Note: Values for hydroxyproline (Hypro), aspartate (Asx), threonine (Thr), serine (Ser), glutamate (GLx), lysine (Lys), and lysinonorleucine (*LNL) are mean and SE of at least 3 tissue samples.

[a] Mean and SE of at least three tissue samples. Boucek, R. J. et al., *Exp. Mol. Pathol.*, 31, 400, 1979.
[b] Mean and SE of at least three tissue samples. Abraham, P. A. et al., *J. Clin. Invest.*, 70, 1245, 1982.
[c] Single-tissue determination. Gunja-Smith, Z., unpublished.

Table 2
DRUGS, ARTERIAL PRESSURE, AORTIC TENSILE STRENGTH, AND MORTALITY IN THE BBW TURKEY

	Arterial pressure			Aortic tensile strength (g/mm²)	Mortality (%)
	Systolic (mmHg)	Diastolic (mmHg)	dp/dt$_{max}$ (mmHg 0.5^{-1})		
Control	189 ± 7	140 ± 3	1736 ± 50	123 ± 3	0
BAPN (0.07%)	196 ± 2	149 ± 2	1642 ± 53	68* ± 2	44* ± 3
+Resperpine	137* ± 4	85* ± 6	2540* ± 52	95* ± 10	2 ± 1
+*dl*-Propranolol	141* ± 2	108* ± 3	1192* ± 55	85* ± 3	1 ± 1
+*d*-Propranolol	198 ± 2	145 ± 3	1685 ± 51	53* ± 30	62* ± 3
+ Hydralazine	157* ± 10	114* ± 7	1903 ± 82	58* ± 3	91* ± 3

Note: Values are means ± SE for 30 animals; * represents values which are significantly different than control ($p < 0.02$); dp/dt, rate of pressure rise per time interval. Data from Simpson, C. F. and Boucek, R. J., *Cardiovasc. Res.*, XVII, 26, 1983.

animals[38] (Table 2). Drugs may reverse or potentiate the BAPN-mediated reduction in aortic tensile strength and increase mortality rate from a ruptured aortic aneurysm. Two drugs, reserpine and *dl*-propranolol, essentially reverse the BAPN toxicity.[39,40]

The mechanism for the reserpine- and *dl*-propranolol-mediated reversal of BAPN toxicity appeared to be a reduction of arterial pressure until the effects of hydralazine were known[41] (Table 2). Hydralazine lowered the arterial pressure to levels comparable to those obtained with reserpine and *dl*-propranolol but hydralazine potentiated rather than reversed the BAPN-mediated reduction of aortic tensile strength and increased the mortality rate from aortic aneurysm with rupture. These seemingly contrary effects on the mechanical properties of the aorta by reserpine (or *dl*-propranolol) and hydralazine suggested a direct rather than an indirect action of drugs on the aorta of the BAPN-fed BBW turkey.

A site of drug action at the level of the lysyloxidase-catalyzed crosslinking reaction was suggested because of the rather selective inhibition of this enzyme by BAPN[36] and because the *l*- and *dl*- but not the *d*-isomer of propranolol reversed the BAPN effect on the aorta from the BBW turkey[42] (Table 2). Subsequently it was learned that chronic propranolol treatment (6 weeks) enhanced the lysyloxidase reaction, the lysyl crosslinking reactions, and the index value for the tensile strength of aortic tissue from the BBW turkey.[43] Propranolol treatment increased the tissue levels of allysine (and hydroxyallysine), lowered the ratio values for chemically unstable (*) and stable crosslinks (*LNL:LNL and *LNL:DES), and increased the density of DES in elastin without raising the elastin concentration (Table 3). These propranolol-mediated modifications of the lysyl crosslink profile supported the relationship between the crosslink ratios or the density of DES in elastin and the mechanical properties of aortic tissue as suggested by the data shown in Table 1.

Subsequently, the propranol-treated BBW turkey was recognized as a unique animal model for identifying a drug effect on the mechanical properties of aortic tissues. For these purposes, a ring (2.9 mm length) was cut from an anatomically standardized region of the abdominal aorta, the region with elastic membrane fragmentation and high vulnerability for aortic aneurysm and rupture. Tissue tension was progressively increased by a motor-driven device and the recorded tension just prior to tissue rupture was related to tissue area and expressed as grams (g/mm^2).[44] Propranolol exerted a dose-related increase in the tensile strength index value which was in part independent of the arterial pressure-lowering effect (Figure 2A).[43]

Table 3
PROPRANOLOL AND LYSYL CROSSLINKING COMPOUNDS OF ELASTIN IN THE ABDOMINAL AORTA OF THE IMMATURE, FEMALE BBW TURKEY

		Reducible crosslinking compounds				
		*HNL		*LNL		DES
	Elastin (μg/mg wet wt)	(Allysine) (cpm/μg hyp)	*MD (cmp/μg hyp)	(cpm/μg hyp)	(cpm/nmol)	(nmol/mg elastin)
Control	13 ± 2 (6)	440 ± 32 (6)	122 ± 7 (6)	86 ± 15 (12)	6624 ± 749 (12)	21 ± 2 (6)
Propranolol (0.12%)	11 ± 3 (6)	701 ± 112 (8)	178 ± 17 (9)	88 ± 7 (12)	4742 ± 309 (12)	32 ± 7 (6)
p	NS[a]	<0.05[b]	<0.02[c]	NS	<0.05[c]	<0.02[c]

Note: Values are means ± SE; number of animals shown in parentheses. Relative proportions of principal amino acids in isolated elastins (valine:alanine:glycine:proline) were 1:0:9:0.5:1.1 for control, 1:0.9:0.5:1.0 for propranolol-treated animals. Abbreviations: hydroxynorleucine, *HNL; hydroxyproline, hyp; merodesmosine, *MD; lysinonorleucine, *LNL; the desmosines, DES.

[a] No significant difference.
[b] Statistical probability by Wilcoxon two-sample rank test.
[c] Statistical probability by Student's *t*-test.

Data from Boucek, R. J. et al., *Biochem. Pharmacol.*, 32, 275, 1983.

A. The Analysis of Elastin Structure, Content, and Organization in Arterial Tissue

Additional support for a relationship between drugs and the passive mechanical properties of arterial tissue came from the studies of the effect of a 6-week course of propranolol or metoprolol on abdominal aortic tissue from the BBW turkey. Both drugs are absorbed and metabolized similarly, and both reduce arterial pressure in the BBW turkey (Figure 2). Metoprolol, however, may have a limited direct effect on arterial tissue since it is not lipophilic, does not stabilize cell membranes, and is a weak antagonist of arterial B_2-adrenergic receptors. As shown in Figure 2B, a four-times larger daily dose of metoprolol failed to raise the tensile strength index values to a level comparable to the propranolol effect; and the density of DES (nmol/mg elastin) was lower in abdominal aortic tissue from the metoprolol- than from the propranolol-treated BBW turkeys (24 ± 1 vs. 32 ± 7n,[5] $p < 0.05$).

The aortic studies on the propranolol-treated BBW turkey model and on patients with Marfan's syndrome suggest a diverse requirement for analyses of the contributions of elastin organization to the mechanical properties of arterial tissue. The analyses are outlined as follows:

- Histology using an elastic stain — identify the integrity of elastic lamellae
- Quantificaton of elastin by
  - — Autoclave: alkali extraction and gravimetric measurement
  - — Chemical: level of a specific elastin-pentapeptide in bovine, porcine, and rabbit tissue[45]
  - — Conversion of hydroxyproline (hypro) or DES to elastin equivalences is a questionable means for quantifying elastin in tissues
- Measure of elastin maturation from
  - — Level of polar amino acid residues in purified elastin
  - — Ratio between chemically unstable and stable crosslinks
  - — Lysine:desmosine ratio in the elastin
  - — Number of desmosine residues in the elastin
  - — Content of desmosines in the elastin

A.

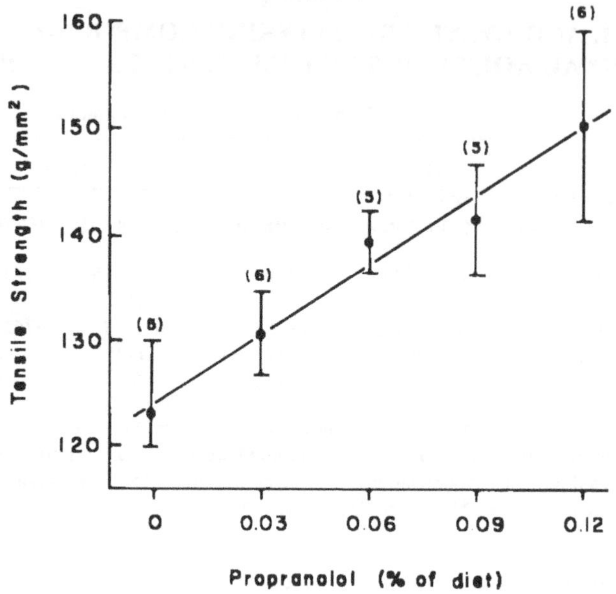

FIGURE 2. Metoprolol or propranolol and the tensile strength of turkey aortic tissue.
FIGURE 2A.

	Propranolol (1% of diet)				
	0	**0.03**	**0.06**	**0.09**	**0.12**
Heart rate	286	242++	240++	235++	232++
(beats/min) ± SE	± 3	± 7	± 7	± 7	± 5
Arterial pressure	1777	1362++	1159++	1301++	1108++
(dp/dt max ± SE)	± 69	± 83	± 43	± 140	± 99

Note: ++ indicates theses values are significantly different for animals receiving propanolol vs. untreated animals. Plasma propranolol, mean value of three determinations, was 10 ng/mℓ at 0.03% and 100 ng/mℓ at 0.12% of the diet. dp/dt, rate of pressure rise per time interval.

IV. COLLAGEN SYNTHESIS AND ORGANIZATION INTO FIBERS AND BUNDLES

Like elastin, collagen synthesis and extracellular organization into fibers and bundles and subsequent aging are multi-stepped processes. Once the collagen molecules are expressed from the cell, the ε-amine group of specific lysyl and hydroxylysyl residues is oxidatively deaminated by lysyloxidase to form the reactive aldehydes, allysine and hydroxyallysine.[13,14] Reactions between the reactive aldehydes or between the reactive aldehyde and neighboring ε-amine groups produce covalent intra- and intermolecular crosslinks which stabilize the developing collagen polymer.

The process of collagen organization can be assessed by biochemical indexes in newly formed connective tissue around a polyvinyl sponge implant in rats.[46] As collagen synthesis and accumulation begin, the most common crosslinks are the dehydro forms of hydroxylysinonorleucine (*HLNL) and histidinohydroxymerodesmosine (*HHMD) (Table 4). As fibers thcken and bundles form, the number of the dehydro forms of dihydroxylysinonorleucine (*DHLNL) rapidly increases, the increase involving the diglycosylated and the nonglycosylated species. The sequence of borohydride-reducible crosslinks as fibers and bundles appeared in electron micrographs of the connective tissue is summarized as follows:

B.

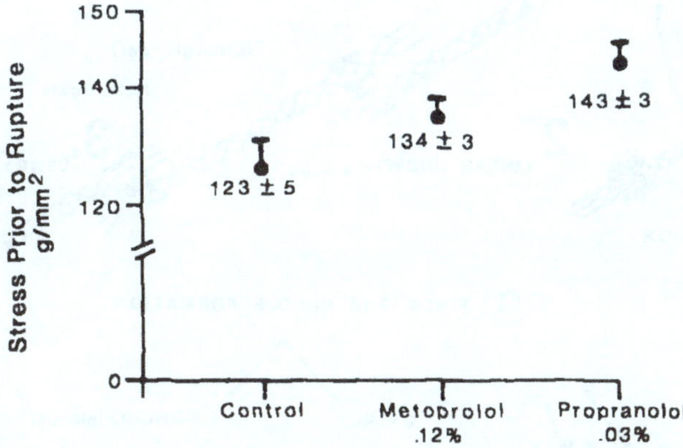

FIGURE 2B. Values are mean ± standard error of a tensile strength index for 6 animals in each animal group and tested for differences by the Student's t-test. Control vs. metoprolol and metoprolol vs. propranolol are significantly different, with $p < 0.05$.

	Arterial Pressure		
	Systolic	**Diastolic**	**dp/dt_{max}**
Control	208 ± 3	157 ± 3	1417 ± 52
Propranolol (.03%)	149 ± 3	110 ± 2	784 ± 31
Metoprolol (.12%)	182 ± 2	143 ± 2	1118 ± 27

Control vs. drug-treated animals are significantly different, with $p < 0.05$.
From Boucek, R. J., Simpson, C. F., Noble, N. L., and Smith, Z. G. (unpublished).

Table 4
DISTRIBUTION OF RADIOACTIVITY IN ACID HYDROLYSATES OF [NaB^3H_4]-REDUCED DEVELOPING CONNECTIVE TISSUE (ADULT MALE RAT)

	Tissue age (days)	10^{-3} × radioactivity (cpm/mg of collagen) of reducible crosslinking compounds					
		***HNL**	***HLNL**	***HLNL**	***LNL**	***HHMD**	**Total**
Early collagen	2	7.6 ± 0.9	1.1 ± 0.2	25.9 ± 1.9	4.3 ± 0.4	22.3 ± 0.6	67.4 ± 4.4
Synthesis	3	7.7 ± 1.0	1.8 ± 0.1	29.7 ± 2.4	2.2 ± 0.2	21.6 ± 0.9	67.1 ± 5.0
Fiber and bundle	4	10.3 ± 0.3	5.5 ± 0.7	32.1 ± 0.7	3.9 ± 0.5	21.9 ± 1.8	78.9 ± 2.6
Organization	6	9.9 ± 1.1	12.9 ± 0.2	31.0 ± 1.3	2.5 ± 0.2	18.4 ± 0.4	77.3 ± 3.0

Note: Results are means ± SE for tissues from three animals. Abbreviations: dehydro forms (*) of crosslinking compounds: hydroxynorleucine, *HNL; dihydroxylysinonorleucine, *DHLNL; hydroxylysinonorleucine, *HLNL; lysinonorleucine, *LNL; and histidinohydroxymerodesmosine, *HHMD.

Data from Boucek, R. J. et al., *Biochem. J.*, 177, 853, 1977.

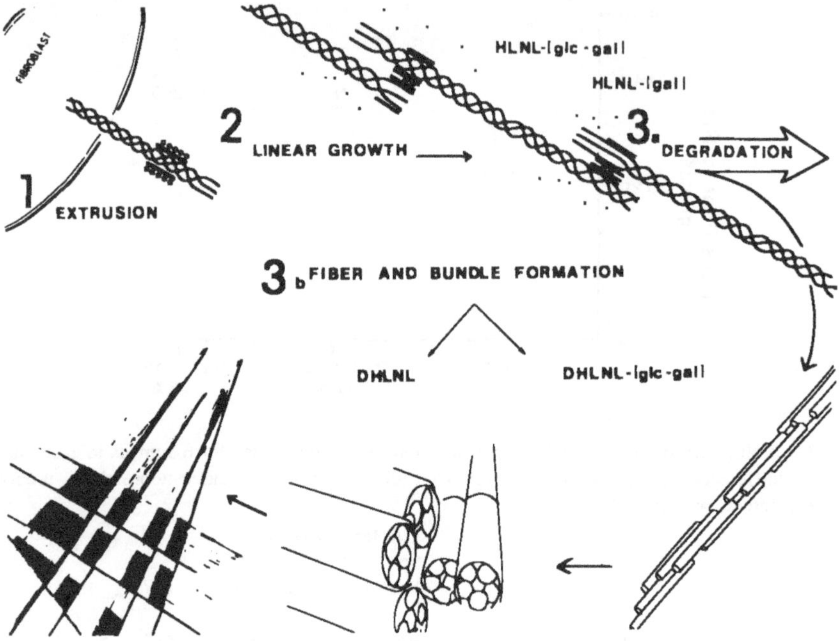

FIGURE 3. A scheme of the intermolecular crosslink formation as monomeric collagen undergoes linear growth (2), fiber and bundle organization (3b), and internal rearrangement to a more stable crosslink, *DHLNL, as bundles form and the susceptibility to bacterial collagenase digestion decreases.

- Highest radioactivity occurs with *HLNL and *HHMD and lowest with *LNL, and the amounts remain constant as fibers and bundles appear
- *DHLNL amounts are low initially but rise sharply with the development of fibers and bundles
- The bulk of *HLNL and *DHLNL are glycosylated, the former with galactosyl or glucosylagalactosyl residues and the latter with Glc-Gal residues.
- Only nonglycosylated *DHLNL undergoes an internal rearrangement as fibers and bundles form
- The changing ratios between *HLNL/Glc-Gal *HLNL and *DHLNL/nonglycosylated *DHLNL suggest three post-translational lysyl modifications as collagen fibers and bundles form in developing connective tissue

Relating the electron microscopic images of fiber and bundle formation, the profiles of crosslinks, and the susceptibility of collagen to bacterial collagenase digestion as a biochemical index of collagen organization, suggest the stepwise process involving crosslink formation that is stylized in Figure 3. The final stabilization step as bundles are organized appears to be an internal rearrangement from an aldimine to a keto-amine form of *DHLNL.

Most of the crosslinks identified with the initial steps of collagen fiber and bundle formation are chemically unstable (*) and their levels decline with more advanced collagen organization.[47] With age-related organization of collagen, chemically stable cross-links appear and, in some tissues, increase in numbers. For example, in human and bovine dentin, levels of *DHLNL and *HLNL decrease while levels of a nonreducible and polyfunctional crosslink, hydroxypyridinium (pyridinoline), increase. *DHLNL, considered to be the precursor crosslink of pyridinoline (Pyr)[48] provides the *DHLNL:Pry ratio value as a biochemical index of age-related organization of dentine collagen. That Pyr formation reflects organi-

Table 5
REDUCIBLE COLLAGEN CROSSLINING COMPOUNDS IN SKIN SPECIMENS FROM CONTROLS AND MARFAN PATIENTS

Group	*HNL	*ACP:*HNL	*HLNL:*HNL	*DHLNL:*HLNL
Controls	4.23 ± 0.93 (8)	20.1 ± 0.22 (4)	1.02 ± 0.21 (4)	0.18 ± 0.01 (4)
Marfan Patients	5.43 ± 0.77 (7)	1.95 ± 0.10 (4)	1.07 ± 0.48 (4)	0.10 ± 0.02[a] (4)

Note: Values are means ± SE, given as cpm/μg of collagen. Figures in parentheses denote the number of tissue samples analyzed from four subjects in each group. Abbreviations for the crosslinking compounds are given in the legen for Table 4.

* Signifcantly different from control values ($p < 0.05$), Student's *t*-test.

Data from Boucek, R. J. et al., *N. Engl. J. Med.*, 305, 988, 1981.

zational changes in another collagen-rich tissue was supported by a parallel refractoriness of the bone collagen to pepsin digestion.[49]

Pyr, a fluorescent material in collagen fibers, was first isolated and chemically characterized by Fujimoto and collegues.[50] They reported a Pyr in collagens from human and rat costal cartilage and Achilles tendon. In rat tissues, the Pyr content increased even after the animal reached maturity, while the content reached a maximum at 30 years of age in human tissues.[51] In adult bovine articular cartilage, one Pyr residue per collagen molecule is reported by Eyre and Oguchl[52] and the concentration of Pyr is estimated to be more than 40 times the content of all of the unstable (reducible) cross-links.[53]

Contributions of Pyr to the intramicrofibrillar organization of bovine articular collagen were suggested by Fujimoto and Moriguchi[51] and identified by Robins and Duncan.[54] Fractionation of CN Br-derived peptides identified two peptides containing stoichiometric amounts of Pyr. From their amino acid composition, the peptides arose from two C-terminal nonhelical chains and from the helical region of an α 1(II) chain. The molecular localization of Pyr suggested a functional role in stabilizing the 4D stagger of adjacent molecules and may act as an intra- and an intermicrofibrillar crosslink.

V. RECOGNITION OF DEFECTIVE COLLAGEN ORGANIZATION IN SKIN AND AORTIC TISSUES

A. Intermolecular Crosslink Profiles

Using the crosslink profiles for assessing collagen organization in human skin (types I and III, 75 and 25% respectively) from control and patients with Marfan's syndrome suggested defects in fiber and bundle organization (Table 5).[55] Levels of reducible lysyl oxidase products, allysine and hydroxyallysine (*HNL) and the initial intramolecular crosslink *ACP, and intermolecular crosslink *HLNL, were similar in the control and Marfan skin samples. But the amounts of *DHLNL relative to *HLNL were significantly lower in the Marfan than in the control skins. If the organizational sequence of the collagens in Marfan's skin is similar to the organization of types I and III collagens in newly formed connective tissue around a sponge implant in the rat,[46] then a defective fiber and bundle organization in the collagens of Marfan skin connective tissue is suggested by the lower *DHLNL levels.

The profile of crosslinking compounds in aortic tissue suggested a defective collagen organization in aortic tissue in the Marfan syndrome.[54] Unlike the skin, *DHLNL is a prominent collagen crosslinking compound in aortic collagen (Table 6). While the *HLNL and *DHLNL levels are probably similar, the Pyr density is considerably lower (50%) in

Table 6
REDUCIBLE COLLAGEN CROSSLINKS IN THE ABDOMINAL AORTA FROM CONTROLS AND A PATIENT WITH THE MARFAN SYNDROME

Group	*HLNL[a]	*DHLNL[a]	*DHLNL:*HLNL[a]	Pyridinoline[b]
Controls				
42-year-old man	4.6, 4.7	6.0, 6.3	1.30, 1.34	1.05, 1.37, 1.22
30-year-old man	—	—	—	1.10
Patient				
32-year-old man	4.5, 6.0	7.5, 8.6	1.67, 1.43	0.52, 0.76, 0.54

[a] Values for two tissue samples are given (as cpm/μg of collagen). Abbreviations for the cross-linking compounds are given in the legend for Table 4.

[b] Values for three tissue samples are given (as nmol of leucine equiv./collagen molecule).

Data from Boucek, R. J. et al., *N. Engl. J. Med.*, 305, 988, 1981.

the Marfan than in the control tissues. Converting the Pyr from nmole leucine equivalents to mol/mol collagen suggests one Pyr residue per two molecules of collagen in the aorta from human males 30 to 40 years of age while the aorta from the Marfan patient had one Pyr residue for four molecules of collagen. A normal *DHLNL and a low Pyr density (high *DHLNL:Pyr ratio value) suggest a defective or an attenuated organization of aortic collagen in this patient with Marfan's syndrome.

B. Thermal Contraction and Bacterial Collagenase Digestion

An indirect measure of collagen organization is the rate of heat denaturation of collagen, a well-characterized feature of collagen in solutions. Using optical rotation to follow the heat-related phase transition of soluble type I collagen, the transition is noted to be time dependent and the melting curve, characterized by the midpoint (T_m), appears to be "regulated by the proline plus the hypro content".[56,57]

Tissue collagen organized into fibers and bundles also undergoes a time and temperature dependent heat denaturation. Thermal denaturation of insoluble collagens may be characterized by a temperature response, the shrinkage temperature, T_s. The T_s is considerably higher than the T_m "reflecting the added stability contributed by molecular packing and tissue structure".[58] Relationships between the T_s and collagen organization at the molecular level are now emerging. For example, Allain and colleagues[59] characterized the heat-labile and stable crosslinks in rat skin collagen as the aldimine and the thermally stable crosslinks as the keto-amine species of intermolecular crosslinks. Differential scanning colorimetry suggested at similar conclusions.[60]

In order to study the T_s of the collagens in a complex tissue such as an artery, the collagens must first be isolated without denaturing the macropolymer. Elastase digestion[61] appears to be satisfactory in this regard, since the procedure does not modify the content or the profile of the chemically stable intermolecular collagen crosslinks.[79] The T_s of aortic collagens can be assessed by recording the tension generated during thermal contraction. After elastase digestion of an abdominal aortic tissue ring, the resultant ring of collagen is mounted on two hooks of a universal tensil testing apparatus (Instron) and suspended in a 0.1 *M* phosphate buffer (pH 7.0). After preconditioning with four length-tension exercises, the ring is extended and held to a constant tension by applying a fixed load. The temperature of the buffer is raised by increments of 1.8°C and the isometric tension generated by thermal contraction recorded. After completing the thermal contraction study, the ring is hydrolyzed and the hypro per tissue ring area is determined. The isometric tension, g, as an expression of T_s

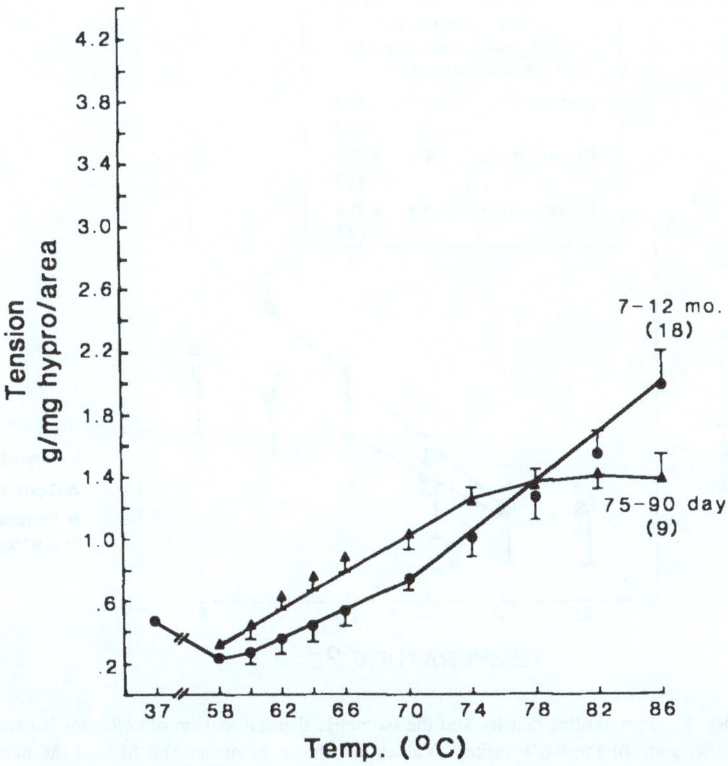

FIGURE 4. The isometric thermal tension of collagen rings from abdominal aorta of 75 to 90-day-old and 7 to 12-month-old rats. At both ages, the collagens generate tension as the bath temperature increases from 60 to 74°C. At temperatures exceeding 74°C, however, the collagens from young rats no longer generated tension while collagens from older rats continued to contract. Apparently collagen organization involving thermally stable intermolecular crosslinks occurred from 3 to 7 months in male Sprague-Dawley rats.

is related to the hypro content of the tissue ring and plotted as a function of temperature (Figure 4).

The density of Pyr appears to be related to the T_s of the collagens from the abdominal aorta of the BBW turkey. This relationship is suggested by the isometric thermal tension of aortic collagen from control and propranolol- or metoprolol-treated BBW turkeys. Both drugs reduced the density of Pyr in the collagens to an equivalent extent (inset Figure 5) and lowered the isometric thermal tension from 70 to 80°C (Figure 5).

Another assessment of collagen organization in arterial tissue is provided from the kinetics of collagen disgestion by collagenase. Differences in the rate of bacterial collagenase (B-collagenase) digestion identifies collagen fiber and bundle organization in the Ivalon sponge implant in rats[46] and age-dependent changes in human collagens.[62] Harris and Farrell[63] crosslinked purified guinea pig collagen in solution with small concentrations of buffered formaldeyde and reported the introduction of "as few as three methylene crosslinks per molecule of collagen produced gels" that resisted B as well as synovial collagenase digestion. More recently, Holton et al.[64,65] summarized the kinetics of collagenase digeston of human collagen of various ages and hypothesize "that the collagen fragments swell at the micro-fibrillar level to permit enzyme penetration in a relatively short time period compared to the overall digestion rate. During the initial rapid interaction, the N-terminal portion is removed and the enzyme penetrates the collagen polymer. A slow cleavage of helical regions then

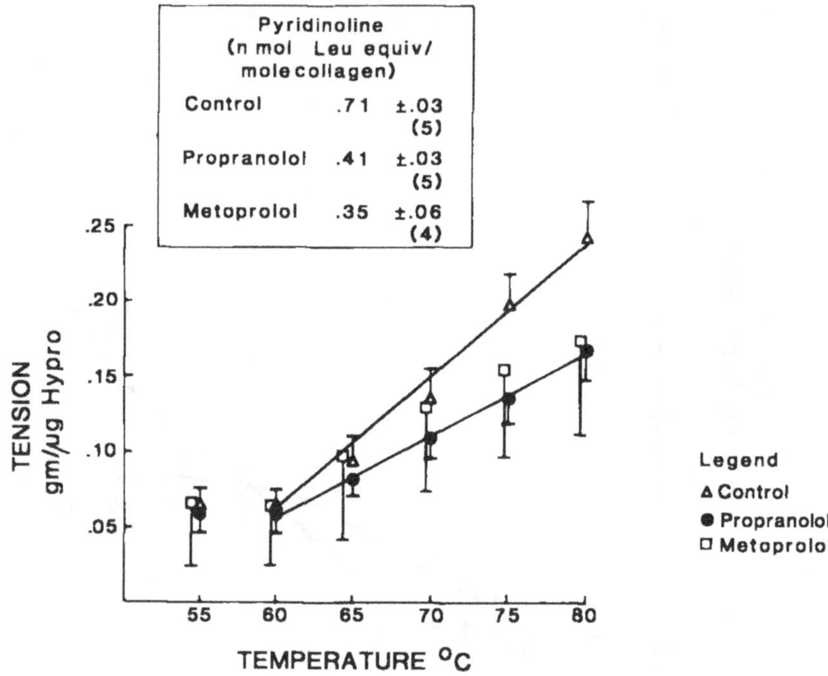

FIGURE 5. Pyridinoline density and the isometric thermal tension of collagens isolated from abdominal aorta of the BBW turkey. Values are shown as means and SE and the number of animals is given in parentheses. The tension (g/μg hypro) as a function of temperature was significantly lower in chronically treated (6 weeks) propranolol and metoprolol as compared to the control.

follows which generates amino-terminated chains which retard digestion as a kinetic blocker. As digestion proceeds, the digestible sequence generated per internal cleavage gets shorter. Since the collagen remaining to participate in the internal cleavage reaction also decreases with conversion, the rate at which acid groups are liberated is best described as a second-order reaction. Human tendon collagens from individuals of 20 to 70 years of age display a 3 to 4 times increase in kinetic blockages of B-collagenase digestion with age. No age-related changes in the B-collagenase digestion of the N-terminal region of collagen could be identified.

If the Holton hypothesis is correct, then complementary information regarding collagen organization may come from studies of thermal contraction and B-collagenase digestion: Thermal contraction, reflecting contributions from N-terminal (and probably C-terminal as well) regions to collagen organization, i.e., intermolecular crosslinks and B-collagenase digestion characterizing intermolecular organization identified with the helical region. Such a combined approach suggests an effect of a diet rich in monounsaturated fatty acids (olive oil) on both collagen crosslinks identified with the teleopeptides and those identified with the helical region in the organization of rat aortic collagens (Figure 6).

A flow chart for the analyses of collagen organization in arterial tissue might be summarized as follows:

- Morphology of collagen bundle orientation (light microscopy, scanning electron microscopy, or X-ray diffraction)
- Quantification of collagen concentration
  - Hydroxyproline (Hypro) concentration in autoclave-alkali-soluble or in B. collagenase-soluble fractions

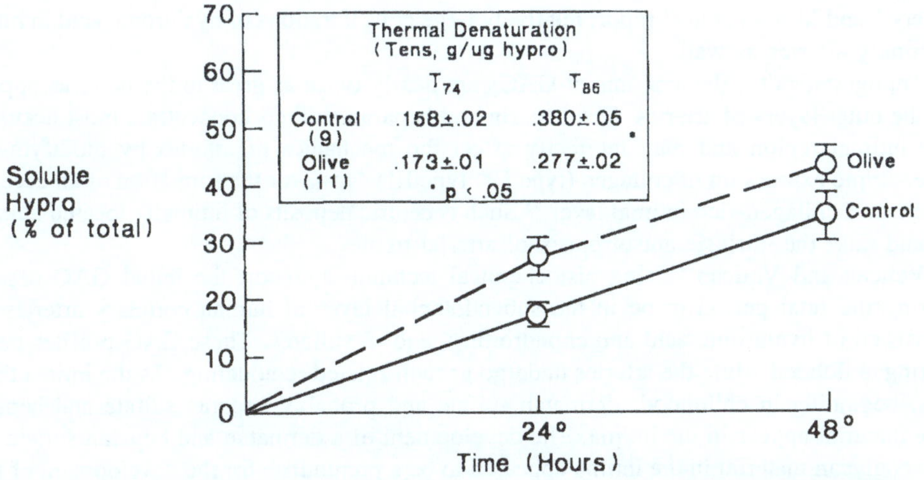

FIGURE 6. Dietary olive oil and the thermal denaturation and susceptibility to collagen digestion of aortic collagen (rat). Groups of male Sprague-Dawley rats were raised postweanling to 7 months of age on a formulated diet containing olive oil (15 energy%) as the sole source of dietary fat or Purina® chow (6 energy%). The saturated:unsaturated fatty acids ratio value of olive oil (10:85) was lower than the Purina chow (25:75). Without affecting arterial pressure or elastin and collagen concentration of the aortic tissues, the olive oil-containing diet mediated a significant reduction in the isometric thermal contraction of aortic collagen at 86°C compared to collagens from chow-fed rats (inset). The collagen solubilized by B-collagenase digestion of acetone-ether defatted abdominal aortic tissue at 24 and 48 hr was significantly greater in tissues from olive oil-fed compared to control rats.

- Estimate of collagen organization
  - — Ratio between chemically unstable and stable crosslinks (newly formed collagens)
  - — Concentration of pyridinoline in the collagen
  - — Thermal contraction at temperatures >60°C of isolated collagens
  - — Rate of B. collagenase digestion in fat-free (and probably) proteoglycan-free tissues

VI. ROLE OF GLYCOPROTEINS AND PROTEOGLYCANS IN ELASTIN AND COLLAGEN ORGANIZATION AND IN THE MECHANICAL PROPERTIES OF ARTERIAL TISSUE

A number of structural glycoproteins isolated from arterial tissues are probably important in positioning cells in the intracellular matrix. The glycoprotein-rich microfibrils of elastin[15] are critical elements in the initial stages of elastin organization (Figure 1). But these glycoproteins may have limited contributions to the mechanical properties of arterial tissue.

Proteoglycan material, on the other hand, as it relates to tissue hydration or to collagen organization, may contribute to the mechanical properties of arterial tissue. The proteoglycan material consists of a protein core with covalently attached polysaccharide side chains made up of glycosaminoglycans (GAGs) which are then associated by a terminal part of a core protein with hyaluronic acid and link-proteins.[66] Arterial proteoglycans contain the following GAGs: heparan sulfate, heparin, dermatan-sulfate, chondroitin-sulfates (-4 and -6), and hyaluronic acid.[67]

An important contribution of proteoglycans to the mechanical properties of arteries may come from their hydration capabilities. For example, proteoglycan material rich in hyaluronate may take up 1000 times the space occupied by the unhydrated chain.[68] However, proteoglycans and GAGs are heterogeneous among arterial tissue.[69] For example, the canine coronary arteries synthesize considerably larger amounts of hyaluronic acid than the carotid artery[70] and Murata et al.[71] report relatively large concentrations of hyaluronic acid in human coronary arteries as well.

Topographically, the amounts of GAGs are nearly twice as great in the inner as opposed to the outer layers of arteries. The heparin and heparan sulfates concentrate most heavily in the intimal region and may indirectly affect the mechanics of arteries by modifying the phenotypic expression of collagen (type I < type III)[72] to favor the formation of an eccentric and type I collagen-rich intimal layer.[73] Such eccentric deposits of intimally located collagen would raise the intrinsic anisotropism of arterial tissues.

Velican and Velican[74] using histochemical techniques, report the initial GAG deposits during the fetal period to be in the subendothelial layer of human coronary arteries and consisted of hyaluronic acid and chondroitin-4 and -6 sulfates. These GAG profiles persist during childhood while the arteries undergo growth-related remodeling. As the intima thickens, beginning in childhood, dermatan sulfate and probably heparan sulfate and heparin-like material appear in the intima. The development of a dermatan and heparan sulfate-rich proteoglycan material in the intima appeared to be a prerequiste for the development of fatty dots and streaks.

Minns et al.[8] evaluated contributions of proteoglycans (ground substance) to the mechanical properties of ascending aorta from a 64-year-old male. The interfibrillar material was removed (α-amylase *[Bacillis subtilis]* or a chelating agent, ethylene diamine tetra-acetic acid [EDTA] and a decrease in stress level, stiffness, relaxation, hysteresis, and other time-dependent mechanical properties were noted. Oxlund and Andreassen,[75] however, could find no changes in the mechanical properties of arterial tissue after removal of 38% of the uronic acid by hyaluronidase digestion.

Whether proteoglycan material makes a direct contribution to the passive mechanical properties of arterial tissue is presently uncertain, but hydration or water content should be part of the anaylses of contributors to the passive mechanical properties of arteries.[76]

VII. ELASTIN AND COLLAGEN ORGANIZATION AND THE PASSIVE MECHANICAL PROPERTIES OF AORTIC TISSUE

Identifying contributions of elastin and collagen organization to the mechanical properties of arterial tissue is exceedingly complex and requires correlative studies involving the chemical and biophysical indexes outlined above and biomechanical descriptors of the passive mechanical properties of the tissue. Relationships between the biochemical and biophysical indexes of elastin and collagen organization and the tensile strength indexes, time dependent length-tension responses to loading (stress-strain characteristics), and tissue hysteresis with loading and unloading after uniaxial incremental loading are presented below.

Uniaxial passive mechanical responses of aortic tissue rings are measured with the Instron. The apparatus is equipped with an upper hook attached to a load cell with strain gauges and a lower hook attached to a moveable cross-head containing a load weighing system with a full scale electronically controlled range of 0 to 20 g. Tissue rings from fixed anatomical sites, measured for width, thickness, and length are mounted on the two hooks and immersed in a temperature-controlled bath of 0.1 *M* phosphate in buffered saline (pH 4.71, 37°C). Length-tension responses to uniaxial incremental loading and unloading are recorded and length-tension values at ten uniformly spaced intervals are determined. The length-tension values are entered into a computer program written to convert the width and thickness of the tissue ring into area (centimeters, cm^2), the tension into mass (g), the mass into "stress" equivalence expressed in kilopascals (kPa) (mass [g] × 0.0981 [acceleration factor] ÷ area [cm^2]), and the percent length change into strain ($length_t - length_o \div length_o$). One thousand computer generated and the experimentally derived data are expressed as a cubic spline function[77] and displayed as a "stress"-strain curve for loading and unloading. The mean

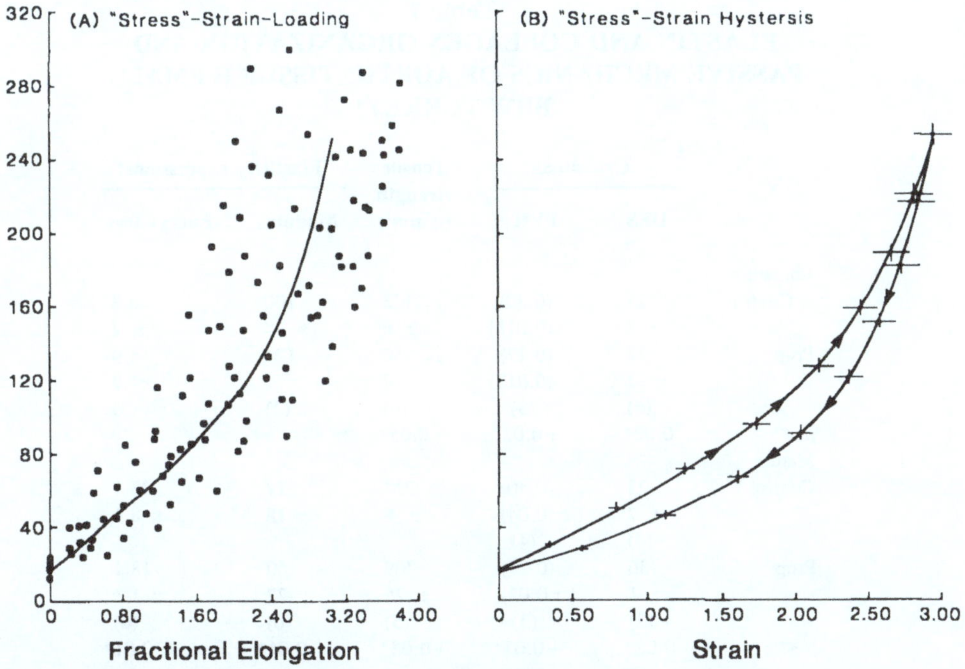

FIGURE 7. Uniaxial passive mechanical response of abdominal aorta tissue (7 month old rat). The "stress"-fractional elongation response of 18 abdominal aortic rings from 6 animals. The tension-length data were converted to "stress"-fractional elongation or strain equivalences and the data points related to the computer-generated curve (A). The mean ± SE of the converted tension-length data for loading and unloading of the tissue ring displays the hysteresis of tissue rings from the group of rats (B).

and standard error (SE) of the experimental points are graphed on the load and unload response. Such a display for 18 abdominal aortic tissue rings from a group of six male rats is shown in Figure 7A. Additionally the display of the loading and unloading response provides a graphic image of tissue hysteresis (Figure 7B).

Key in identifying elastin and collagen organizational contributions to the passive mechanical properties is an experimental animal model where the anatomy and the concentration of the scleroproteins are held constant while the organization is modified. The propranolol-treated BBW turkey provides such an animal model. In the propranolol-treated BBW turkey, elastin organization (maturation) is apparently accelerated (increase DES content in the elastin) while age-related collagen organization is attenuated (reduced Pyr from 0.3 to 0.2 in immature and from 0.4 to 0.3 mol/mol of collagen in adult BBW turkeys) (Table 7). Abdominal aorta tissue with proportionately large amounts of mature elastin (high DES density in elastin) and smaller amounts of age-related collagen organization (low Pyr density in the collagens) have higher tensile strength index values, a lower modulus of elasticity with loads approximating high physiologic levels of arterial pressure,* and a low hysteresis with loading and unloading (expressed as energy loss

VIII. SUMMARY AND CONCLUDING REMARKS

Structure and function considerations for arterial tissues identify unique active and passive mechanical requirements. Key in these considerations are the determinants of the passive mechanical properties of arterial tissue, the anatomic distribution, the concentration, and the

* The systolic aortic pressure of the BBW turkey ranges from 180 to 220 mmHg.[78]

Table 7
ELASTIN AND COLLAGEN ORGANIZATION AND PASSIVE MECHANICS OF AORTIC TISSUE (FEMALE BBW TURKEY)

	Crosslinks		Tensile strength	Elasticity expressions[a]	
	DES	PYR	(g/mm²)	Modulus	Energy loss
Immature					
Control	21	(0.30)	123	200	6.3
	± 2	(0.013)	± 6	± 40	± 7
Prop	32	(0.17)	150	124	4.9
	± 7	(0.013)	± 7	± 1	± .9
	(6)	(5)	(6)	(3)	(3)
$p <$	0.02⁺	+0.02⁺	0.05⁺		
Mature					
Control	23	(0.40)	214	217	33.7
	± 7	(± 0.04)	± 8	± 18	± 8.6
	(4)	(4)			
Prop	36	(0.28)	309	160	18.1
	± 1	(±0.02)	± 25	± 22	± 1.3
	(3)	(3)	(5)	(6)	(6)
$p <$	0.02⁺	+0.02⁺	+0.05⁺	0.01*	0.01*

Note: Abbreviations: DES, desmosines nmol/mg elastin; PYR, pyridinoline mol/mol collagen; Prop., propranolol.

[a] Modulus, the tangent of the stress-strain curve for loading at 40 kPa stress; energy loss, the integral of the stress-strain curves for loading minus unloading divided by loading. Values are means ± SE with number of animals given in parentheses; statistical probability, p: +, by Student's t-test or *, Wilcoxon two-sample rank test for the combined data from immature and mature animals.

physical properties of the structural proteins, elastin and collagen. The anatomic arrangement of elastin- and collagen-rich membranes or fascia and the intrinsic physical properties of elastin and collagen are remarkably complementary. Elastin in the elastic lamellae is arranged as an elastic three-dimensional macropolymer with elastomeric capabilities while collagen is organized as a two-dimensional polymer with limited or no extensibility in one plane. Collagen bundles in the adventitia of arteries are organized as an interwoven fabric to provide the ultimate support at high mechanical loads.

Identifying the contributions of elastin and collagen organization to the passive mechanical properties of arterial tissue requires the isolation and characterization of polymeric elastin and collagen and an experimental animal model where organizational and structural arteries can be independently examined. While enzymatic procedures seem to isolate collagen in a polymeric form, isolation of native elastin from arteries for physical characterizations is problematic. Consequently, indirect means for assessing collagen and elastin organization in arterial tissue are required. Since organization of the two proteins proceeds in a stepwise manner and is accompanied by the formation of specific intermolecular crosslinking compounds, the profiles of crosslinking amino acids provide a biochemical means for assessing the hierarchal forms of arterial elastin and collagen. Furthermore, collagen organization may be assessed after the scleroprotein has been isolated from arterial tissue by thermal shrinkage and by the susceptibility to bacterial collagenase digestion.

An experimental animal model where structural arrangement and concentrations of elastin and collagen are constant while organization of the scleroproteins are altered (recognized by changes in crosslink profiles and physical properties of the collagens) is the propranolol-treated domesticated turkey. Chronic propranolol treatment of the turkey accelerates elastin maturation and retards age-related organization of collagens in the structurally weak abdominal region of the aorta. Propranolol treatment significantly improves tensile strength and elasticity indexes of this aortic region when compared to the untreated animals.

In order to identify contributions of elastin and collagen organization to the passive mechanical properties of arterial tissues, elaborate characterizations of the structure (histology and ultrastructure), concentration (appropriate biochemical markers), intermolecular cross-linking amino acids and, when possible, the biophysical properties of the two proteins are required. Because in arteries the elastin- and collagen-rich fibers and bundles are part of a fabric, indirect means of characterizing the organizational contributions of these proteins are required to identify their contributions to the static biomechanical properties of the tissue.

REFERENCES

1. **Roy, C. S.,** The elastic properties of the arterial wall, *J. Physiol.*, 3, 125, 1881.
2. **Burton, A. C.,** Relation of structure to function of the tissues of the wall of blood vessels, *Physiol. Rev.*, 34, 619, 1954.
3. **Bergel, D. H.,** The dynamic elastic properties of the arterial wall, *J. Physiol.*, 156, 458, 1961.
4. **Roach, M. R. and Burton, A. C.,** The reason for the shape of the destensibility curves of arteries, *Can. J. Biochem. Physiol.*, 35, 681, 1957.
5. **Wolinsky, H. and Glagov, S.,** Structural basis for the static mechanical properties of the aortic media, *Circ. Res.*, 14, 400, 1964.
6. **Samila, Z. J. and Carter, S. A.,** The effect of age on the unfolding of elastin lamellae and collagen fibers with stretch in human carotid arteries, *Can. J. Physiol. Pharmacol.*, 59, 1050, 1981.
7. **Roach, M. R. and Burton, A. C.,** The effect of age on the elasticity of human iliac arteries, *Can. J. Biochem. Physiol.*, 37, 557, 1959.
8. **Minns, R. J. and Soden, P. D.,** The role of the fibrous components and ground substances in the mechanical properties of biological tissues, *J. Biomech.*, 6, 153, 1973.
9. **Davidson, J. M. and Crystal, R. G.,** The molecular aspects of elastin gene expression, *J. Invest. Dermatol.*, Suppl. 79, 133, 1982.
10. **Burnett, W., Eichner, R., and Rosenbloom, J.,** Correlation of functional elastin messenger ribonucleic acid levels and rate of elastin synthesis in the developing chick aorta, *Biochemistry*, 19(6), 1106, 1980.
11. **Ryhanen, L., Graves, P. N., Bressan, G. M., and Prockop, D. J.,** Synthesis of an elastin component of molecular weight about 70,000 by polysomes from chick embryo aortas, *Arch. Biochem. Biophys.*, 185(2), 344, 1978.
12. **Fahrenbach, W. H., Sandberg, L. B., and Cleary, E. G.,** Ultrastructural studies on early elastogenesis, *Anat. Rec.*, 155, 563, 1966.
13. **Siegel, R. C., Pinnell, S. R., and Martin, G. R.,** Cross-linking of collagen and elastin properties of lysyl-oxidase, *Biochemistry*, 9, 4486, 1970.
14. **Pinnell, S. R. and Martin, G. R.,** The cross-linking of collagen and elastin: enzymatic conversion of lysine in peptide linkage to α-amino-acipic-δ-semialdehyde (allysine) by an extract from bone, *Proc. Natl. Acad. Sci. U.S.A.*, 61, 708, 1966.
15. **Greenlee, T. K., Jr., Roos, R., and Harman, J. L.,** The fine structure of elastic fibers, *J. Cell Biol.*, 30, 59, 1966.
16. **Sandberg, L. B., Weissman, N., and Gray, W. R.,** Structural features of tropoelastin related to the site of cross-links in aortic elastin, *Biochemistry*, 10, 52, 1971.
17. **Franzblau, C., and Lent, R. W.,** Studies on the chemistry of elastin, *Brookhaven Symp. Biol.*, 21, 358, 1968.
18. **Partridge, S. B., Elsden, D. F., and Thomas, J.,** Constituition of the cross-linkages in elastin, *Nature (London)*, 197, 1297, 1963.
19. **Sandberg, L. B., Soskel, N. T., and Leslie, J. G.,** Elastin structure, biosynthesis, and relation to disease states, *New Engl. J. Med.*, 304, 566, 1981.
20. **Davidson, J. M., Hill, K. E., and Alford, J. L.,** Developmental changes in collagen and elastin biosynthesis in the porcine aorta, *J. Biol. Chem.*, submitted, 1966.
21. **Schwartz, E., Adamany, A. M., and Blumenfeld, O. O.,** Isolation and characterization of the internal elastic lamina from calf thoracic aorta, *Exp. Mol. Pathol.*, 34, 299, 1981.

22. **Wolff, E. K.,** Elastic and pseudo-elastic tissue of the aorta, *Virchows Arch. Pathol. Anat. Physiol.*, 270, 37, 1928.
23. **Lehoczky, J. M., Terner, J., and Rodbard, S.,** Spontaneous fragmentation of elastin in the dog carotid artery, *J. Atheroscler. Res.*, 2, 223, 1962.
24. **McDonald, B. E., Bird, H. R., and Lalik, J. J.,** The incidence of spontaneous pathologic changes in aortas of growing turkeys, *Poultry Sci.*, 43, 341, 1964.
25. **Simpson, C. F., Boucek, R. J., and Nobel, N. L.,** Similarity of aortic pathology in Marfan's syndrome, copper deficiency in chicks and beta-aminopropionitrile toxicity in turkeys, *Exp. Mol. Pathol.*, 32, 81, 1980.
26. **Abraham, P. A., Perejda, A. J., Carnes, W. H., and Uitto, J.,** Marfan syndrome. Demonstration of abnormal elastin in aorta, *J. Clin. Invest.*, 70, 1245, 1982.
27. **Boucek, R. J., Nobel, N. L., Gunja-Smith, Z., and Simpson, C. F.,** Considerations of aortic elastin chemistry in the genesis of the intimal plaque (broad-breasted white turkey), *Exp. Mol. Pathol.*, 31, 400, 1979.
28. **Miller, E. J. and Fullmer, H. M.,** Elastin: diminished reactivity with aldehyde reagents in copper deficiency and lathyrism, *J. Exp. Med.*, 123, 1097, 1966.
29. **Hall, D. A. and Czerkawski, J. M.,** The reaction between elastase and elastic tissue, *Biochem. J.*, 80, 128, 1961.
30. **Gerler, A.,** The non-specific electrostatic nature of the absorption of elastase and other basic proteins in elastin, *Eur. J. Biochem.*, 20, 541, 1971.
31. **Kagan, H. M., Crombil, G. D., Jordan, R. E., Lewis, W., and Franzblau, C.,** Proteolysis of elastin-legand complexes, *Biochemistry*, 11, 3412, 1972.
32. **Saruk, M. and Eisenstein, R.,** Aortic lesion in Marfan syndrome: the ultrastructure of cystic medial degeneration, *Arch. Pathol. Lab. Med.*, 101, 74, 1977.
33. **Barnett, B. D., Bird, R. H., Lalich, J. J., and Strong, F. M.,** Toxicity of beta-aminopropionitrile for turkey poults, *Proc. Soc. Exp. Biol. Med.*, 94, 67, 1957.
34. **Simpson, C. F., Pritchard, W. R., Harmes, R. H., and Sautter, J. H.,** Electron microscopy of the cardiovascular system of the normal and B-aminopropionitrile fed turkey, *Exp. Mol. Pathol.*, 1, 321, 1962.
35. **McSherry, B. J., Ferguson, A. E., and Ballantyne, J. A.,** Dissecting aneurysm in internal hemorrhage in turkeys, *J. Am. Vet. Med. Assoc.*, 124, 279, 1954.
36. **Narayanan, A. S., Seigel, R. C., and Martin, G. R.,** On the inhibition of lysyl oxidase by B-amino-propinitrile, *Biochem. Biophys. Res. Commun.*, 46, 745, 1972.
37. **Todd, G. L., O'Dell, N. L., Whitfield, G. B., Walden, M. D., Bernard, G. R., and O'Brien, L. J.,** Functional Effects of Lathyrism Upon the Turkey's Cardiovascular System. II. Length-Tension Studies on the Descending Aorta, Abstract 1—6, Annu. Meet. Soc. Exp. Biol. Med. Southwest. Sec., Medical College of South Carolina, Charleston, S.C., October 17, 1969.
38. **Simpson, C. F., Klinge, H. M., Robbins, R. C., and Harms, R. H.,** B-aminopropionitrile-induced aortic ruptures in turkeys: inhibition by reserpine and enhancement by monamine oxidase inhibitors, *Toxicol. Appl. Pharmacol.*, 12, 48, 1968.
39. **Ringer, R. K.,** The Influence of Reserpine on Early Growth, Blood Pressure and Dissecting Aneurysm in Turkeys, Proc 1st Conf. of the Use of Tranquilizing and Antihypertensive Agent Serpasil in Animal and Poultry Products, Rutger's University, New Brunswick, N.J., May 7, 1959, 21.
40. **Simpson, C. F., Klinge, J. M., and Palmer, R. F.,** B-aminopropionitrile-induced dissecting aneurysms of turkeys: treatment with propranolol, *Toxicol. Appl. Pharmacol.*, 16, 143, 1970.
41. **Simpson, C. F. and Boucek, R. J.,** The B-aminopropionitrile-fed turkey: a model for detecting potential drug action on areterial tissue, *Cardiovasc. Res.*, XVII, 26, 1983.
42. **Simpson, C. F., Boucek, R. J., and Noble, N. L.,** Influence of d-, l- and dl-propranolol and practolol and B-aminopropionitrile-induced aortic rupture of turkeys, *Toxicol. Appl. Pharmacol.*, 38, 169, 1976.
43. **Boucek, R. J., Gunja-Smith, Z., Nobel, N. L., and Simpson, C. F.,** Modulation by propranolol of the lysyl cross-links in aortic elastin and collagen of the aneurysm-prone turkey, *Biochem. Parmacol.*, 32, 275, 1983.
44. **Kimball, D. A., Coulson, W. F., and Carnes, W. M.,** Cardiovascular studies on copper-deficient swine. III. Properties of isolated aortic elastin, *Exp. Mol. Pathol.*, 3, 10, 1964.
45. **Sandberg, L. B. and Davidson, J. M.,** Elastin and its gene, *Peptide Protein Rev.*, 3, 169, 1984.
46. **Boucek, R. J., Noble, N. L., and Gunja-Smith, Z.,** A possible role for dehydrodihydroxylysinonorleucine in collagen fibre and bundle formation, *Biochem. J.*, 177, 853, 1977.
47. **Bailey, A. J. and Schimokomaki, M. S.,** Age related changes in the reducible cross-links of collagen. *FEBS Lett.*, 16, 86, 1971.
48. **Walters, C. and Eyre, D. R.,** Collagen cross-links in human detin: increasing contents of hydroxypyridinium residues with age, *Calcif. Tissue Int.*, 35, 4, 1983.
49. **Uchiyama, A., Inoue, T., and Fujimoto, D.,** Synthesis of pyridinoline during in vitro aging of bone collagen, *J. Biochem.*, 90, 1795, 1981.

50. **Fujimoto, D., Akiba, K., and Nakamura, N.,** Isolation and characterization of a fluorescent material in bovine Achilles' tendon collagen, *Biochem. Biophys. Res. Commun.*, 76, 1124, 1977.
51. **Fujimoto, D. and Moriguchi, T.,** Pyridinoline, a non-reducible cross-link of collagen, *J. Biochem.*, 83, 863, 1978.
52. **Eyre, D. R. and Oguchi, H.,** The hydroxypyridinium cross-links of skeletal collagens: their measurement, properties and a proposed pathway of formation, *Biochem. Biophys. Res. Commun.*, 92, 403, 1980.
53. **Eyre, D. R., Paz, M. A., and Gallop, P. M.,** Cross-linking in collagen and elastin, *Ann. Rev. Biochem.*, 53, 717, 1984.
54. **Robins, S. P. and Duncan, A.,** Cross-linking of collagen: location of pyridinoline in bovine articular cartilage at two sites of the molecule, *Biochem. J.*, 215, 175, 1983.
55. **Boucek, R. J., Noble, N. L., Gunja-Smith, Z., and Butler, W. T.,** The Marfan syndrome: a deficiency in chemically stable collagen cross-links, *N. Engl. J. Med.*, 305, 988, 1981.
56. **Burge, R. E. and Hynes, R. D.,** The thermal denaturation of collagen in solution and its structural implications, *J. Mol. Biol.*, 1, 155, 1959.
57. **Piez, K. A.,** The relation between amino acid composition and denaturation of vertebrate collagens, *J. Am. Chem. Soc.*, 82, 247, 1960.
58. **Piez, K. A.,** Molecular and aggregate structures of the collagens, in *Extracellular Matrix Biochemistry*, Piez, K. A. and Reddi, A. H., Eds., Elsevier, New York, 1984, 1.
59. **Allain, J. C., Le Lous, M., Bazin, S., Bailey, A. J., and Delaunay, A.,** Isometric tension developed during heating of collagenous tissue: relationships with collagen cross-linking, *Biochim. Biophys. Acta*, 533, 147, 1978.
60. **Flandin, F., Buffevant, C., and Herbage, D.,** A differential scanning colorimetry analysis of the age-related changes in the thermal stability of rat skin collagen, *Biochim. Biophys. Acta*, 791, 205, 1984.
61. **Faris, B., Moscaritolo, R., Levine, A., Snider, R., Goldstein, R., and Franzblau, C.,** Isolation of purified insoluble aortic collagen, *Biochim. Biophys. Acta*, 534, 64, 1978.
62. **Kohn, R. R. and Rollerson, E.,** Aging of human collagen in relation to susceptibility to the action of collagenase, *J. Gerontol.*, 15, 10, 1960.
63. **Harris, E. D., Jr. and Farrell, M. E.,** Resistance to collagenase: a characteristic of collagen fibrils cross-linked by formaldehyde, *Biochim. Biophys. Acta*, 278, 133, 1972.
64. **Holton, J. S., Litt, M. H., Kohn, R. R., and Hamlin, C. R.,** Human collagen digestion: I. Nature of the digestion kinetics as a function of age and structure, *Exp. Gerontol.*, 18, 477, 1983.
65. **Holton, J. S., Litt, M. H., Kohn, R. R., and Hamlin, C. R.,** Human collagen digestion: II. Theoretical model and application to data, *Exp. Gerontol.*, 18, 485, 1983.
66. **Gardell, S., Baker, J., Caterson, B., Heinegard, D., and Roden, L.,** Link protein and a hyaluronic acid-binding region as components of aorta proteoglycan, *Biochem. Biophys. Res. Commun.*, 95, 1823, 1980.
67. **Oegema, T. R., Jr., Hascall, V. C., and Eisenstein, R.,** Characterization of bovine aorta proteoglycan extracted with guanidine hydrochloride in the presence of protease inhibitors, *J. Biol. Chem.*, 254, 1312, 1979.
68. **Heinegard, D. and Paulsson, M.,** Structure and metabolism of proteoglycans, in *Extracellular Matrix Biochemistry*, Piez, K. A. and Reddi, A. H., Eds., Elsevier, New York, 1984, 277.
69. **Picard, J., Breton, M., Deudon, E., and Berrou, E.,** Heterogeneity of arterial proteoglycans, *Glycoconjugate Res.*, 11, 737, 1979.
70. **Noble, N. L. and Boucek, R. J.,** Glycosaminoglycan synthesis in arterial tissue in renoprival hypertension, *Fed. Proc.*, 37, 1777, 1978.
71. **Murata, K., Yokoyama, Y., and Horiuchi, Y.,** Acid glycosaminoglycan in human coronary artery connective tissue, *Connect. Tissue (Jpn.)*, 12, 53, 1980.
72. **Majack, R. A. and Bornstein, P.,** Heparin regulates the collagen phenotype of vascular smooth muscle cells: induced synthesis of an MPVrPV 60,000 collagen, *J. Cell. Biol.*, 100, 613, 1985.
73. **Majak, R. A. and Bornstein, P.,** Heparin related glycosaminoglycans modulate the secretory phenotype of vascular smooth muscle cells, *J. Cell. Biol.*, 99, 1688, 1984.
74. **Velican, D. and Velican, C.,** Histochemical study on the glycosaminoglycans of the human coronary arteries, *Acta Histochem.*, 59(2), 190, 1977.
75. **Oxlund, H. and Andreassen, T. T.,** The roles of hyaluronic acid, collagen and elastin in the mechanical properties of connective tissues, *J. Anat.*, 131, 611, 1980.
76. **Cox, R. H.,** Mechanical properties of the coronary vascular wall and the contractile process, in *The Coronary Artery*, Kalsner, S., Ed., Oxford University Press, New York, 1982, 59.
77. **Pearson, C. E., Ed.,** *Handbook of Applied Mathematics. Selected Results and Methods*, Van Nostrand Reinhold, New York, 1977, 994.
78. **Weiss, H. S. and Sheahan, M.,** The influence of maturity and sex on the blood pressure of the turkey, *Am. J. Vet. Res.*, 19, 209. 1958.
79. **Gunja-Smith, Z.,** personal communication.

Chapter 11

CURRENT ADVANCES ON THE STUDY OF THE BIOMECHANICAL PROPERTIES OF TENDONS AND LIGAMENTS

Savio L.-Y. Woo and Terry J. Sites

TABLE OF CONTENTS

I. Introduction 224

II. Recent Findings 224
- A. The Effects of Storage by Freezing on Ligaments and Tendons 224
- B. Temperature-Dependent Behavior 226
- C. Effects of Maturation on Soft Tissue Properties 229
- D. Ligament Failure Modes 231
- E. Functional Adaptation and Homeostasis 232
- F. Healing of Ligaments and Tendons 236

III. Summary and Conclusions 238

Acknowledgments 239

References 239

I. INTRODUCTION

Ligaments and tendons are a form of dense, regular connective tissue comprised primarily of collagenous fibrils oriented in a packed, parallel arrangement. Other constituent materials include elastin, proteoglycans, and water. Ligaments serve as the major static supports between bones of a joint as well as contributing significantly to their dynamic support, in accord with periarticular musculature. The morbidity associated with ligamentous injuries results in a significant personal and socioeconomic impact and as such, heightened interest in soft tissue research has been greatly supported.

Many important contributions concerning the biomechanical behavior of ligaments, tendons, and other soft tissues have been generated during recent decades which have stimulated more advanced investigations and research plans. These tissues exhibit complex rheologic characteristics as portrayed by their nonlinear elastic as well as time and history dependent viscoelastic properties.[1]

Further, irregular shapes and geometry as well as the complex anatomical construction of ligaments and tendons contribute to the difficulties in testing for their biomechanical properties.[2] Periarticular ligaments are attached to bone on either side of a joint and insertions are often broad-based and nondescript so that accurate measurement of ligament length is impossible. Testing of isolated ligaments can lead to fiber retraction, thus, the ligament length does not represent its pre-cut condition. Mounting of the isolated ends into clamps often results in slippage at the grip-specimen interface, as well as inducing damage to the underlying tissue which leads to premature failure. Thus, most testing has been conducted utilizing bone-ligament-bone preparations since the loading conditions can be made to more closely approximate in vivo conditions. However, a distinct disadvantage has been the inability to separate the biomechanical properties of the ligament substance from those properties contributed by its insertion sites. The importance of making these distinctions is enhanced when new and improved techniques are introduced in more recent investigations.[2-4] In order to determine the mechanical behavior of the ligament substance alone, special devices such as the buckle transducer[4] and magnetic field displacement transducer[5] have recently been designed to measure the ligament load and strain during tensile loading. However, these systems depend on direct contact with the ligament and as such, undesired artifact may be imposed upon ligament properties. In our laboratory, we have chosen to use a noncontact video technique (video dimension analyzer, or VDA) to measure the surface strains of the ligament substance when testing a bone-ligament-bone preparation. In this system, a television camera is used to transmit a picture of the ligament specimen to a video monitor (Figure 1). An electronic dimensional analyzer then projects two windows on the specimen image which can be seen on the monitor. These windows can be placed over any two stained lines (gauge length) placed on the ligament. The horizontal scan time between the gauge length lines is converted to an output voltage. As the gauge length increases under uniaxial tensile loading, the output voltage increases proportionately. Therefore, the ligament strain can be automatically determined and recorded. The frequency response of the VDA system is as high as 120 Hz and errors in linearity and accuracy are less than 0.5%[2].

In this chapter, we will concentrate on a series of recent advanced studies in connective tissue research as it pertains to the biomechanics of ligaments and tendons. The readers may refer to our previous work for background information concerning the detailed methodology used to measure the biomechanical properties of ligaments and tendons.[2,3]

II. RECENT FINDINGS

A. The Effects of Storage by Freezing on Ligaments and Tendons

Biomechanical testing of ligaments and tendons has evolved into very complex metho-

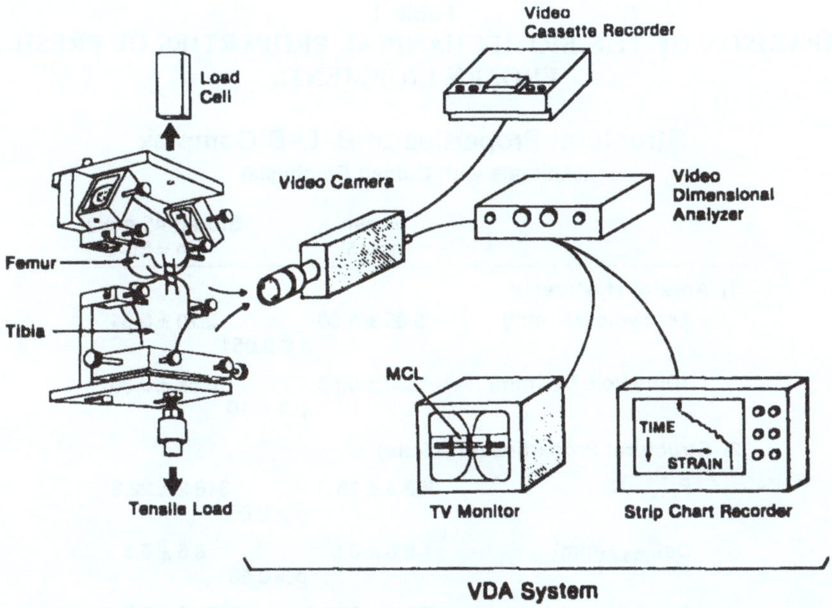

FIGURE 1. The experimental apparatus designed to measure the tensile characteristics of the ligament-bone complex as well as the ligament substance.

dologies. This often requires a considerable period of time for each individual test, and as such, freezing of the specimens becomes a necessity. Also, in the case of human specimens, cadaveric samples have generally been frozen. Therefore, the effects of frozen storage on the mechanical properties of these tissues must be addressed. Furthermore, the results obtained will aid the current clinical interest of freezing cadaver tendons and ligament allografts for transplantation.

Several biomechanical studies comparing the properties of fresh soft tissues with those following storage have been conducted, but conflicting results have been reported. Wertheim[6] found no difference in the elastic modulus of dog tendons between those tested at sacrifice and the contralateral ones tested 5 days later. Smith[7] found that rabbit anterior cruciate ligaments (ACL) became less extensible within 1 hr after death and suggested that data on ligaments can be valid only if tests are performed within 30 min following sacrifice. Viidik et al. showed differences in ACL tensile properties following storage at 18 to 20°C from 2 to 96 hr.[8] Matthews and Ellis reported a decrease in the elastic modulus of cat extensor tendons following storage for 2 weeks at −10°C.[9] Most recently, Noyes and Grood examined the effect of freezing storage for 4 weeks (at −15°C) on monkey ACL and found no changes in the structural properties of bone-ligament-bone specimens or in the ligament cross-sectional area.[10] Dorlot et al. reported a slight increase in stiffness of human ACL stored at −18°C from 5 to 60 days.[11] Barad et al. found no significant differences in structural or mechanical properties between monkey ACL stored at 4°C overnight and those frozen at −80°C for 3 to 5 weeks.[12]

In our laboratory, a study was designed to evaluate possible changes in the mechanical properties of the rabbit medial collateral ligaments (MCL) substance and/or of the structural properties of the bone-ligament unit following 3 months of limb storage at −20°C.[13] Fresh contralateral limbs were dissected immediately at sacrifice and tested as controls. Biomechanical characteristics relating to the structural properties of the bone-ligament-bone (B-L-B) complex included cyclic stress relaxation, area of hysteresis, and load deformation. Stress-strain curves were also obtained to represent the mechanical properties of the

Table 1
COMPARISON OF THE BIOMECHANICAL PROPERTIES OF FRESH AND FROZEN LIGAMENTS

Structural Properties of B–L–B Complex
(Animals with Closed Epiphysis)

	Fresh (n = 5)		Stored 45 days (n = 5)
1) Area of Hysteresis			
1st cycle (N · mm)	5.86 ± 1.60	$p < 0.05$*	2.20 ± 0.54
10th cycle (N · mm)	1.36 ± 0.50	$p > 0.10$	0.58 ± 0.30
2) Structural Properties (at failure)			
P_{max} (N)	368.4 ± 15.0	$p > 0.05$	316.2 ± 22.3
$Def._{max}$ (mm)	6.6 ± 0.6	$p > 0.50$	6.6 ± 0.5
A_{max} (N · mm)	1330.0 ± 200.0	$p > 0.50$	1170.0 ± 200.0

* paired t-test, statistically significant

Note: P_{max} is the ultimate load in Newtons. $Def._{max}$ is the deformation at failure. A_{max} is the ultimate energy absorbed.

ligament substance. We found no significant differences between fresh and frozen samples in most of the parameters measured (Table 1) except for area of hysteresis, where the frozen samples demonstrated significant decreses during the first few cycles of loading and unloading, when compared to fresh, contralateral controls. These differences diminished and became insignificant with further cycling. Thus, the area of hysteresis may be a sensitive indicator of minor changes in the bone-ligament complex secondary to storage by freezing. These changes may result from some insult to cellular integrity or associated ground substance[8] or from changes in ligament fluid dynamics.[14] Freezing did not appear to affect the ligament insertion sites as the modes of failure were not altered between the experimentals and the controls.

We conclude that there are no appreciable effects of storage by freezing up to 3 months on biomechanical properties of ligaments, especially if the properties are measured following preconditioning. However, crucial to the results is the care taken in preparing the tissue sample prior to freezing and ample care must be taken to protect the sample from dehydration. The ligaments were stored with muscle and other periarticular connective tissues left in place rather than in the completely dissected state. Each sample was then double-wrapped in saline-soaked gauze and sealed in airtight plastic bags. Thawing was carried out at refrigerator temperatures (4°C) overnight, and specimens were tested soon after thawing.

B. Temperature-Dependent Behavior

An important consideration when testing tendons and ligaments is what influence does the environmental temperature have on their biomechanical behavior? The concern for controlling the environmental condition during testing of soft tissue has stemmed from the realization that there has been significant disparity between reported results. Obviously, any

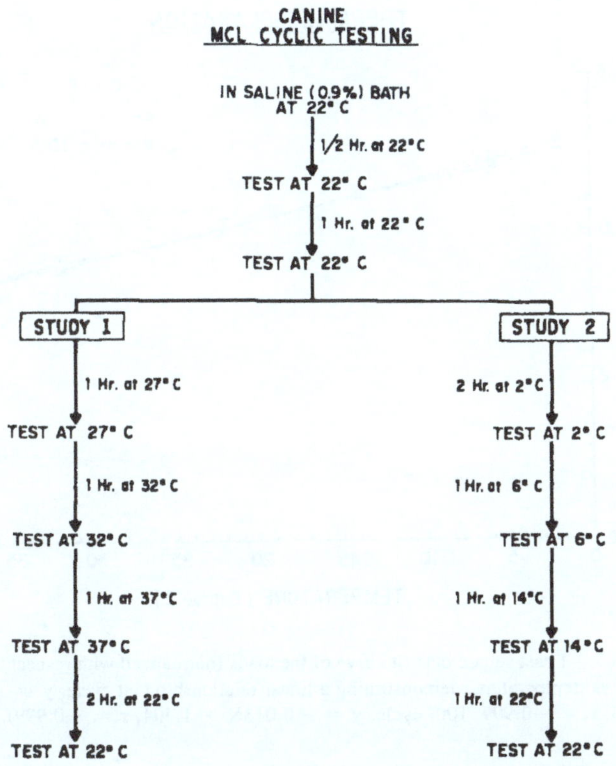

FIGURE 2. Algorithm demonstrating the testing protocol to study the effects of temperature on the canine MCL. Repeated tests at 22°C at the beginning and end are for checking precision of the testing procedure.

differences between testing protocols may cause variance in results reported, but little attention has been given to the effects of temperature on soft-tissue biomechanical behavior. Most tests have been performed by immersing the specimen in a bath, such as an isotonic solution, where pH and temperature can be closely controlled. Rigby et al. suggested that no changes occur in the mechanical properties of ligaments between 0 and 37°C.[15] Apter reported that collagen has a negative temperature-elastic modulus relationship from 0 to 70°C, with the higher modulus appearing at low temperatures.[16] Similar testing of elastin revealed a positive temperature-modulus relationship. Hunter and Williams found an inverse relationship between joint stiffness and temperature.[17] Kuei et al. also revealed a similar temperature dependence for articular cartilage.[18] Further studies by our laboratory were conducted to quantitate the temperature-dependent behavior of canine MCL between 2 and 37°C.[19]

The specimen used in this study was a canine femur-MCL-tibia preparation. The specimen was clamped and submerged into a 0.9% saline bath contained in a specially designed tank (approximately 8 ℓ in volume) equipped with a heating and cooling system. The testing tank was then mounted onto the Instron testing machine for cyclic tensile testing. The temperature was monitored by a thermostat. The accuracy of the temperature control was within 0.5°C. The study was performed in two parts because of the length of time to complete each test.

Study 1 — Temperatures ranged from 22 to 37°C (Figure 2). Each specimen was sequentially tested at 22, 22, 27, 32, 37, and 22°C. After applying 2 *N* of preload, the specimen was cycled between 0 to 2 mm extension (approximately 0 to 3% ligament strain) for one cycle at an extension rate of 2 cm/min, and then cycled an additional nine times between 1

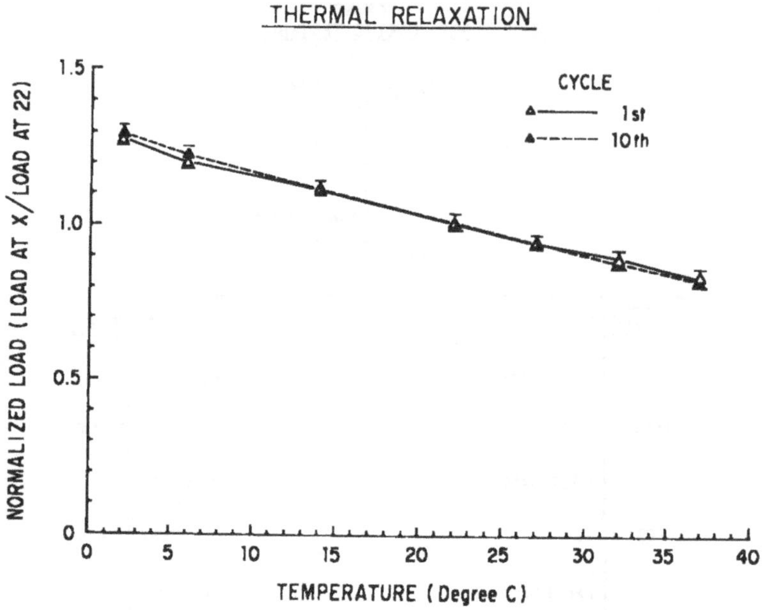

FIGURE 3. Least square best fit curve of the loads (normalized with respect to the load at 22°C) vs. temperature, demonstrating a linear relationship (1st cycle; y = −0.0.12X + 1.283, r = −0.999; 10th cycle: y = −0.013X + 1.304, r = −0.999).

to 2 mm extension (approximately 1 to 3% ligament strain) at the same extension rate. The specimen was then unloaded for 1 hr at 22°C and the same cyclic test was performed. A minimum of 1 hr was required to allow recovery of the initial, untested condition due to the viscoelastic nature of the ligaments. Thereafter, the specimen was unloaded for 1 hr between each cyclic test performed at 27, 32, and 37°C. For the final test at 22°C, a 2-hr recovery period was allowed.

Study 2 — Temperature ranged from 2 to 22°C (Figure 2). Each specimen was sequentially tested at 22, 22, 2, 6, 14, and 22°C. The specimen was first submerged in a 0.9% saline bath at 22°C for 1/2 hr in its unloaded state. Thereafter, an identical testing procedure (as in Study 1) was applied at each temperature except the recovery period between tests ranged from $1^1/_2$ to 2 hr. Note that the repeated cyclic tests at 22°C for the beginning and end of each study were to check the repeatability and reliability of our experimental technique.

Loads for each cycle at each temperature were recorded and normalized with the respective loads at 22°C to illustrate the ligaments' temperature-dependent behavior in terms of cyclic relaxation. The area of hysteresis was determined by normalizing with respect to the value at 22°C. In addition, at any specific test temperatures, the loads for each cycle were normalized with the load of the first cycle at that specific temperature (Figure 3).

The results demonstrate that the ligament reflects increasing stiffness in terms of cyclic loading as the temperature declines, and this can be expressed by a simple linear relationship (Figure 3). For cyclic stress, the relaxation behavior plateaued to lower values as the testing temperature was increased. The area of hysteresis reflected a similar temperature dependency. Another important contribution from this study was the finding that between temperature changes, the ligament requires between 1 to $2^1/_2$ hr to return to its untested, resting characteristics. This may be of particular importance in explaining why our findings differ from those obtained by others. It also demonstrates how critical the use of a standardized procedure is to obtain dependable data and that temperature must be reported and controlled, as it can have profound effects on soft tissue behavior.

C. Effects of Maturation on Soft Tissue Properties

The effects of age (maturation) on tendons, ligaments, and their insertions have been demonstrated. Many authors, using rat tail tendons, have shown age-dependent increases in collagen fibril size, ultimate load, and tensile strength from puberty to adulthood;[20-22] afterwards, no changes were observed until senescence, where decreases in these properties may result. There exists limited information on changes in ligament properties as a function of age, however Booth and Tipton described the changes in bone-ligament junction strength for male and female rats during growth from 15 to 90 days.[23] Males had significantly higher junction strength than females in terms of absolute values, but when normalized to body weight, females had consistently higher values through adulthood.

Noyes and Grood compared the tensile properties of human ACL between young adults (16 to 26 years of age) and those over 50 years of age.[10] These authors found that the strength and stiffness of the bone-ligament complex was significantly decreased with advanced age. Furthermore, the mode of failure changed from ligament substance in the younger groups to that of ligament avulsion at its insertion sites for those samples beyond 50 years of age. In our laboratory, age-related changes on the biomechanical properties of the ligament substance and the bone-ligament complex were evaluated.[24]

The ligament chosen for this study was the MCL from the rabbit knee and the test specimen was a femur-MCL-tibia (FMT) preparation. Four groups of male New Zealand white rabbits were utilized. Group I contained animals $1^1/_2$ months of age (open epiphysis by radiologic examination), group II animals were sacrificed at 4 to 5 months of age (open epiphysis), group III animals were sacrificed at 6 to 7 months of age, and group IV animals were sacrificed at 12 to 15 months of age. All animals in groups III and IV had closed epiphisis. The testing protocol was as follows: each clamped FMT specimen was submerged in a 37° saline bath which was then mounted on an Instron testing machine. A small preload of 0.5 *N* was applied. The specimen was then subjected to 10 cycles of loading-unloading between deformations of 0 to 1.0 mm, at an extension rate of 1.0 cm/min. Following cyclic testing, the specimen was stretched to failure at a rate of 1 cm/min (ligament strain rate of approximately 0.3%/sec). The structural properties of the FMT complex were represented by area of hysteresis, load-deformation curves, ultimate load, and energy absorbed. The mechanical properties of the ligament were expressed by stress-strain curves.

The structural properties of the FMT complex increased dramatically from 1 to 6 months of age, at which time the magnitude of differences between groups diminished. Statistically significant differences ($p < 0.05$) in the area of hysteresis for the 1st cycle were seen between groups I and II, I and III, and I and IV. The mechanical properties of the ligament substance demonstrated relatively early maturation, in that by 4 to 5 months of age the stress-strain curves, in the functional range, were similar to the adults (Figure 4). However, the structural properties of the FMT complex did not reflect plateau values until closure of the epiphyses, at about 7 months. It was also noted by histological examination in our laboratory that the insertion site of the MCL on the tibial side is affected by its proximity to the growth plate, where (1) osteoclastic activity in this region weakens the subperiosteal attachment and (2) part of the ligament insertion is at the area of the metaphysis. These findings are supported by the failure modes as the rabbits with open epiphyses all failed by tibial avulsion and 62% failed by ligament substance tear.

Thus, it seems that the age of the animals has a significant effect on the relative strength of the various components in the FMT cmplex. For young animals, the ligament tissues (stress-strain) reflect adult properties prior to epiphyseal closure (Figure 5). However, the insertion sites are directly affected by growth plate activity and remain structurally inferior until epiphyseal closure. Much interest has been generated on the changes in soft tissue with senescence. A return to weakness at the insertion sites is felt to be secondary to a loss of bone mass. Hormonal influence with age may affect many of these changes in bone as well

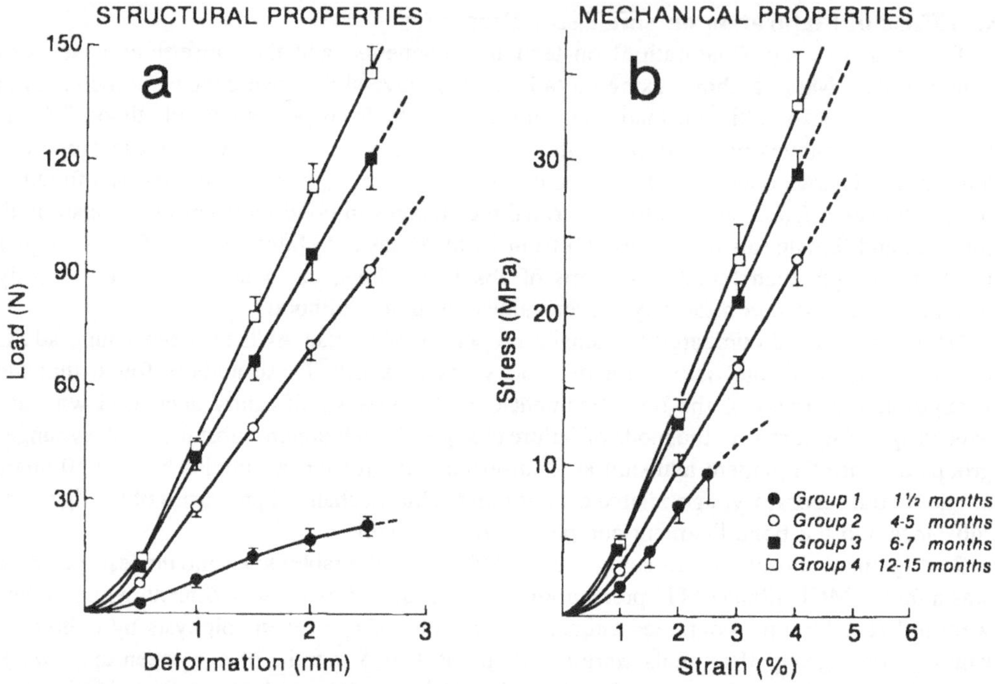

FIGURE 4. The structual properties of the FMT complex for all four age groups are shown in (a). There are rapid changes in the stiffness characteristics between the two younger groups with smaller changes between the older groups. The mechanical properties of the MCL substance for all four age groups are shown in (b). Note that the age-related changes were much less, as compared to those for the structural properties.

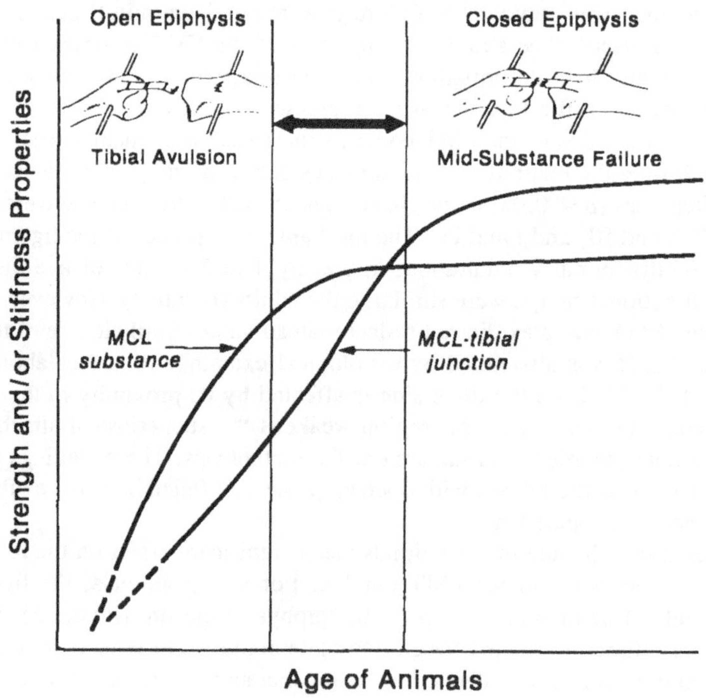

FIGURE 5. A schematic diagram hypothesizing the asynchronous rates of maturation between the FMT complex and the MCL substance.

as in ligament biochemistry and biomechanics. Additional studies must be performed in order to substantiate these hypotheses.

D. Ligament Failure Modes

The functional support provided by a ligament is as a bone-ligament-bone unit. Clinically, failure in adults is more common by substance tear rather than by avulsion. However, in the laboratory, many investigators using cadavers and experimental animals have commented on the difficulty of producing ligament substance injuries and have traditionally stated that the bone is the weakest component of the system.[25-29] This discrepancy between clinical observations and experimental results has stimulated interest in determining what factors influence failure types and injury mechanisms. Identifiable factors include species and ligament type, axis of loading, activity levels, strain-rate, and age.

Although ligaments serve primarily as static stabilizers of joints, each ligament type (i.e., MCL, ACL, ligamentum flavum, etc.) is unique in anatomic location and structural orientation. As such, one must always be aware that experimental results are based upon the particular ligaments utilized for the investigation and that caution must be exercised in formulating generalizations concerning ligament behavior. Investigations by our laboratory have been concerned with several specific factors that are described below.

(1) Axis of force application: Applying the load along the longitudinal axis of the ligament or along the longitudinal axis of the tibia.[30] In the case of the rabbit ACL, if the load was directed along the ligament axis, knee flexion, which varied between 0 and 90°, had no significant effect upon the ultimate load values, and most failures were by bony avulsion. However, when forces were directed along the tibial axis, progressively inferior structural properties were found as knee flexion increased, and failures occurred primarily by progressive fiber failure of the ligament substance. Thus, when the load was applied along the ligament axis, the ligament fibrils were oriented so that even distribution of forces occurred and maximum strength resulted. Force application along the tibial axis resulted in uneven loading along the ligament and the fibrils resisting most of the load failed in a progressive fashion.

(2) Effects of animal activity on ligament strength: Findings in our laboratory and others[31,32] have demonstrated that immobilization of a joint for clinically relevant periods leads to a precipitous decline in ligament-bone junction strength; particularly for those ligaments that insert via the periosteum and do not pass through well-defined zones of fibrocartilage (i.e., the tibial insertion of the MCL). Most failures occur by bony avulsion, and histologic observation identifies the compromised bone-ligament junction as subperiosteal resorption. Detailed discussion can be found in Section II.E.

(3) Effects of strain rate on failure modes: Considerable attention has been given to the effects of rate of stretch on the failure mode of a bone-ligament complex. Some investigators feel that the reason others have been unsuccessful in obtaining ligament substance failure is primarily due to employing slow strain rates.[20,33] Noyes,[32] in a study on ACL of primates, found that at a slow stretch rate (0.00662 sec^{-1}) there were 29% ligamentous failures and 57% failures by tibial avulsion. However, following a 100-fold increase in the stretch rate (0.662 sec^{-1}), the percentage of failures by ligament tear increased to 66% and the percentage of tibial avulsions declined to 28%. Other research on ligaments and tendons had similar results.[34,35]

Recent investigations in our laboratory, however, have shown that the effects of maturation on ligament insertion strength can be even more significant than strain rate. For example, using a moderate strain rate of 0.3/sec, we have noticed that for animals with open epiphyses, all failures occurred by tibial avulsion, whereas, for those animals with closed epiphyses, there was only 12% failure by tibial avulsion.

The effect of strain rate on the MCL midsubstance material properties as well as its effects on mode of failure of the FMT complex have been investigated by our laboratory. Two groups of New Zealand white rabbits were selected: (1) open epiphysis (2 to 3 months old) and (2) closed epiphysis (7 to 8 months old). The FMT complex was subjected to uniaxial tensile tests at five different strain rates (crosshead speeds of 10 cm/sec to 0.01 mm/sec). In the case of high strain rates, a high speed video recording system* was used to record at 2000 frames per second (normal videotaping speed is 60 frames per second).

In group I, the ultimate load, energy absorbed, elastic modulus, and stress-strain increased with increasing strain rate. For group II, only the ultimate load, tensile strength, and energy absorbed increased with increasing strain rate. However, it should be noted that significant differences were obtained only between the values resulting from the slowest and fastest strain rate tests, a range of 4 decades. Failure modes, on the other hand, were independent of strain rate. All failures in the animals with open epiphyses occurred by tibial avulsion; whereas, all failures in the animals with closed epiphyses occurred by ligament disruption. Therefore, we concluded that the age of the animals was the most influential factor determining failure mode (Figure 5).

E. Functional Adaptation and Homeostasis

The effects of immobilization on soft connective tissue are profound, as exemplified by the many works showing significant deteriorations of cartilage, tendon and ligament morphology, biochemistry, and biomechanics.[36-41]

A number of biomechanical studies have been conducted utilizing knees from various animal models, but it seems evident that the ultimate load exhibited by ligaments, following approximately 2 months of immobilization, is less than 50% of the load at failure for contralateral controls.[32,41] Even more concerning was the discovery that following cessation of immobilization, prolonged recovery times were required for the bone-ligament-bone complexes to acquire strength equal to normal controls.[32] It should be noted that all results presented have been concerned with ultimate loads, and little or no information has been presented on the mechanical properties of the ligament substance.

In our laboratory, rabbit knees were immobilized for varying periods: group I, 9 weeks of immobilization; group II, 12 weeks of immobilization; group III, 9 weeks of immobilization followed by 9 weeks of remobilization; and group IV, 12 weeks of immobilization followed by 9 weeks of remobilization.[41] Significant reductions in the structural properties of the FMT complex following immobilization were found. For group I, the ultimate loads and energy absorbing capabilities of the experimental complexes were only 29 and 16%, respectively, of the contralateral controls ($p < 0.01$). Group II showed still further reduction. Stress-strain curves revealed a significant softening of the ligament substance after 9 weeks of immobilization, with continued decline at 12 weeks so that a steep, time-dependent, negative effect was illustrated (Figure 6).

Following stress deprivation, the modes of failure of the FMT complex were predominantly by bony avulsion, indicating disruption at the attachment sites. Thus, the integrity of the ligaments and their respective insertions were violated.

Biomechanical testing of the FMT complex following remobilization showed that the ligament mechanical properties rapidly returned to normal control values after 9 weeks (Figure 7), but the structural properties of the bone-ligament complex remained inferior to controls. Modes of failure for groups III and IV reflect this continued disruption at the bony insertion sites. It should be mentioned that this result was for the medial collateral ligament and may not be generalized to other ligaments. For example, the ACL is positioned in a "hostile" synovial fluid environment, thus immobilization may be of greater detriment to the ACL physiological balance, while remobilization may not be as effective in restoring homeostatic conditions.

* Spin Physics, SP 2000®, Motion Analysis System, Eastman Kodak Co., Rochester, N.Y., 1984.

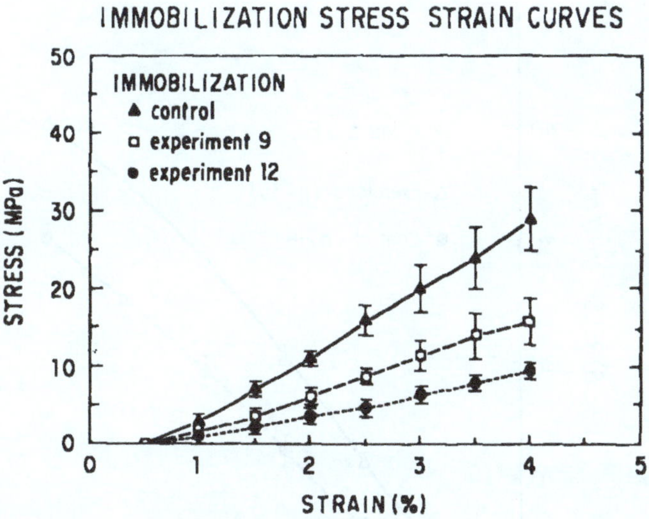

FIGURE 6. The time-dependent effect of immobilization on the mechanical properties of the ligaments.

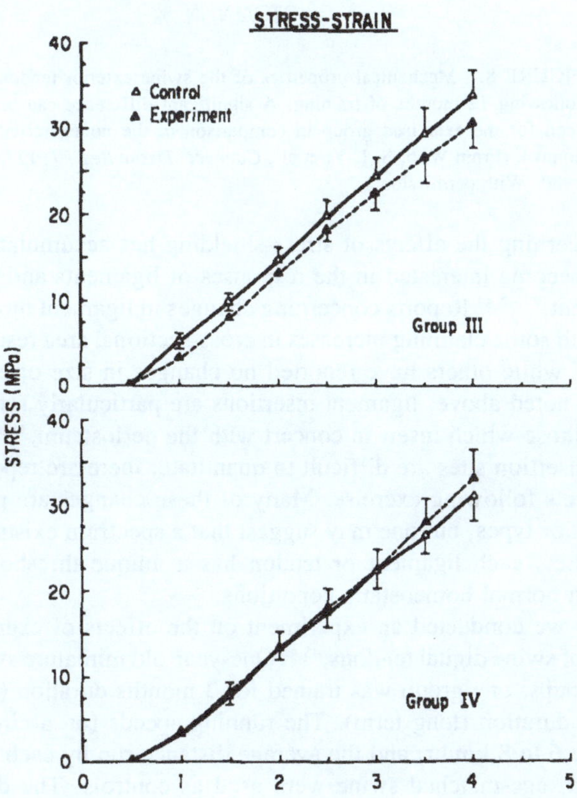

FIGURE 7. Curves demonstrating the complex recovery of the mechanical properties of the ligament in the functional range following remobilization.

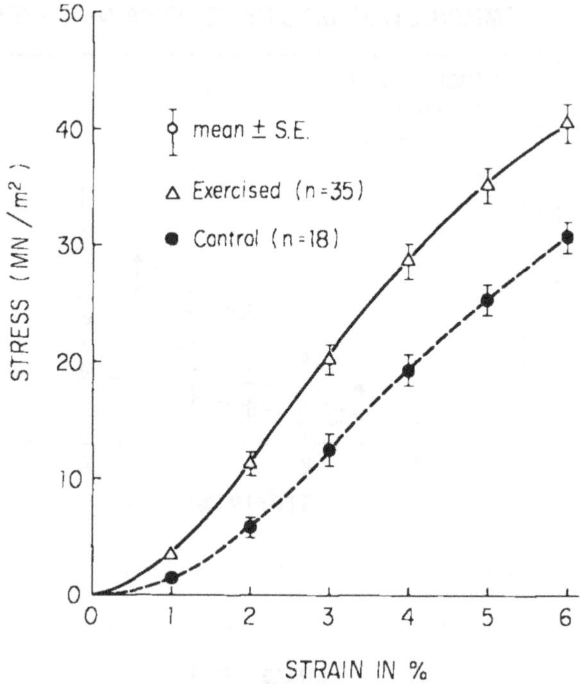

FIGURE 8. Mechanical properties of the swine exterior tendon following 12 months of training. A significant difference can be seen for the exercised group in comparison to the nonexercised control. (From Woo, S. L-Y. et al., *Connect. Tissue Res.*, 7, 177, 1980. With permission.)

As evidence concerning the effects of stress-shielding has accumulated, more and more investigators have become interested in the responses of ligaments and tendons to exercise or stress-enhancement.[29,42-47] Reports concerning changes in ligament morphology have been in disagreement, with some claiming increases in cross-sectional area resulting from collagen fiber hypertrophy,[47] while others have reported no changes in size or even atrophy due to losses of water. As noted above, ligament insertions are particularly sensitive to stress and motion, especially those which insert in concert with the periosteum.[31,32] Although subperiosteal changes at insertion sites are difficult to quantitate, there are reports of a reversal of immobilization effects following exercise. Many of these changes are probably influenced by specific ligament or types, but one may suggest that a spectrum exists for tissue response to activity levels, i.e., each ligament or tendon has a unique threshold level of activity required to maintain normal homeostatic conditions.

In our laboratory we conducted an experiment on the effects of exercise on the biomechanical properties of swine digital tendons.[45,46] One-year-old miniature swine were randomly divided into two groups: one group was trained for 3 months duration (short term) and the other for 12 month duration (long term). The running speeds (on a circular dirt track and on a treadmill) were 6 to 8 km/hr, and the average distance run by each animal was 40 km/week. Nonexercised, age-matched swine were used as controls. The digital extensor and flexor tendons of the two middle toes of the four limbs were studied. The effects of exercise on the mechanical properties of the extensor tendons are detailed in Figure 8. Short term (3 month) exercise had little or no effect on the tendon properties. However, for the long term (12 month) exercise group there were positive changes. The tensile strength of the extensor tendons was 22% higher. Long term exercise also resulted in an increase in the extensor tendon mass. Specifically, the cross-sectional area increased 21% ($p < 0.005$).

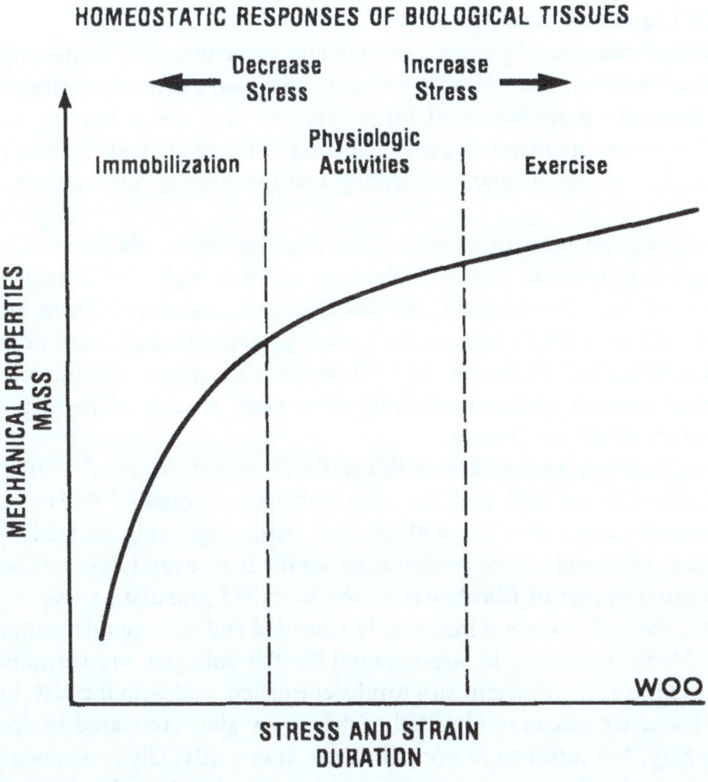

FIGURE 9. A hypothetical curve to describe the nonlinear effects of stress and motion of the homeostatic responses of soft connective tissues.

For the flexor tendons, the mechanical properties of the tendon substance exhibited no statistical difference between the exercised and the controls, even after 12 months of training. However, the ultimate load of the exercised flexor-tendon complexes increased 19%. Since the mode of failure in both exercised and controls was by bony avulsion, it was felt that the increased values of the experimentals were secondary to changes at the bony insertion sites. Tipton et al. also noted that FMT complexes of caged, exercised dogs had higher strength/body weight ratios than those from caged, control, nonexercised dogs.[47]

From these findings we can propose a hypothetical curve to describe the homeostatic responses for connective tissues such as tendons and ligaments (Figure 9). The relationship between the levels and durations of stress and motion, and the resulting changes in tissue properties and tissue mass could be represented by a highly nonlinear, concave upward-shaped curve. The normal range (physiological activities) is bracketed in the middle portion of the curve. With stress and motion deprivation, a rapid reduction in tissue properties (and/or mass) may occur since the slope of the curve in this region is steep. For example, a short period of immobilization of the rabbit knee resulted in a significant alteration of the mechanical properties of the MCL substance, as well as the structural properties of the MCL-bone complex. On the contrary, with exercise training or increased levels of stress and motion, the resulting changes may not be as pronounced since the slope of the curve in this region is relatively flat. For example, with three months of exercise training, the bone and tendon had little change in mechanical properties and mass. Only with long term, 12-month training were the positive responses on the mechanical properties (and mass) seen. Similarly, the positive effect of exercise training on the ligament-bone junction was not as pronounced.

F. Healing of Ligaments and Tendons

The treatment of damaged ligaments and tendons is a common but challenging orthopedic problem. The approach in determining the most adequate treatment protocols has yet to be determined. Accurate descriptions of other types of soft tissue healing have provided a springboard for current interests in ligament healing.[48] Those interests include (1) nonrepaired ligament healing v. repaired ligament healing and (2) activity level effects on the healing ligament.

In our laboratory we have used male New Zealand white rabbits (3.0 to 3.5 kg) with complete surgical injuries to the midsubstance of their right MCL (experimentals) and exposures only of their contralateral, left MCL (sham controls).[49] There was no repair or immoblization and all animals had unrestricted cage activity until sacrifice in groups at 10 days, 3 weeks, 6 weeks, 14 weeks, and 40 weeks after injury. Additional age, sex, and activity matched animals (normal controls) were used at each of these time intervals to provide a baseline of all parameters.

Histologic evaluation revealed that healing of the disrupted, nonrepaired ligament is similar to the classic descriptions provided for other soft tissue wounds.[50,51] Hemorrhage and an acute inflammatory response ensue with the soft tissue gap being bridged by a fibrin clot. The inflammatory cells and edema predominate for the first several days followed by vascular ingrowth and proliferation of fibroblasts — the so-called granulation tissue. By the end of the first month, the inflammation has largely subsided and collagen deposition continues at a rapid pace. Many more months are required for the collagen and surrounding matrix to organize into mature scar. Biochemistry results correlated well with the histologic findings.[49] Water content was significantly elevated at the injury site, compared to controls, up to 6 weeks postsurgery, but returned to normal values afterwards. Glycosaminoglycans (GAGs) content, as estimated by hexosamine content, was elevated early and continued to be elevated above normal even at 40 weeks. Collagen content dropped significantly by 3 weeks postinjury, and then steadily increased over the ensuing weeks, but never reached control values. The healing MCL also contained significant amounts of type III collagen which is normally present in only minute quantities. Repairing of the ligament ends primarily with sutures results in a similar healing scenario as above[52] except for some important differences revealed in our study. The defect was reduced, thus less granulation and collagen was required for bridging the gap. This resulted in less edema, less scar formation, and less time for healing — similar to "healing by first intention."

The mechanical and structural properties of the untreated MCL scars were persistently inferior to the sham controls even at 40 weeks postinjury. MCL scars (formed without treatment) were larger, weaker, and less stiff in tension than normal MCL substance and, based upon their persistent changes in composition, along with their decreasing rate of improvement over the course of this study, it is highly unlikely that untreated MCL scars will ever approach normal ligament characteristics. Cyclic and stress-relaxation properties of the MCL scars, however, did show more complete recovery, indicating that the MCL healing may be adequate for the functional, low loading range.

For the canine model, adult mongrel dogs with an average weight of 221 $\pm$ 7 N were separated into three groups of ten animals each.[53,54] In group I, the early mobilization group, the MCL of the left knee was transected at the joint line and not repaired. The right knee was sham operated, leaving the MCL intact. The animals were allowed 6 and 12 weeks of cage activity. For group II, the shorter term immobilization group, the MCL was repaired after transection. The knee joint was immobilized for 3 weeks and then remobilized either for 3 weeks or 9 weeks. In group III, the longer term immobilization group, the MCL was lacerated and repaired as before. Immobilization was extended to 6 weeks plus an additional 6 weeks of remobilization. At sacrifice, the knees of each animal were attached to a specially designed laxity device, and the varus-valgus (V-V) knee laxity was determined. In addition,

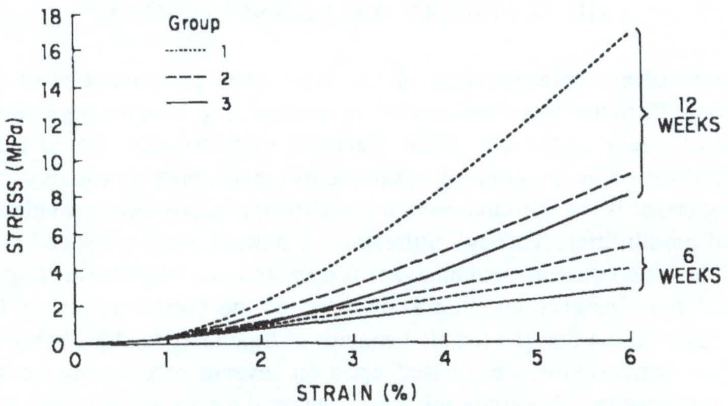

FIGURE 10. Mechanical properties of the healing canine MCL.

the bone-ligament-bone specimen was subjected to tensile tests to determine the mechanical properties of the MCL substance and the structural properties of the bone-ligament complex following reports by Woo et al.[4]

At 6 weeks, the knees from all experimental legs had significantly higher V-V laxity values. Animals from group II showed the least amount of V-V laxity yet were still twice that of the controls. At 12 weeks, experimental knees from group I achieved the best results as their values were similar to those of the controls. An identical trend was observed for the mechanical properties of the healing MCL (Figure 10). However, both the stress-strain curves and tensile strength of the healing MCL were significantly inferior as compared to the controls.

The results obtained for both the V-V knee laxity and the mechanical properties of the healing MCL suggest that the currently preferred clinical practice, i.e., no primary surgical repair for patients with isolated tears of their MCL, may be appropriate. However, nonrepair may result in a more compliant ligament if lengthening occurs during healing. Clayton and others reported that sutured and unsutured MCL healed with equal length at rest, but that the unsutured ligaments may be more compliant with loading.[51]

Further concerns are that early mobilization of the young scar will result in scar lengthening and increased laxity. However, in our study, stress application following only a few weeks of immobility has not revealed any increase in laxity.

Even more important to healing may be the dependence of the reorganizing tissue on stress and motion. Experiments conducted on animal models utilizing repaired ligaments and tendons treated by immobilization, cage activity, and exercise have demonstrated a spectrum of results: immobilized animals having the weakest and exercised animals portraying the strongest properties.[55-57] Thus, it seems that early and active mobilization of the repaired ligaments or tendons may lead to improved qualities. However, the end point of these effects is uncertain and will require studies of longer follow-up.

It is important to note that healing is also dependent on ligament type. For example, the ACL is not known to heal unless approximated with sutures, and even then, healing is often incomplete, leading some surgeons to challenge the use of primary repair of the cruciates.[56] This has lead to much recent interest on autograft, allograft, and xenograft, as well as synthetic replacement of the torn ACL. The important question is whether there is a mechanical difference between repaired or nonrepaired ligaments which heal. Most studies in this area are encouraging as well as studies on the mediators of stress effects such as blood supply and hormones. Hopefully, additional research will result in a complete understanding of its function and nutritional source so that the dismal sequelea associated with ACL injuries will be of only historic interest.

III. SUMMARY AND CONCLUSIONS

Much of our current understanding of the biomechanical properties of ligaments and tendons has come from the development and application of new technologies such as accurate determination of stress and strain of the ligament midsubstance. These technical breakthroughs have afforded us the ability to examine old ideas more thoroughly and accurately. Even more important is the stimulation that these examinations have provided for exploring new ideas and possibilities. Through utilization of normal and high speed video recording together with the video dimensional analyzer system and other bioengineering advances, we have presented new insights into the knowledge of the biomechanics of ligaments and tendons. We have shown that the tensile behavior of ligaments (and probably all soft tissue) is dependent on temperature, i.e., based upon an inverse relationship between ligament stiffness and temperature. This study reinforces the need for a standardized testing procedure since any change in environmental conditions may influence the results.

Other studies in this chapter have utilized the VDA sytem to separate the differences between the ligament substance and its insertion sites under various conditions. The investigations concerning the strain-rate sensitivity of ligaments have reinforced the concept that tensile properties of the ligaments, like most other soft tissues, are not largely sensitive to rates of stretch, but are more dependent on the animal age. Through 4 decades of strain rates, those animals with open epiphyses all failed by tibial avulsion, but those animals that had reached skeletal maturity failed by tearing of the ligament substance. Additional differences between the younger and older animals have also been revealed. The differences in mechanical properties of the MCL between the younger and older groups were not nearly as large as those for structural properties of the FMT complex, and the asynchronous rate of maturation in the FMT complex has been demonstrated, i.e., the ligament substance matures earlier than the ligament-bone junction.

Our results concerning the homeostatic adaptation of ligaments and tendons to varying activity levels have revealed that each component of the bone-ligament-bone complex has unique time-dependent responses to stress. Ligament properties can completely recover following short-term remobilization, but recovery of the FMT complex structural properties is incomplete due to persistent weakness at the insertion sites. The response of tendons and ligaments to exercise seems to be type specific. The mechanical properties of the flexor tendons underwent no significant change following prolonged exercise, when compared to nonexercised controls. However, the extensor tendons demonstrated significant increases, as well as tendon hypertrophy. Animal studies conducted by Tipton et al. have demonstrated an increase in junction strength of bone-ligament-bone preparations following exercise.[47] They also reported that the strength of repaired ligaments is increased with training and decreased following immobilization.

Studies concerning tendon and ligament healing continue to generate much enthusiasm. Utilizing the VDA system, we have been able to isolate the mechanical properties of the healing site as it undergoes continual reorganization. From a clinical standpoint, an important observation has been that for an isolated MCL tear, primary surgical repair seems to offer no lasting benefit in terms of V-V laxity and of mechanical properties at the healing site. Prolonged immobilization proves to be detrimental to the healing process as well as to the overall bone-ligament-bone strength. Obviously, longer term animal studies are needed to determine the functional outcome. Since many of these functional adaptations of ligament and tendon healing are type-specific, continued investigations are necessary to understand the unique responses of all ligaments (ACL, PCL, spinal ligaments, etc.) and tendons (Patellar, Achilles, etc.). Additional developments in biomedical engineering and technology will be required to expand on our fundamental understandings of ligament and tendon function.

ACKNOWLEDGMENTS

The authors gratefully acknowledge the financial support of NIH Grants GM24900 and AM14918, the RR & D of the Veterans Administration, and the Malcolm and Dorothy Coutts Institute for Joint Reconstruction. We would also like to recognize the important contributions and technical assistance of students, residents, and fellows of the Orthopaedic Bioengineering laboratories during the course of this work.

REFERENCES

1. **Fung, Y. C. B.,** Biorheology of soft tissues, *Biorheology,* 10, 139, 1973.
2. **Woo, S. L-Y., Gomez, M. A., Woo, Y. K., and Akeson, W. H.,** Mechanical properties of tendons and ligaments *Biorheology* 19, 385, 1982.
3. **Woo, W. L-Y., Gomez, M. A., and Akeson, W. H.,** Mechanical behaviors of soft tissues: measurements, modifications, injuries and treatment, in *Biomechanics of Trauma,* Nahum, A. N. and Melvin, J., Eds., Appleton-Century-Crofts, New York, 1985, 109.
4. **Lewis, J. L. and Bhybut, G. T.,** In vivo forces in the collateral ligaments of canine knees, *Trans. Orthop. Res. Soc.,* 6, 4, 1981.
5. **Arms, S. W., Pope, M. H., Boyle, J. B., Davignon, P. J., and Johnson, R. J.,** Knee medial collateral ligament strain, *Trans. Orthop. Res. Soc.,* 7, 47, 1982.
6. **Wertheim, M. G.,** Memoirs sur l'elasticite et la cohesion des principaux tissues de corps humain, *Ann. Chim. Phys.,* 21, 385, 1847.
7. **Smith, J. W.,** The elastic properties of the anterior cruciate ligament of the rabbit, *J. Anat.,* 88, 369, 1954.
8. **Viidik, A., Sanquist, L., and Magi, M.,** Influence of postmortem storage on tensile strength characteristics and histology of rabbit ligaments, *Acta Orthop. Scand.,* Suppl. 79, 1965.
9. **Matthews, L. S. and Ellis, D.,** Viscoelastic properties of cat tendon: effects of time after death and preservation by freezing, *J. Biomech.,* 1, 65, 1968.
10. **Noyes, F. R. and Grood, E. S.,** The strength of the anterior cruciate ligament in humans and rhesus monkeys, *J. Bone Joint Surg.,* 8, 1074, 1976.
11. **Dorlot, J. M., Ait ba Sidi, M., Gramblay, G. M., and Drouin, G.,** Load-elongation behavior of the canine anterior cruciate ligament, *J. Biomech. Eng.,* 102, 190, 1980.
12. **Barad, S., Cabaud, H. E., and Rodrigo, J. J.,** Effects of storage at −80°C as compared to 4°C on the strength of rhesus monkey anterior cruciate ligaments, *Trans. Orthop. Res. Soc.,* 7, 378, 1982.
13. **Woo, S. L-Y., Camp, J. F., Gomez, M. A., and Akeson, W. H.,** Comparison of the biomechanical properties of fresh and frozen ligaments, in *Advances in Bioengineering,* American Society of Mechanical Engineers, New York, 1983, 98.
14. **Stouffer, D. C. and Butler, D. L.,** An analysis of crimp unfolding, fluid expulsion and fiber failure in collagen fiber bundles, in *Advances in Bioengineering,* Spilker, R. L., Ed., American Society of Mechanical Engineers, New York, 1984, 46.
15. **Rigby, B., Hirai, N., Spikes, J., and Eyring, H.,** The mechanical properties of rat tail tendon, *J. Gen. Physiol.,* 43, 265, 1958.
16. **Apter, J.,** Influence of composition on thermal properties of tissues, in *Biomechanics: Its Foundations and Objectives,* Fung, Y. C. B., Perrone, N., and Anliker, Eds., Prentice-Hall, Englewood Cliffs, N.J., 1972, 217.
17. **Hunter, J. and Williams, M. G.,** A study of the effect of cold on joint temperature and mobility, *Can. J. Med. Sci.,* 29, 255, 1951.
18. **Kuei, S., Woo, S. L-Y., Gomez, M. A., and Akeson, W. H.,** The viscoelastic, thermoelastic, and time dependent properties of the knee ligaments, *Trans. Orthop. Res. Soc.,* 4, 25, 1979.
19. **Woo, S. L-Y., Orlando, C. A., Sato, S., Camp, J. F., Gomez, M. A. and Akeson, W. H.,** The Effects of Aging, Temperature and Postmortem Storage on Ligament Tensile Behavior, 1984 Am. Soc. Mech. Eng. Winter Annu. Meet., 1984, 337.
20. **Haut, R. C.,** Age-dependent influence of strain rate on the tensile failure of rat-tail tendon, *J. Biomech. Eng.,* 105, 296, 1983.
21. **Morein, G., Goldgefter, L., Kobyliansky, E., Goldschmidt-Nathan, M., and Nathan, H.,** Change in mechanical properties of rat tail tendon during postnatal ontogenesis, *Ant. Embryol. (Berlin),* 154(1), 121, 1978.

22. **Nathan, H., Goldgefter, L., Kobyliansky, E., Goldschmitt-Nathan, M., and Morein, G.,** Energy absorbing capacity of rat tail tendon at various ages, *J. Anat.*, 127, 589, 1978.
23. **Booth, F. W. and Tipton, C. M.,** Ligamentous strength measurements in prepubescent and pubescent rats, *Growth*, 34, 177, 1970.
24. **Woo, S. L-Y., Orlando, C. A., Frank, C. B., Gomez, M. A., and Akeson, W. H.,** Tensile properties of medial collateral ligament as a function of age, *Trans. Orthop. Res. Soc.*, 10, 38, 1985.
25. **Horwitz, M. T.,** Injuries of the ligaments of the knee joint. An experimental study, *Arch. Surg.*, 38, 946, 1939.
26. **Miltner, L. J., Hu, C. H., and Fang, H. C.,** Experimental joint sprain. Pathologic study, *Arch. Surg.*, 35, 234, 1937.
27. **Smith, J. W.,** The elastic properties of the anterior cruciate ligament of the rabbit, *J. Anat.*, 88, 369, 1954.
28. **Viidik, A.,** Biomechanics and functional adaptation of tendons and joint ligaments, in *Studies on the Anatomy and Function of Bone and Joints*, Evans, F. G., Ed., Springer-Verlag,Berlin, 1966, 17.
29. **Zuckerman, J. and Stull G. A.,** Effects of exercise on knee ligament separation force in rats, *J. Appl. Physiol.*, 26, 716, 1969.
30. **Roux, R. D., Hollis, J. M., Gomez, M. A., Inoue, M., Kleiner, J. B., Akeson, W. H., and Woo, S. L-Y.,** Tensile testing of the anterior cruciate ligament (ACL): a new methodology, *Trans. Orthop. Res. Soc.*, 11, 237, 1986.
31. **Laros, G. S., Tipton, C. M., and Cooper, R. R.,** Influence of physical activity on ligament insertions in the knees of dogs, *J. Bone Joint Surg.*, 53A, 275, 1971.
32. **Noyes, F. R.,** Functional properties of knee ligaments and alterations induced by immobilization. A correlative biomechanical and histological study in primates, *Clin. Orthop.*, 123, 210, 1977.
33. **Noyes, F. R., DeLucas, J. L., and Torvik, P. J.,** Biomechanics of anterior cruciate ligament failure. An analysis of strain-rate sensitivity of mechanisms of failure in primates, *J. Bone Joint Surg.*, 56A, 236, 1974.
34. **Abrahams, M.,** Mechanical behavior of tendon in vitro, *Med. Biol. Eng.*, 5, 433, 1967.
35. **Crowninshield, R. D. and Pope, M. H.,** The strength and failure characteristics of rat medial collateral ligaments, *J. Trauma*, 16(2), 99, 1976.
36. **Akeson, W. H., Woo, S. L-Y., Amiel, D. and Frank, C. F.,** The biology of ligaments, in *Basic Principles of Knee Rehabilitation*, Funk, J., Ed., C. V. Mosby, St. Louis, 1983.
37. **Amiel, D., Woo, S. L-Y., Harwood, F. L., and Akeson, W. H.,** The effect of immobilization of collagen turnover in connective tissue: biochemical-biomechanical correlation, *Acta Orthop. Scand.*, 53, 325, 1982.
38. **Evans, E. B., Eggers, G. W. N., Butler, J. K., and Blumel, J.,** Experimental immobilization and remobilization of rat knee joints, *J. Bone Joint Surg.*, 42A, 737, 1960.
39. **Klein, L., Payer, J. S., Heiple, K. G., Bahnuik, E., and Goldberg, V. M.,** Isotopic evidence for resorption of soft tissues and bone in immobilized dogs, *J. Bone Joint Surg.*, 64A, 225, 1982.
40. **Thaxter, T. H., Mann, R. A., and Anderson, C. E.,** Degeneration of immobilized knee joints in rats, *J. Bone Joint Surg.*, 47A, 567, 1965.
41. **Woo, S. L-Y., Gomez, M. A., Woo, Y-K., and Akeson, W. H.,** Mechanical properties of tendons and ligaments. II. The relationships of immobilization and exercise on tissue remodeling, *Biorheology*, 19, 397, 1982.
42. **Cabaud, H. E.,** Exercise effects on the strength of the rat anterior cruciate ligaments, *Am. J. Sports Med.*, 8, 79, 1980.
43. **Tipton, C. M., Schild, R. J., and Tomanek, R. J.,** Influence of physical activity on the strength of knee ligaments in rats, *Am. J. Physiol.*, 212(4), 783, 1967.
44. **Viidik, A.,** The effect of training on the tensile strength of isolated rabbit tendons, *Scand. J. Plastic Reconstr. Surg.*, 1, 141, 1967.
45. **Woo, S. L-Y., Gomez, M. A., Amiel, D., Ritter, M. A., Gelberman, R. H., and Akeson, W. H.,** The effects of exercise on the biomechanical and biochemical properties of swine digital flexor tendons, *J. of Biomech. Eng.*, 103(1), 52, 1981.
46. **Woo, S. L-Y., Ritter, M. A., Amiel, D., Sanders, T. M., Gomez, M. A., Garfin, S. R., and Akeson, W. H.,** The biomechanical and biochemical properties of swine tendons. Long term effects of exercise on the digital extensors, *Connect. Tissue Res.*, 7, 177, 1980.
47. **Tipton, C. M., James, S. L., Mergner, W., and Tcheng, T-K.,** Influence of exercise on the strength of the medial collateral ligaments of dogs, *Am. J. Physiol.*, 218, 894, 1970.
48. **Brookis, J. G.,** Wound healing: suture materials, *The Scientific Fundamentals of Surgery*, Appleton-Century-Crofts, New York, 1972, 186.
49. **Frank, C. B., Woo, S. L-Y., Amiel, D., Harwood, F. L., Gomez, M. A., and Akeson, W. H.,** Medial collateral ligament healing. A multidisciplinary assessment in rabbits, *Am. J. Sports Med.*, 11(6), 379, 1983.

50. **Jack, E. A.,** Experimental rupture of the medial collateral ligament of the knee, *J. Bone Joint Surg.*, 32B, 396, 1950.
51. **Clayton, M. L., Miles, J. S., and Abdulla, M.,** Experimental investigations of ligamentous healing, *Clin. Orthop.*, 61, 146, 1968.
52. **O'Donoghue, D. H., Rockwood, L., Zarecznyj, B.,** Repair of knee ligaments in dogs. I. The lateral collateral ligament, *J. Bone Joint Surg.*, 43A, 1167, 1961.
53. **Gomez, M. A., Woo, S. L-Y., Sato, S., Amiel, D. Harwood, F. L., and Akeson, W. H.,** The Mechanical Properties of Healing Canine MCL Ligament, 37th Alliance for Eng. Med. Biol., 1984, 269.
54. **Inoue, M., Gomez, M. A., Hollis, J. M., Roux, R. D., Lee, E. B., Burleson, E. M., and Woo, S. L-Y.,** Medical collateral ligament healing: repair vs. non-repair, *Trans. Orthop. Res. Soc.*, 11, 78, 1986.
55. **Vailas, A., Tipton, C. M., and Matthes, R. B.,** Physical activity and its influence on the repair process of medial collateral ligaments, *Connect. Tissue Res.*, 9, 25, 1981.
56. **O'Donoghue, D. H., Frank, G. R., and Jeter, G. L.,** Repair and reconstruction of the anterior cruciate ligament in dogs: factors influencing long term results *J. Bone Joint Surg.*, 53A, 710, 1971.
57. **Woo, S. L-Y., Gelberman, R. H., Cobb, N. G., Amiel, D., Lothringer, K., and Akeson, W. H.,** The importance of controlled passive motion on flexor tendon healing. A biomechanical study, *Acta Orthop. Scand.*, 52, 615, 1982.

Chapter 12

THE COLLAGEN FRAMEWORK OF ARTICULAR CARTILAGE: ITS PROFOUND INFLUENCE ON NORMAL AND ABNORMAL LOAD-BEARING FUNCTION

Neil D. Broom

TABLE OF CONTENTS

I. Introduction 244

II. Limitations of a Purely Morphological Approach to Understanding Structure/Function Relationships in Articular Cartilage 244

III. The Response of Articular Cartilage to Stress 246
- A. Mechanical Anisotropy of the General Matrix — A Clue to its Ultrastructural Arrangement 246
- B. Rupture Characteristics of the General Matrix — A Further Clue to its Ultrastructural Arrangement 247

IV. Articular Cartilage Ultrastructure 250
- A. Ultrastructural Changes in the General Matrix Associated with Directional Stress 250
- B. Current Ultrastructural Findings and Previously Accepted Structural Ideas 250
- C. A New Concept of Fibrillar Structure in Cartilage 251
- D. A Functioning Physical Model of Articular Cartilage 254
- E. Fibril-to-Fibril Interactions in Articular Cartilage 256
  - 1. TEM Evidence for Fibril-to-Fibril Associations 256
  - 2. Low Temperature SEM Evidence for Fibril-to-Fibril Associations 257
- F. The Role of the Proteoglycan Component 257

V. The Relationship Between Fibrillar Architecture and Abnormal Function 257
- A. Structural Characteristics of Mechanically Soft Articular Cartilage 260
- B. The Origin of Softening in Articular Cartilage 260

VI. Stress-Induced Structural Transformation in Articular Cartilage 262

VII. Concluding Remarks 263

Acknowledgments 264

References 264

I. INTRODUCTION

The primary role of articular cartilage is to facilitate the efficient transmission of mechanical stress across the articulating joint. Any alteration in the structural configuration of this tissue will inevitably influence its ability to perform this functional role. It is therefore important to have a thorough appreciation of the structural principles governing the behavior of normal articular cartilage if there are to be further advances in our understanding of those processes which lead ultimately to its malfunction.

Until recently there has been little attempt to develop a meaningful synthesis between the fine structure of cartilage and its mechanical properties. Previous studies by other investigators have for example sought to explain the load-bearing ability of cartilage by interpreting structural features as revealed by ultrastructural techniques.[1] Conversely, others have carried out detailed examination of the mechanical response of cartilage and then attempted to correlate this description with known structural and biochemical properties.[2,3] Still others have developed rheological models of cartilage using measured physicochemical data but without seriously addressing the question of how these models relate to the actual structural configuration of the tissue.[4]

In some recent work the present author has attempted to define more rigorously some of the biomechanical principles governing the behavior of articular cartilage and their relation to the degenerative process. These studies have required the development of new experimental approaches which allow microscopic observation of the fresh hydrated tissue with simultaneous micromechanical manipulation, as well as the interpretation of data derived from transmission electron microscopy (TEM) and scanning electron microscopy (SEM) with cryopreservation.

This chapter therefore presents a critical re-evaluation of the generally accepted biomechanical concepts for cartilage in the light of several new ideas arising from these recent experiments. The material presented deals principally with the structural architecture of the intercellular areas of the general matrix, leaving aside the highly specialized pericellular lacunar regions. The term "general matrix" in this chapter refers collectively to the middle and deeper regions of the cartilage thickness where there is a relatively common structural configuration (see following section).

II. LIMITATIONS OF A PURELY MORPHOLOGICAL APPROACH TO UNDERSTANDING STRUCTURE/FUNCTION RELATIONSHIPS IN ARTICULAR CARTILAGE

Only limited structural information is revealed by light microscopy when normal articular cartilage is examined either in its fully hydrated states or as prepared histological sections. This is because the diameter of the principal structural element, namely the collagen fibril, is too small to be resolved by optical means. It is possible however to obtain some hint of the general fibrillar arrangement in cartilage, using differential interference contrast (DIC) light microscopy, from both the variation in optical texture of the matrix and the arrangement of the chondrocytes. In the articular surface there is an obvious delineation parallel to the surface in the first few microns of depth with the chondrocytes appearing as nearly flattened disks (Figure 1a). The immediate underlying region exhibits no discernible matrix texture and contains rounded chondrocytes of variable orientation (Figure 1b). Deeper still, any texture that is visible in the matrix is always in the radial direction (i.e., perpendicular to the articular surface) and the chondrocytes tend to be aligned as radial columns (Figure 1c, d). Further elucidation of the structural layout of normal articular cartilage requires ultrastructural techniques.

Most ultrastructural studies of articular cartilage have been limited to describing the

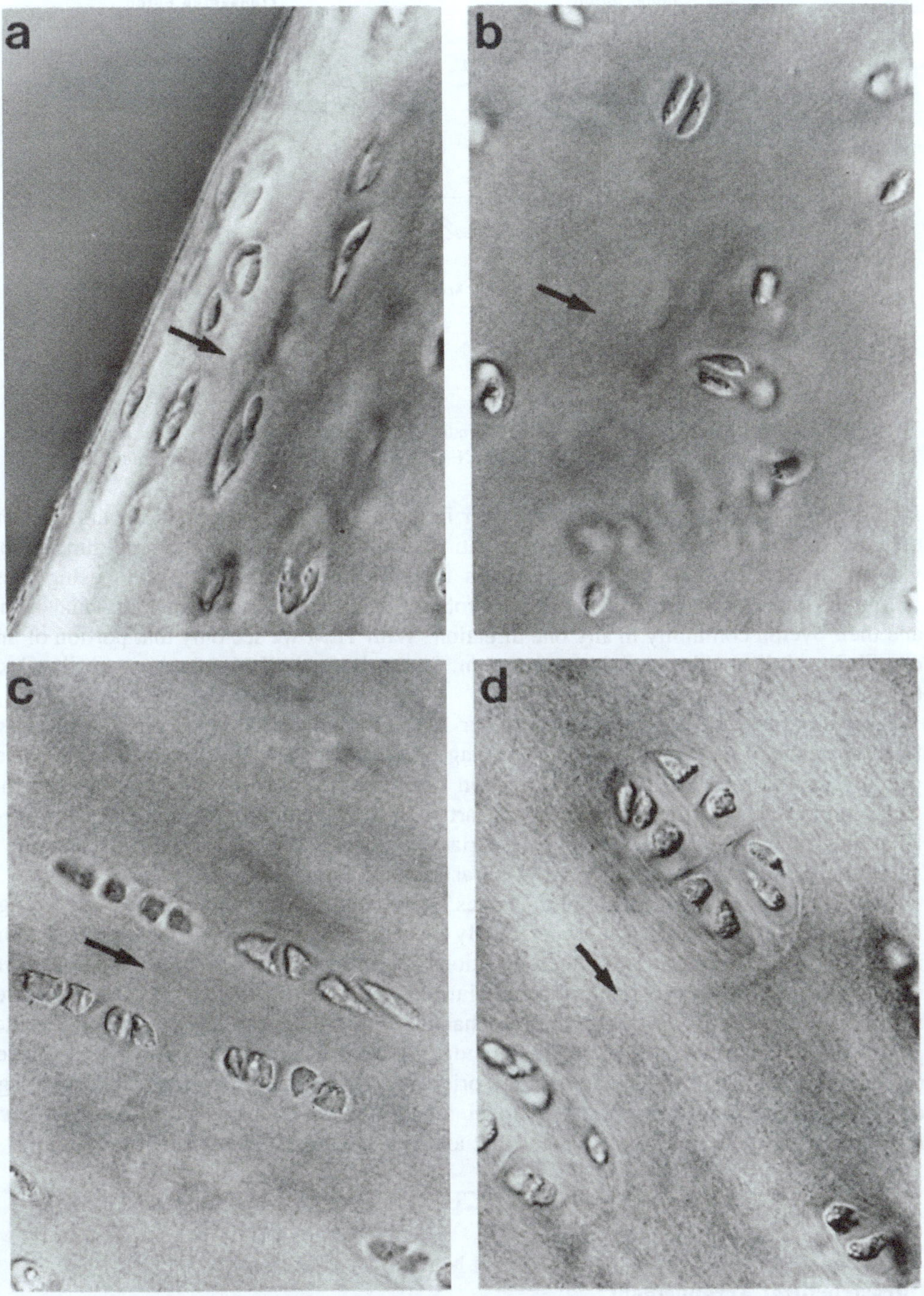

FIGURE 1. Appearance of hydrated articular cartilage when imaged using differential interference contrast (DIC) light microscopy. (a) Superficial layer, (b) Intermediate zone, and (c) deep zone. Note fine radial texture in vicinity of chondrocyte cluster in (d). Arrows indicate radial direction. Magnification × 460.

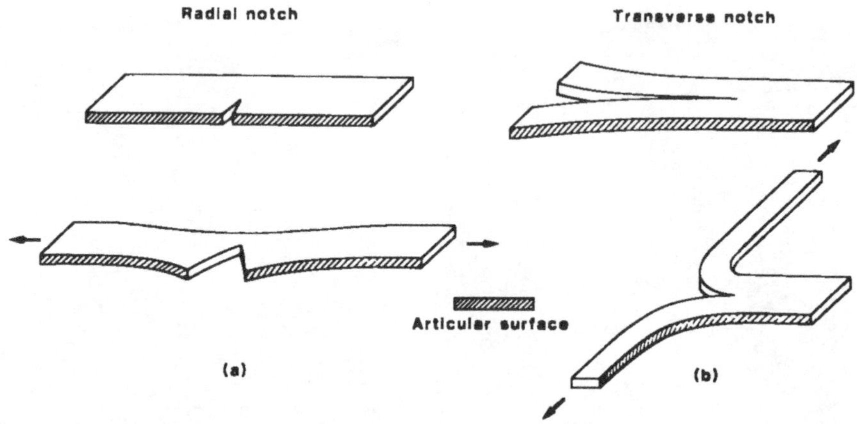

FIGURE 2. Schematic representation of (a) radial and (b) transverse notch geometries and their different loading configurations. (From Broom, N. D., *J. Anat.*, 139, 275, 1984. With permission.)

arrangement of fibrils as they appear, either in the restricted plane of a TEM thin section, or in the fractured surface of a SEM preparation. Neither provides a truly three-dimensional view of the collagenous architecture. It is therefore impossible to resolve, using these techniques alone, a number of structural problems such as the length of individual fibrils and their overall continuity in any one direction. With TEM we see only that portion of any one fibril as it lies within the plane of section. Similarly, with SEM only those fibrils which lie in the plane of fracture are revealed.

Such ultrastructural studies have however led to the now widely accepted zonal classification for the fibrillar arrangement in cartilage, i.e., the superficial, intermediate, and deep or radial layers in which there is a change in fibril orientation from tangential through to a more radial one on proceeding from the articular surface through to the calcified layer adjacent to the subchondral bone.[5] The variations in matrix texture as shown in Figure 1 are entirely consistent with this general ultrastructural classification.

But how are these different structural zones interrelated? Is a continuum achieved through fibrils traversing large distances interzonally or with more shorter elements bound more intimately to neighboring ones? Such questions cannot be answered simply by examining ultrastructure in the conventional manner. Further, although distinctive structural features are closely associated with degenerative changes in articular cartilage the zonal fibrillar picture described above is of little help in predicting how such features might arise. It makes no attempt to explain how the range of fibril orientations are achieved throughout the cartilage depth and consequently is of little help in identifying mechanisms likely to have been responsible for these structural changes that are clearly associated with degeneration.

III. THE RESPONSE OF ARTICULAR CARTILAGE TO STRESS

A. Mechanical Anisotropy of the General Matrix — A Clue to its Ultrastructural Arrangement

Significant structural anisotropy or texture will be evident in those materials which exhibit large differences in extensibility in different directions in response to the application of load. The strongly anisotropic nature of the general matrix of cartilage is readily demonstrated in thin hydrated slices by comparing the extension of the matrix at and around the root of a predetermined notch cut either perpendicular or parallel to the articular surface.[6] These two notch configurations and their response to stress are shown schematically in Figure 2. It should be noted that the radial notch geometry involves the severing of the strongly tangential

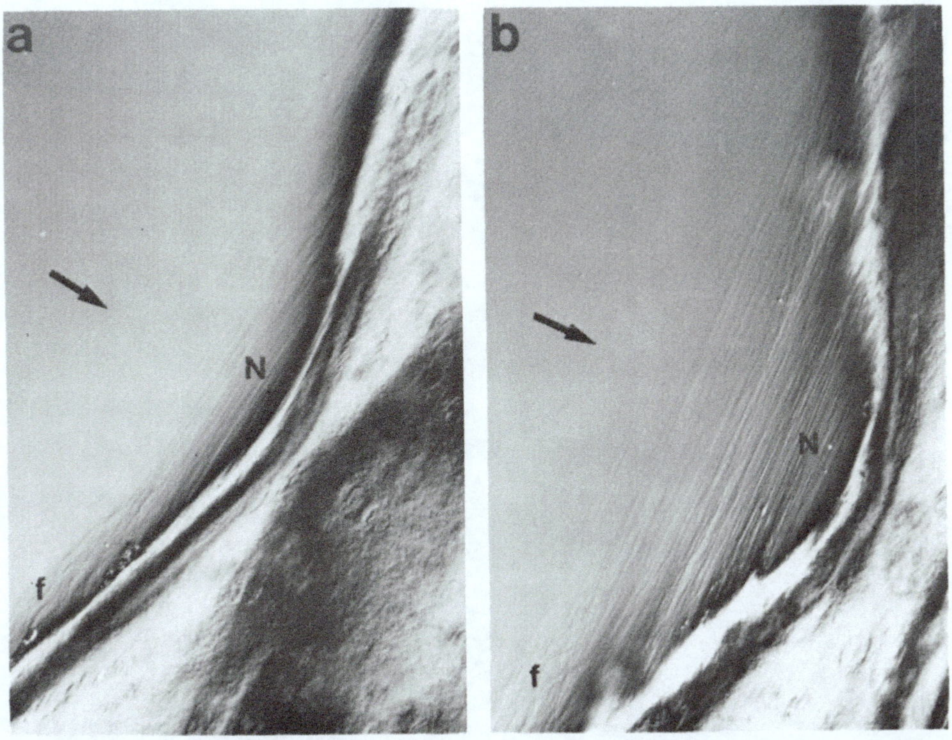

FIGURE 3. Propagation of a radial notch through (a) middle and (b) deep matrix of normal articular cartilage maintained in its fully hydrated condition. Note progressive fracture of fibers (f) most distant from notch root (N). Radial direction indicated with arrow. DIC × 135.

arrangement of fibrils comprising the articular surface. This enables the general matrix to respond to the applied stress in a manner not masked by this strain-limiting structure. By simultaneously observing the tissue matrix using DIC light microscopy it has been demonstrated that in the middle and deep zones the transverse strains can be greater than 100% and as much as an order of magnitude higher than the surface-to-deep or radial extensivity.[7] Clearly then, the arrangement of fibrils in the general matrix creates a pronounced strain-locking effect in the radial direction only.

B. Rupture Characteristics of the General Matrix — A Further Clue to its Ultrastructural Arrangement

When the notch geometries described in the previous section are loaded to the onset of matrix rupture a fundamental difference is observed in the mode of propagation between the radial and transverse directions.[6] Under a steady load the radial notch advances radially in a stable manner, the rupture process accompanied by fibril bundles being drawn out across the notch tip and progressively strained to fracture (Figure 3). By contrast the rupture process resulting from propagation of the transverse notch is characteristically unstable. It is in fact impossible to extend the original notch root any further in the transverse direction. Instead the onset of matrix rupture always involves an immediate sideways deflection of the notch tip into a radial direction either down into the deep zone or up towards the articular surface (Figure 4a to c).

Either of these secondary rupture directions in effect represents a reversion to a radial mode of propagation. This proceeds in the normal manner in the downwards direction, but in the upwards direction, as the radially deflected notch approaches the upper layers incor-

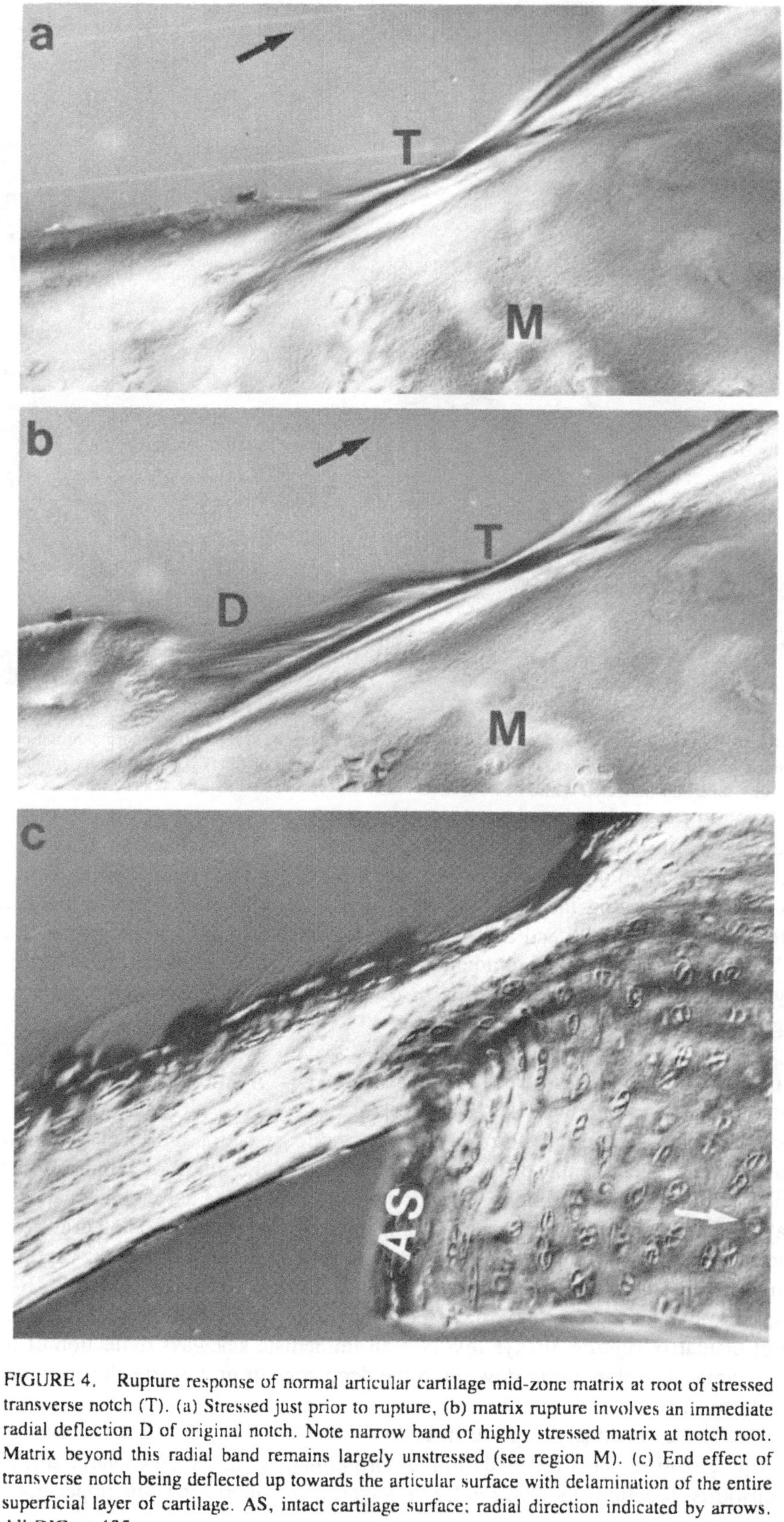

FIGURE 4. Rupture response of normal articular cartilage mid-zone matrix at root of stressed transverse notch (T). (a) Stressed just prior to rupture, (b) matrix rupture involves an immediate radial deflection D of original notch. Note narrow band of highly stressed matrix at notch root. Matrix beyond this radial band remains largely unstressed (see region M). (c) End effect of transverse notch being deflected up towards the articular surface with delamination of the entire superficial layer of cartilage. AS, intact cartilage surface; radial direction indicated by arrows. All DIC × 135.

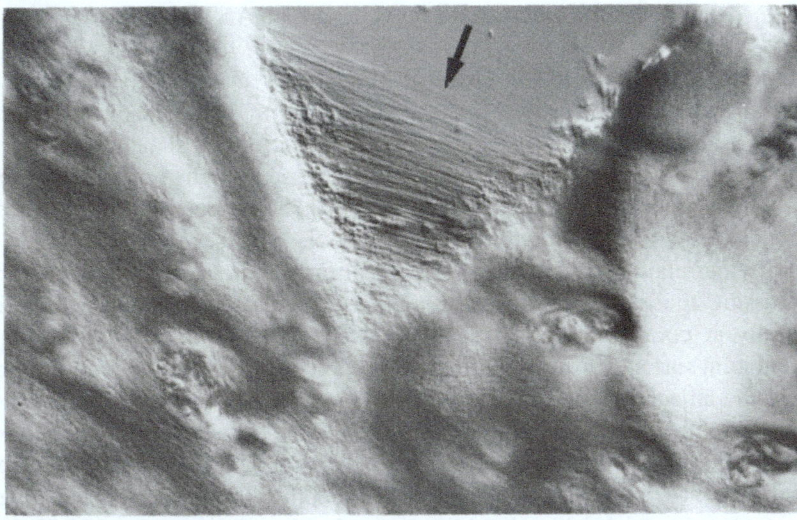

FIGURE 5. Acute geometry of radial notch propagating through general matrix of aged human articular cartilage. Radial direction indicated by arrow. DIC × 135.

porating the articular surface it is again deflected into a transverse direction tracking parallel to but at a relatively uniform depth just below the articular surface. In effect, the final stages of rupture of the upwards deflected notch involve a highly repeatable process of delamination of the entire articular surface (Figure 4c).

Several important structural principles can be deduced from these notch experiments. First, there is a fundamental difference in the mechanical strength of the general matrix in the radial and transverse directions. An obvious structural grain or texture exists in the radial direction along which it is relatively easy to divide the matrix. By contrast, across this "grain" there is an extremely high resistance to rupture, hence the repeated deflection of the original transverse notch. Second, the delamination phenomenon suggests that a major structural discontinuity is present just below the superficial layer and parallel to it.[7] And although in the thin experimental slice of hydrated articular cartilage this delamination process is consistent with there being a zone of weakness in this upper region it has been argued elsewhere[7] that this will not necessarily be so in the full continuum of cartilage in the functioning joint.

It should be clear to the reader that these matrix extensivity measurements and rupture observations conducted using the different notch geometries provide a unique means of assessing the functional properties of articular cartilage, whether in its healthy, aged, or degenerate condition. Although Section V of this chapter deals with the problem of cartilage degeneration it is appropriate at this point to comment briefly on the relationship between aging and the response of the general matrix to stress. Figure 5 shows the propogation of a radial notch through the general matrix of intact aged cartilage. The geometry of the propagating notch is considerably more acute than that observed in the healthy cartilage (see Figure 3). That actual rupture of the aged matrix occurs at such low matrix extensivities, as indicated by this notch geometry, clearly demonstrates its greatly reduced compliance. Although not considered in this presentation, an analysis of the level of stress at which such notches propagate through a given matrix, and their relative acuity, has provided additional insights into the structural integrity of cartilage.[6]

IV. ARTICULAR CARTILAGE ULTRASTRUCTURE

A. Ultrastructural Changes in the General Matrix Associated with Directional Stress

Using this notch propagation technique combined with glutaraldehyde fixation of the stressed tissue it is possible to permanently capture the deformed state of the general matrix in both the radial and transverse directions.[7] The stressed notch root and adjacent matrix represent a whole gradation of deformed states. Further, the observation of ultrastructural changes associated with the variation in matrix strain in these regions provides data that helps to clarify the relationship between cartilage ultrastructure and its mechanical response. Before discussing these data it should be emphasized that considerable variability exists ultrastructurally at comparable zonal depths between samples of articular cartilage taken from different joint surfaces. Any ultrastructural response to stress must always be considered in relation to the adjacent undeformed or control matrix.

In the vicinity of the radial notch immediately below the ruptured region the ultrastructural response to a transversely directed load can be observed. The fibrils exhibit a pronounced transverse alignment (Figure 6a). With increasing depth this transverse arrangement gives way to a structure comprising largely oblique acutely kinked segments of fibrils, persisting in this general form throughout the remaining depth of matrix below the notch (Figures 6b, c). This acutely kinked fibrillar structure should be constrasted with the unstressed control matrix at an equivalent depth in which there is a predominantly radial arrangement (Figure 6d).

With the transverse notch the ultrastructural response to a radially directed stress can be observed. There is a very marked radial alignment of fibrils in the matrix in the immediate vicinity of the notch root (Figure 7a). This aligned structure, corresponding to the narrow band of acutely stressed matrix visible at the light microscope level (see Figures 4a, b), persists for only a short distance beyond the notch root before reverting to a structure similar in appearance to that observed in the unstressed matrix (Figure 7b).

B. Current Ultrastructural Findings and Previously Accepted Structural Ideas

From both the nature of these ultrastructural responses and the matrix extensibility and rupture characteristics discussed earlier it is now possible to develop some basic biomechanical principles from which the general matrix of articular cartilage derives its unique functional properties. First, the very marked mechanically anisotropic properties of the general matrix, as demonstrated by the fundamentally different response to extension in the radial and transverse directions, is consistent with the ultrastructural findings of other investigators who have shown that the general matrix, while containing fibrils of varying obliquity, has a predominantly radial texture.[8,9] This general description, however, does not adequately account for the ultrastructural responses to stress described above. For example, the oblique component of fibrils has been interpreted by some authors as fulfilling a type of bracing role complementing the more radial component,[8,10] the assumption presumably being that the architecture is developed from fibrils individually possessing distinct overall orientations within the matrix (see schematic in Figure 8) and that varying degrees of anisotropy arise from lesser or greater degrees of departure from the radial direction.

The nature of the ultrastructural response at the root of the radial notch is clearly at odds with the above interpretation. First, the rapid gradation from a distinctly transverse to an oblique, repeatedly kinked fibrillar arrangement persisting throughout the remaining depth of matrix below the notch (Figures 6a to c) is not consistent with that expected in a structure developed from obliquely oriented fibrils possessing long range directional continuity. If this were the case, as the matrix is stressed in the transverse direction the oblique fibrils would tend to become increasingly aligned in the direction of stress. They would in turn impose a significant restriction on the amount of transverse extension permissible in the

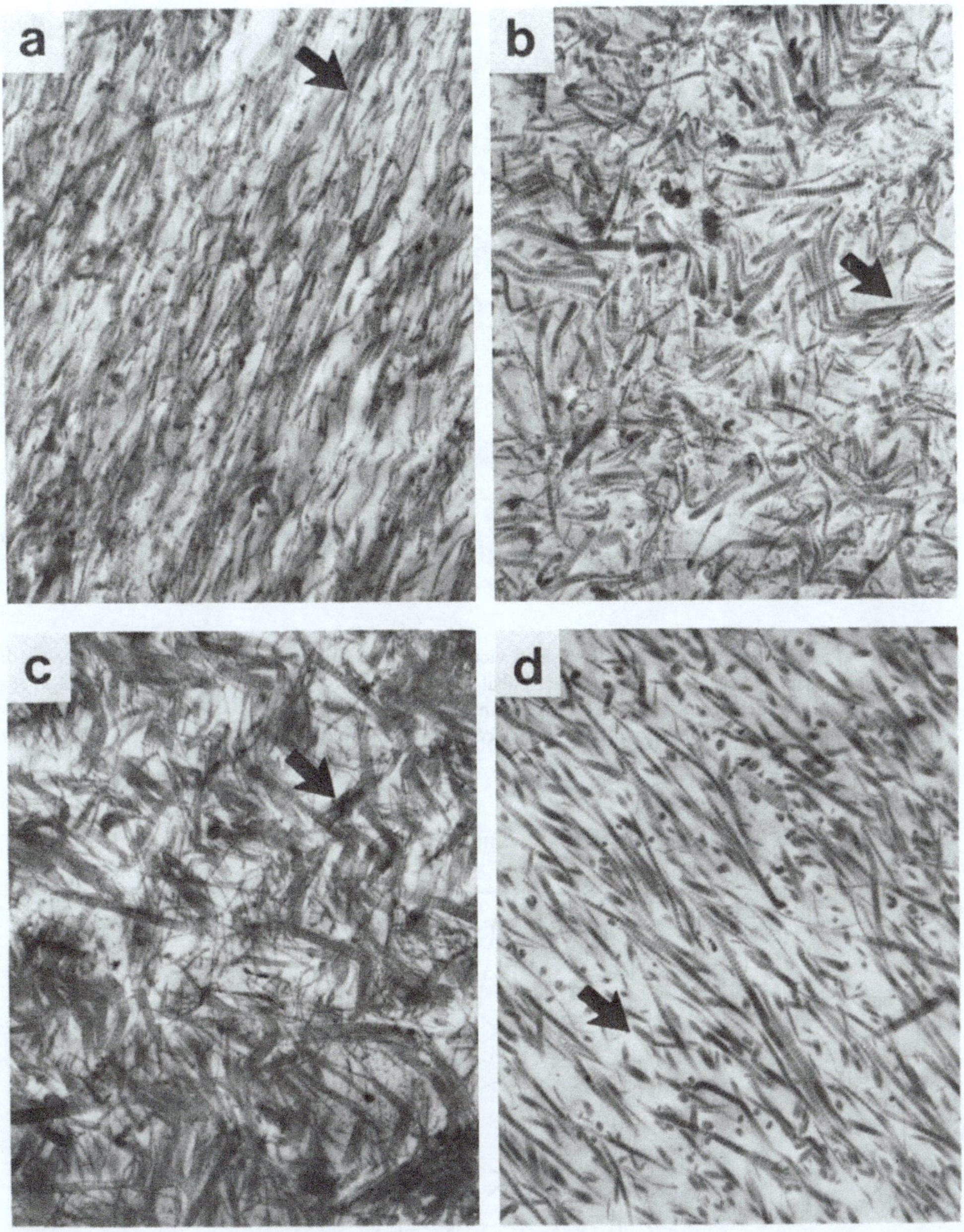

FIGURE 6. TEM, ultrastructural response in vicinity of radial notch. Radial direction indicated by arrows. (a) Near transversely aligned taut fibrils immediately below notch root (bovine patellar groove); magnification × 10,040. (b) Oblique, acutely kinked fibrillar structure developed approximately 35 μm below aligned structure shown in (a); magnification × 10,040. (c) Similar kinked structure to that in (b) but induced in human femoral head articular cartilage; magnification × 11,900. (d) Unstressed control general matrix ultrastructure for direct comparison with (b); magnification × 10,040. (From Broom, N. D., *J. Rheum.*, 13, 142, 1986. With permission.)

matrix. This however, is clearly inconsistent with measured transverse strains exceeding 100% in the matrix below the radial notch root.[7]

C. A New Concept of Fibrillar Structure for Cartilage

An alternative structural interpretation can however be found which satisfactorily accounts

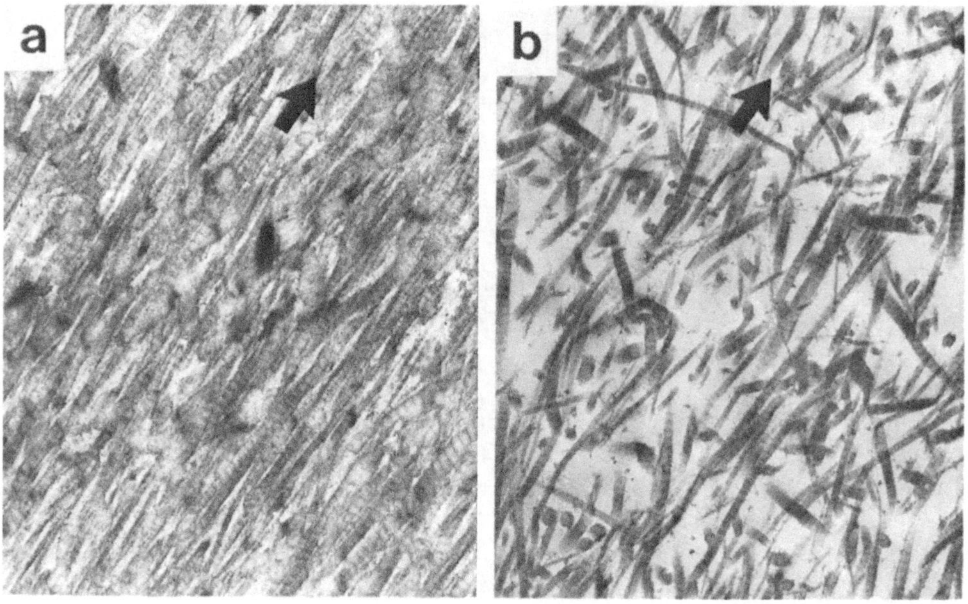

FIGURE 7. (TEM) ultrastructural response in vicinity of transverse notch. Radial direction indicated by arrow. Magnification × 19,000. (a) Compact radial array of nearly straightened fibrils in region of intense stress similar to that revealed optically in Figures 4a, b. (b) Ultrastructure approximately 100 μm beyond region shown in (a) and at same depth below articular surface. (From Broom, N. D., *J. Rheum.*, 13, 142, 1986. With permission.)

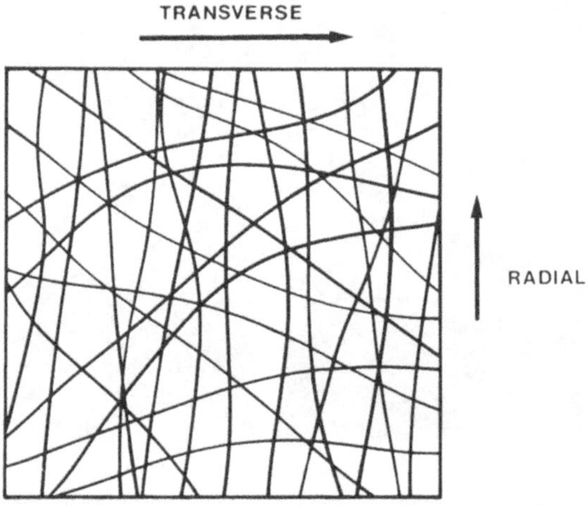

FIGURE 8. Idealized schematic showing possible development of fibrillar framework in general matrix from elements exhibiting large scale directional continuity and a range of orientations about a radial mean. (From Broom, N. D., *J. Rheum.*, 13, 142, 1986. With permission.)

for the observed extensivity and ultrastructural data.[7,11] In this model the oblique fibrillar component is generated from a primary array of radial fibrils. Each individual fibril is repeatedly deflected laterally along its length somewhat in the form of a crude zig-zag arrangement as shown schematically in Figure 9. If a more idealized schematic of this model is considered (see Figure 10) it is possible to see more clearly the probable relationship between matrix mechanical properties and its ultrastructure. First, the very large transverse

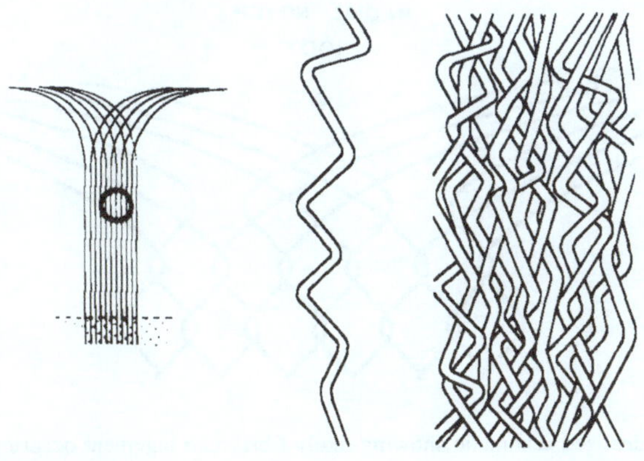

FIGURE 9. Idealized schematic depicting manner in which an array of fibrils possessing a primary radial orientation (see (a)) can, at the ultrastructural level create a pseudorandom framework (b) and (c). Although in (a) the fibrils are shown to turn from a radial to a transverse orientation in the superficial layer, whether such continuity actually exists in cartilage is still open to question. (From Broom, N. D., *Arthritis Rheum.*, 25, 1209, 1982. With permission.)

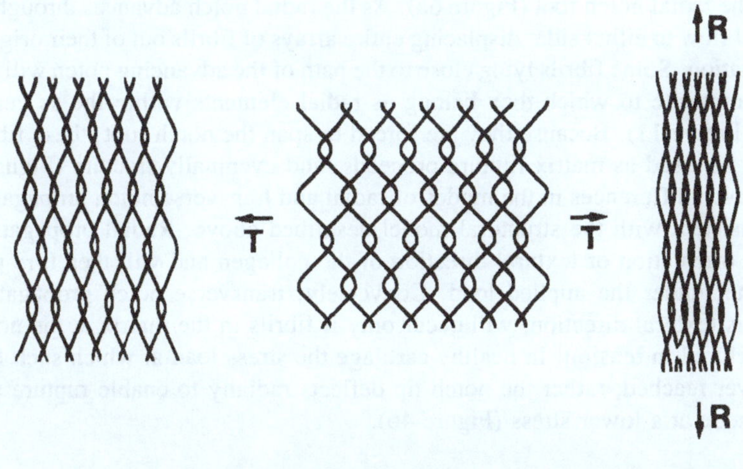

FIGURE 10. Idealized schematic illustrating transverse and radial strain response of the repeatedly deflected radial array. (A) Unstrained network; (B) relatively large transverse extension of network A; (C) strain limiting response of network A in radial direction. T, transverse loading direction, R, radial loading direction. (From Broom, N. D., *J. Rheum.*, 13, 142, 1986. With permission.)

strains observed in the matrix below the radial notch are consistent with the transverse extensions permitted geometrically in the repeatedly deflected radial array (compare A and B in Figure 10). The radial contraction accompanying such an extension will, as a consequence, create from the original zig-zag radial geometry a more obliquely kinked structure as is evident ultrastructually (Figures 6b, c). Conversely, the model predicts a pronounced strain-locking behavior in the radial direction (compare A and C in Figure 10) as is clearly seen ultrastructually in the matrix at the transverse notch root (Figure 7a). The rapid straightening of the repeatedly deflected radial fibrils will minimize the amount of radial strain that can be induced in the matrix any significant distance beyond the transverse notch root. The

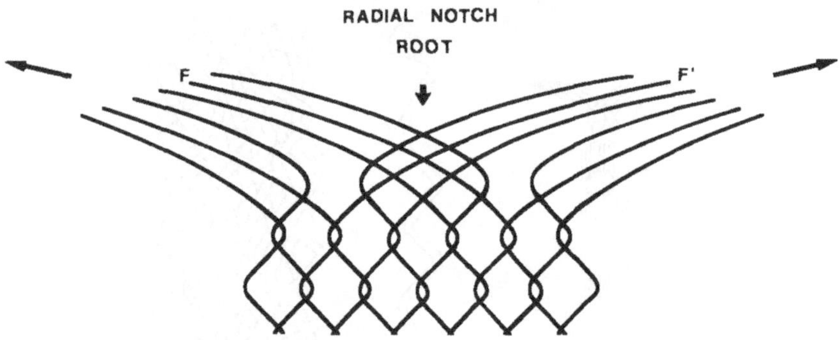

FIGURE 11. Idealized schematic showing likely fibril rearrangement occuring at radial notch root following propagation. Fibrils F and F' have been deflected to the side of the advancing notch opposite to which they belong as radial elements within the intact matrix. (From Broom, N. D., *J. Rheum.*, 13, 142, 1986. With permission.)

narrow band of intensely stressed matrix at the root of the transverse notch, visible with DIC light microscopy (Figure 4a and b), is a consequence of this strain-locking role of the radial arrangement of fibrils.

The model also explains the presence of the transversely aligned fibrils in the immediate vicinity of the radial notch root (Figure 6a). As the radial notch advances through the matrix, material will flow to either side, displacing entire arrays of fibrils out of their original primary radial orientation. Some fibrils lying close to the path of the advancing notch will be deflected to the side opposite to which they belong as radial elements within the as yet unruptured matrix (see Figure 11). Because they are forced to span the notch root, these fibrils become increasingly stressed as matrix rupture proceeds, and eventually fracture (Figure 3).[6,7]

The observed differences in the modes of radial and transverse notch propagation are also entirely consistent with the structural model described above. Radial propagation is along the primary orientation or textural direction of the collagen and will therefore proceed in a stable manner under the applied load. Conversely, transverse notch propagation at right angles to this textural direction, will occur only if fibrils in the matrix at the notch root are actually fractured in tension. In healthy cartilage the stress load at which such fracture will occur is never reached; rather the notch tip deflects radially to enable rupture to occur via the radial mode at a lower stress (Figure 4b).[7]

D. A Functioning Physical Model of Articular Cartilage

While matrix coherency in the radial direction is easily explained with this structural concept by the continuity of fibrils in the radial direction, transverse matrix cohesion must be provided by some form of fibril-to-fibril interconnection. Broom and Marra,[12] using a working physical model, have recently demonstrated the functional viability of such a concept. The model comprises a system of radial elements each repeatedly deflected into oblique orientations and appropriately crosslinked with neighboring elements to create a three-dimensional network. The lattice spaces are then packed with partially inflated air balloons to simulate the highly deformable proteoglycan component trapped within the collagen mesh-work (Figure 12a). The integration of the cartilage general matrix with the in-plane fibrillar structure of the articular surface was also very effectively incorporated in this system by turning remnant lengths of the radial elements into a surface plane and tying each at random with others to create a tension resisting membrane analogous to the superficial layer (see Figure 12b).

Although the above physical model represents a gross oversimplification of the actual tissue architecture, functional studies conducted with it clearly demonstrate that a system of

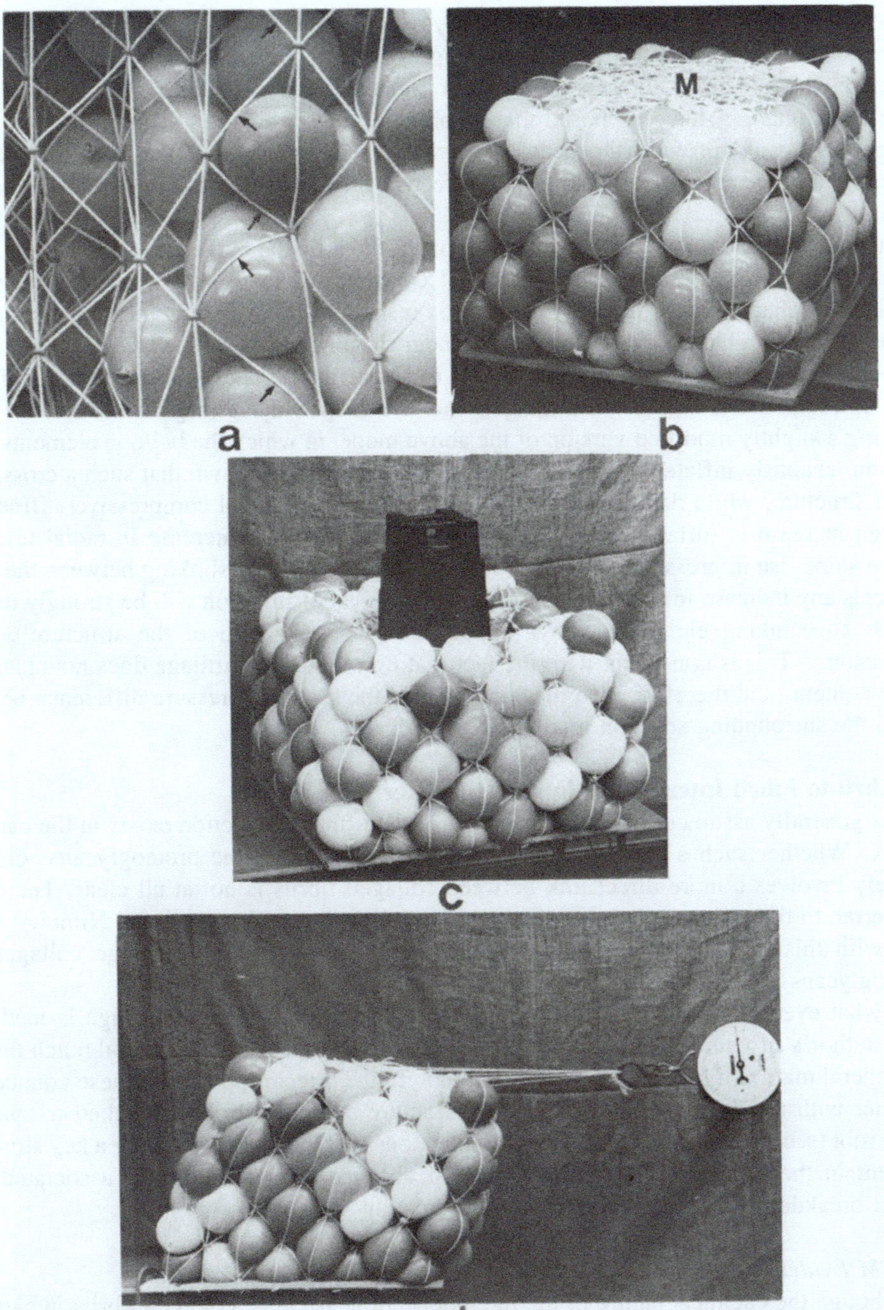

FIGURE 12. Mode of construction and functional response of a 3-dimensional physical model of articular cartilage (a) 3-Dimensional network generated from the repeated crosslinking of the primary radial elements. Small arrows identify oblique segments comprising single radial fibril. The deformable component comprises air-filled rubber balloons packed into the interfibrillar spaces. (b) The repeatedly deflected radial elements turn over on reaching the model surface and are tied at random with each other to form a tension-resisting fibrous membrane (M) representing the articular surface. (c) The model sustaining a concentrated 80-kg compressive load. (d) The model sustaining a shear load of 40 kg applied by means of 4 guy ropes attached to surface membrane. (Figures 12b to d from Broom, N. D. and Marra, D. L., *Connect. Tissue Res.* 14, 1, 1985. With permission.)

radial fibrils, appropriately deflected laterally over short distances and systematically cross-linked to form a captive meshwork for the highly deformable component, can function most effectively both in direct compression (Figure 12c) and in shear (Figure 12d).[12] Thus, whereas other investigators[10] have proposed that an oblique set of fibrils, as distinct from the radial component would provide the necessary bracing elements within the general matrix of articular cartilage, this model shows that effective bracing can in fact be achieved with the radial arrangement.

The model also demonstrates the crucial role of the articular surface in load bearing. With compression the local "punching" effect of concentrated loads down into the underlying matrix is mechanically buffered by the tension resisting fibrous membrane acting somewhat in the manner of a hammock (Figure 12c). Under shear conditions this same tension resisting membrane counters any tendency for tensile forces acting parallel to the articular surface to open up radial clefts that might propagate into the deeper layers (Figure 12d).[12]

Using a slightly modified version of the above model in which the balloon elements could be simultaneously inflated to varying degrees it has also been shown that such a crosslinked radial structure, while demonstrating a twofold increase in radial compressive stiffness for a given increase in inflation pressure, shows only a fractional increase in radial thickness for the same rise in pressure. Thus, for a given frequency of crosslinking between the radial elements any increase in pressure beyond a certain level of inflation will be strongly resisted by the crosslinking elements with a consequential stabilization of the structure's radial dimension.[12] This is consistent with the fact that normal intact cartilage does not change its water content (and therefore its dimensions) when the swelling pressure difference between it and the surrounding solution is altered.[13]

E. Fibril-to-Fibril Interactions in Articular Cartilage

It is generally assumed that some form of fibril-to-fibril interaction exists in the cartilage matrix. Whether such a relationship is mediated via some of the proteoglycans, or alternatively involves a more direct link between collagen fibrils is not at all clear. The reader is referred to the comprehensive reviews by Muir,[14] Lash and Vasan,[15] and Nimni,[16] which deal with this problem of the nature of the interactions between cartilage collagen and proteoglycans.

In what ever way the collagen-to-collagen connection in articular cartilage is mediated, if the author's structural concept is in principle correct, propagation of a radial notch through the general matrix of articular cartilage will require the direct rupturing of these connections together with the drawing out and eventual fracture of those fibrils compelled to span the notch root (see Figures 3 and 11). These connections will therefore constitute a key structural element in the fibrillar architecture, and their absence should be clearly associated with matrix breakdown. This will be discussed in Section V.

1. TEM Evidence for Fibril-to-Fibril Associations

Although the chemical nature of the interconnections between collagen fibrils in cartilage remains largely speculative some very recent ultrastructural studies by Broom and Marra[17] have revealed a variety of fibril-to-fibril association, in the cartilage matrix that appear to have escaped the attention of other investigators. These associations, most clearly revealed by TEM, ranged from a relatively loose transient clustering of several fibrils over short distances along their lengths (Figure 13a) through to fibrils so intimately bound as to suggest a specific chemical affinity (Figures 13b, c). And, although some of these structures are at first sight difficult to identify in the restrictive two-dimensional TEM thin section, they can be recognized with careful observation in most thin sections of general matrix incorporating the radial direction.

2. *Low Temperature SEM Evidence for Fibril-to-Fibril Associations*

That these distinctive groupings of fibril segments represent a type of nodal point or transient association of much longer radial fibrils is largely confirmed by low temperature SEM.[17] Using cryopreservation of fresh hydrated cartilage, followed by fracture and sublimation to remove frozen water from the plane of fracture,[18,19] it is possible to obtain a more extensive ultrastructural view of the undisturbed fibrillar architecture. These structures exhibited a well-developed interconnecting arrangement with fibrils anastomosing and aggregating with each other to achieve a coherent three-dimensional structure clearly developed from a primary radial array (Figure 14). It would seem then that the nodal points which are crucial to the satisfactory functioning of the physical model of cartilage (see Figure 12a) are a simplified representation of these fibril-to-fibril associations visible at the ultrastructural level.

Another consideration relating to the low temperature SEM studies is that a significant depth below the fractured matrix surface can be revealed by sublimation. Any long range fibril obliquity, if present, should be easily revealed; yet this is not the case. Examination of large numbers of cryopreserved cartilage specimens indicates that only occasionally do fibrils appear to extend any significant distance in directions oblique to the primary direction.[17] It is therefore suggested that the preferred orientation studies conducted by Aspden and Hukins[8] using high angle X-ray diffraction, in which they describe a range of fibril orientations about a radial mean, be interpreted as quantifying the angular spread of the short segments of much longer radial fibrils that have been repeatedly deflected into oblique orientations as indicated in the various schematic representations in this chapter (Figures 9 and 10). Further, the evidence from the radial notch propagation studies suggests that the length of individual fibrils is some reasonable fraction of the full cartilage depth (see Figure 3 and Broom[6] and Clark[20]). And, whereas it has not yet been verified experimentally that the primary radial arrangement of fibrils do in fact turn into a transverse surface orientation as shown in both the schematic in Figure 9a and the physical model in Figure 12b, this is certainly a plausible concept.

F. The Role of the Proteoglycan Component

Within the three-dimensional structure developed from the primary radial array of fibrils the proteoglycan component is immobilized. The macromolecular structure of the proteoglycan complex is such that it presents an extremely large and highly ordered array of negativity, hence its very strong affinity with the dipolar water molecule. In effect, the proteoglycan binds water so successfully that it becomes, by volume, the principle constituent in articular cartilage.[21] Caplan[22] refers to this as the "structuring of water."

Mechanically the proteoglycan gel is weak in shear.[23] Thus, with respect to articular cartilage its physicochemical properties can only be harnessed in cooperation with the tension-bearing elements of the collagenous mesh-work. This principle is of course demonstrated in a simplified way by the physical model shown in Figure 12. Scott[24] has examined theoretically the concept of "fit" between the filler component and the interfibrillar spaces or domains in soft connective tissues. Small or very flexible filler polymers would be permitted to move freely through the mesh-work resulting in a highly compliant matrix. In articular cartilage the hyaluronic acid/proteoglycan complex is an extensively branched and aggregated large cross-sectioned molecule.[25] Such a polymer would be effectively trapped within the interlocking mesh-work, hence explaining the comparative rigidity of this tissue.[24]

V. THE RELATIONSHIP BETWEEN FIBRILLAR ARCHITECTURE AND ABNORMAL FUNCTION

Having now established what are believed to be some of the important structural principles

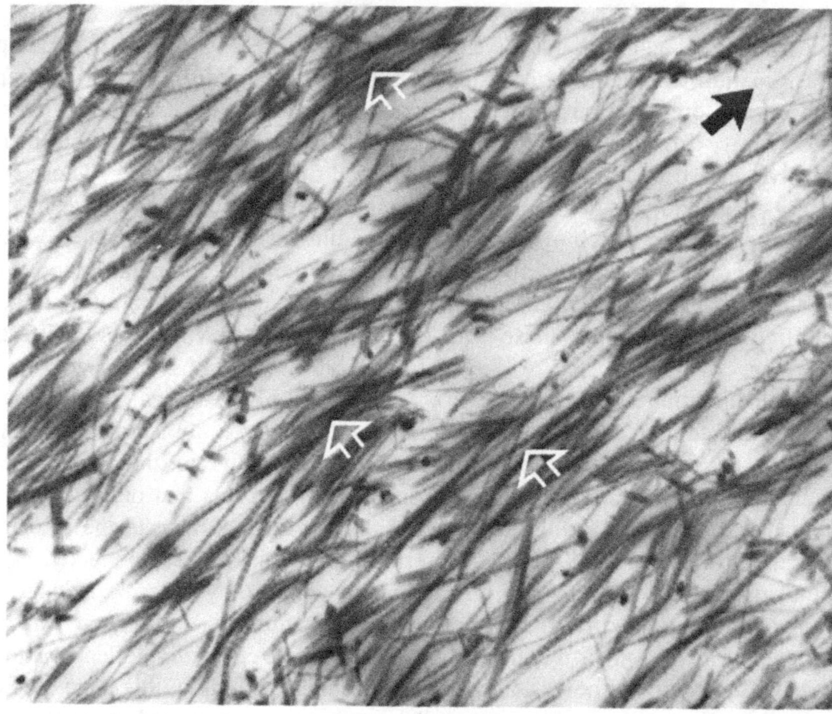

A

B

FIGURE 13. TEM showing varied fibril-to-fibril associations in articular cartilage. Radial direction indicated by solid arrows. (a) Relatively loose transient clustering of fibrils in general matrix of normal bovine articular cartilage (see small open arrows); magnification × 20,000. (b) More intimate fibril-to-fibril associations (e.g., see open arrows) extending over greater distances than those in (a). Normal human head of femur; magnification × 9,100. (c) Parallel aggregates of intimately associated fibrils oriented about the radial direction in human head of femur cartilage; magnification × 20,000. (Figures 13a and b from Broom, N. D. and Marra, D. L., *J. Anat.*, 146, 185, 1986. With permission.)

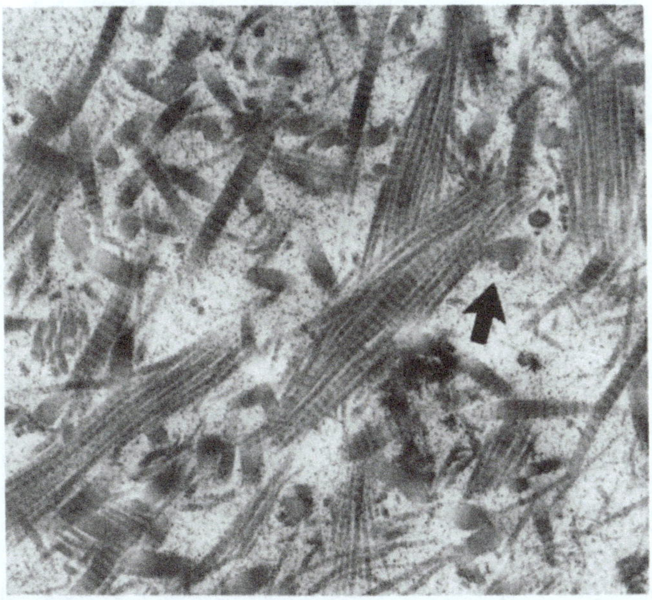

FIGURE 13C.

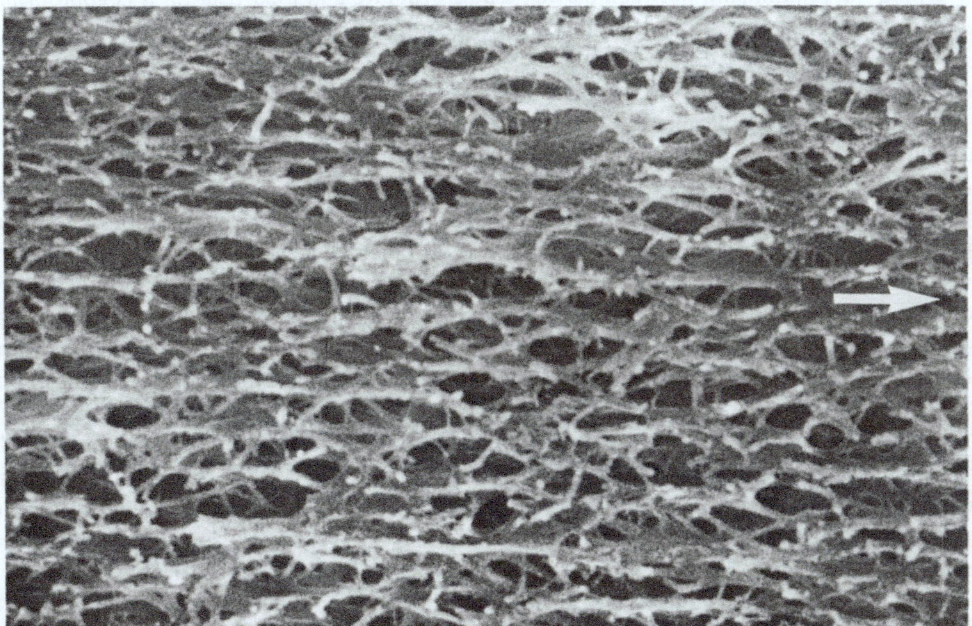

FIGURE 14. SEM fractured and etched appearance of hydrated, frozen bovine articular cartilage showing an interconnecting primary radial arrangement of fibrils in middle zone. Radial direction see arrow; magnification × 11,700. (From Broom, N. D. and Marra, D. L., 146, 185, 1986. With permission.)

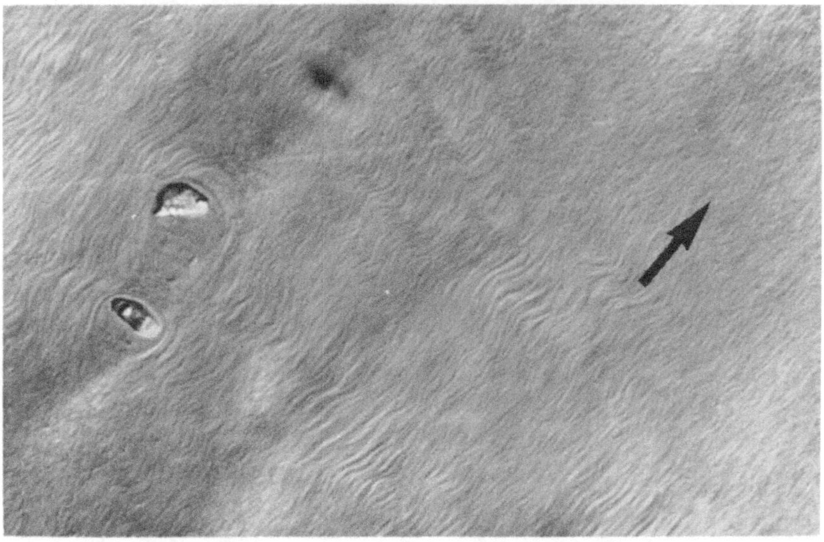

FIGURE 15. DIC; crimped fibrous texture in general matrix of bovine articular cartilage exhibiting nonprogressive softening. Radial direction indicated by arrow; magnification × 415. From Broom, N. D., *Anat.*, 139, 275, 1984. With permission.)

governing the relationship between cartilage ultrastructure and load-bearing it is appropriate to examine briefly some aspects of cartilage degeneration.

A. Structural Characteristics of Mechanically Soft Articular Cartilage

Almost invariably abnormal softening in articular cartilage is associated with a distinct structural alteration evident even at the light microscope level.[11,26] In cartilage exhibiting both nonprogressive softening and osteoarthritic changes there are local or extensive regions of matrix containing a well-developed radial fibrous texture frequently incorporating a variable amount of waveform or crimp (Figure 15). Biomechanical studies of the response of these degenerate tissues in their fresh hydrated condition using simultaneous micromechanical compression and DIC light microscopy demonstrate that matrices containing such radial arrays deform under extremely low compressive stresses via a mechanism involving an in-phase structural collapse (Figure 16). Radial and transverse notch propagation studies similarly confirmed a significant loss of mechanical strength in these visibly fibrous regions.[26] Propagation of a radial notch through the softened arrays requires substantially lower loads than those required in normal matrix; i.e., suggesting that the interfibril bond strength is considerably reduced in such regions. Also a transverse notch can be propagated further in the transverse direction in the general matrix containing such arrays thus indicating a reduced tensile strength of the fibrils themselves.[26]

Ultrastructurally these regions of prominant radial texture in the softened cartilage lack the interconnecting fibrillar structure seen in the normal matrix.[7,27] The radial texture visible at the light microscope level (Figure 15) is seen to be an aggregation of a number of near-parallel individual collagen fibrils (Figures 17a, b). Such arrays, in cooperation with the proteoglycan component, will function as an unbraced structure and therefore be deficient in compressive strength.

B. The Origin of Softening in Articular Cartilage

What then is the origin of these structurally weakened regions in the degenerate cartilage matrix? It is conceivable that their presence is a consequence of a modified cellular expression

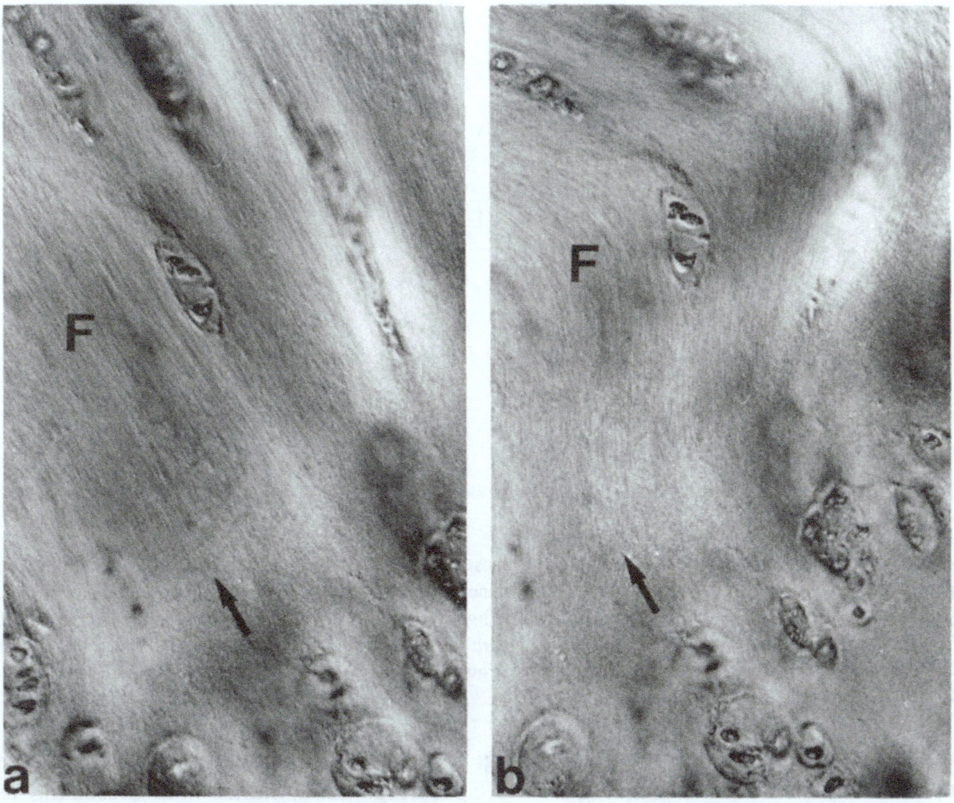

FIGURE 16. DIC, prominent radially aligned fibrous arrays (F) in general matrix of human osteoarthritic cartilage undergoing in-phase collapse with direct compression. (a) Zero load (b) 35 g/mm². Direction of compression is radial, see arrow. Magnification × 200. (From Broom, N. D., *Arthritis and Rheum.*, 27, 1028, 1984. With permission.)

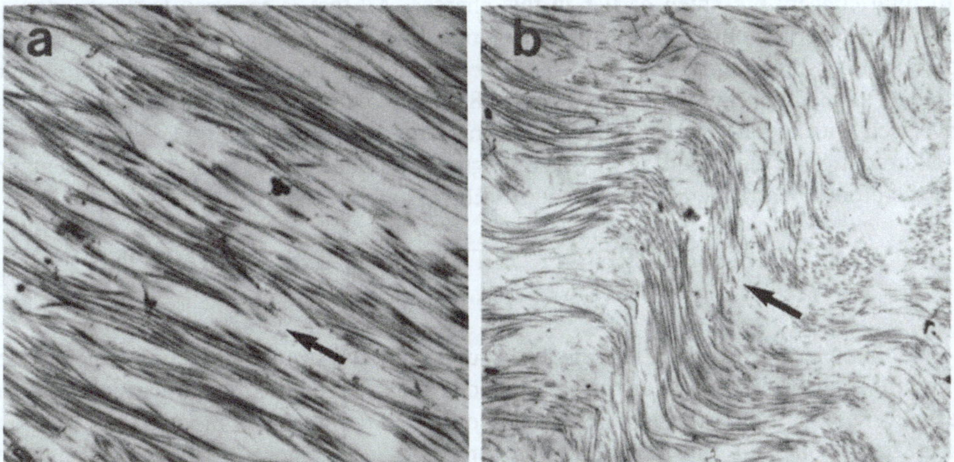

FIGURE 17. Ultrastructural appearance of general matrix of softened cartilage similar to that shown optically in Figure 15. (a) Parallel aggregates of radial fibrils. Arrow indicates radial direction; magnification × 8,200. (b) Similar to (a) but with a superimposed waveform or crimp. Arrows indicate radial direction; magnification × 6,100.

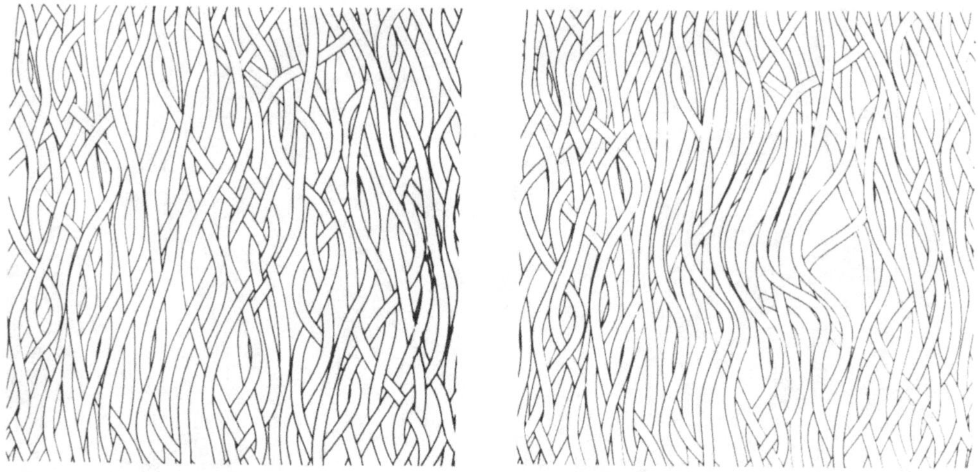

FIGURE 18. Idealized schematic illustrating proposed fibrillar relationships between normal and softened general matrix. (a) Fibrils in normal matrix constrained in a high energy interconnecting arrangement hence yielding a structure that is largely texture-free at the light microscope level of observation (see Figure 1c). (b) Matrix as in (a) but containing a central region where the interfibril constraints have been degraded; a more parallel, aligned configuration therefore results (see Figures 15 and 16). (From Broom, N. D., *Arthritis Rheum.*, 27, 1028, 1984. With permission.)

as part of the reparative response of cartilage.[28,29] Structurally however, they blend almost imperceptibly into the surrounding normal-appearing matrix rather than presenting as a discordant structure. In fact it has been pointed out by Meachim and Osborne[30] that repair tissue generally forms distinct plugs and is readily identified histologically by its coarsely streaked and whorled morphology.

In the author's view it is more plausible to view these softened regions as arising from a local degradative modification of the collagen network originally comprising the normal matrix.[7] This view is consistent with structural changes that can be predicted from the structural concept presented earlier in this chapter. All that is required is a breakdown in the forces constraining the original primary radial array in its high energy interconnecting configuration, thus permitting short-range reorienting of the fibril segments into a more parallel, aligned configuration (see Figure 18). Such a structural transformation cannot be envisaged with the alternative model based on large scale fibril obliquity (Figure 8). This would require the repositioning of entire lengths of oblique fibrils into a radial orientation, and therefore a rearrangement of the entire fibrillar architecture.

VI. STRESS-INDUCED STRUCTURAL TRANSFORMATIONS IN ARTICULAR CARTILAGE

One final piece of experimental evidence supporting the structure/function concepts dealt with in this chapter concerns the effect of mechanical trauma on articular cartilage. In some very recent experiments the author[31] has demonstrated that repeated in vitro trauma of fresh cartilage-on-bone can transform the interconnecting arrangement of fibrils comprising the general matrix of normal cartilage into a strongly aligned radial configuration incorporating a variable waveform or crimp (Figures 19 and 20). In fact, the structure resulting from this stress-induced transformation exhibits characteristics almost identical to those observed in articular cartilage displaying nonprogressive softening (Figures 15 and 17).

Again this stress-induced structural transformation is consistent with that predicted from

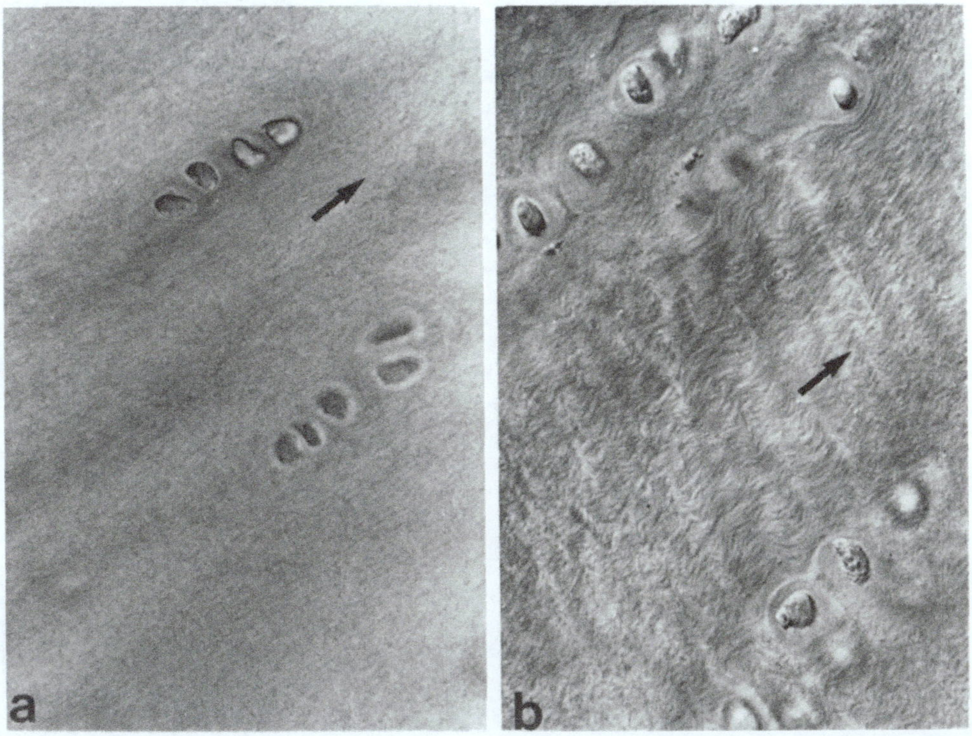

FIGURE 19. Stress-induced structural transformation in bovine articular cartilage. Arrows indicated radial direction. DIX × 455. (a) Appearance of control matrix. (b) Matrix adjacent to (a) following repeated cartilage-on-bone traumatization of approximately 300 impacts at an energy level of 2.7 J.

the radially deflected fibrillar model[7,12] High energy compression impact would tend to disrupt first the interfibril linkages by over-extension of the collagenous network in the transverse direction, presumably as a consequence of the time-dependent rheological properties of the hydrated cartilage matrix. Partial or total destruction of the pseudo-random fibrillar arrangement would therefore occur and fibril segments now no longer constrained in a high energy, short-range, interconnected configuration could then revert to a lower energy parallel radial arrangement. The total length of each fibril previously accommodated within the zig-zag configuration of repeatedly deflected segments would, as a consequence, incorporate a degree of redundancy in its length, this being absorbed within the now parallel arrays by means of an in-phase collapse. Again this is consistent with the presence of a distinctive waveform or crimp in the radial arrays comprising the traumatized matrix (Figures 19 and 20).

VII. CONCLUDING REMARKS

This chapter has attempted to clarify first some of the important principles governing the load-bearing function of normal articular cartilage. It has endeavored to show that the collagenous fibrillar architecture of the general matrix, although overtly variable in appearance, does possess a unique structural form. In the author's view this structural form is created from repeated crosslinking of a primary radial array, and is able to explain in a consistent manner structural/functional changes involved in the development of a variety of abnormal matrix characteristics associated both with degeneration and mechanical trauma.

There is substantial evidence that the structural integrity of healthy articular cartilage

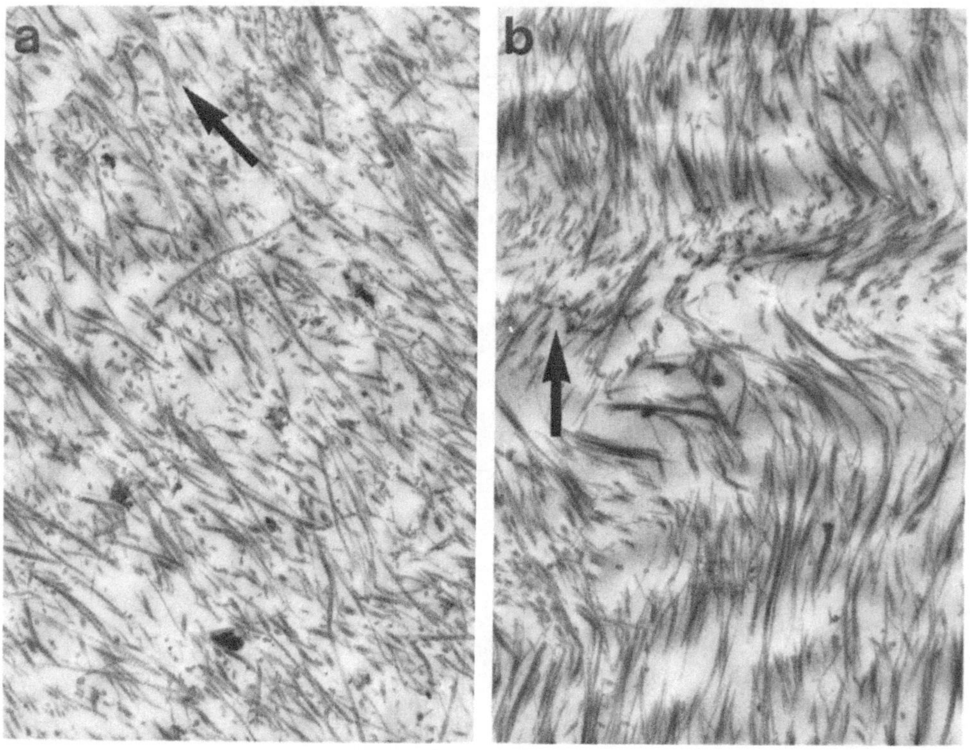

FIGURE 20. TEM showing (a) control general matrix (magnification × 11,000) and (b) adjacent traumatized matrix (magnification × 10,300) comparable with the structures shown in Figures 19a, b respectively. Arrows indicate radial direction.

requires that there be specific interactions between the collagenous elements comprising the three-dimensional fibrillar arrangement. The biochemical nature of these interactions remains unclear at present but it is considered that further studies of this problem could advance significantly our knowledge of the degenerative mechanisms involved in cartilage breakdown.

ACKNOWLEDGMENTS

This chapter was written while the author was supported by a Project grant from the Medical Research Council of New Zealand. He would like to thank Nicola Hopper for typing the manuscript.

REFERENCES

1. **Clarke, I. C.,** Articular cartilage: a review and scanning microscope study, *J. Bone Joint Surg.*, 53B, 732, 1971.
2. **Weightman, B. and Kempson, G. E.,** Load carriage, in *Adult Articular Cartilage*, Freeman, M. A. F., Ed., Pitman, Kent, England, 1979, chap. 5.
3. **Kempson, G. E.,** Mechanical properties of articular cartilage, in *Adult Articular Cartilage*, Freeman, M. A. F., Ed., Pitman, Kent, England, 1979, chap. 6.
4. **Mow, V. C., Holmes, M. H., and Lai, W. M.,** Fluid transport and mechanical properties of articular cartilage: a review, *J. Biomech.*, 17, 377, 1984.

5. **Meachim, G. and Stockwell, R. A.,** The matrix, in *Adult Articular Cartilage,* Freeman, M. A. F., Ed., Pitman, Kent, England, 1979, chap. 1.
6. **Broom, N. D.,** Further insights into the structural principles governing the function of articular cartilage, *J. Anat.,* 139, 275, 1984.
7. **Broom, N. D.,** The collagenous architecture of articular cartilage — a synthesis of ultrastructure and mechanical function, *J. Rheum.,* 13, 142, 1986.
8. **Aspden, R. M. and Hukins, D. W. L.,** Collagen organisation in articular cartilage determined by x-ray diffraction, and its relationship to tissue function, *Proc. R. Soc. London, Ser. B,* 212, 299, 1981.
9. **Boyde, A. and Jones, S.,** Scanning electron microscopy of cartilage, in *Cartilage,* Vol. 1, Hall, B. K., Ed., Academic Press, New York, 1983, chap. 5.
10. **Bullough, P. and Goodfellow, J.,** The significance of the fine structure of articular cartilage, *J. Bone Joint Surg.,* 50B, 852, 1968.
11. **Broom, N. D.,** Abnormal softening in articular cartilage: its relationship to the collagen framework, *Arthritis Rheum.,* 25, 1209, 1982.
12. **Broom, N. D. and Marra, D. L.,** New structural concepts of articular cartilage demonstrated with a physical model, *Connect. Tissue Res.,* 14, 1, 1985.
13. **Maroudas, A. and Venn, M.,** Chemical composition and swelling of normal and osteoarthritic femoral head cartilage, *Ann. Rheum. Dis.,* 36, 399, 1977.
14. **Muir, I. H. M.,** Biochemistry in *Adult Articular Cartilage,* Freeman, M. A. R., Ed., Pitman, Kent, England, 1979, chap. 3.
15. **Lash, J. W. and Vasan, N. S.,** Glycosaminoglycans of cartilage, in *Cartilage,* Vol. 1, Hall, B. K., Ed., Academic Press, New York, 1983, chap. 8.
16. **Nimni, M. E.,** Collagen: structure, function, and metabolism in normal and fibrotic tissues, *Sem. Arthritis Rheum.,* 13, 1, 1983.
17. **Broom, N. D. and Marra, D. L.,** Ultrastructural evidence for fibril-to-fibril associations in articular cartilage and their functional implication, *J. Anat.,* 146, 185, 1986.
18. **Echlin, P.,** Low temperature scanning electron microscopy: a review, *J. Microsc.,* 112, 47, 1978.
19. **Gardner, D. L., O'Connor, P., and Oates, K.,** Low temperature scanning electron microscopy of dog and guinea-pig hyaline articular cartilage, *J. Anat.,* 132, 267, 1981.
20. **Clark, J. M.,** The organisation of collagen in cryofractured rabbit articular cartilage: a scanning electron microscope study, *J. Orthop. Res.,* 3, 17, 1985.
21. **Maroudas, A.,** Physicochemical properties of articular cartilage, in *Adult Articular Cartilage,* Freeman, M. A. R., Ed., Pitman, Kent, England, 1979, chap. 4.
22. **Caplan, A. I.,** Cartilage, *Sci. Am.,* 251, 82, 1984.
23. **Hukins, D. W. L. and Aspden, R. M.,** Composition and properties of connective tissue, *Trends Biochem. Sci.,* July, 260, 1985.
24. **Scott, J. E.,** Composition and structure of the pericellular environment, *Phil. Trans. R. Soc. London Ser. B,* 271, 235, 1975.
25. **Rosenberg, L., Hellman, W., and Kleinschmidt, A. K.,** Macromolecular models of proteinpolysaccharides from bovine nasal cartilage based on electron microscope studies, *J. Biol. Chem.,* 245, 4123, 1970.
26. **Broom, N. D.,** The altered biomechanical state of human femoral head osteoarthritic articular cartilage, *Arthritis Rheum.,* 27, 1028, 1984.
27. **Bullough, P. G., Yawitz, P. S., Tafra, L., and Boskey, A. L.,** Topographical variations in the morphology and biochemistry of adult canine tibial plateau articular cartilage, *J. Orthop. Res.,* 3, 1, 1985.
28. **Nimni, M. and Deshmukh, K.,** Differences in collagen metabolism between normal and osteoarthritic human articular cartilage, *Science,* 181, 751, 1973.
29. **Eyre, D. R., McDevitt, C. A., Billingham, M. E. J., and Muir, H.,** Biosynthesis of collagen and other matrix proteins by articular cartilage in experimental osteoarthritis, *Biochem. J.,* 188, 823, 1980.
30. **Meachim, G. and Osborne, G. V.,** Repair of the femoral articular surface in osteoarthritis of the hip, *J. Pathol.,* 102, 1, 1970.
31. **Broom, N. D.,** Structural consequences of traumatizing articular cartilage, *Ann. Rheum. Dis.,* 45, 225, 1986.

Chapter 13

VISCOELASTIC PROPERTIES OF ARTICULAR CARTILAGE AND MENISCUS

Elizabeth R. Myers, Wenbo Zhu, and Van C. Mow

TABLE OF CONTENTS

I. Introduction 268

II. The Multiphasic and Composite Nature of Cartilage 268
- A. Tissue Composition 268
- B. Matrix Organization 269

III. The Material Properties and the Visoelastic Behavior of Cartilage 271
- A. Definitions of Elasticity and Viscoelasticity 271
- B. The Roles of Collagen, Proteoglycan, and Water in the Material Properties of Cartilage 274
  - 1. Tensile Properties 274
  - 2. Compressive Properties 277
  - 3. Shear Properties 280
    - a. Dynamic Shear Experiment 281
    - b. Stress Relaxation Experiments 281
    - c. The Roles of Collagen, Proteoglycan, and Water 283

IV. Summary and Conclusions 285

Acknowledgment 286

References 286

I. INTRODUCTION

Articular cartilage is a thin layer of soft connective tissue that overlies the articulating bony ends in a diarthrodial joint. The menisci are crescent-shaped structures of soft tissue found between the plateau of the tibia and the condyles of the femur. Both types of cartilage play an important role in providing the excellent lubrication and wear characteristics of the joint and in maintaining the smooth, efficient function of the bearing system for the body. Joints in which the cartilage layer has worn away show bone erosion and eburnation leading to clinical arthritic symptoms. The ability of cartilage to perform the important mechanical functions in vivo depends on the mechanical properties of the tissue. These mechanical properties in turn depend on the composition of the tissues, the microstructural forms of the constituents, and the anatomical shape of the tissues.

The two common structural macromolecules of all connective tissues are collagen and proteoglycan. For each type of tissue, a specific compositional blend of materials, a specific ultrastructural arrangement of the collagen fibrillar network, and a certain amount and form of proteoglycan exist. In addition, water, the major component of all connective tissues by volume and weight, plays a determinant role in the mechanical properties of many connective tissues. These properties are the manifestation of the frictional resistance generated by the flow of this water through the interstitium of the tissue. In this chapter, we present a description of how the composition, including water, and how the microscopic arrangement of the structural molecules might influence the viscoelastic properties of connective tissues. Specifically, we wish to describe the mechanical behavior of articular cartilage and meniscus under various loading conditions and to explain how one may separate the frictional effects of interstitial fluid flow through the tissue from the intrinsic mechanical properties of the porous collagen/proteoglycan solid matrix.

II. THE MULTIPHASIC AND COMPOSITE NATURE OF CARTILAGE

Cartilaginous tissue can be thought of as a material composed of two distinct phases: a fluid phase (the interstitial water and the various inorganic salts dissolved in it) and a solid phase (the organic solid matrix which is made of cells, collagen fibrils, proteoglycans, and other glycoproteins). The solid phase itself behaves like a fiber-reinforced composite material. Thus, each phase and each distinct constituent of both the solid and fluid phases play roles in the functional behavior of cartilage. It is therefore necessary to know the composition and structure of cartilaginous tissues and the composite multiphasic nature of the tissue in order to understand their mechanical and functional properties. Although the various cartilaginous tissues (articular cartilage, meniscus, nasal cartilage, etc.) of the mammalian body perform a number of different functions, their compositions nevertheless are remarkably similar. The structural organization of each type of cartilage, however, varies a great deal, and it provides, along with the gross anatomical form of the tissue, the distinctive features differentiating the various tissues. These structural features also have a strong influence on the mechanical and functional behavior of the tissue.

A. Tissue Composition

Presented in Table 1 are the basic components of articular cartilage and meniscal cartilage. Collagen composes the majority of the dry content of articular cartilage, making up to 50 to 80% of the dry weight or about 20% of the wet weight.[1,2] In the meniscus, collagen also composes the largest portion of the solid matrix, ranging up to 90% of the dry weight in human material.[3] The content of collagen in articular cartilage varies with depth from the articulating surface; it drops from approximately 80% of the dry weight in the surface zone to about 60% in the middle region.[4] Collagen in cartilage is organized into fibrils, which

Table 1
APPROXIMATE AMOUNTS OF WATER, COLLAGEN, AND PROTEOGLYCAN IN ARTICULAR CARTILAGE AND MENISCUS (% WET WEIGHT)

Tissue	Water (%)	Collagen (%)	Proteoglycan (%)
Articular cartilage	70—80	15—25	5—10
Meniscus	60—70	20—25	≈1

are long, ropelike aggregates of collagen molecules, and sometimes into fibers, which are collections of fibrils. (Please see Volume I, Chapters 1 and 5, and see Chapters 9 and 12 of this volume for a complete discussion of collagen composition and structural hierarchy).

Proteoglycans constitute the second largest portion of the solid material in cartilage, accounting for 5 to 10% of the wet weight in human articular cartilage.[1] There are fewer proteoglycans in meniscal cartilage than in articular cartilage; about 2% of the solid matrix or approximately 1% of the wet weight is proteoglycan in canine articular cartilage.[5] The amount of proteoglycan in articular cartilage increases with depth, reaching the highest content by weight in the middle zone of the tissue.[1,2,6-8] Cartilage proteoglycans have been extensively studied by Muir,[9,10] Hardingham and co-workers,[11,12] and Rosenberg and co-workers.[13,14] In brief, proteoglycans are composed of chondroitin sulfate (having a molecular weight of 20 to 30 $\times$ 10^4 daltons) and keratan sulfate (having a molecular weight of 20 to 30 $\times$ 10^4 daltons), both of which form chains attached to a linear protein core making a proteoglycan subunit with a molecular weight of approximately 1 $\times$ 10^6 daltons. The globular end of the protein core of the proteoglycan subunit is attached to a filamentous hyaluronate chain with a molecular weight of approximately 5 $\times$ 10^5 daltons; in this manner a proteoglycan aggregate is formed with a molecular weight that may be as high as 200 $\times$ 10^6 daltons. Each proteoglycan subunit and aggregate has a highly complex molecular conformational state, and it is generally believed that these complex molecular conformations are necessary for the organization of the extracellular matrix of cartilage[9,10] and for providing the necessary material properties of cartilage.[12,15]

Cartilaginous tissues have a water content, including both interstitial and intracellular fluid, of approximately 60 to 80% by weight.[1,2,5] Within the interstitial fluid are electrolytes such as sodium, chloride, potassium, and calcium.[16] Most of the interstitial fluid within the tissue is freely exchangeable by diffusion with the bathing solution surrounding the tissue.[17,18] In addition, the interstitial fluid may be extruded from the tissue by applying a pressure gradient across the tissue[19,20] or by compressing the tissue.[16,21,22] The size of the open connected pores within which the interstitial fluid resides is very small and has been estimated to range from 20 to 70 Å.[15,18,19] The frictional resistance against the flow of the interstitial fluid through the porous solid matrix is very high, and thus the permeability of the tissue is very small.[19-22]

B. Matrix Organization

Each cartilaginous tissue differs according to the way in which the components are organized in the extracellular matrix. (see Volume I, Chapter 1 and 5, and see Chapters 9 and 12 of this volume for a more complete discussion of collagen composition and structural hierarchy.) Figures 1 and 2 give schematic illustrations of the fibrillar structure of articular cartilage and meniscus.

In the superficial zone of articular cartilage, most collagen fibrils are tangent to the articular surface and are closely spaced (Figure 1). The cells are oblong in shape with the long cell

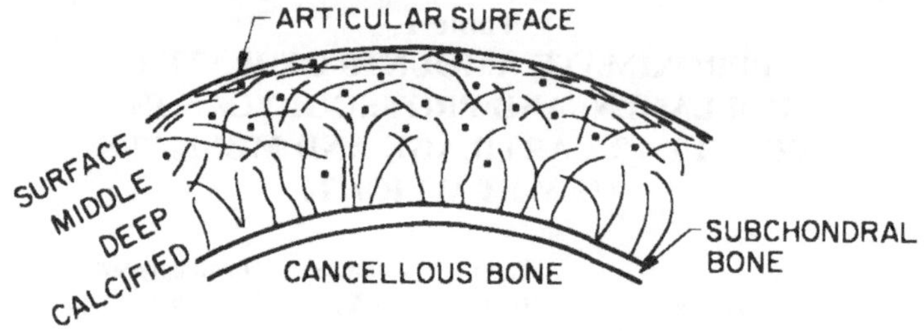

FIGURE 1. Schematic illustration of the fibrillar structure of the articular cartilage layer. There are four zones of note: the surface or superficial layer, the middle or intermediate layer, the deep zone, and the calcified zone.

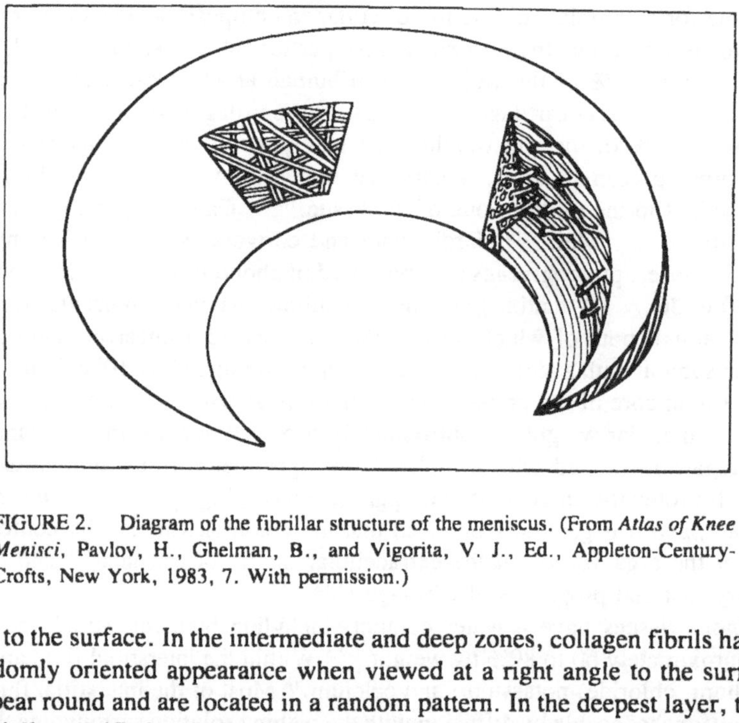

FIGURE 2. Diagram of the fibrillar structure of the meniscus. (From *Atlas of Knee Menisci*, Pavlov, H., Ghelman, B., and Vigorita, V. J., Ed., Appleton-Century-Crofts, New York, 1983, 7. With permission.)

axis parallel to the surface. In the intermediate and deep zones, collagen fibrils have a widely spaced, randomly oriented appearance when viewed at a right angle to the surface. Chondrocytes appear round and are located in a random pattern. In the deepest layer, the cartilage may be partially calcified and the fibrils are normal to the articular surface appearing to insert into the underlying bone.[4,23-25]

Within the plane that is tangent to the articular surface, a split-line pattern is discernible when the surface is punctured with a round, pointed instrument.[26] The elongated splits created by the round instrument may reflect a predominant collagen fibril arrangement in the plane of the surface along the axis parallel to the split lines.[25,27] Many biomechanical tensile tests have used this split-line pattern as the local coordinate system to provide the appropriate orientation for harvesting tissue specimens.[28-32]

The fibrous structure of the meniscus also has a layered appearance (Figure 2), but it differs from that of articular cartilage. The menisci of the knee have a semilunar shape and are found between the femoral condyles and the tibial plateau. At both surfaces there appears to be a thin layer, approximately 50 μm thick, of randomly oriented collagen fibrils of small diameter in the plane of the surface; underneath are large, coarse, circumferentially oriented collagen bundles. Interposed between the large fiber bundles are small, more radially arranged collagen fibrils, possibly serving to connect one large circumferential bundle to another.[33,34]

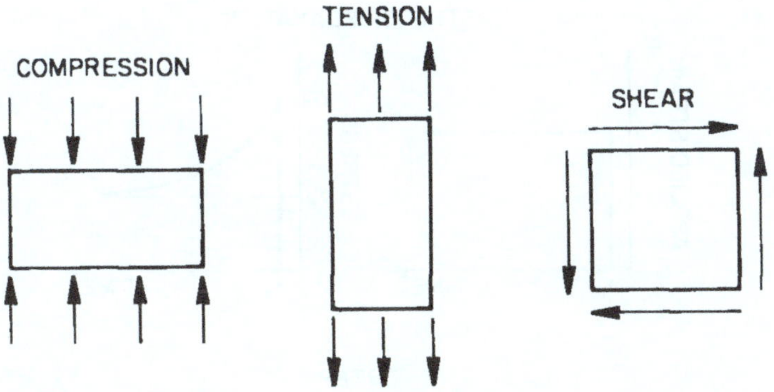

FIGURE 3. Representation of the actions of tension, compression, and shear on a material body.

The layered morphology of the collagen network (i.e., inhomogeneity) and the preferred orientation of collagen fibers (i.e., anisotropy) of both articular and meniscal cartilage provide a convenient means to examine how the collagen determines the strength and stiffness of the overall tissue. This may be accomplished by simply cutting specimens out of each layer oriented in a specific direction relative to the split-line pattern and determining the tensile properties of the specimens as a function of the layered inhomogeneity and anisotrophy.[28-32]

III. THE MATERIAL PROPERTIES AND THE VISCOELASTIC BEHAVIOR OF CARTILAGE

A. Definitions of Elasticity and Viscoelasticity

Many solid materials such as steel and titanium exhibit a linearly elastic behavior when deformed or placed under load. To a materials scientist or engineer, a material is elastic in a linear sense if it possesses the following properties: (1) the strain vanishes when the applied stress is removed, (2) the stress and strain are related in a linear manner, and (3) the material is insensitive to the rate at which stress or strain is applied. In addition, many linearly elastic materials have equal mechanical properties in all directions; this property is known as isotropy. This particular property exists when the characteristic features of the structural organization of the material are (1) randomly distributed in the material and (2) at a dimensional scale that is small compared with the scale of the material as a whole.

In theories describing the continuum nature of elastic materials, coefficients called moduli are used to determine the relationship between stresses and strains. When a stress is applied at right angles to the ends of a rectangular piece of material in a uniaxial test (i.e., a normal stress such as tension or compression, see Figure 3), the resulting proportionality coefficient between stress and strain is called Young's modulus. Often the Young's modulus for a piece of material has the same value in tension or in compression. The two can be differentiated, however, by speaking of one as the tensile modulus and the other as the compressive modulus. If a stress is applied parallel to the ends of a rectangular piece of material (Figure 3), it is known as a shear stress and the coefficient relating this shear stress and the resulting shear strain is known as the modulus of rigidity or the shear modulus.

This idealized elastic material model is adequate for many solids, including bone, but rarely does it completely describe the mechanical behavior of polymeric or soft biological materials. For these materials, a time dependency is found in the mechanical response to constant applied loads or deformations. Linearly elastic materials do not exhibit such time dependencies when subjected to a constant load or deformation nor are the moduli dependent

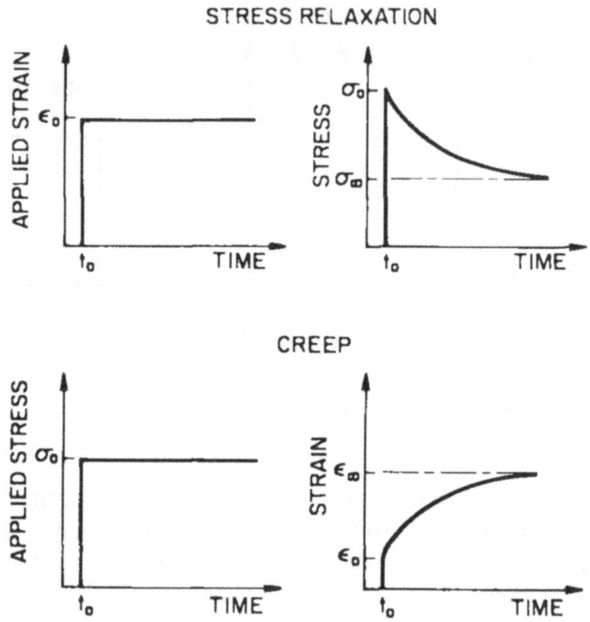

FIGURE 4. Viscoelastic responses of a polymer or soft connective tissue. The upper portion of the figure depicts stress relaxation under a constant imposed strain. The lower portion shows creep under a constant applied stress.

on the rate at which the deformation is imposed. Figure 4 illustrates two time-varying or viscoelastic responses typical of both polymers and soft connective tissues such as cartilage under the action of a constant load or deformation. In the upper portion of Figure 4, a phenomenon known as stress relaxation for viscoelastic solids is depicted. When a constant deformation (either tensile, compressive, or shear) is applied to a sample at t_0, a peak stress (σ_0) is registered followed by a diminution of the stress with time until a nonzero equilibrium value (σ_∞) is reached. This means that over time it takes less stress than the initial stress generated in the sample to maintain a constant stretch or constant twist. The lower portion of Figure 4 shows a second important viscoelastic phenomenon known as creep. In this case, when a constant stress is applied to a sample at t_0, it does not simply deform to some initial strain (ϵ_0) and remain there but rather the deformation increases in magnitude over time until an equilibrium strain (ϵ_∞) is reached. These two phenomena, stress relaxation and creep, are typical behaviors of viscoelastic solids and the detailed characteristics of these behaviors are often used to describe the mechanical response of viscoelastic materials.

In general, for single-phase polymeric materials, the motion of the molecular chains and the intermolecular friction associated with this motion within the stressed material are the mechanisms responsible for the visoelastic response of high polymers.[35] But what mechanisms are responsible for the time-dependent viscoelastic behavior of soft connective tissues? At least two distinct mechanisms give rise to the viscoelastic response of cartilage. First, there is the intrinsic viscoelasticity of the solid phase arising from the motion of polymeric chains such as collagen and proteoglycans during deformation, very much like the mechanisms responsible for the viscoelastic behavior of single-phase polymeric materials. This type of viscoelasticity could result from: (1) the intrinsic viscoelastic behavior of proteoglycans,[12,36] (2) the intrinsic viscoelastic behavior of collagen (see Chapter 11), and (3) the intrinsic viscoelastic behavior of the fiber-reinforced composite collagen-proteoglycan matrix (see Section III.B.3). The last mechanism could also include such effects as the retarding action associated with the movement of proteoglycans through the fine collagen meshwork when the tissue is deformed. Second, viscoelasticity may also arise from the biphasic nature of the tissue due to the flow of the interstitial fluid though the porous solid matrix in response to a pressure gradient or an applied load.[15,22] This mechanism would only be seen in biphasic materials consisting of a solid phase and a fluid phase where a portion of the fluid is free to move when the material is deformed as a whole.

Recall that the material properties of a linearly elastic material were relatively simple to define because there was a linear proportionality constant known as a modulus relating stress and strain. How does one model a viscoelastic material such as cartilage? One general viscoelastic law used by investigators to model the viscoelasticity of soft tissues, when the interstitial fluid flow effects are negligible, is the quasilinear viscoelastic theory developed by Fung.[37,38] This theory is applied to cartilage when it is subjected to infinitesimal pure shear. Since no interstitial flow occurs under this condition,[39] the intrinsic viscoelastic properties of the solid phase may be determined. The quasilinear viscoelastic law assumes that the stress in a material is dependent not only upon the magnitude of the strain at the time of observation but also upon the cumulative effects over time of the strains imposed on the material, i.e., the history of strain. The cumulative effect of strain depends on the stress-relaxation function which must be determined experimentally. In addition, the quasilinear viscoelastic theory postulates that the elastic stress-strain behavior of the material may be nonlinear. The stress at the time of observation is given by the time integral of the product of the experimentally determined stress-relaxation function and the nonlinear elastic stress-strain law.

A second model that has been used to describe the biphasic viscoelastic nature of cartilage is one that takes into account the large fluid content of the tissue (see Table 1) and considers the viscoelastic response to be due to frictional drag caused by the flow of fluid through the porous solid matrix. This theory was developed by Mow and co-workers and is called the KLM biphasic theory for cartilage.[15,22] It is important to clearly delineate the implications of the biphasic theory; in its simplest form, it postulates that the porous solid matrix of cartilage (but not the whole biphasic solid-fluid mixture making up cartilage) is a linearly elastic material. Therefore, it is possible to find a linear proportionality constant or elastic modulus describing the relationship between stress and strain for the solid matrix in an appropriate mechanical test. This modulus can only be determined by measuring the stress and strain behavior of cartilage at equilibrium. Before equilibrium, there is a viscoelastic response caused directly by the flow of the interstitial fluid through the porous solid matrix of cartilage. The frictional drag resulting from the resistance offered by the porous solid matrix to the flow of the interstitial fluid is described by the hydraulic permeability of cartilage.

So far we have discussed the application of stress or strain to a sample of material in a stepwise manner only (Figure 4). This means that stress and strain are applied suddenly and then held constant as illustrated in the left-hand side of Figure 4. Stresses or strains in material tests are applied commonly in several ways: in a stepwise manner, at a constant rate with respect to time, in sinusoidal waves etc. These functions are illustrated in Figure 5. Each type of test can reveal information about the nature of viscoelastic materials.

Because cartilage is undoubtedly a viscoelastic material, little is gained by trying to apply a linearly elastic model to fit the stress-strain response of cartilage unless one is dealing with the equilibrium response. At equilibrium, when all fluid flow has ceased, the linearly elastic material model does seem to describe adequately the stress-strain behavior of cartilage as long as the strains are small. To model the time-dependent behavior of cartilage, however, either the biphasic fluid-solid nature of the material must be taken into account when fluid-flow effects are important or the quasilinear viscoelastic material model must be used when the fluid-flow effects are not important.

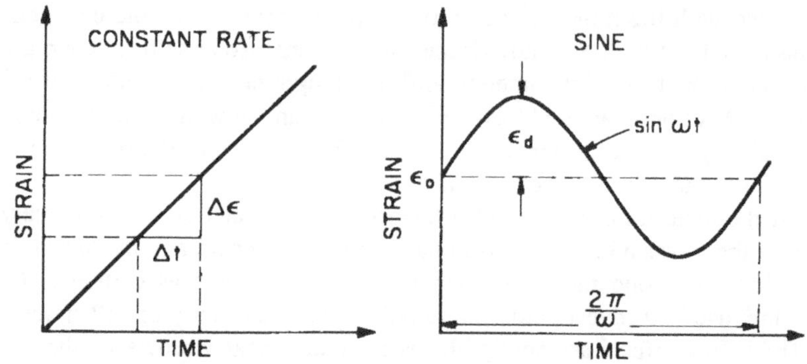

FIGURE 5. Illustrations of two inputs used commonly in material tests. The constant rate test imposes a constant rate of strain with respect to time ($\Delta\epsilon/\Delta t$). The steady sinusoidal test imposes a fluctuating strain in the form of a sine wave over an initial strain of ϵ_0. The symbol ω stands for the frequency of the input expressed in radians per second. This ω is related to the frequency f (see Figures 16 to 19) by $\omega = 2\pi f$.

In the above, we briefly described how the quasilinear viscoelastic model may be used to describe the intrinsic viscoelasticity of the porous solid matrix and how the biphasic nature of hydrated tissues like articular cartilage may give rise to a viscoelastic response of the tissue as a whole. In the remainder of Section III we describe the actual experimental determinations of the viscoelasticity of cartilaginous tissues, including the measurement of the equilibrium moduli and hydraulic permeabilities, and the response of cartilaginous tissues in tension, compression, and shear.

B. The Roles of Collagen, Proteoglycan, and Water in the Material Properties of Cartilage

1. Tensile Properties

The response of cartilage to applied tension is strongly viscoelastic and strongly dependent on the composition and organization of the matrix (see also chapter 11). Shown in Figure 6 is a typical stress-strain curve obtained under a slow constant strain-rate test (quasistatic response) for a strip of articular cartilage taken from the surface zone parallel to the local split-line direction and pulled in uniaxial tension under a constant rate. The input for such a constant strain-rate experiment is illustrated schematically in Figure 5, where the rate is given by $\Delta\epsilon/\Delta t$. There are several regions of the stress-strain curve attributed to certain structures and changes in the structures of articular cartilage. Early in the response is the "toe region", and it is believed that the increasing slope of the stress-strain curve in the toe region of connective tissues is due to the uncrimping and straightening or aligning of the collagen fibrils in the specimen as it is being stretched.[28,40] Next on the stress-strain diagram is the "linear region" in which the stress and strain appear to be related in a linear manner by a proportionality constant defined by the slope of the line ($\Delta\sigma/\Delta\epsilon$). It is thought that this region reflects the stiffness of the straightened collagen fibrils as they are pulled in uniaxial tension. Finally, beyond the linear region, the collagen fibrils of articular cartilage fracture. The tensile failure stress is therefore a measure of the strength of the collagen fibrils and of the fiber-reinforced composite matrix.

Human and bovine menisci have also been studied in a tensile constant strain-rate test[41] and several typical curves are shown in Figure 7. The shape of the curves is similar to that for articular cartilage. The magnitude of the tensile stress, however, is greater for meniscus specimens prepared in a direction parallel to the fiber bundles than for articular cartilage (compare the ordinates of Figures 6 and 7). This result can be explained in part by the fact that the meniscus may contain a larger portion of collagen then articular cartilage and by

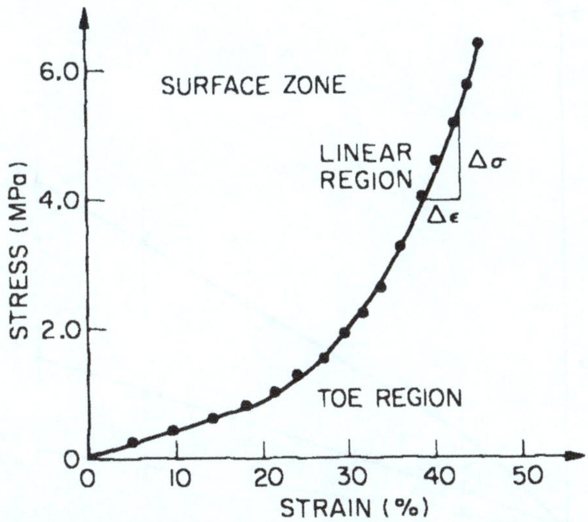

FIGURE 6. Stress-strain curve for a bovine articular cartilage specimen cut from the surface zone parallel to the split lines. The toe region is marked by an increasing slope, whereas the linear region appears to be a straight line with a slope of $\Delta\sigma/\Delta\epsilon$.

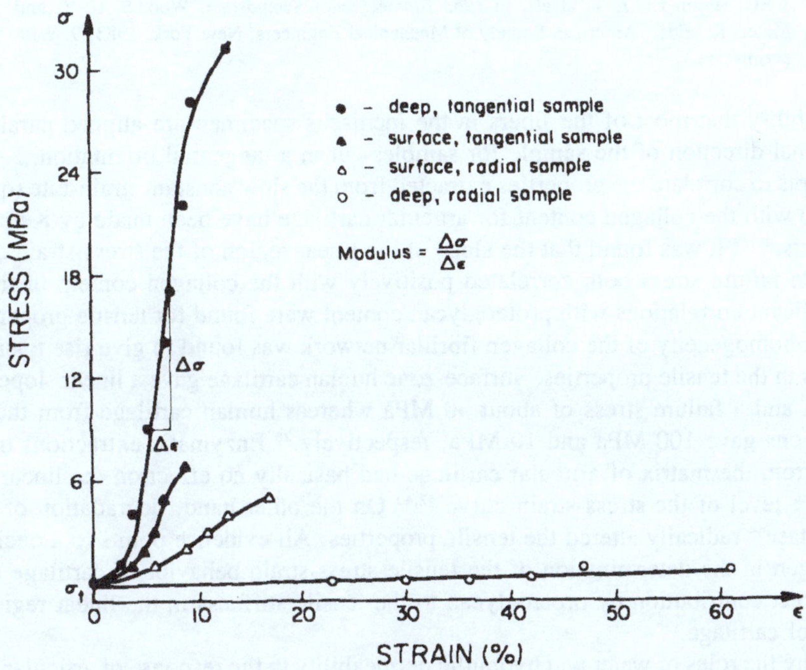

FIGURE 7. Stress-strain curves for samples of meniscus cut from four regions of a bovine medial mensicus. There are linear regions in each curve with a slope of $\Delta\sigma/\Delta\epsilon$. (From Whipple, R. et al., in *1984 Advances in Bioengineering*, Spilker, R. L., Ed., Am. Soc. Mech. Eng., New York, 1984, 32. With permission.)

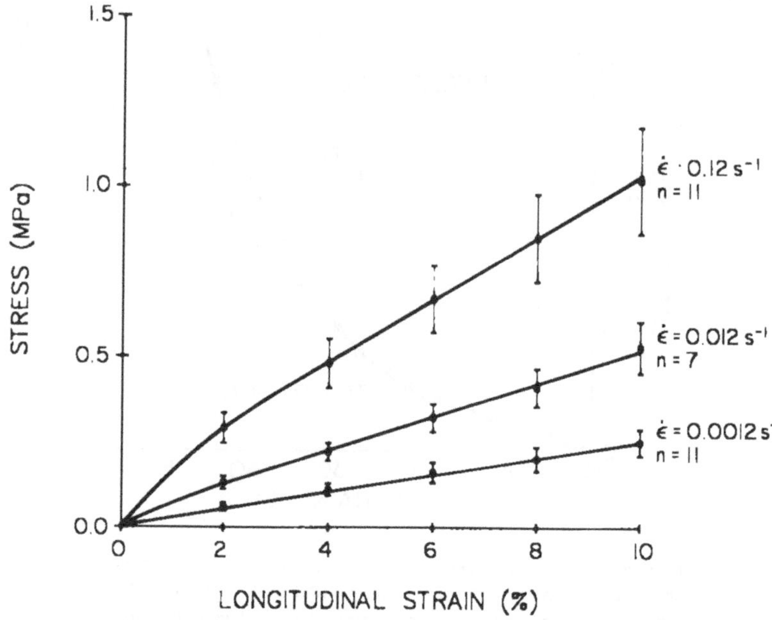

FIGURE 8. The toe region of the stress-strain curves of bovine articular cartilage tested at three different strain rates ($\dot{\epsilon}$). The symbols are mean values and the bars represent ± 1 SD. (From Li, J. T. et al., in *1983 Biomechanics Symposium*, Woo, S. L.-Y. and Mates, R., Eds., American Society of Mechanical Engineers, New York, 1983, 9. With permission.)

the possibility that most of the fibers in the meniscus specimen are aligned parallel to the longitudinal direction of the sample for samples cut in a tangential orientation.

Attempts to correlate the properties extracted from the slow constant strain-rate (quasistatic response) with the collagen content for articular cartilage have been made by Kempson and co-workers.[28,29] It was found that the slope of the linear region of the stress-strain curve and the tensile failure stress both correlated positively with the collagen content of the tissue. No significant correlations with proteoglycan content were found for tensile properties. The layered inhomogeneity of the collagen fibrillar network was found to give rise to significant variances in the tensile properties. Surface-zone human cartilage gave a linear slope of about 300 MPa and a failure stress of about 30 MPa whereas human cartilage from the middle/deep regions gave 100 MPa and 10 MPa, respectively.[28] Enzymatic extractions of proteoglycans from the matrix of articular cartilage had basically no effect on the linear slope or the failure level of the stress-strain curve.[42,43] On the other hand, degradation of collagen with elastase[44] radically altered the tensile properties. All evidence points to a decisive role for collagen in the determination of the tensile stress-strain behavior of cartilage and little if any direct contribution by proteoglycan to the tensile stiffness in the linear region or the strength of cartilage.

What are the roles of water and hydraulic permeability in the response of articular cartilage and meniscus to a constant strain-rate tensile test? The question can be considered by examining the effect of strain rate on the overall shape of the stress-strain curve. If the strain rate is relatively high, then the concave toe region disappears and a convex region is seen instead.[45] This response is shown in Figure 8. The biphasic nature of articular cartilage becomes important in this case. A fast rate means that fluid has no time to flow through the matrix and becomes pressurized, there is little energy dissipation, and the result is a higher level of stress at a given strain than would be found in a slow strain-rate test where fluid exudation has had time to occur. Pressurization of the interstitial fluid can occur only

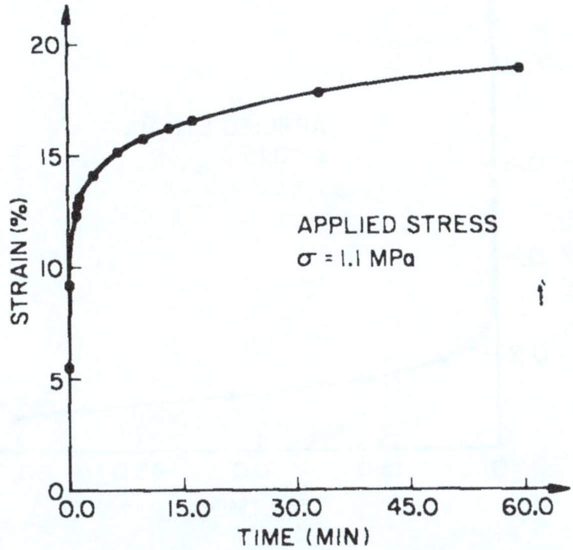

FIGURE 9. Creep curve for a bovine articular cartilage sample taken from the middle zone and held under a tensile stress of 1.1 MPa.

if the tissue behaves as a biphasic medium where the collagen and proteoglycan act as a porous fiber-reinforced composite material and where pull out and alignment of the collagen fibrils are unimportant, such as in the early stages of the stretch in these experiments (see Figure 8). It is important to note that under physiological conditions, the tensile strain within articular cartilage cannot be more than a few percent.[46] In this way, proteoglycans do contribute to the transient tensile viscoelastic response of cartilage in tension.

Creep and stress relaxation take place in cartilage when tensile stresses or strains are applied as a step function. As opposed to a constant strain-rate test (unless the strain rate is very small), the true elastic modulus of the solid matrix can be determined experimentally in either a creep or stress-relaxation test at equilibrium. This equilibrium tensile modulus differs from the linear slope of the constant stress-strain curves shown in Figures 6 and 7 because no effects of fluid flow and intrinsic matrix viscoelasticity can occur at equilibrium. Figures 9 and 10 show typical creep and stress-relaxation curves, respectively, for articular cartilage in uniaxial tension. The tensile moduli of human articular cartilage obtained at equilibrium at low strain levels are in the 1 to 10 MPa range.[47]

The roles of proteoglycan and collagen in the normalized tensile creep response can be examined by considering the effects of enzymatic extraction of proteoglycans from the cartilage solid matrix. Figure 11 is a graph of normalized tensile strain vs. time from a creep test showing that articular cartilage samples treated with chondroitinase "jumped" initially to a much higher relative strain level compared with control samples. It seems that the chondroitin sulfate chains on the proteoglycans provide a strong retarding or resisting action against collagen fibrillar arrangement along the axis of tension. When chondroitin sulfate is removed from the proteoglycans within the matrix, the alignment of collagen fibrils takes place immediately upon application of load, and then very little tensile creep takes place after the first second. This is not true in untreated cartilage, where the unaltered proteoglycans appear to offer great resistance against fibrillar alignment and a comparatively slow tensile creep phenomenon takes place.[48]

2. *Compressive Properties*

The response of cartilage to compression or compressive loads is also strongly viscoelastic.

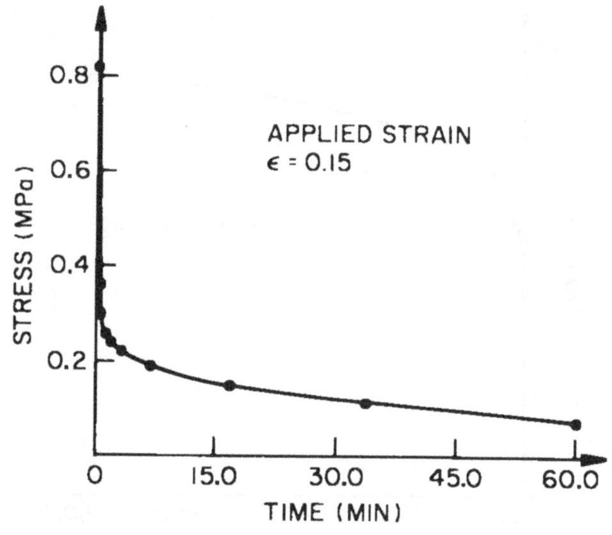

FIGURE 10. Stress-relaxation curve for a bovine articular cartilage specimen from the midzone. The applied tensile strain was 15%.

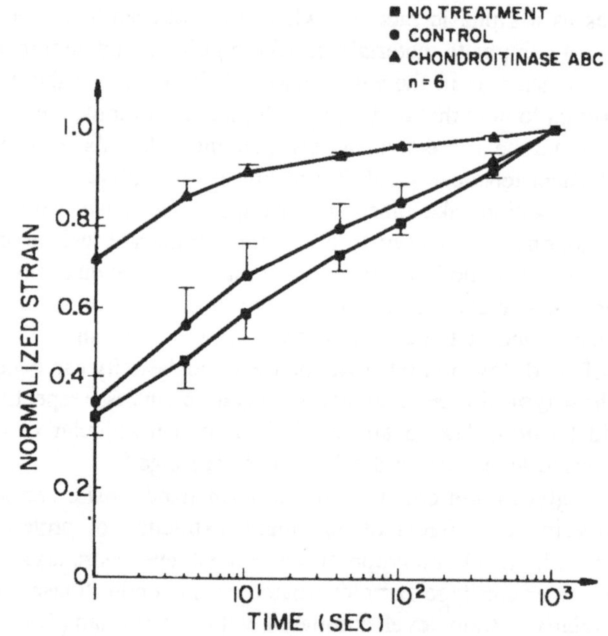

FIGURE 11. Effects of treatment with chondroitinase ABC on the tensile creep response of bovine articular cartilage. The normalized strain is the strain at a given time divided by the strain measured at 1000 sec. The symbols each represent the mean value for six samples from one of three data pools: samples that received no treatment, control samples that were incubated in buffer, and enzyme-treated samples that were incubated in buffer plus chondroitinase ABC. (From Schmidt, M. B. et al., *Trans. Orthop. Res. Soc.*, 11, 1986. With permission.)

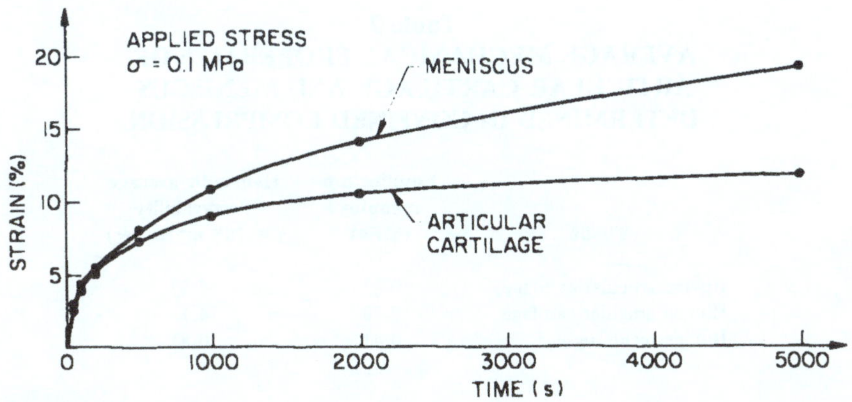

FIGURE 12. Creep curves for bovine cartilage samples tested in confined compression. Results are shown for both a meniscus specimen and an articular cartilage specimen.

In contrast to tension, however, it is the frictional resistance to fluid flow that provides the dominant mechanism in the overall viscoelastic response of the tissue.[15,22] The compressive creep phenomenon for articular cartilage and meniscus has been studied extensively in the confined compression test, wherein a circular cylindrical specimen is enclosed in a smooth, impermeable, rigid confining ring that prevents radial expansion, focusing all motion to take place in the axial direction of the load. The compressive load is applied via a fine porous filter so that all the load is exerted onto the solid phase. The pore pressure in the filter is ambient, which allows fluid exudation from the tissue to occur unimpeded by any external pressure. The confinement and the manner of loading greatly simplify the mathematical analysis of the problem and permit values for an intrinsic equilibrium modulus and the hydraulic permeability to be extracted from the material tests.[15,22,49]

Figure 12 shows typical creep curves for articular cartilage and meniscus obtained from the confined compression experiment. Note that there is no instantaneous deformation; this is because both the fluid phase and solid phase of cartilage are incompressible and there has not been sufficient time for a significant amount of fluid exudation to occur. As time progresses, however, compression takes place because fluid is being squeezed out from the tissue; this creep continues to occur until equilibrium is achieved. Thus, the rate of creep in compression is governed by the ease with which fluid may be squeezed out from the tissue. In other words, the permeability of the tissue determines the rate of creep in compression. At equilibrium the frictional resistance caused by fluid flow no longer exists and the applied compressive load is carried by the solid phase throughout the tissue. Thus, the equilibrium stress-strain behavior of the tissue determines the intrinsic elastic modulus of the solid phase. Figure 12 also shows that the meniscal cartilage exhibits greater equilibrium deformation under the same applied load than articular cartilage, indicating that the solid matrix of mensicus is more compliant than that of articular cartilage. Typical values for the equilibrium modulus and hydraulic permeability of articular cartilage and meniscus, obtained from this confined compression protocol, are given in Table 2.

The equilibrium modulus of articular cartilage determined in the confined compression creep test correlates inversely with water content of the tissue[49] and directly with uronic acid content.[50] If an enzyme such as trypsin is used to remove most of the proteoglycan content of articular cartilage, the compressive equilibrium modulus is decreased by an order of magnitude compared with the control value.[51] Thus, it is evident that proteoglycans are the main components of the tissue that offer the resisting force to compressive loads and deformations in cartilage at equilibrium.

Without including any intrinsic viscoelasticity arising from the motion of the macromol-

Table 2
AVERAGE MECHANICAL PROPERTIES OF ARTICULAR CARTILAGE AND MENISCUS DETERMINED IN CONFINED COMPRESSION

Tissue	Equilibrium modulus (MPa)	Hydraulic average Permeability (× 10^{15} $m^4/N \cdot sec$)
Bovine articular cartilage	0.85	4.67
Human articular cartilage	0.79	4.7
Bovine meniscus	0.41	0.81

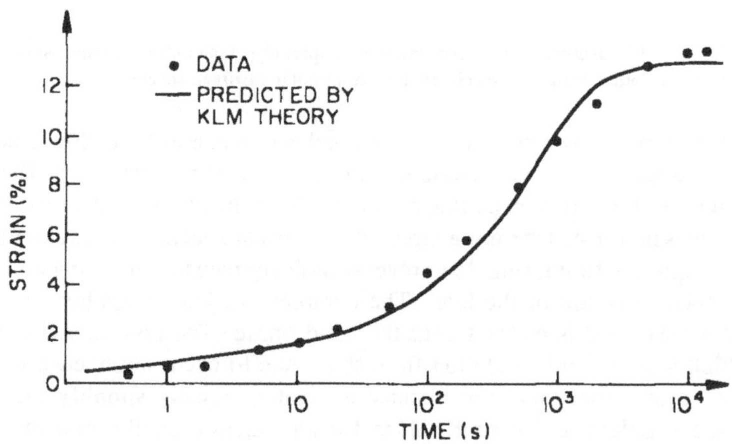

FIGURE 13. Predicted and experimental results of creep in confined compression for a plug of bovine articular cartilage. The solid line represents the creep response predicted by the KLM biphasic model. The results for the curve fit yielded a confined compression modulus of 0.86 MPa and an average hydraulic permeability of 1.6 × 10^{-15} $m^4/N \cdot sec$.

ecules composing the solid matrix, the KLM biphasic theory for cartilage[22] can quite adequately predict the transient creep response during the confined compression test. This means that the predominant mechanism responsible for the viscoelastic behavior of cartilage in compression and its load-carrying capacity is the frictional drag associated with interstitial fluid flow through the porous solid matrix during compression. This result is depicted in Figure 13, where both data and theoretical predictions are given, and a good fit between the two is seen. The relative motion between the fluid phase and the solid phase in cartilage may also be used to account for the stress-relaxation behavior in compression.[15,22] Stress relaxation due to fluid exudation in cartilage provides an excellent means by which the tissue may relieve excessive stress under physiological loading conditions.[52]

3. *Shear Properties*

In order to determine the intrinsic viscoelastic properties of the solid phase alone, cartilage can be tested in a pure shear experiment under infinitesimal strain conditions. No fluid flow occurs under these conditions because no pressure gradients or volume changes are developed within the tissue.[39] The determination of the intrinsic viscoelasticity of the solid phase provides a means to assess the relative contribution of the two mechanisms responsible for the viscoelastic behavior of cartilage as a whole. It is worth repeating this concept to clarify this powerful technique that permits the uncoupling of the two distinct mechanisms responsible for the viscoelastic behavior of cartilaginous tissues. As stated previously, the intrinsic properties of the solid phase as determined in a tensile or compressive material test can only be determined at equilibrium or by fitting transient data with an appropriate model that includes the fluid-flow effects on the mechanical response of the tissue. This is not necessary in the infinitesimal pure shear test because no fluid flow takes place in the interstitium. Any property that is determined under these conditions, therefore, whether under dynamic or equilibrium conditions, reflects the intrinsic viscoelastic character of the solid phase (i.e., the properties of the collagen/proteoglycan solid matrix) alone. The intrinsic viscoelastic behavior of the cartilage matrix has been observed in both the dynamic shear test and the stress-relaxation test.

a. Dynamic Shear Experiment

The viscoelasticity of the collagen/proteoglycan solid matrix may be determined by measuring the storage and loss moduli of cartilage through a dynamic pure shear experiment where a thin (thickness = 800 μm) circular (diameter = 6.35 mm) wafer of tissue is subjected to a steady sinusoidal torsional strain (see Figure 5) of small magnitude over a range of frequencies.[53-55] The torque generated by the sinusoidal torsional shear displacement is used to calculate the storage modulus G′ and the loss modulus G″ as a function of frequency.[37,53-54] There are two sets of quantities: G′ and G″ or equivalently the magnitude of the dynamic shear modulus (|G*|) and the phase shift angle (δ). These quantities are related by the following equations:

$$|G^{*}| = \mathrm{sqrt}[(G')^2 + (G'')^2] \tag{1}$$

$$\tan \delta = G''/G' \tag{2}$$

Either of these pairs of coefficients (G′ and G″ or |G*| and δ) can be used. The phase shift angle is defined as the angle between the applied steady sinusoidal torsional strain and the steady sinusoidal torque response; it characterizes the energy dissipation within a viscoelastic material. The material is purely elastic when δ = 0° and purely viscous when δ = 90°, whereas δ lies somewhere between 0 and 90° for a general viscoelastic material. The magnitude of the shear modulus (|G*|) reflects the overall stiffness of the tissue in shear. The collagen/proteoglycan solid matrix of normal bovine articular cartilage is found to exhibit a slightly viscoelastic behavior.[39,53-55] The phase shift angle lies between 9 and 20° over the physiological frequently range of 0.01 to 20 Hz when the tissue is subjected to a small shear strain (0.1%). The value of the dynamic shear modulus varies from 0.2 to 2.5 MPa.[53-55]

b. Stress Relaxation Experiments

Another manifestation of the viscoelastic behavior of the collagen/proteoglycan solid matrix is stress relaxation in shear.[55] If the specimen in the torsional pure shear test undergoes a sudden step change of angular displacement (Figure 4) and the time variation of the resulting torque generated by the tissue is measured, then the stress-relaxation function of the solid phase is completely determined. A typical stress-relaxation curve is shown in Figure 14. The shear stress calculated from the measured torque reaches a peak value at the instant the shear strain is applied. The time for the tissue to reach a nonzero equilibrium varies from tissue to tissue, and it depends upon the magnitude of the applied shear strain. The equilibrium shear modulus of the tissue can be obtained as the slope of the linear regression on the equilibrium stress-strain relation when the steps of shear displacement are applied in an incremental manner. The average equilibrium shear modulus for human patellar cartilage is 0.23 MPa.[55] It has been found that the instantaneous stress-strain relation exhibits a slightly nonlinear pattern over a range of shear strain of 0.2 to 2.0%. Using the quasilinear viscoelastic

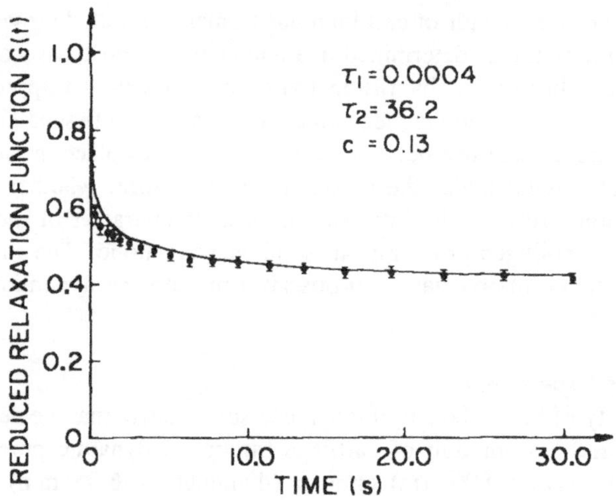

FIGURE 14. A typical stress-relaxation curve upon a step change of shear strain, expressed in terms of the mean of ten cycles of stress relaxation normalized by the initial stress. The solid line represents the theoretical prediction. The spectrum parameters τ_1, τ_2, and c are determined from the stress-relaxation experiment.[55,56]

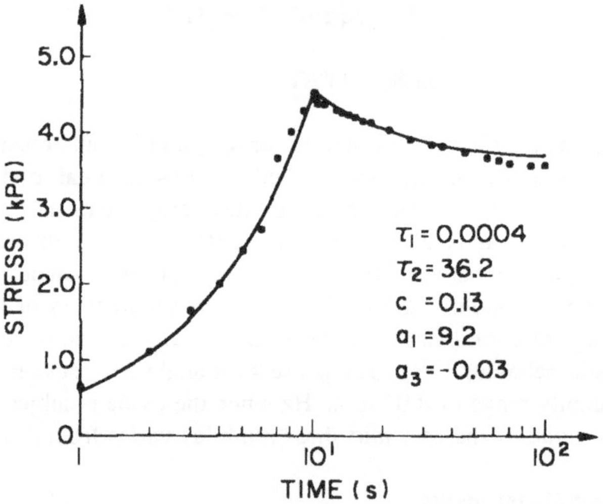

FIGURE 15. A strain-rate controlled stress-relaxation curve in shear. Shown is a comparison of the theoretical predicted results (solid line) and the experimental results. The spectrum parameters τ_1, τ_2, and c and the elastic coefficients a_1 and a_3 are determined from the stress-relaxation experiment.[55,56]

material model proposed by Fung[37,38] (see Section III.A), the strain-rate controlled stress-relaxation experiment has been analyzed for articular cartilage in shear.[55] The theoretical prediction is shown by the solid line in Figure 15; there is reasonable agreement between the predictions of the model and the experimental results. This result indicates that the quasilinear viscoelastic theory is adequate for analysis of the intrinsic "flow-independent" viscoelastic behavior of the collagen/proteoglycan solid matrix of cartilage in shear.

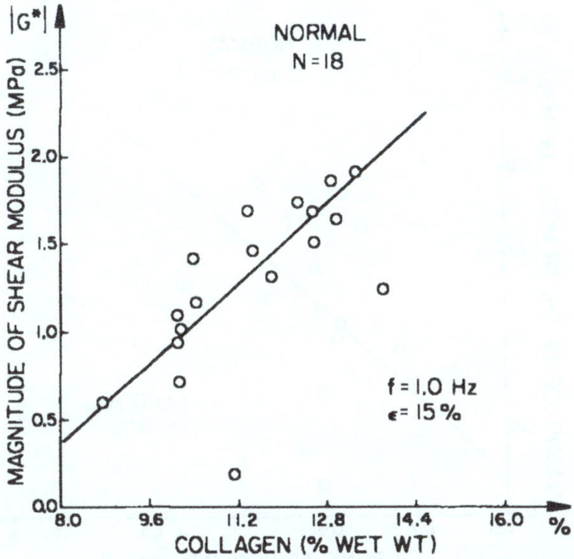

FIGURE 16. A direct correlation between the collagen content (by wet weight) and |G*| for bovine articular cartilage. The compressive clamping strain is 15%.

c. The Roles of Collagen, Proteglycan, and Water

The collagen network plays an active mechanical role in contributing to the shear properties of cartilage. When a steady sinusoidal torsional shear deformation is applied to the middle zone of cartilage, the randomly distributed collagen network is stretched. This stretching provides the mechanism for storage of elastic energy within the tissue. In other words, the tension in collagen acts to strengthen the shear properties of the matrix. This phenomenon is confirmed by the fact that an increase in collagen content results in an increase in the magnitude of the shear modulus, as shown in Figure 16. The change in collagen content does not affect the phase shift angle significantly at a moderate frequency and a small magnitude of strain. For the most part, collagen stores energy like an elastic material. The slight viscoelastic nature of collagen only becomes evident when the tissue is under significant deformation. Based upon the work of Woo et al.,[56] the energy dissipation of collagen in a tensile test in terms of the phase shift angle is about 3.6° for canine medial collateral ligament, which is composed mainly of collagen. Therefore, energy dissipation of collagen is very small compared with that of cartilage ($9° \leq \delta \leq 20°$). The energy dissipation is probably due to frictional effects within the network formed by proteoglycans.

According to the complex shear modulus measurements made on proteoglycan solutions at a concentration of 50 mg/mℓ (a concentration similar to that found in cartilage), the magnitude of the complex shear modulus (|G*|) and the phase shift angle are about 1.0 Pa and 70° respectively[12,36] while the values for the cartilage solid matrix are about 10^6 Pa and 15°.[39,53-55] Hence, proteoglycans themselves do not seem to contribute to the overall rigidity of the tissue. There are direct correlations, however, between the uronic acid content and |G*| (see Figure 17) and the uronic acid content and the phase shift angle (δ) for articular cartilage samples (see Figure 18). Apparently, proteoglycans entrapped in a collagen meshwork interact with the collagen fibrils in such a way that the fibrils are separated and maintained in their proper spatial orientations. Under such an arrangement, the collagen fibrils are likely to be stretched during shear even if the tissue is under compressive load as in the physiological situation in the joint. Thus, this organization of collagen fibrils within the collagen/proteoglycan solid matrix appears to provide cartilage with its strength and stiffness characteristics not only in tension but in shear as well.

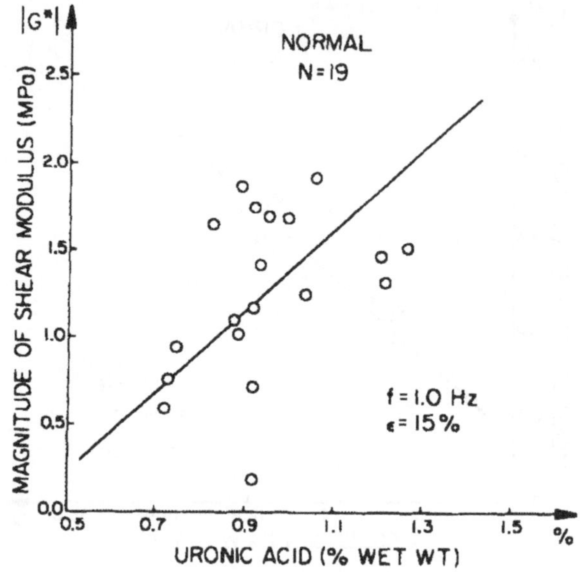

FIGURE 17. A direct correlation between the uronic acid content (by wet weight) and |G*| for bovine articular cartilage. The compressive clamping strain is 15%.

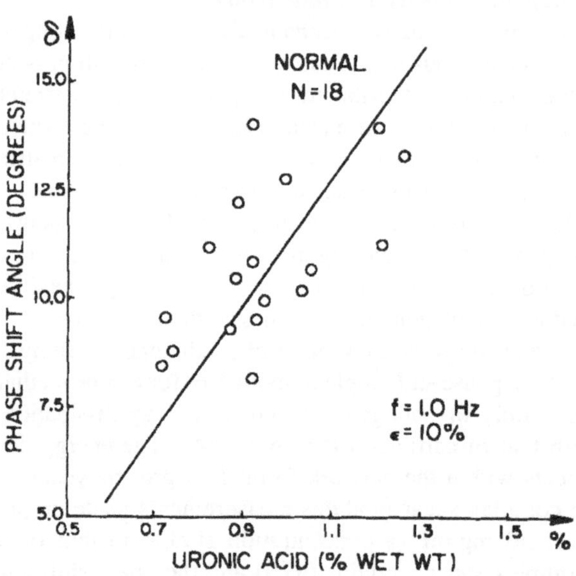

FIGURE 18. A direct correlation between the uronic acid content (by wet weight) and the phase shift angle (δ) for bovine articular cartilage. The compressive clamping strain is 10%.

When proteoglycans are degraded and depleted from cartilage with chrondroitinase ABC or hyauluronidase, the physical and chemical interactions between collagen and proteoglycan are altered and greatly reduced. This is reflected by the decrease of |G*| in treated tissues. Figure 19 shows a comparison of |G*| in proteoglycan-extracted tissues with |G*| in controls. In the proteoglycan-extracted tissue, the collagen network is probably collapsed. When such a tissue is subjected to the action of shear, instead of being stretched, the collagen fibrils

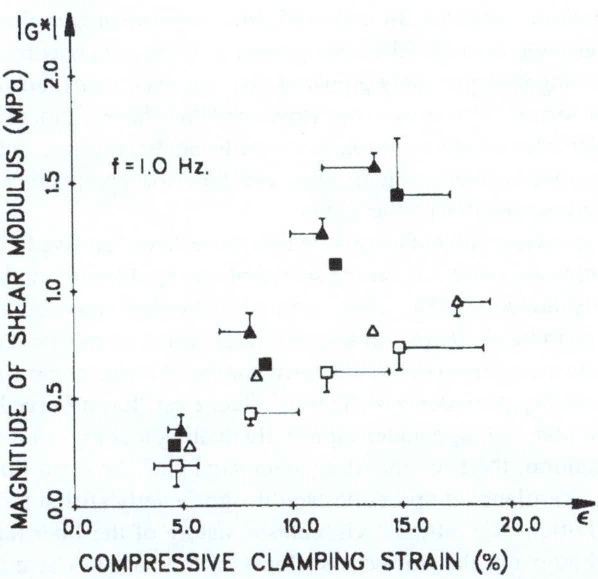

FIGURE 19. A comparison of |G*| in proteoglycan-extracted tissues with |G*| of controls. Solid and open squares represent control and chondroitinase ABC treated specimens respectively. Solid and open triangles represent control and hyaluronidase treated samples. The bars represent 1 SD. Twelve samples were tested within each group. The |G*| decreases in treated tissues.

are probably sliding relative to one another, thus defeating the fiber-reinforced composite character of the collagen/proteoglycan solid matrix of normal cartilage.

IV. SUMMARY AND CONCLUSIONS

Cartilage strips pulled in uniaxial tension at high strains show both stiffness and strength characteristics that depend upon the nature of the collagen framework. The frictional resistance to interstitial fluid flow and the frictional retarding action of proteoglycans to movement of the collagen network within the fiber-reinforced composite solid matrix are important factors governing the transient behavior of cartilage if it is subjected to high strain rates in tension. In compression, the proteoglycan component of the solid matrix is the major contributing factor to the equilibrium compressive modulus of cartilage and the frictional resistance to interstitial fluid flow is the determinant factor of the transient compressive response. Permeability controls the rate of fluid flow in cartilage and it is inversely related to the proteoglycan content. Under small pure shear conditions, the shear stiffness of cartilage depends mostly on the tensile stiffness of collagen fibrils because the fibrils are arranged in an orientation within the tissue such that they are stretched when the tissue is sheared. The dissipation of energy within the solid matrix depends largely upon the proteoglycans present within the solid matrix. The fiber-reinforced composite solid matrix is only slightly viscoelastic in shear because in shear it is the relatively elastic collagen fibrils within the collagen/proteoglycan matrix that support the stress generated within the tissue.

It is relatively easy, both intuitively and in the laboratory, to consider the material properties of isolated cartilage specimens (such as moduli and permeability) as they relate to and are influenced by composition and structure. A more important and difficult consideration, however, is the influence of the components on the ability of cartilage to perform its biomechanical functions in vivo. No longer can one consider the simple loading modes shown in Figure 3; there are three-dimensional, time-dependent complex loading conditions on the surface of cartilage in vivo. Also, the geometry of the joint becomes important because one is no longer dealing with precisely prepared cartilage specimens such as circular cylinders or thin rectangular strips. The stress, the flow, and the deformational fields in the entire cartilage layer under normal joint motion have yet to be determined, but hypotheses can be put forward as to what is happening in vivo and how the components of cartilage might influence the overall mechanics of the joint.

Stresses across the knee joint during function have been assessed,[57] and may range as high as 10 MPa, yet from Table 2 it can be seen that the equilibrium modulus of the cartilage solid matrix is only about 1 MPa. Thus, some mechanism must be at play that stiffens articular cartilage as these dynamic stresses are transmitted across the knee during motion. One such mechanism stems from the viscoelastic nature of articular cartilage; under dynamic conditions viscoelasticity provides a stiffening effect over that observed under equilibrium conditions. For articular cartilage under rapidly fluctuating loading conditions, such as those found during locomotion, the frictional drag associated with the rapid flow of the interstitial fluid accompanying cartilage compression would significantly stiffen the transient behavior of the tissue. In addition, the intrinsic viscoelastic nature of the fiber-reinforced composite solid matrix would augment the transient stiffness of the tissue. These fundamental mechanisms have been clearly elucidated in the laboratory. Exactly how these mechanisms come into play for the articular and meniscal cartilages of the knee joint under physiological loading conditions is not known. Due to the extreme theoretical and experimental difficulties encountered in this subject, very little work has been done in this area. Thus, this area continues to offer significant challenges for future investigators in the field.

ACKNOWLEDGMENT

The research that made this work possible was supported by National Science Foundation grant MSM82-11968, National Institute of Arthritis, Diabetes, Digestive and Kidney Diseases grants AM26440, AM19094, and AM35926, and the John A. Clark and Edward T. Crossan Endowment at Rensselaer Polytechnic Institute. Any options, findings, and conclusions or recommendations expressed in this publication are those of the authors and do not necessarily reflect the views of the National Science Foundation. We wish to thank Ms. Rose A. Boshoff for her assistance in editing this manuscript.

REFERENCES

1. **Maroudas, A., Bayliss, M. T., and Venn, M. F.,** Further studies on the composition of human femoral head cartilage, *Ann. Rheum. Dis.*, 39, 514, 1980.
2. **Venn, M. F.,** Variation of chemical composition with age in human femoral head cartilage, *Ann. Rheum. Dis.*, 37, 168, 1978.
3. **Peters, T. J. and Smillie, I. S.,** Studies on the chemical composition of the menisci of the knee joint with special reference to the horizontal cleavage lesion, *Clin. Orthop. Rel. Res.*, 86, 245, 1972.
4. **Muir, H., Bullough, P., and Maroudas, A.,** The distribution of collagen in human articular cartilage with some of its physiological implications, *J. Bone Joint Surg.*, 52-B, 554, 1970.
5. **Adams, M. E. and Muir, H.,** The glycosaminoglycans of canine meniscus, *Biochem. J.*, 197, 385, 1981.
6. **Stockwell, R. A. and Scott, J. E.,** Distribution of glycosaminoglycans in human articular cartilage, *Nature (London)*, 215, 1376, 1967.
7. **Maroudas, A., Muir, H., and Wingham, J.,** The correlation of fixed negative charge with glycosaminoglycan content of human articular cartilage, *Biochim. Biophys. Acta*, 177, 492, 1969.
8. **Frazen, A., Inerot, S., Hejderup, S.-O., and Heinegard, D.,** Variations in the composition of bovine hip articular cartilage with distance from the articular surface, *Biochem J.*, 195, 535, 1981.

9. **Muir, I. H. M.,** Biochemistry, in *Adult Articular Cartilage*, 2nd ed., Freeman, M. A. R., Ed., Pitman Medical, Kent, England, 1979, 145.
10. **Muir, H.,** Proteoglycans as organisers of the extracellular matrix, *Biochem. Soc. Trans.*, 11, 613, 1983.
11. **Hardingham, T. E.,** The role of link protein in the structure of cartilage proteoglycan aggregates, *Biochem J.*, 177, 237, 1979.
12. **Hardingham, T. E., Muir, H., Kwan, M. K., Lai, W. M., and Mow, V. C.,** Viscoelastic properties of proteoglycan solutions with varying proportions present as aggregates, *J. Orthop. Res.*, 5, 36, 1987. publication.
13. **Rosenberg, L. C., Hellmann, W., and Kleinschmidt, A. K.,** Electron microscope studies of proteoglycan aggregates from bovine articular cartilage, *J. Biol. Chem.*, 250, 1877, 1975.
14. **Rosenberg, L. C.,** Structure of cartilage proteoglycans, in *Dynamics of Connective Tissue Macromolecules*, Burleigh, P. M. C. and Poole, A. R., Eds., North-Holland, Amsterdam, 1975, 105.
15. **Mow, V. C., Holmes, M. H., and Lai, W. M.,** Fluid transport and mechanical properties of articular cartilage: a review, *J. Biomech.*, 17, 377, 1984.
16. **Linn, C. C. and Sokoloff, L.,** Movement and composition of interstitial fluid of cartilage, *Arthritis Rheum.*, 8, 481, 1965.
17. **Torzilli, P.,** Influence of cartilage conformation on its equilibrium water partition, *J. Orthop. Res.*, 3, 473, 1985.
18. **Maroudas, A.,** Physicochemical properties of articular cartilage and the intervertebral disc, in *The Joints and Synovial Fluid*, Sokoloff, L., Ed., Academic Press, New York, 1980, 95.
19. **McCutchen, C. W.,** The frictional properties of animal joints, *Wear*, 5, 1, 1962.
20. **Mansour, J. M. and Mow, V. C.,** The permeability of articular cartilage under compressive strain and at high pressures, *J. Bone Joint Surg.*, 58-A, 509, 1976.
21. **Edwards, J. and Smith, A. U.,** The uptake of fluid by living cartilage after compression, *J. Physiol.*, 183, 5P, 1966.
22. **Mow, V. C., Kuei, S. C., Lai, W. M., and Armstrong, C. G.,** Biphasic creep and stress relaxation of articular cartilage in compression: theory and experiments, *J. Biomech. Eng.*, 102, 73, 1980.
23. **Weiss, C., Rosenberg, L., and Helfet, A. J.,** An ultrastructural study of normal young adult human articular cartilage, *J. Bone Joint Surg.*, 50-A, 663, 1968.
24. **Clarke, I. C.,** Articular cartilage: a review and scanning electron microscope study, *J. Bone Joint Surg.*, 53-B, 732, 1971.
25. **Aspden, R. M. and Hukins, D. W. L.,** Collagen organization in articular cartilage determined by X-ray diffraction, and its relationship to tissue function, *Proc. R. Soc. London Ser. B*, B212, 299, 1981.
26. **Hulkrantz, W.,** Ueber die spaltrichtungen der gelenknorpel, *Verh. Dsch. Anat. Ges.*, 12, 248, 1898.
27. **Bullough, P. and Goodfellow, J.,** The significance of the fine structure of articular cartilage, *J. Bone Joint Surg.*, 50-B, 852, 1968.
28. **Kempson, G. E.,** Mechanical properties of articular cartilage, in *Adult Articular Cartilage*, 2nd ed., Freeman, M. A. R., Ed., Pitman Medical, Kent, England, 1979, 333.
29. **Kempson, G. E., Muir, H., Pollar, C., and Tuke, M.,** The tensile properties of the cartilage of human femoral condyles related to the content of collagen and glycosaminoglycans, *Biochim. Biophys. Acta*, 297, 456, 1973.
30. **Roth, V. and Mow, V. C.,** The intrinsic tensile behavior of the matrix of bovine articular cartilage and its variation with age, *J. Bone Joint Surg.*, 62-A, 1102, 1980.
31. **Myers, E. R., Lai, W. M., and Mow, V. C.,** A continuum theory and an experiment for the ion-induced swelling behavior of articular cartilage, *J. Biomech. Eng.*, 106, 151, 1984.
32. **Woo, S. L.-Y., Akeson, W. H., and Jemmott, G. F.,** Measurements of nonhomogeneous directional mechanical properties of articular cartilage in tension, *J. Biomech.*, 9, 785, 1976.
33. **Bullough, P. G., Munuera, L., Murphy, J., and Weinstein, A. M.,** The strength of the menisci of the knee as it relates to their fine structure, *J. Bone Joint Surg.*, 52-B, 564, 1970.
34. **Pavlov, H., Ghelman, B., and Vigorita, V. J.,** *Atlas of Knee Menisci*, Appleton-Century-Crofts, New York, 1983.
35. **Ferry, J.,** *Viscoelasticity of Polymers*, 2nd ed., John Wiley & Sons, New York, 1970.
36. **Mow, V. C., Make, A. F., Lai, W. M., Rosenberg, L. C., and Tang, L.-H.,** Viscoelastic properties of proteoglycan subunits and aggregates in varying solution concentrations, *J. Biomech.*, 17, 325, 1984.
37. **Fung, Y. C.,** Stress-strain-history relations of soft tissues in simple elongation, in *Biomechanics: Its Foundations and Objectives*, Prentice-Hall, Englewood Cliffs, N.J., 1972, 181.
38. **Fung, Y. C.,** Quasi-linear viscoelasticity of soft tissues, in *Biomechanics: Mechanical Properties of Living Tissues*, Springer-Verlag, New York, 1981, 226.
39. **Layes, W. C. and Bodine, A. J.,** Flow-independent viscoelastic properties of articular cartilage matrix, *J. Biomech.*, 11, 407, 1978.

40. **Viidik, A., Danielsen, C. C., and Oxlund, H.,** On fundamental and phenomenological models, structure and mechanical properties of collagen, elastin and glycosaminoglycan complexes, *Biorheology*, 19, 437, 1982.
41. **Whipple, R., Wirth, C. R., and Mow, V. C.,** Mechanical properties of the meniscus, in *1984 Advances in Bioengineering*, Spilker, R. L., Ed., American Society of Mechanical Engineers, New York, 1984, 32.
42. **Kempson, G. E., Tuke, M. A., Dingle, J. T., Barrett, A. J., and Horsfield, P. H.,** The effects of proteolytic enzymes on the mechanical properties of adult human articular cartilage, *Biochim. Biophys. Acta*, 428, 741, 1976.
43. **Li, J. T., Mow, V. C., Koob, T. J., and Eyre, D. R.,** Effect of chondroitinase ABC treatment on the tensile behavior of bovine articular cartilage, *Trans. Orthop. Res. Soc.*, 9, 35, 1984.
44. **Mow, V. C., Eyre, D. R., Schoonbeck, J. M., and Chun, L.,** unpublished data, 1985.
45. **Li, J. T., Armstrong, C. G., and Mow, V. C.,** The effect of strain rate on mechanical properties of articular cartilage in tension, in *1983 Biomechanics Symposium*, Woo, S. L.-Y., and Mates, R., Eds., American Society of Mechanical Engineers, New York, 1983, 9.
46. **Armstrong, C. G., Bahrani, A. S., and Gardner, D. L.,** In vitro measurement of articular cartilage deformations in the intact human hip under load, *J. Bone Joint Surg.*, 61-A, 744, 1979.
47. **Akizuki, S., Mow, V. C., Muller, F., Pita, J. C., Howell, D. S., and Manicourt, D. H.,** The tensile properties of human knee joint cartilage I. Influence of ionic conditions, weight bearing, and fibrillation on the tensile modulus, *J. Orthop Res.*, 4, 379, 1986.
48. **Schmidt, M. B., Schoonbeck, J. M., Mow, V. C., Eyre, D. R., and Chun, L. E.,** Effects of enzymatic extraction of proteoglycans on the tensile properties of articular cartilage, *Trans. Orthop. Res. Soc.*, 11, 1986.
49. **Armstrong, C. G. and Mow, V. C.,** Variations in the intrinsic mechanical properties of human articular cartilage with age, degeneration, and water content, *J. Bone Joint Surg.*, 64-A, 88, 1982.
50. **Armstrong, C. G., Mow, V. C., and Eyre, D. R.,** Correlations between the intrinsic mechanical properties of human articular cartilage and its composition, in Proc. Eighth Can. Congr. Appl. Mech., Moncton, Canada, 1981, 173.
51. **Stahurski, T. M., Armstrong, C. G., and Mow, V. C.,** Variation of the intrinsic aggregate modulus and permeability of articular cartilage with trypsin digestion, in *1981 Biomechanics Symposium*, Van Buskirk, W. C. and Woo, S. L.-Y., Eds., American Society of Mechanical Engineers, New York, 1981, 137.
52. **Holmes, M. H., Lai, W. M., and Mow, V. C.,** Singular perturbation analysis of the nonlinear, flow-independent compressive stress-relaxation behavior of articular cartilage, *J. Biomech. Eng.*, 107, 206, 1985.
53. **Roth, V., Schoonbeck, J. M., and Mow, V. C.,** Low frequency dynamic behavior of articular cartilage under torsional shear, *Trans. Orthop. Res. Soc.*, 7, 150, 1982.
54. **Mow, V. C., Lai, W. M., and Holmes, M. H.,** Advanced theoretical and experimental techniques in cartilage research, in *Biomechanics: Principles and Applications*, Huiskes, R., van Campen, D., and de Wijn, J., Eds., Martinus Nijhoff, The Hague, 1982, 47.
55. **Zhu, W. B., Lai, W. M., and Mow, V. C.,** Intrinsic quasi-linear viscoelastic behavior of the extracellular matrix of cartilage, *Trans. Orthop. Res. Soc.*, 11, 1986.
56. **Woo, S. L.-Y., Gomez, M. A., and Akeson, W. H.,** The time and history-dependent viscoelastic properties of the canine medial collateral ligament, *J. Biomech. Eng.*, 103, 293, 1981.
57. **Ahmed, A. M. and Burke, D. L.,** In vitro measurement of static pressure distribution in synovial joints. I. Tibial surface of the knee, *J. Biomech. Eng.*, 105, 216, 1983.

INDEX

A

Abalone, 33
Abnormal dermal repair, 123—124
Acetylcholinesterase, 26
Acid mucopolysaccharides. see Glycosaminoglycans
ACP, see Aldol condensation product
Adhesions, 165—166, see also specific types
Aging, 2, see also Maturation
  aorta and, 104—109
  atheroma as complication of, 108—109
  dermal connective tissue and, 96—103
  dermis and, 96
  fibrils and, 178
  fibroblasts and, 99—102
  lesions of, 103
  ligaments and, 229
  tendon and, 229
ALC, see Anterior lens capsule
Alcian Blue, 61—63
Aldehydes, 208
Aldol condensation product (ACP), 202
Altricial animals, 4, see also specific types
Alveolar basement membrane, 88
Alveolar macrophages, 126
Amelida, 34
Amelid cuticle collagens, 32
Amino acids, 76, 142
β-Aminopropionitrile (βAPN), 128, 162, 164—171, 204
  cross-linking inhibition by, 166
  effectiveness of, 167—171
  percutaneous resorption of, 167—169
  safety of, 167—171
  topical, 167—171
β-Aminopropionitrile (βAPN)-base, 167, 170
β-Aminopropionitrile (βAPN)-fumarate, 167, 169—172
Aneurysms, 204, see also specific types
Animals, 4, 29, 128—130, 170—171, see also specific types
Anisotropy, 250, 271
Annelida, 32—33, see also specific types
Annulus fibrosus, 190
Anterior lens capsule (ALC), 74, 88
Anthozoa, 27
Antibodies, 54, 80, 85, see also specific types
Antifibrogenic agents, 151
Antigen, 54, 85, see also specific types
Aorta, 33, 104—109, 203, 204
Aortic tissue
  cross-linking compounds in, 211
  defective collagen organization in, 211—215
  mechanical properties of, 204—208
  passive mechanical properties of, 216—217
βAPN, see β-Aminopropionitrile
Apolar interactions between collagen molecules, 10
Arm muscle, 33, 34
Arterial tissue
  elastin in, 207—208
  elastin synthesis in, 202—204
  lamellar formation in, 202—204
  mechanical properties of, 215—216
Arterial wall, 15, 104—106
Arteries, 53, 108, 203, 215
Arteriosclerosis, 108
Arthropoda, 34—35, see also specific types
Articular cartilage, 244
  composite nature of, 268—271
  fibril-to-fibril interactions in, 256—257
  hydraulic permeability of, 276
  matrix organization of, 269—271
  multiphasic nature of, 268—271
  origin of softening in, 260—262
  physical model of, 254—256
  proteoglycans and, 257
  softening in, 260—262
  stress and, 246—249
  stress-induced structural transformations in, 262—263
  structural characteristics of, 260
  structure-function relationships in, 244—246
  tissue in, 268
  ultrastructure of, 250—257
Artifacts, 54
Aschelminthes, 29—32, see also specific types
Asparagine, 79, 83
Atheroma, 108—109
Atherosclerosis, 108
Axons, 128
Azo dye, 43

B

Bacterial collagenase, 79, 212—215
Basement membrane, 2, 34, 35, 45, 74, 119
  alveolar, 88
  glomerular, 77
  intestinal, 32
  in liver, 147
  molecular organization of proteoglycans in, 87—88
  noncollagen components of, 81—88
    entactin, 85—86
    laminin, 81—85
    proteoglycans, 86—88
  supramolecular organization in, 88—89
  type IV collagen in, 74—81
    domain structure and organization, 76—80
    general aspects, 74—76
    molecular models, 80—81
Biochemical structure of laminin, 82—83
Bioecosystem aspect of liver, 138, 151—153
Biomechanical properties, 224
Biosynthesis, see Synthesis
Biphasic theory for cartilage, 273, 280
Birefringence intensities, 45

Blood coagulation, 116
Blood vessels, 2, 52
Bone-ligament-bone preparations, 2, 17, 224, 225
Bradykinin, 116
Bundles, 208—211
Burn scar contractures, 167

C

Carp, 36
Cartilage, 2, 14, 36, see also specific types
  articular, see Articular cartilage
  biphasic theory for, 273, 280
  composite nature of, 268—271
  compressive properties of, 277—280
  cranial, 33, 34
  degeneration of, 260
  elasticity of, 271—274
  fibrillar structure for, 251—256
  fibrils in, 17
  fluid in, 273
  hyaline, 34
  hydraulic permeability of, 273, 279
  immobilization and, 166, 232
  invertebrate, 34
  lathyritic, 10
  meniscal, 268
  multiphasic nature of, 268—271
  proteoglycans in, 55, 269
  shear properties of, 280—285
  squid, 33
  tensile properties of, 274—277
  viscoelasticity of, 271—274
cDNA hybridization analysis, 74, 115, 130
Cell-binding domain, 85
Cells
  endodermal, 85
  endothelial, see Endothelial cells
  epithelial, see Epithelial cells
  foam, 107
  interactions between, 138, 148—151
  ito, see Ito cells
  Kupffer, 138, 146
  matrix interactions with, 138, 148—151
  Schwann, 86, 87, 128
  smooth muscle, see Smooth muscle cells (SMC)
  surface receptors on, 85
Cellstic, 130
Cell substrate attachment antigen (CSAT), 85
Cephalopoda, 33, see also specific types
Chalones, 120
Characterization of types of collagen, 45
Chemoattractants, 118, see also specific types
Chemotaxis, 116
Chick metatarsal tendons, 4
Chondrichthyes, 36
Chondroitinases, 61, 277, 284
Chondroitin sulfates, 11, 12, 55, 58, 87, 143, 215, 277
Chromophore, 60
Circulating monocytes, 146, 148
Cirrhosis of liver, 138, 140, 144, 146, 151
Clones, 32
Cnidaria (Coelenterata), 27—29, see also specific types
Cockroach, 35
Codfish, 36
Coelenterata (Cnidaria), 27—29
Colchicine, 151
Collagenases, see also specific types
  bacterial, see Bacterial collagenases
  digestion by, 44—45, 212—215
  dynamics of activity of, 121
  fibroblast, 116
  inhibitors of, 124
  in liver, 148
  production of, 146
Collagenolytic activity in liver, 148
Collagenous antigen, 54
Complement, 26, 116
Compliance mechanism, 18
Compression properties, 271, 277—280
Conchiodine, 33
Connective tissue, 2, see also specific types
  aging and, 96
  dermal, see Dermal connective tissue
  fetal, 4
  fibril growth in, 2—9
  immature, 4
  mechanical role of, 14—19
  metabolism of components of, 144—146
  soft, 178
Contractures, 121, 166, 167
Copolymerization of collagen types, 14
Coral, 29
Core protein, 86
Cornea, 2, 4, 10, 11, 17
Corneal stromata, 4
Crab, 35
Crack propagation, 17
Cranial cartilage, 33, 34
Creep, 15, 272, 277, 279
Crimp, 181
Critical electrolyte concentration, 61—63
Crohn's disease, 126, 128
Cross-linking, 12, 34, 76, 121, 162
  βAPN and, 166
  intermolecular, 79, 208, 211—212
  intramolecular, 79, 208
  joint stiffness and, 166
  lysyl-derived, 79
  lysyloxidase-catalyzed, 206
  stable, 206
  unstable, 206
Crustacea, 34, see also specific types
CSAT, see Cell substrate attachment antigen
Cuticle collagen, 29, 32
Cuttlebone, 33
Cuttlefish, 33
Cyclostomes, 36
Cysteines, 78, 79
Cytoskeleton, 98

D

Defective collagen organization, 211—215
Degradation of collagen in liver, 146
Dehydrolysinonorleucine, 202
Dermal collagen matrix, 117—121
Dermal connective tissue aging, 96—103
  changes in fibroblast appearance, 102—103
  cultures of young and aging fibroblasts, 99—102
  dermis in aging subjects, 98—99
  dermis in young subjects, 96—98
  interpretation of aging lesions, 103
Dermal wound repair
  abnormal, 123—124
  inflammatory events in, 115—117
  maturation in, 121—123
Dermatan sulfate, 12, 59, 87, 143, 215
Dermatosparaxis, 11
Dermis, 52, 96—99, see also Skin
Desmosines, 202, 207
Deuterostomia, 26, 27, 36
DHLNL, see Dihydroxylysinonorleucine
DIC, see Differential interference contrast
Differential interference contrast (DIC) light microscopy, 244
Differential scanning calorimetry (DSC), 186
Digestion
  bacterial collagenase, 79, 212—215
  collagenase, 44—45
  elastase, 212
  hyaluronidase, 216
  papain, 64
  pepsin, 76, 80, 140
Digestive tube, 53
Digital tendons, 4, 234
Dihydroxylysinonorleucine (DHLNL), 34, 35, 208
Dimers, 74
Dimethylsulfoxide (DMSO), 168, 170
Distribution of collagen, 51, 53
Disulfide bonds, 32, 34, 35
DMSO, see Dimethylsulfoxide
DNA, 33, 74, 130
D period, 2, 10
DSC, see Differential scanning calorimetry
Dyes, 43, see also specific types
Dynamic shear, 281

E

Echinodermata, 35—36, see also specific types
E collagen, see Type VI collagen
EGF, see Epidermal growth factor
Elastase digestion, 212
Elastic artery, 108
Elastic fibers, 2, 14, 15, 98, 103
Elasticity, 14, 18, 271—274
Elastic laminae, 106
Elastic membrane, 203, 204
Elastic modulus, 182
Elastin, 14, 103, 104, 118 202—208
Elastoidin, 36
Elaunin, 98
Electron microscopy, 53, 59—60, 74
Embryonic development, 35
Encapsulation, 129
Endodermal cells, 85
Endoneurium, 52, 128
Endothelial cells, 104, 119, 138, 141, 144—146
Energy, 15, 281
Entactin, 74, 85—86
Enzymes, 34, see also specific types
Epibody, 120
Epidermal collagen matrix, 117—121
Epidermal growth factor (EGF), 118, 123
Epineurium, 52, 128
Epithelial cells, 86, 119, 120
Epithelium, 120
Eumetazoa, 27
Exoskeleton, 33, 34
Extensibility of tissue, 14
Extracellular matrix in liver, 138—144, 146—148
  collagens, 138—141
  glycosaminoglycans and proteoglycans, 143—144
  noncollagenous proteins, 143
Extrafibrillar matrix, 178, 186—187

F

Fascicle of tendon, 181
Fast Green, 43, 140
Fertilization processes, 35
Fetal connective tissue, 4
Fibers, 26, 53, 98, 208—211, see also specific types
Fibrillar structure for cartilage, 251—256
Fibrillogenesis, 10, 11, 13, 45
Fibrils, 45, 178
  aging and, 178
  architecture of, 257—262
  associations between, 256, 257
  in cartilage, 244
  diameter of, 4, 15, 178, 185
  fibroblasts and, 125
  growth of, 9—14
  growth of in connective tissues, 2—9
  interactions among in articular cartilage, 256—257
  kinetics of formation of, 10
  length of, 17, 246
  mass-average diameter of, 15
  orientation of, 17, 18
  polydisperse distribution of sizes of, 15
  polymerization of, 116
  radial, 256
  reconstituted, 27
  resilience of, 15
  stability of, 12
  structure of, 179
  transverse aspects of structure of, 2
Fibroblasts, 85, 96, 98, 105, 117, 146
  in aging, 99—102
  changes in appearance of during life, 102—103
  collagenase in, 116
  cultures of, 99—102

fibrils and, 125
keloid, 123
proliferation of, 118, 146
Fibronectin, 12, 85, 98, 103
fibrillogenesis and, 13
hepatocyte attachment to, 148
in liver, 143, 144
production of, 146
SMC secretion of, 104
Fibrosis, 114, 138, 144, 147
hepatic, 126
pharmacology of, 163
topical treatment for, 165—167
Fibrous plaques, 106
Fibrous scars, 107
Fish, 36, see also specific types
Flavobacterium heparinum, 61
Flexor digital tendon, 4, 235
fMLP, see *N*-Formylmethionylleucylphenylalanine
Foam cells, 107
N-Formylmethionylleucylphenylalanine (fMLP), 116
Fragmentation of collagen, 99
Fragmented elastic membrane, 204
Freezing, 224—226
Fruitfly, 35
Functional adaptation, 232—235
Function of collagen, 2

G

GAG, see Glycosaminoglycans
Gap region, 18
Gastropoda, 33
Genomic clones, 32
Gingiva, 13
Gladius, 33
Glomerular basement membrane, 77
Glomerulonephritis, 80
Glucocorticosteroids, 128
Glucosylgalactosyl units, 27
Gamma-glutamyltranspeptidase, 151
Glycopeptides, 34, see also specific types
Glycoproteins, see also specific types
in collagen, 14
in collagen organization, 215—216
in elastin organization, 215—216
kidney-derived, 151
in liver, 143
in mechanical properties of arterial tissue, 215—216
noncollagenous, 81, 140
vs. proteoglycans, 55
from pulmonary alveoli, 26
structural, 215
sulfated, 85
Glycosaminoglycans (GAG), 2, 11, 14, 215
functions of, 55
general characteristics of, 55—58
light microscopy of, 60—61
in liver, 143—144
noncollagenous, 140
sulfated, 57—59
o-Glycosidically linked substitutents, 32
Glycosylated residues on collagen molecule, 14
Goodpasture syndrome, 78, 80
Gore-Tex, 130
Ground substance, 56
Growth factors, 116, 118, 121, 123, see also specific types
Growth of fibrils, 9—14

H

Half-life of collagen, 146, 148
Head-to-head aggregation, 80
Healing, 103, 236—237
Hemostasis, 116
Heparan sulfate, 58, 59, 74, 143, 144, 215
Heparan sulfate proteoglycan (HSPG), 86, 87
Heparin, 58, 59, 143, 215
Hepatic fibrosis, 126
Hepatocytes, 126, 138, 141—144, 146
attachment of to proteins, 148
collagen synthesized by, 144
fibronectin production by, 146
laminin synthesis by, 146
regenerating, 147
HHMD, see Histidinohydroxymerodesmosine
Hierarchical structure of collagen, 178, 188—191
High performance liquid chromatography (HPLC), 115, 130
High resolution ultrasound, 115
Histamine, 116, 124
Histidinohydroxymerodesmosine (HHMD), 208
Histochemical concepts, 53—54
Histochemical methods, 59—63, see also specific types
HLNL, see Hydroxylysinonorleucine
Homeostasis, 152—153, 232—235
Homology, 76
Homotrimer, 33
Hormones, 120, see also specific types
Horses, 4, 171
HPLC, see High performance liquid chromatography
HSPG, see Heparan sulfate proteoglycan
Hyaline cartilage, 34
Hyaluronic acid, 12, 57—59, 87, 143, 215
Hyaluronidase, 216, 284
Hydralazine, 206
Hydranth, 29
Hydration, 186—187, 215
Hydraulic permeability of cartilage, 273, 276, 279
Hydroxylation, 33, 35
Hydroxylysine, 26
Hydroxylysinonorleucine (HLNL), 208
Hydroxyproline, 27, 32
Hydroxypyridinium (pyridinoline), 79, 210, 211
Hydrozoa, 27
Hypertrophic scars, 123, 165, 167
Hysteresis, 216, 217

I

Imitation step, 10
Immature connective tissue, 4
Immobilization, 166, 232
Immunoassays for procollagen peptide, 151
Immunocytochemical techniques, 143
Immunofluorescence, 96
Immunohistochemistry, 54
Immunohistology, 54
Inflammation, 115—117
Infrared (IR), 186
Injury, 115, 124—128, 165, see also Wound
Insects, 34, see also specific types
Interference colors, 45
Interleukins, 118, 151
Intermolecular cross-links, 79, 208, 211—212
Internal elastic membrane, 203
Interstitial collagens, 27, 76
Intervertebral disk, 178, 189—191
Intestinal injury, 126—128
Intestine, 32, 128, 178, 188—189
Intracellular SLS aggregates, 10
Intracytoplasmic granules, 102
Intramolecular cross-links, 79, 208
Invertebrates, 26—27, 34, see also specific types
In vitro protein synthesis, 115, 130
Ionic interactions between collagen molecules, 10
IR, see Infrared
Isometric tension, 212
Ito cells, 141, 146

J

Jellyfish, 27, 29
Joint contractures, 166
Joint stiffness, 166—167

K

Keloids, 123, 130, 165
Keratan sulfate, 58, 59
Keratinocytes, 119
Kidney, 149, 151
Kinetics of fibril formation, 10
Kupffer cells, 138, 146

L

Lamellar formation, 202—204
Lameness, 171
Laminae, 106
Laminin, 74, 80
  antibodies to, 85
  in basement membrane, 81—85
  binding properties of, 85
  binding of to type IV collagen, 85
  biochemical structure of, 82—83
  hepatocyte attachment to, 148
  kidney and, 149
  lipocyte production of, 144
  in liver, 143, 144
  receptors for, 85
  synthesis of, 83—85
  synthesis of in culture, 146
  type IV collagen and, 89
Lamprey, 36
Lateral aggregation, 80
Lateral growth phase, 10
Lathyrogens, 10, 162, 164—165, see also specific types
LDL, see Low-density lipoproteins
Lens epithelium, 86
Lesions of aging, 103
Leukocytes, 116
Leukotrienes, 116
Ligaments, see also specific types
  aging and, 229
  biomechanical properties of, 224
  failure modes of, 231—232
  freezing of, 224—226
  healing of, 236—237
  immobilization and, 232
  periodontal, 14
  storage of, 224—226
  strain-rate in, 231—232
  strength of, 231
Ligamentum nuchae, 14
Light microscopy, 60—61, 96, 244
Linear elastic region, 18
Linear growth phase, 10
Lipocytes, 144
Lipoproteins, 96, 103, 104, see also specific types
Liver
  as bioecosystem, 138, 151—153
  cirrhosis of, see Cirrhosis of liver
  collagen degradation in, 146
  collagenolytic activity in, 148
  collagens in, 138—143
  collagen synthesis in, 145
  composition of, 138—144
  extracellular matrix in, 138—144, 146—148
  half-life of collagen in, 146, 148
  homeostasis in, 152—153
  injury to, 126
  regeneration of, 146—148
  repair of, 146—148
Liver diseases, 138, see also specific types
Load bearing, 256
Lobster, 34
Locust, 35
Loss modulus, 281
Low-density lipoproteins (LDL), 96, 104
Lung injury, 125—126
Lymphocytes, 117
Lymphokines, 117
Lysyl-derived cross-links, 79
Lysyl oxidases, 162, 164, 172, 206

M

Mackerel, 36
Macrophages, 116, 126
Macroscopic crimp, 17
Mallory's method, 42, 43

Mantle skin, 33, 34
Marfan's syndrome, 203, 204, 211, 212
Mass-average fibril diameter, 15
Masson's method, 42
Matrix resiliency, 67
Maturation, 121—123, 163—164, 185—186, 229—231, see also Aging
Mechanical anisotropy, 246—247
Mechanical properties, 36, see also specific types
of aortic tissue, 204—208
of arterial tissue, 215—216
of tendon, 182—185
of tissue, 268
Mechanical role of connective tissue, 14—19
Mechanical trauma, 262
Membrane, 35, 143, 203, 204
Meniscal cartilage, 268, 269
Mesogloea, 27, 29
Metabolism of connective tissue components, 144—146
Metal impregnation techniques, 43
Metatarsal tendons, 4
Metoprolol, 207
Microfibrils, 4, 6, 178
Microsomes, 144
Minerals, 2, 14
Miniature wound implants, 115
Moduli, 271, see also specific types
Molecular models of type IV collagen, 80
Molecular organization of proteoglycans in basement membranes, 87—88
Molecular packing, 6
Mollusca, 33—34, see also specific types
Monocytes, 116, 146—148
Mouse tumor, 74, 76—78
mRNA, 123, 130
Mucopolysaccharidases, 61
Murine schistosomiasis, 148
Muscle, 33, 34, 166, see also specific types
Myelin, 128
Myofibroblasts, 103, 120, 146, 167

N

Nematocyst, 29
Nematoda, 29, 32, see also specific types
Nerve ends, 166
Nerve growth factor, 118
Nerves, 53, 128
NMR, see Nuclear magnetic resonance
Noncollagenous components, 81—88, 146
Noncollagenous domains of type IV collagen, 76
Noncollagenous proteins, 81, 140, 141, 143, see also specific types
Notochord, 36
Nuclear magnetic resonance (NMR), 115, 186
Nucleus pulposus, 189

O

Octopus, 33
Oligonucleotide, 33
Oligosaccharides, 83, see also specific types
Osteichthyes, 36
Overlap region, 18
Oxytalan fibers, 98
Oyster, 34

P

Packing, 6, 9
Panniculus carnosus, 129
Papain digestion of tissue sections, 64
Papillary zone of dermis, 96
Parasitic worms, 29
Parenchyma/stroma ratio, 146
Parietal yolk sac (PYS), 86, 87
PAS, see Periodic acid-Schiff
Passive mechanical properties of aortic tissue, 216—217
pC-collagen, 10
PDGF, see Platelet-derived growth factor
Pearl oyster, 34
D-Penicillamine, 164
Pepsin digestion, 76, 80, 140
Peptides, 36, 118, 151, see also specific types
Percutaneous resorption of βAPN, 167—169
Perineural adhesions, 166
Perineurium, 128
Periodic acid-Schiff (PAS) technique, 63
Periodontal ligament, 14
Peritendinous adhesions, 165—166
Peritoneal adhesions, 12, 125
Peritoneal injury, 124—125, see also specific types
PF-HR9, 74
Phase shift angle, 281
Physical model of articular cartilage, 254—256
Physical properties, 178, see also specific types
Picrosirius-polarization method, 43—51
characteristics of, 43—44
collagen types aiding, 45—51
procedures for, 44—45
Plaques, 106
Plasma membrane, 143
Platelet-derived growth factor (PDGF), 116, 118, 121
Platelets, 114
PMN, see Polymorphonuclear leukocytes
pN-collagen, 10
Polychaete, 32
Polydisperse distribution of fibril sizes, 15
Polymerization of collagen, 163—164
Polymers, 208
Polymorphonuclear leukocytes (PMN), 116
Polyvinyl alcohol (PVA), 129, 130
Poststreptococcal glomerulonephritis, 80
Precipitation by salt, 34
Precocial animals, 4, see also specific types
Primitive animals, 29, see also specific types
Primitive interstitial collagens, 27
Procollagen, 10, 11, 35, 123, 130, 151
Proline, 32—33, 145
Prolylhydroxylase, 123, 144
Propranolol, 206, 207
Prostanglandins, 116

Proteins, see also specific types
complement, 26
core, 86
in vitro synthesis of, 130
noncollagenous, 140
synthesis of in arteries, 203
Proteoglycans, 11, 54—63, see also specific types
aggregate of, 56
arterial, 215
articular cartilage and, 257
of basement membranes, 86—88
in cartilage, 11, 87
chondroitin sulfate, 87
collagen interactions with, 63—68
in collagen organization, 215—216
in elastin organization, 215—216
electron microscopy of, 59—60
general characteristics of, 55—57
vs. glycoproteins, 55
heparan sulfate (HSPG), 86, 87
histochemical methods and, 59—63
hyaluronic acid and, 57—59
in liver, 143—144
molecular organization of in basement membranes, 87—88
shear properties of cartilage and, 283—285
structure of, 55—57
in tensile creep response, 277
viscoelastic behavior of, 272
Protostomia, 26, 27, 33
Pulmonary alveoli, 26
PVA, see Polyvinyl alcohol
Pyr, see Pyridinoline
Pyridinoline (hydroxypyridinium), 79, 210, 211
PYS, see Parietal yolk sac

Q

Quasi-hexagonal packing, 6, 9
Quasilinear viscoelastic theory, 273, 281

R

Radial fibrils, 256
Rat skin, 13
Rat tail tendon, 12
Recombinant DNA, 33
Regeneration, 114, 138
Reinforcement efficiency, 17
Reserpine, 206
Resilience, 14, 15
Resolution, 114
Restriction maps, 33
Reticulin fibers, 53
Retinal epithelium, 86
RNA, 123, 130
Rotary shadowing electron microscopy, 74
Rupture behavior, 19
Ruthenium Red, 59
R values, 29

S

Safranin O, 59
Salt precipitation, 34
Scanning electron microscopy (SEM), 244, 246, 257
Scars, 107, 162—163, 165—167
Schiff base, 164
Schistosomiasis, 148
Schwann cells, 86, 87, 128
Scyphozoa, 27
Sea anemone, 29
Sea-cucumber, 35
Sea mussel, 34
Sea pen, 27
Secretory vacuoles, 10
Segment long spacing (SLS) crystallites, 27, 32
Segment long spacing (SLS) type aggregates of procollagen, 10
SEM, see Scanning electron microscopy
Seven-S domain of type IV collagen, 78—80
Shark, 36
Shear, 280—285
dynamic experiment, 281
modulus, 281
properties of cartilage, 283—285
stress, 14, 16, 271, 281—282
Shrinkage temperature, 212
Sirius Red, 43, 140
Skin, 2, 53, see also Dermis; specific types
codfish, 36
defective collagen organization in, 211—215
fibrils in, 17
mackerel, 36
mantle, 33, 34
rat, 13
squid, 33
SLS, see Segment long spacing
SMC, see Smooth muscle cells
Smooth muscle cells (SMC), 86, 96, 104—106, 128
Soft tissue, 178, 229—231
Spiny lobster, 34
Squid, 33, 34
Stable cross-linking, 206
Stains, 42—43, see also specific types
Storage, 224—226, 281
Strain energy, 15
Strain-locking behavior, 253
Strain-rate in ligaments, 231—232, 274
Strain-stress characteristics, 18, 216
Stress
articular cartilage and, 246—249
experiments on, 281—282
relaxation, 272, 277
shear, 14, 16, 271
structural transformations in articular cartilage and, 262—263
Stress-shielding, 234
Stress-strain characteristics, 18, 182, 216
Stroma/parenchyma ratio, 146
Structural glycoproteins, 215
Structure of collagen, 2, 26—27, see also specific types

Structure-function relationships in articular cartilage, 244—246
Subcutaneous granuloma, 170
Subcuticular membrane, 35
Subfibrils, 178
Sulfated GAGs, 57—59
Sulfated glycoprotein, 85
Supramolecular organization in basement membranes, 88—89
Swine digital tendons, 234
Synthesis
  of collagen, 118, 144, 145
  of elastin, 202—204
  of HSPG, 87
  in vitro protein, 115, 130
  of laminin, 83—85, 146
  of noncollagenous components of liver, 146
  of procollagenase, 123
  of proteins, 115, 130, 203
  types in collagen, 119

T

Tapetum, 17
TEM, see Transmission electron microscopy
Temperature-dependent behavior, 226—228
Tendinitis, 171
Tendon, 2, 178
  adhesions of, 165
  aging and, 185—186, 229
  anatomy of, 179—182
  biomechanical properties of, 224
  chicken foot sublimis, 171
  chick metatarsal, 4
  fascicle of, 181
  flexor, 4, 235
  freezing of, 224—226
  healing of, 236—237
  hierarchical structure of collagen in, 188
  horse, 4
  immobilization and, 232
  injury to, 124, 165
  lathyritic, 10
  maturation and, 185—186
  mechanical properties of, 182—185
  rat tail, 12
  sheath of, 124
  storage of, 224—226
  stress-strain of, 182
  swine digital, 234
Tensile creep response, 277
Tensile modulus, 277
Tensile properties, 15, 129, 274—277
Tension, 271
Tetramers, 74
Teuthoidea, 33
TGF, see Transforming growth factors
Thermal contraction, 212—215
Thermal stability of proline, 33
Tissue, see also specific types
  aortic, see Aortic tissue
  in articular cartilage, 268
  connective, see Connective tissue
  destruction of, 114
  distribution of collagen in, 51
  hydration of, 215
  mechanical properties of, 268
  papain digestion of sections of, 64
  physical properties of, 178
  soft, see Soft tissue
Tissue hysteresis, 216, 217
Tissue repair, 114, 128—130, see also Wound repair
Topical treatment, 165—167, see also specific types
Transforming growth factors (TGF), 118
Transmission electron microscopy (TEM), 244, 246, 256
Transverse aspects of fibril structure, 2
Trauma, 262
Trichrome stains, 42—43
Triple helix, 76
Tropocollagen, 178
Tropoelastin, 118
Tumors, 74, 76—78
Turban, 33
Type I collagen, 2, 14, 45
  chemotactic, 118
  in dermis, 96
  distribution of, 53
  in epineurium, 128
  hepatocyte attachment to, 148
  in intestine, 189
  in liver, 138, 140, 141, 148
  proteoglycan interactions with, 65
  SMC secretion of, 104
  synthesis of, 119
  type III ratio to, 123, 140
Type II collagen, 2, 45, 65, 118, 128
Type III collagen, 2, 14, 45
  chemotactic, 118
  in dermis, 96
  distribution of, 53
  in intestine, 189
  in liver, 138, 140—142, 148
  proteoglycan interactions with, 65
  SMC secretion of, 104
  synthesis of, 119
  tendon healing and, 236
  type I ratio to, 123, 140
Type IV collagen, 2
  antibodies against, 80
  in basement membrane, 74—80
  hepatocyte attachment to, 148
  laminin binding to, 85, 89
  in liver, 138, 140, 142
  molecular models of, 80
  noncollagenous domains of, 76
  organization of, 76
  production of, 119
  seven-S domain of, 78—80
  structure of, 76—80
Type V collagen, 138, 140, 142, 147
Type VI collagen, 143

U

Ultrasound, 115
Ultrastructural appearances, 29
Ultrastructural arrangement, 246—249
Ultrastructural correlations, 51—53
Ultrastructure of articular cartilage, 250—257
  current/previous findings, 250—251
  fibril-to-fibril interactions, 256
  functional physiology of models, 254—256
  new concept of fibrillar structure, 251—254
  role of proteoglycan component, 257
  ultrastructural changes in general matrix, 250
Unstable cross-linking, 206

V

Van Gieson's method, 42
Vertebrata, 15, 36, see also specific types
Viscoelasticity, 224, 271—274
Viscose cellulose sponge, 129
Vitreous humor, 2, 17

W

Water, 14, 283—285
Waveform, 181
Worms, 29, see also specific types
Wound, 120—123, see also Injury
Wound healing, see Wound repair
Wound implants, 115
Wound repair, 114, 128—130, see also Tissue repair

X

X-ray diffraction, 4, 6, 8, 181, 186

Y

Yield behavior, 19
Yield and rupture region, 18